Early Days in Complex Dynamics

A History of Complex Dynamics in One Variable During 1906–1942

HISTORY OF MATHEMATICS
VOLUME 38

Early Days in Complex Dynamics

A History of Complex Dynamics in One Variable During 1906–1942

Daniel S. Alexander
Felice Iavernaro
Alessandro Rosa

AMS
American Mathematical Society
Providence, Rhode Island

The London Mathematical Society
London, England

2010 *Mathematics Subject Classification.* Primary 01A55, 30–03, 01A60.

For additional information and updates on this book, visit
www.ams.org/bookpages/hmath-38

Library of Congress Cataloging-in-Publication Data

Alexander, Daniel S.
Early days in complex dynamics : a history of complex dynamics in one variable during 1906–1942 / Daniel S. Alexander, Felice Iavernaro, and Alessandro Rosa.
p. cm. — (History of mathematics ; v. 38)
Includes bibliographical references and index.
ISBN 978-0-8218-4464-9 (alk. paper)
1. Iterative methods (Mathematics) 2. Analytic functions. I. Iavernaro, Felice, 1967– II. Rosa, Alessandro, 1974– III. Title.

QA297.8.A439 2011
518′.26—dc23

2011014903

∞ The paper used in this book is acid-free and falls within the guidelines
established to ensure permanence and durability.
The London Mathematical Society is incorporated under Royal Charter
and is registered with the Charity Commissioners.
Visit the AMS home page at `http://www.ams.org/`

10 9 8 7 6 5 4 3 2 1   17 16 15 14 13 12

Contents

List of Figures

Preface

Our goal is to tell the story of the development of complex dynamics in the first half of the 20th century. We introduce the reader to the mathematics we cover through its origins in the 19th century. We then provide additional context for our work through discussions of differential equations, in particular, the relation of Henri Poincaré to the study of complex dynamics and the problem of small divisors.

The works of Pierre Fatou and Gaston Julia, and the controversial events surrounding their work, take up the middle third of our story. But the study of complex dynamics in the first half of the 20th century did not stop with Fatou and Juila's exploration of rational functions, and so we cover subsequent developments in the last third of our narrative, including the beginnings of transcendental and algebraic dynamics, as well as a detailed examination of the center problem, which culminated in 1942 with Carl Ludwig Siegel's successful solution of a small divisors problem that links complex dynamics to KAM theory.

The conclusion of our own narrative, however, does not signal the end of our book: we include numerous appendices, the bulk of which are written by mathematicians currently involved in the development of complex dynamics. Our hope is that they underscore the connections between current research and its history.

Our book ends with the usual back matter—a glossary, a detailed index and an exhaustive bibliography—but we also include four appendices of our own devising: Two contain extended biographical sketches of Fatou and Julia, and the next consists of capsule biographies of many of the other mathematicians whose work you encounter along the way. Our last appendix discusses the computer graphics we use to illustrate the works we discuss. Appendices where the author's name is not given were written by us.

Some brief comments regarding our methodology: Unless explicitly noted, translations are our own. For the sake of notational coherence, we have made very inconsequential changes in notation in some of our direct quotations. For example, most functions will be referred to as f whether or not that was the name used by the author. However, when we refer to an equation number in our quotations, you may assume the author did likewise, although obviously using a different number.

Most of the technical terms we use are introduced in the first chapter and can be located through the index where their first reference will be to their definition. Definitions of terms introduced later are indexed similarly. Bibliographical citations are keyed by the publication date and a name is added only if the citation is unclear.

We hope you enjoy our work. In the words of the great Stan Lee, we produced the kind of story we ourselves would enjoy reading.*

*The actual quote is, "For just once, I would do the type of story I myself would enjoy reading..." (Lee [**1974**, p. 17]).

Acknowledgments

First we would like to thank our families whose support during this project has meant more to us than they will ever know. In particular, Dan wishes to thank Rebecca, Caroline and Elise, and he dedicates this book to them. Felice wishes to dedicate the present work to his uncle Felice, to his parents Tina and Pietro, to his wife Miriam, to Vittorio D'Andria, and to Donato Trigiante whose invaluable and steadfast advice unveiled the innermost beauty of mathematics. Sandro wishes to dedicate this book to his family, and to the memory of the beloved Alfonso Farina, who took a little student and taught him to love mathematics.

For their generous contributions to this work, we wish to thank the authors of the appendices: Mark Elin and David Shoikhet (Ort Braude College, Karmiel, Israel) Lawrence A. Harris (University of Kentucky, U.S.), Aimo Hinkkanen (University of Illinois at Urbana-Champaign, U.S.), Xinhou Hua and Rémi Vaillancourt (University of Ottawa, Canada), Katsuhiko Matsuzaki (Waseda University, Japan), Peter Mercer (Buffalo State University, New York, U.S.), John W. Milnor (Stony Brook University, New York, U.S.), Simeon Reich (Technion-Israel Institute of Technology, Israel), Joel L. Schiff (University of Auckland, New Zealand) and Dierk Schleicher (Jacobs University Bremen, Germany). We mourn the passing Sanford L. Segal (University of Rochester, U.S.), and we extend our thanks to his wife Rima Segal, who graciously allowed us to use his contribution. Although he wrote no essay for this work, the encouragement and support of Benoit B. Mandelbrot meant very much to us, and we mourn his passing.

We wish thank the following people for their help along the way: Gerard Alberts (Centrum voor Wiskunde en Informatica, Amsterdam, Netherlands), Giuseppe Angiuli (Università di Bari, Italy), Toma Albu (Koc University, Sariyer-Instanbul, Turkey), Pierluigi Amodio and Francesca Mazzia (Università di Bari, Italy), Nataly Ampilova (University of St. Petersburg, Russia), Akio Arimoto (Musashi Institute of Technology, Tokyo, Japan), David Aubin (Université Pierre-et-Marie-Curie, Paris, France), June Barrow-Green (The Open University, Milton Keynes, U.K.), Bruno Belhoste (Institut National de Recherche Pédagogique, Paris, France), Filippo Bracci (Università di Roma—Tor Vergata, Italy), Walter Bergweiler (Mathematisches Seminar, CAU Kiel, Germany), Claudine Billoux (École Polytechnique, France), Paul Blanchard and Robert Devaney (Boston University), Daniel Bogert and Paola Rosa, François Boisson (Lycée Charlemagne, Paris, France), Luigi Brugnano and Donato Trigiante (Università di Firenze, Italy), Jean-Luc Chabert (Université de Picardie Jules Verne, Amiens, France), Arnaud Chéritat (Université Paul Sabatier, Toulouse, France), Joseph Cima (University of North Carolina, U.S.), Roberto Comunale and Maurizio Guadalupi, Pierre Crépel (Université de Lyon, France), John Dann (Monash University, Australia), Vassileios Drakopoulos (University of Athens, Hellas), Jean Charles Dufait (Lycée François 1er, Le Havre,

France), Gregory Febraro (Des Moines, Iowa, U.S.) Danièle Ghesquier-Pourcin (CNRS-Paris, France), Leon Fishlyn, Patricia Gillet and Armelle LeGoff (Centre Historique des Archives Nationales, Paris, France), Jerry, Terry W. Gintz, Florence Greffe (Académie des Sciences de Paris, France), Christian Henriksen (Technical University, Lyngby, Denmark), Tom Iacono, Giorgio Israel (Università di Roma "La Sapienza", Italy), Armand Lattès (Université Paul Sabatier, Toulouse, France), Sandrine Malotaux (Université Paul Sabatier, Toulouse, France), Anna Maria Masciale, Christian Mira (Institut National des Sciences Appliquées, Toulouse), Alan Norton (National Center for Atmospheric Research, Colorado, U.S.), Brynjulf Owren (Institutt for Matematiske Fag, Trondheim, Norway), Hiroshi Okumura (Maebashi Institute of Technology, Japan), Carsten Lunde Petersen (Roskilde University, Denmark), Christine Phili (National Technical University, Athens, Greece), Kin-Keung Poon (Hong Kong Institute of Education, Tai Po, Hong Kong), Gabriel T. Pripoæ (University of Bucharest, Romania), João Filipe Queiró (Universidade de Coimbra, Portugal), Mary Rees (University of Liverpool, England), André Renaud (Institut Elie Cartan, Nancy, France), David Renfro (ACT, Iowa City, Iowa U.S.), Michael von Renteln (Univerität Karlsruhe, Germany), Sergio Plaza Salinas (Universidad de Santiago de Chile), Cesario Santoro, Donald Sarason (University of California, Berkeley, U.S.), Sabine Starita (Institute "Henri Poincaré", Paris, France), Doru Stefanescu (University of Bucharest, Romania), John Stillwell (College of Arts and Sciences, San Francisco, U.S.), Jennifer Ulrich (Columbia University, New York, U.S.), Jussi Vaisala (University of Helsinki, Finland), Pierre Valiron (Université Joseph Fourier, Grenoble, France), Antonietta Vierucci, Massimo Villarini (Università di Modena, Italy), David J. Wright (Oklahoma State University, U.S.)

Our gratitude goes to Lazare Georges Vidiani (Fontaine Les Dijon, France) for his strenuous efforts on our behalf, his patience and the kindness he extended towards us. Special thanks to Bertrand Fatou (Plessis-Robinson, France), for sharing his father Robert's touching account of the time he spent with his uncle Pierre. We also thank Colette Rey (Saint-Raphael dans le Var, France) for helping us with our genealogical research about the Julia family, and Jennifer Wright Sharp and Cristin Zannella (American Mathematical Society, Providence, Rhode Island, U.S).

We also thank the following individuals who helped us with our archival work: Martine Cornede (Archives Nationales, Centre des Archives d'Outre-Mer, Aix-en-Provence); François Bordes and Claude Gonzalez (Archives Municipales de la Mairie de Toulouse); Dominique Demangel and Christine Celier (Archives Municipales de la Mairie de Nice); Cécile Espine (Archives Municipales de la Mairie de Versailles); Véronique Guitton (Archives Municipales de la Mairie de Nantes); Annie Gualla, Valérie Jan, Patricia Le Bris and Jean-François Noblet (Archives Municipales de la Mairie de Lorient); Bruno Le Gall (Archives Municipales de la Mairie de Quimper); Daniel Skreko (Archives Municipales de la Mairie de Pornichet); Corinne Maréchal (Ministère de la Défense—Direction de la Mémoire, du Patrimoine et des Archives).

We wish to extend our thanks to those who keep the web sites devoted to history of mathematics and the preservation of scientific journals up and running, especially those at the American Mathematical Society at `http://www.ams.org`, Gallica (Bibliothèque Nationale de France) at `http://gallica.bnf.fr/`, the Göttinger Digitalisierungszentrum at `http://gdz.sub.uni-goettingen.de`, the Jahrbuch Project of the European Mathematical Society at http://www.emis.de, and the NUMDAM project at http://www.numdam.org. We regret the passing of the Historia Matematica Forum (maintained by Julio Gonzalez Cabillon and Fred Rickey) whose archives are available at http://mathforum.org/library/view/6952.html.

We are indebted to the organizations that have supported us. We wish to recognize the College of Arts and Sciences at Drake University whose generous support included funding for travel and the purchase of photographs; the Drake University Humanities Center whose support provided us with time, that most precious of resources; the Mathematics Department at Drake University for its graciousness and helpful ideas; the Dipartimento di Matematica, Università di Bari, Italy, for its support and generous use of its resources; and finally, the Istituto Nazionale di Alta Matematica "Francesco Severi" Roma, Italy, which has provided all manner of support for our work.

We wish to express our deep gratitude to two individuals who have supported our project from its onset, Edward Dunne at the American Mathematical Society and Jeremy Gray of The Open University. The spirit of this publication reaffirms our belief in Paul Erdős's conviction that mathematics is an expression of collaborative activity.

Finally, all royalties from our work will be split between the Albert Leon Whiteman Memorial Prize Fund at the American Mathematical Society and the AMS Epsilon Fund.

Part 1

Preliminaries

CHAPTER 1

A Complex Dynamics Primer

The mutability of its fundamental concepts is not an argument for rejecting Science. Each new scientific theory preserves a hard core of the knowledge provided by its predecessor and adds to it. Science progresses by replacing old theories with new. But an age as dominated by Science as ours does need a perspective from which to examine the scientific beliefs which it takes so much for granted, and history provides one important source of such perspective.

Thomas S. Kuhn, *The Copernican Revolution*, [**1976**], p. 3.

What is complex dynamics? At the risk of giving a reductive definition, at its most elemental level it concerns the character of the set

$$\{z, f(z), f^2(z), \dots, f^n(z), \dots\},$$

where z is a complex number and f is a given complex function. The quantity f^n, generally referred to as the nth *iterate* of f, denotes the n-fold composition of f.

Our book gives a history of complex dynamics in the first half of the 20th century. The seeds for complex dynamics, however, lie in the last third of the 19th century, when several mathematicians studied the iteration of an arbitrary analytic function f in the vicinity of a *fixed point*, that is, a point λ satisfying $f(\lambda) = \lambda$.

Our goal in the present chapter is to lay the mathematical and historical foundations for the developments of the 1900s that we discuss in later chapters by tracing the arc of the 19th century study of iteration.[1]

1.1. Signs of life

It is ultimately futile to attempt a definitive dating of the genesis of a branch of mathematics. Sometimes it is best simply to call attention to early signs of its life: In 1870 a young *Gymnasium* teacher from Pforzheim, Germany, Ernst Schröder, published a paper in *Mathematische Annalen* with the intriguing title "On infinitely many algorithms for solving equations".[2] It was followed quickly by a second paper, "On iterated functions".[3] Together, the two papers examine iterative equation solving algorithms on the complex plane.

To properly study the algorithms he developed, Schröder realized that he needed to better understand the process of iteration, and so he examined the following question: Given an analytic function f, what is the nature of the infinite

[1]There are several excellent contemporary introductions to complex dynamics. See for example, Blanchard's expository paper [**1984**] or books by Milnor [**2006a**] and Steinmetz [**1993**]. Other excellent references are Beardon [**1991**] and Carleson and Gamelin [**1993**].

[2]Translation of "*Ueber unendlich viele Algorithmen zur Auflösung der Gleichungen*" [**1870**].

[3]"*Ueber iterirte Funktionen*" [**1871**].

sequence $\{z, f(z), f^2(z), \ldots, f^n(z), \ldots\}$, where z is an arbitrary point in the complex plane?

Even though his are the first papers that we know to treat the iteration of complex functions systematically, it is difficult to say why he chose to study the subject of iterative algorithms. It is not even clear whether the idea was even his since he credited the underpinnings of his study to an obscure German mathematician, Heinrich Eggers, remarking,

> *I was inspired in this work in 1867 by reports from Dr. Heinrich Eggers of Meklenburg, who was formerly Professor at the Gymnasium at Schaffhusen and has now emigrated to America.* [**1870**, p. 371][4]

Eggers' role is mysterious. We could not find these reports, and there appear to be no papers by him prior to 1875 when he published two very brief notices[5] [**1876a, 1876b**] that treat iterative algorithms similar to those found in Schröder's papers. Perhaps because of their brevity, these works are nowhere near as deep or as rich as Schröder's study, and it is difficult to argue with Stewart's characterization of them as "inconsequential" [**1993**, p. 1]. Moreover, as we will see, one of the strengths of Schröder's paper are his attempts to understand the iteration of complex functions from a theoretical viewpoint. That perspective is absent from Eggers' work.

Perhaps Schröder felt that no other explanation for his study was necessary. Solving equations, after all, is one of the oldest activities to engage mathematicians. Given the procedures involved for solving cubic and quartic equations, and the impossibility of finding roots of general higher degree polynomials via radicals, it was natural to look towards numerical methods, not just for polynomials, but for arbitrary equations as well. One such method is what is commonly known as Newton's method, and, given the intense interest in the solutions of equations in the first half of the 19th century, it is perhaps no accident that this interest was accompanied by a resurgence of interest in Newton's method.[6]

1.2. Newton's method: The inspiration for complex dynamics?

Newton's method for finding a root of $q(z) = 0$ is commonly given by the recursion formula

$$z_{k+1} = z_k - \frac{q(z_k)}{q'(z_k)}.$$

If all goes well, an appropriately chosen initial point z_0 generates a sequence $\{z_k\}$ converging to a root of q. The *Newton's method function* for a given q is the

---

[4]Schröder's first paper [**1870**] was translated into English by G. W. Stewart in [**1993**], and while all translations in this work are generally our own, the translation of this quotation is Stewart's, modulo a very few minor changes. Stewart's introduction to his translation contains a very nice analysis of Schröder's first paper in the context of numerical analysis. Alexander [**1994a**, Chap. 1] discusses the role of Schröder's investigation in the development of complex analysis.

[5]These notices appeared in *The Analyst*, a well-regarded amateur mathematics journal published in Des Moines, Iowa, between 1874 and 1883 that was a precursor to the *Annals of Mathematics.* See Deborah Kent's article [**preprint**, pp. 13–14, 19] for more information about the *Analyst.*

[6]See Alexander [**1996a**], Kollerstrom [**1992**], or Ypma [**1995**] for accounts of the history of Newton's method.

functional representation of the algorithm

$$N_q(z) = z - \frac{q(z)}{q'(z)},$$

and application of Newton's method is tantamount to the iteration of N_q, that is, $\{z_k\} = \{N_q^k(z_0)\}$. When analyzing the convergence of Newton's method to the roots of q, it is convenient to use the functional representation N_q. For example, it is trivial, yet nonetheless very useful, to see that the zeros of q become fixed points of N_q.

The standard formulation of Newton's method is not due to Newton, who used a version of this algorithm to invert power series by solving equations of the form $f(y) - x = 0$, where $f(y)$ is the series to be inverted and y is its inverse.[7] Rather, it was evidently first given in a then contemporary paper—from the point of view of Schröder—by Joseph Fourier [**1831**] that appeared posthumously. Indeed, among the infinitely many algorithms referred to in the title of Schröder's first paper are Newton's method and a family of its generalizations.

One issue with Newton's method is the selection of an initial value, or seed, z_0. Determining when "all goes well", that is, when the selection of z_0 leads to a sequence $\{z_n\}$ which converges to a root of $q(z) = 0$, as well as when it does not, is a highly nontrivial question. Generally speaking, while an initial value "near" a root of q gives rise to a convergent sequence, determining precisely what "near" means can be quite difficult.

For example, the function

$$f(z) = z^3 - 4z^2 - 3z + 5 \tag{1.1}$$

has three real roots: one, z_1 near $z = -1.3$; another, z_2 near $z = 0.87$; and the third, z_3 near $z = 4.22$. Surrounding each root is a small disc where Newton's method generates sequences which converge rapidly to that particular root. However, there are regions where strange behavior occurs. For example, in the disc D of radius 0.005 centered at $z = -0.115$, an arbitrary value may converge to any of the three roots under Newton's method, and reasons for this do not emerge from routine calculation. Moreover, there are some initial values in this interval (though they may be hard to find by hand) which lead to periodic sequences of the form $\{z_0, z_1, \ldots, z_{p-1}, z_0, z_1, \ldots, z_{p-1}, \ldots\}$ where a finite string of values endlessly repeats.

1.3. Schröder: The cornerstones of a theory of iteration

Schröder evidently decided that if one is to study algorithms such as Newton's method, one needs to understand iteration; that solutions of equations are not in general real valued may well account for his choice to place iteration on the complex plane.

This desire to unravel the mysteries of iteration led him to articulate several important concepts, perhaps the most fundamental of which is the following from [**1870**, §2], one of the sections inspired by Eggers, which we state in contemporary terms:

---

[7] Solving for y in the series for the arcsine,

$$y + \frac{1}{6}y^3 + \frac{3}{40}y^5 + \frac{5}{112}y^7 + \frac{35}{1152}y^9 + \frac{63}{2816}y^{11} + \cdots - x = 0,$$

Newton obtained the standard series for sine.

THEOREM 1.1 (Schröder's Fixed Point Theorem). *If f is an analytic function with $f(\lambda) = \lambda$ and $|f'(\lambda)| < 1$, then there is a neighborhood D of the fixed point λ on which all points converge to λ under iteration by f; that is, for any $z \in D$,*

$$\lim_{n\to\infty} f^n(z) = \lambda.$$

Schröder's theorem gives rise to the following definition—although the terminology is not Schröder's, he certainly articulated the concepts it describes:

DEFINITION 1.2 (Fixed point classification I). If a fixed point λ satisfies $|f'(\lambda)| < 1$, it is said to be an *attracting fixed point*. If the modulus $|f'(\lambda)|$ is zero, it is *superattracting*. The quantity $f'(\lambda)$ is called the *multiplier* of the fixed point, and $|f'(\lambda)|$ is the *modulus of the multiplier.*

Schröder's proof of Theorem 1.1 is heuristic and relied on the Taylor series of f expanded about the attracting fixed point λ. In the case where $0 < |f'(\lambda)| < 1$, he chose points z that he described as "infinitely close" to λ, hence the degree one term of the series "outweighs all succeeding terms" [**1870**, p. 321]. He reasoned that he could ignore higher order terms, thus $f(z) - \lambda = f'(\lambda)(z - \lambda)$, effecting an implicit linearization of iteration, an idea we will return to.[8] Implicitly replacing z by $f(z)$, he deduced that $f^n(z) - \lambda = (f'(\lambda))^n(z - \lambda)$, from whence he concluded that $f^n(z) - \lambda$ converges to zero as $n \to \infty$.

The treatment of the superattracting case proceeded similarly with the consideration of $f(z) - \lambda = a_n(z - \lambda)^n$, where a_n is the first nonzero Taylor coefficient, in effect, reducing iteration to the power mapping $(z - \lambda) \mapsto a_n(z - \lambda)^n$.

Although Schröder did not discuss other kinds of fixed points in any detail, they are often classified as follows in the present-day literature:

DEFINITION 1.3 (Fixed point classification II). If the modulus of its multiplier is strictly greater than one, the fixed point is called *repelling*. If $|f'(\lambda)| = 1$, then λ is a *rationally neutral* (or *rationally indifferent*) fixed point if $|f'(\lambda)|$ is a root of unity and *irrationally indifferent* (or neutral) if it is not.

The heuristic Schröder used in his proof of Theorem 1.1 explains why Newton's method works: if λ is a simple root for a given function q, routine calculations reveal not only that $N_q(\lambda) = \lambda$, but also that $N_q'(\lambda) = 0$. Thus, simple roots of $q(z) = 0$ are superattracting fixed points of N_q, and Newton's method converges very quickly, quadratically[9] in fact, on a neighborhood D of λ.

If λ is a root of multiplicity r of $q(z) = 0$, then it remains an attracting fixed point, but is no longer superattracting since the derivative of N_q at λ equals $\frac{r-1}{r}$. In order to speed up convergence, Schröder used a modified version of Newton's method,

$$M_q = z - r\frac{q}{q'},$$

that preserved quadratic convergences [**1870**, §3]. This tinkering with the nature of Newton's method was at the heart of his development of a family of algorithms (one of the families referred to in the title of his paper) that increase the rate of

[8]Schröder did use the equal sign, but wrote ϵ instead of $(z - \lambda)$; however, just a few lines earlier he used ϵ^k to represent $(z - \lambda)^k$ in the Taylor series expansion of f around λ.

[9]A sequence $\{z_k\}$ converges (at least) quadratically to a point λ if there exist positive constants C and K such that if $k > K$, then $|z_{k+1} - \lambda| < C|z_k - \lambda|^2$. The convergence is of order r if $|z_k - \lambda|^2$ is replaced by $|z_k - \lambda|^r$.

convergence of the traditional Newton's method N_q. In the next few sections he constructed a family of Newton's method-like algorithms whose convergence is of arbitrary order r.[10]

Schröder cited Eggers' role in the genesis of the material he presented in §3, but not for the sections where he constructed algorithms of an arbitrary order of convergence. Interestingly, Eggers' article [**1876b**] describes not only the generalization of Newton's method for roots of multiplicity greater than one, but it also indicates, via an example, how one might construct an algorithm whose convergence is cubic, noting that,

> *As it would require considerable space to prove* [the generalization to arbitrary order], *I suppress the demonstration here, but I may be permitted to add one example.* [**1876b**, p. 150]

The example he provided involved an iterative algorithm for finding the roots of the real function $f(x) = (x-2)^3(x-4)$ proceeding from the initial value $x = 2.5$. Eggers' justification for the convergence of his algorithm employed ideas very similar to the heuristic Schröder used in his proof of Theorem 1.1, but which is apparently restricted to the real line, so perhaps the idea to extend the heuristic—and the iterative methods—to the complex plane was Schröder's. That this might be the case is suggested by the fact that nowhere in his articles did Eggers explicitly treat the case of complex iteration.

Schröder's heuristic begs a fundamental question, and not just for Newton's method: How far away can an arbitrary point be from an attracting fixed point λ of a function f and still converge to it under iteration? Schröder did not know the answer to this question, and, indeed, it emerged as one of the deepest unanswered questions in the study of iteration until independently answered by Pierre Fatou and Gaston Julia near the end of World War I in works that are among the central subjects of our book.

In contemporary terms, Schröder's perspective was *local*, meaning it focused on behavior in the vicinity of fixed points, rather than *global*, where the focus is on a classification of the convergence behavior of the *iterative family* $\mathcal{I} = \{f^n\}$ on the entire Riemann sphere. In other words, while Schröder knew that points converge to an attracting fixed point under iteration on a neighborhood of an attracting fixed point, he did not know what happened globally, except in special cases.

1.4. Schröder and functional equations

Functional equations, that is, equations with a function for an unknown, have proved very useful in the study of iteration because they provide a means to reduce the iteration of a function to a simpler mapping. Schröder introduced two functional equations that have become fundamental. Our goal here is not so much to review Schröder's approach and methods (although we will do that to some extent), but rather to indicate the utility of these functional equations in the study of iteration.

1.4.1. The Schröder functional equation. The first, called the Schröder functional equation, seeks to linearize iteration by conjugating f to the mapping $z \mapsto hz$:

[10] See Stewart [**1993**, Intro.] and Alexander [**1994a**, §1.4] for more details about Schröder's algorithms.

DEFINITION 1.4 (Schröder functional equation).

$$S \circ f = h \cdot S,$$

where f is a known analytic function and h is an unspecified complex constant.

If a solution S to the Schröder equation can be found that is defined on D, then the following diagram commutes and f is conjugate to the mapping $z \mapsto hz$ on $N = S(D)$,

$$(1.2) \qquad \begin{array}{ccc} D & \xrightarrow{f} & D \\ {\scriptstyle S}\big\downarrow & & \big\downarrow{\scriptstyle S} \\ N & \xrightarrow[z \mapsto hz]{} & N. \end{array}$$

Should D contain a neighborhood of a fixed point λ whose multiplier $f'(\lambda)$ is nonzero, it follows from routine calculations that $h = f'(\lambda)$ (and $S(\lambda) = 0$), in which case we arrive at what we will refer to as the *canonical Schröder equation*:

DEFINITION 1.5 (Canonical Schröder functional equation).

$$(1.3) \qquad S \circ f = f'(\lambda) \cdot S,$$

where λ is a fixed point of f with $f'(\lambda) \neq 0$ and D contains a neighborhood of λ.[11]

The linearization, provided by an invertible solution S to the canonical Schröder equation on D, greatly simplifies iteration on a neighborhood of λ since it follows from (1.3) not only that iteration is conjugate to the mapping $z \mapsto f'(\lambda) \cdot z$ on a neighborhood of the origin, but that $f^n = S^{-1} \circ (f'(\lambda)^n \cdot S)$ on D. Because $S(\lambda) = 0$, we have $S^{-1}(0) = \lambda$. It therefore follows immediately from the expression for f^n that if $0 < |f'(\lambda)| < 1$, then

$$\lim_{n \to \infty} f^n(z) = \lambda$$

for any z near λ; that is, all z in D converge to the attracting fixed point under iteration by f.

The heuristic Schröder used in his proof of Theorem 1.1 naturally leads to the canonical Schröder equation. Indeed, as we will see below, about 15 years after Schröder's investigation, it was shown that the existence of an attracting, but not superattracting, fixed point is equivalent to a solution of the canonical Schröder equation. Thus, iteration near a fixed point whose multiplier has a modulus strictly between one and zero is always linearizable.

Surprisingly, Schröder did not suggest the canonical equation. His examples of solved Schröder equations were not constructed with the goal of conjugating iteration to multiplication by the derivative of an attracting fixed point. This was due in part to the fact that he was unable to solve arbitrary Schröder equations directly. More will be said about these matters shortly. Perhaps he did not bring up the canonical equation because it made little sense to him to suggest an approach he could not carry out. Moreover, he had already made the point that iteration was tantamount to multiplication by a derivative with his proof of Theorem 1.1.

[11] Solutions are generally normalized so $S'(\lambda) = 1$.

1.4.2. The Abel functional equation. The second functional equation that Schröder introduced is known as the Abel functional equation, and it reduces iteration to addition:[12]

DEFINITION 1.6 (Abel functional equation).

$$A \circ f = A + h,$$

where h is a complex constant and f is known.

The canonical Abel equation reduces iteration to a unit translation:

DEFINITION 1.7 (Canonical Abel functional equation).

$$S \circ f = S + 1. \tag{1.4}$$

Sometimes points converge to a fixed point from one direction only, and the Abel equation can be useful in modeling iteration in this situation. The effect of (1.4) is to translate the dynamics of a function f to that of the elementary mapping $g(z) = z + 1$, whose iteration takes place on the Riemann sphere, and the point at infinity is viewed as a fixed point. In the infinite strip $\Re(z) > C$ union $\Re(z) < -C$ for large C, points are both attracted to ∞ and repelled from it.

There are functions f whose iteration is modeled by g in the sense that there is a region D with a fixed point of f on its boundary such that

$$\begin{array}{ccc} D & \xrightarrow{\;f\;} & D \\ {\scriptstyle A}\downarrow & & \downarrow{\scriptstyle A} \\ \tilde{D} & \xrightarrow[z \mapsto z+1]{} & \tilde{D}, \end{array} \tag{1.5}$$

where $A(\lambda) = \infty$ and $\tilde{D}$ contains the point of infinity on its boundary. Since this diagram commutes, $A \circ f = A + 1$, and there is a solution to the Abel equation for f on D. The behavior we describe occurs if $f(\lambda) = \lambda$ and $f'(\lambda)$ is a root of unity. We will discuss the iteration of such functions further in Sections 1.7.3 and 7.2.5.

In our example of $g(z) = z+1$ the point at infinity does not act like an attracting fixed point since there are points near infinity that are repelled from infinity under iteration by g. In other situations the point at infinity can be an attracting fixed point. In fact, this occurs with any polynomial of degree greater than one and is one reason why iteration of complex functions is generally viewed as taking place on the Riemann sphere $\overline{\mathbb{C}}$.

To make the notion of iteration near infinity rigorous, one can analyze the behavior of f on the set $D = \{z : |z| > R\}$, which is said to be a *neighborhood of*

[12]Niels Abel examined this functional equation—but not in the context of iteration—in a fragment of uncertain date published posthumously which is found in an edition of his collected works [**1881**] that appeared 10 years after Schröder's investigation. Copies of this fragment were scarce prior to this edition, so it is doubtful that Schröder was aware of it. We do not know the origins of the Schröder equation but, since linearization is useful in a wide variety of contexts, we suspect its roots go back a long way.

infinity, via the coordinate change

$$\begin{array}{ccc} D & \xrightarrow{\ f\ } & D \\ {\scriptstyle \frac{1}{z}}\downarrow & & \downarrow{\scriptstyle \frac{1}{z}} \\ \tilde{D} & \xrightarrow[\ \tilde{f}\ ]{} & \tilde{D}, \end{array} \tag{1.6}$$

which maps D to a neighborhood $\tilde{D}$ of zero. If zero is indeed a fixed point of $\tilde{f}$, infinity is said to be a fixed point of f and is attracting, repelling, or neutral according to the nature of $\tilde{f}'(0)$. The complete set of points which converge to infinity is called the *basin of infinity*.

1.4.3. An early application of functional equations. As he freely admitted, Schröder was unable to develop general solutions to either functional equation and thought that finding a direct solution to a functional equation would be a "difficult task" [**1871**, p. 302]. Because of this, he demonstrated their utility via the construction of examples, one of which involves an application of the Schröder equation to Newton's method, which will be discussed shortly.[13]

Due to the difficulties he found with a direct approach to the solution of functional equations, he instead suggested an "opposite approach" [**1871**, p. 302] that for the Schröder equation is as follows: rather than solving for S directly, fix h, and instead consider the family

$$\mathcal{F}_S = \{S^{-1} \circ h \cdot S\} \tag{1.7}$$

for various strategically chosen S in hopes that relatively simple functions f would emerge.

One such strategy is to let S be a periodic function. The expression $S^{-1} \circ h \cdot S$ often results in a rational identity. Although Schröder succeeded in finding formulae for the iterates of several functions in this manner, it was not an approach susceptible to a wide generalization. Nonetheless, in particular circumstances it can be quite useful, and other mathematicians followed this approach, as we will see in Section 7.5.1.

Schröder had considerable success applying this "opposite approach" to the study of Newton's method for the quadratic. One of the tasks he set for himself was the characterization of Newton's method for polynomials of arbitrary degree. He began with the quadratic case, most likely in hopes that he would discover techniques that would help him understand Newton's method for functions of arbitrary degree.

He assumed that the roots of an arbitrary quadratic are distinct, mapped them to ± 1, and focused his attention on $q(z) = z^2 - 1$. Fixing $h = 2$ and choosing $S(z) = \arctan(iz)$, he manipulated a standard trigonometric identity to obtain the relation

$$f(z) = \frac{2z}{1+z^2} = -i\tan(2 \cdot \arctan(iz)), \tag{1.8}$$

which he in turn interpreted as stemming from the Schröder equation

$$S \circ f = 2 \cdot S.$$

[13]We will give another example from Schröder's paper in Section 7.5.3.

The function f is the reciprocal of the Newton's method function N_q for $q(z) = z^2 - 1$, and in a somewhat baroque fashion Schröder used this information to iterate the Newton's method function for $q(z)$, determining that $N_q^n(z) \to \pm 1$ depending on which side of the imaginary axis z lay.[14]

Arthur Cayley also attempted to classify the dynamics of Newton's method for complex polynomials in [**1879a**], [**1879b**], and [**1880**]. Using very different methods, he nonetheless duplicated Schröder's success in the degree two case but was likewise unable to extend his results to higher degree polynomials.[15] Cayley was evidently unaware of Schröder's work, and despite its being oft cited, Cayley's work is probably tangential to the development of complex dynamics. It is unclear what drew Cayley to the problem of Newton's method. It could have been the imperative to explore means to solve equations. Reflective of his interest in Newton's method was his inclusion of a question pertaining to its convergence in the quadratic case on the mathematics exam for 1879 Smith Prize at the University of Cambridge.

1.5. Fundamental concepts of iteration

Although Schröder did not articulate things in this manner, one way to describe the convergence of arbitrary points to the fixed point is via the concept of an *orbit* of a point under iteration:

DEFINITION 1.8 (Forward orbit of a point). Let f be a complex function and $z \in \overline{\mathbb{C}}$. The set $O^+(z)$, called the *forward orbit of* z under f, is defined by $O^+(z) = \{f^n(z)\}$ where n is a nonnegative integer and $f^0(z) = z$.

To extend the definition of an orbit, we need the concept of the *total inverse* of f:

DEFINITION 1.9 (Total inverse of f.). By the *total inverse* of f we mean the following multifunction: $f^{-1}(z) = \{w : f(z) = w\}$.

The backward orbit of a point z under f is thus the orbit of z_0 under the multifunction f^{-1}:

DEFINITION 1.10 (Backward orbit of a point). The *backward* orbit of z is the set $O^-(z) = \{f^{-n}(z)\}$ where, again, n is a nonnegative integer. Thus, the set $O^-(z)$ coincides with $\{w : f^n(w) = z\}$ for some nonnegative integer n. An equivalent way to define the backward orbit of a point z under f is thus to say it consists of the set of *preimages* of z, where a *preimage* of z is a point w satisfying $f^n(w) = z$, for some positive index n. One can of course also consider the orbit of a point under a single branch of the inverse. The *total orbit of* z is the union $O(z) = O^+(z) \cup O^-(z)$.

Another important concept related to the inverse of a function is that of a *critical point* and a *critical value* of f:

DEFINITION 1.11 (Critical value). A *critical point* of a function f is a point z such that $f'(z) = 0$. The point $w = f(z)$ is called a *critical value* of f and is a branch point of the inverse. For this reason, aa critical value of f is sometimes referred to as a *critical point of the inverse of* f.

---

[14]See Alexander [**1994a**, §1.7].

[15]See Alexander [**1994a**, §1.8].

If f is a rational function of degree d,[16] then

$$w \text{ is not a critical value} \Longleftrightarrow f^{-1}(w) = \{z_1, \ldots, z_d\},$$

namely $f^{-1}(w)$ consists of d distinct points. When w is not a critical value, there are neighborhoods D_i of z_i and D of w such that $f|_{D_i} : D_i \to D$, is one-to-one and onto. Moreover, there is a well-defined branch of the inverse of f at w, f_i^{-1}, mapping D to D_i.

Related to this is the concept of the *post-critical set.*

DEFINITION 1.12 (Post-critical set). Let C denote the set of critical points of a function f. The *post-critical set* is the union of the orbits $O^+(c)$ where $c \in C$.

We can use the terminology of orbits to describe the dynamics of Newton's method for the quadratic: if z is in the right half-plane, then $O^+(z)$ converges to 1; to the left of the imaginary axis, $O^+(z)$ converges to -1. Points on the imaginary axis, which form the boundary of the convergence regions for Newton's method, do not converge to either root under the iteration of N_q and in fact remain on the axis. For this reason, the imaginary axis is said to be *invariant* under iteration by N_q, a fact Schröder understood.

Indeed, Schröder briefly described the behavior of the iteration of N_q on the imaginary axis, which will be discussed in more detail in Section 7.5.3. We will, however, mention two of his observations here. The first requires a definition of a *periodic point*:

DEFINITION 1.13 (Periodic point of order p). If, for $p \in \mathbb{N}$, a point z satisfies $f^p(z_0) = z_0$, with $f^k(z_0) \neq z_0$ for $0 < k < p-1$, then z_0 is said to be a periodic point of f of order p.[17] The orbit $O^+(z_0) = \{z_0, z_1 = f(z), \ldots, z_{p-1} = f^{p-1}(z)\}$ is said to be a *periodic orbit.* If $p = 1$, then z_0 is a fixed point. An orbit is attracting, repelling, or neutral depending on whether

$$\left| \frac{d}{dz} f^p(z)\Big|_{z=z_k} \right|$$

is less than, greater than, or equal to one. The quantity inside the absolute value is called the multiplier of the orbit.[18]

Schröder observed, first, that there were periodic points of N_q of every order on the imaginary axis [**1871**, p. 320]. Although he made no mention of the fact, these periodic orbits are either repelling or neutral. His second observation was that if z is on the imaginary axis but not eventually periodic, then $N_q^k(z)$ takes on infinitely many values [**1871**, p. 320]; in other words, the orbit $O^+(z)$ consists of infinitely many distinct points. The theory developed in the early 20th century tells us that the orbit of a nonperiodic point in the boundary J between two convergence regions is dense in J, so its closure equals J. Schröder's observation begs a question, how

[16] A complex rational function is a function of the form p/q where p and q are polynomials. Its degree is $d = \max\{\deg p, \deg q\}$.

[17] Our definition of a periodic orbit suggests that the concept of "orbit" can be viewed as an infinite sequence of points ordered by the index of iteration, or as the set of distinct values in the sequence, as in the case of a periodic orbit. In general, if we do not use the term "periodic", the reader should assume that the orbit under question consists of infinitely many distinct values ordered by the index of iteration.

[18] An application of the chain rule shows that the multiplier is the same for all points in the orbit.

did he view the character of the infinite set formed by the $N_q^k(z)$? Aside from this second observation, Schröder was silent on this matter. It should be pointed out that at the time of Schröder's investigation, the mathematical language to describe denseness or closure was not in use.

1.6. Newton's method on the Riemann sphere: Two examples

1.6.1. Example I: Newton's method for the cubic. Schröder and Cayley each hoped to understand Newton's method for higher degree polynomials. However, they failed in their attempts to understand Newton's method for the cubic, which following Schröder's lead for the quadratic case, we will express as $c(z) = z^3 - 1$.

One roadblock they encountered was that the techniques they each used for the quadratic case were ad hoc and simply do not generalize. Perhaps the greatest obstacle was that due to the sheer complexity of the behavior of Newton's method for the cubic c, the mathematical tools available to them were not up to the task. Unlike Newton's method applied to the quadratic, which induces a partitioning of the sphere into two convergence regions bounded by a simple closed curve, iteration by the Newton's method function for the cubic, N_c, partitions $\overline{\mathbb{C}}$ into infinitely many convergence regions. It would be over 45 years after Schröder's work before the behavior he confronted with N_c was well understood, as will be discussed in Chapter 7.

The reader wanting a glimpse into the difficulties Schröder and Cayley faced should refer to Color Plate Figure 5.[19] The regions of convergence are color-coded in order to depict the dynamics of N_c from a global perspective. The blue region is the set of points which converge under N_c to $z = 1$, points in the red region converge to $z = e^{i\frac{2\pi}{3}}$, and those in the green to $z = e^{i\frac{4\pi}{3}}$. As one can see, not only does the set of points which converge to a particular root of $c(z) = 0$ consist of infinitely many disjoint connected components, but the boundary between them borders intricate braids of convergence regions that extend from the point at infinity to the origin.

These images also provide insight into the behavior we described in equation (1.1) regarding Newton's method for the function $f(z) = z^3 - 4z^2 - 3z + 5$, where very small variations in the initial conditions of points in the disc of radius 0.005 centered at $\alpha = -0.115$ cause points to converge to different roots: three distinct convergence regions for N_c intersect this disc, analogous to what occurs on a disc in Color Plate 5 intersecting one the braids that form along the edge of the larger convergence regions.

The partitioning of the sphere that occurs with N_c leads to additional definitions: *(or basin) of convergence*

DEFINITION 1.14 (Domain of convergence). Given a function f, the set of all z for which $f^n(z)$ converges to a fixed point λ as $n \to \infty$ is called the *total domain (or basin) of convergence* of λ and is often denoted $A(\lambda)$. It is possible that this set has more than one connected component (if so, as we will see later, it must have infinitely many), in which case the component containing λ is called the *immediate domain (or basin) of convergence.* If λ is the point at infinity, then the total domain

[19]See page 143 for more information about the images displayed in Color Plate 5.

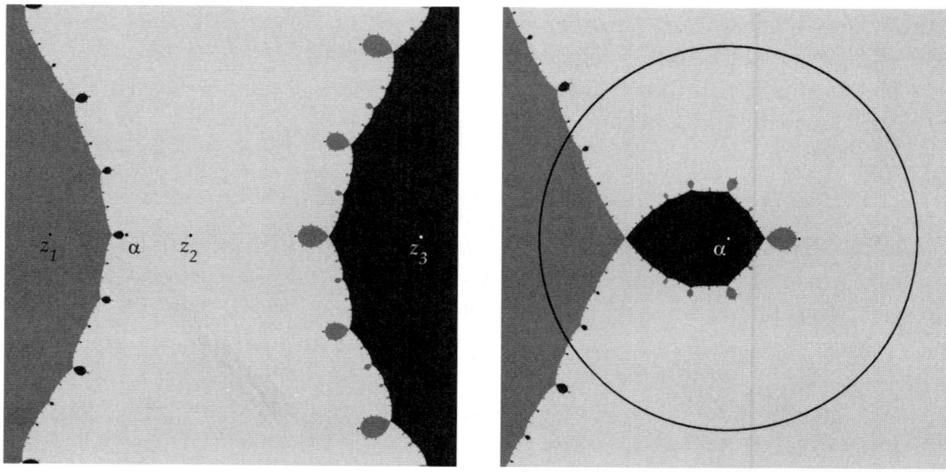

FIGURE 1.1. The dynamics of the Newton's method function N_f for $f(z) = z^3 - 4z^2 - 3z + 5$. The domains of convergence for the three roots of $f(z) = 0$ are pictured, as is the disc D centered at α. Similar to what occurs with Newton's method for the cubic $c(z) = z^3 - 1$, there are innumerable distinct convergence regions within D.

of convergence for λ is called *the basin of infinity*.[20] The domain of convergence of an attracting periodic orbit $\{z_0, \ldots, z_{p-1}\}$ of f is the union of the domains of convergence of f^p to the points z_k.

In Color Plate 5 the union of the blue regions is the total domain of convergence for $z = 1$; the large blue region in the left image represents the immediate domain of convergence of $z = 1$.

The union of the domains of convergence of N_c is an example of what is known as the Fatou set. The complement of the Fatou set—typically the boundary of the union of the domains of convergence—is known as the Julia set. A more precise definition of the Fatou and Julia sets is given the next section, and another in Chapter 7. We will often refer to the Julia and Fatou sets in our book, but the reader should understand that such use is anachronistic since this terminology did not come into common usage until the early 1980s.

1.6.2. The Fatou set, the Julia set, and rotation domains. Repelling and rationally neutral periodic points are in the Julia set. Indeed, for rational and transcendental functions,[21] the Julia and Fatou sets can be defined as follows:[22]

[20]For polynomials, the boundary of the basin of infinity is the what is known as the Julia set, to be discussed in the next section.

[21]Fatou and Julia's initial work focused on the iteration of rational functions of degree two or higher. We will explore the reasons for their choice in subsequent chapters. Fatou later extended his study to the iteration of other kinds of functions, as we will discuss in Chapter 10.

[22]There are historical and mathematical issues associated with this definition. It is the one Julia chose in his monograph on rational functions [**1918a**]. The standard definition, which we will postpone until Chapter 7 when we introduce Montel's theory of normal families, is to define the Fatou set as the domain of normality of the iterative family $\{f^n\}$ and the Julia set as its complement. This is the definition that Fatou used in his monograph on rational functions [**1919-1920**], and it extends naturally to transcendental functions. While Julia's definition also

DEFINITION 1.15 (Julia and Fatou sets). Given a rational or transcendental function f, the Julia set of f, denoted $J(f)$ or J, is the closure of the set of repelling and neutral periodic points of f. The Fatou set, denoted $F(f)$ or F, is the complement of the $J(f)$ in $\overline{\mathbb{C}}$.

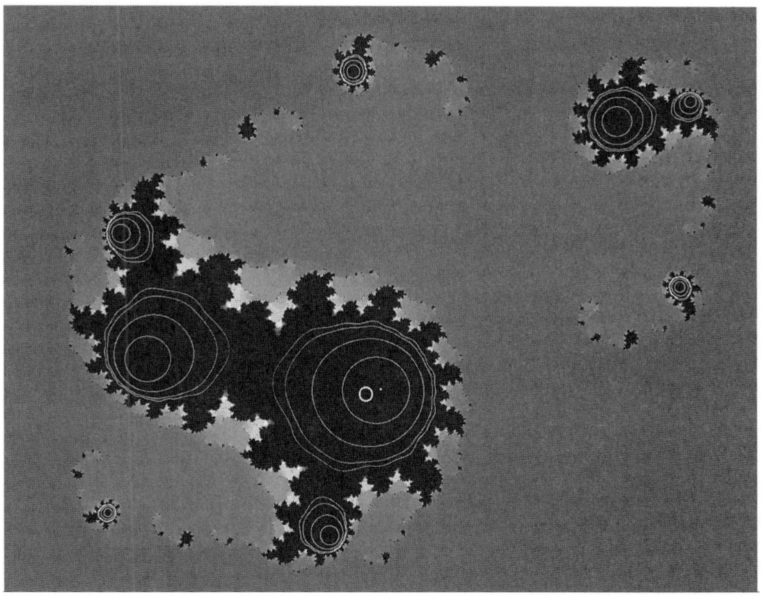

FIGURE 1.2. The origin O is a center for $f(z) = e^{i2\pi\sqrt{3}}z + z^2 + iz^3$. Four orbits around O are indicated, and the Siegel disc D is the component of the Fatou set containing the center and the four orbits. The Siegel disc has three preimages under f, itself, and the regions with two largest sets of orbits, each of which in turn have three preimages. The large grey region surrounding the preimages is the basin of infinity.

The Fatou set of a rational complex function contains all attracting periodic orbits, as well as their domains of convergence. It can also contain sets called *rotation domains* on which iteration is conformally equivalent via a Schröder equation, to an irrational rotation of a disc or annulus through $2\pi\theta$, where $\theta \in [0, 1)$.[23]

In the event the rotation domain D is conformally equivalent to a disc, it contains in its interior an irrationally neutral fixed point λ, called a *center*. The domain D is also referred to as a *Siegel disc*, and its angle of rotation θ is given by the multiplier $f'(\lambda) = e^{i2\pi\theta}$ of λ. Figure 1.2 illustrates a Siegel disc.

When D is a rotation domain conformally equivalent to an annulus, D is called a *Herman ring* or an *Arnold-Herman ring*. The rotation does not correspond to a

---

extends, it takes considerably more effort to show this (and indeed took many years), and for this reason it is nowadays generally stated as a consequence of the standard definition of the Fatou set as the domain of normality. We discuss issues involved with the different definitions at the beginning of Section 7.2.2 and on page 223; also see Bergweiler [**1993**, §3.4].

[23]The possible configurations for components of the Fatou set are discussed beginning near the bottom of p. 332 and continuing to the next page.

multiplier of a fixed point, however, and is not as straightforward to compute as it is for a center.[24] We depict a Herman ring in Color Plate 1.[25]

Caption for Color Plate 1: (See p. 63.) This image depicts a Herman ring for the function

$$f(z) = \frac{e^{i2\pi\theta} z^2(z-4)}{1-4z},$$

where θ is chosen so that iteration by f is conjugate to the rotation $z \mapsto ze^{i2\pi\phi}$ where $\phi = (\sqrt{5}-1)/2$.[26] The grayish region containing the origin O, denoted D_0 is the immediate domain of convergence of O, which is a superattracting fixed point. The Herman ring A surrounds D_0, and the paths in A along which the first 1000 points of $O^+(z)$ lie for $z = 0.4, 0.7, 1, 1.3, 1.6$, are indicated. Note that the orbit for $z = 1$ lies on the unit circle, which f fixes.

Also pictured are the myriad preimages of A and D_0. As we will see later, when there is rotary behavior, the degree of connection of a component of the Fatou set is either one or two, hence the Julia set—which is not depicted—snakes through the image, dividing the Fatou set into infinitely many components. Each preimage of A or D_0 is its own distinct component.

While the existence of centers for rational functions of degree greater than one was speculated for quite some time, they were not shown to exist until 1942. It took another 40 years to establish the existence of Herman rings, although there was an anticipation of them in Germany in the early 1930s, which we will discuss in Chapter 11.

1.6.3. Example II: Newton's method for the quadratic. The dynamics of the Newton's method function N_q for $q(z) = z^2 - 1$ are infinitely less complex than those of the cubic, but it is nonetheless worthwhile to explain the local behavior of N_q from a contemporary perspective.

The points $z = \pm 1$ are the fixed points of

$$N_q(z) = \frac{z^2+1}{2z};$$

since $N_q'(\pm 1) = 0$, they are superattracting fixed points. Theorem 1.1 guarantees that both $z = \pm 1$ are surrounded by neighborhoods $D(\pm 1)$ on which all points converge to the corresponding fixed point under iteration by N_q. The question then arises, how far does the convergence of N_q extend beyond $D(\pm 1)$? The local theory developed by Schröder does not answer this question, but, by dint of calculation and good fortune, he and Cayley were independently able to show that the domains of convergence for N_q extend to the left half-plane for $z = -1$ and the right half-plane for $z = 1$.

---

[24]One, in essence, averages the angular distance between consecutive points in an orbit. See Milnor [**2006a**, pp. 164–166] or Steinmetz [**1993**, Chap. 4] for details.

[25]While color images have been separated from the text, we place their captions within our exposition.

[26]For a rational function, the angle of rotation of a center is simply the multiplier of the fixed point. Finding the angle of rotation for a Herman ring is considerably more complicated. In our example, it cannot be gleaned trivially from the $e^{i2\pi\theta}$ term in f, although the angle of rotation is a continuous function of θ. See Steinmetz [**1993**, pp. 102ff.].

Each half-plane thus comprises both the immediate and complete domain of convergence for the respective fixed points. Such domains are said to be *completely invariant* under iteration since the complete orbit $O(z)$ of a point z remains in the half-plane of its origin. The union of the right and left half-planes forms the Fatou set, and the boundary between the convergence regions, the imaginary axis, is the Julia set.

1.7. Towards a deeper understanding of the local theory: The work of Kœnigs and his students

Despite articulating several fundamental ideas concerning the iteration of complex functions, Schröder's understanding of the local theory had several significant gaps. He did not investigate the nature of periodic points or attempt to classify fixed points; nor did he develop a unified approach to the Schröder and Abel equations. These tasks would be left to the French mathematician Gabriel Kœnigs and his students, Auguste Grévy and Léopold Leau.

1.7.1. A rigorous approach to the Schröder functional equation. Kœnigs' three papers from the mid-1880s [**1883, 1884b, 1885b**] represent the most significant exploration of complex dynamics of the 19th century. Although Kœnigs cited Schröder in [**1884b**] (but not in his first paper), it is not clear how much of an influence Schröder was on Kœnigs; the lack of any reference to Schröder in Kœnigs' first paper suggests it may have been minimal.

ERNST SCHRÖDER

GABRIEL KŒNIGS

FIGURE 1.3. Photo of Kœnigs courtesy of Académie des sciences—Institut de France.

Kœnigs developed a rigorous, general approach to the behavior of iteration near attracting fixed points. He produced initial proofs of Theorem 1.1 and the existence of solutions to the Schröder and Abel functional equations when $f(\lambda) = \lambda$, where

$0 < |f'(\lambda)| < 1$ in [**1883**], but later realized they were lacking somewhat in rigor. Prompted by Gaston Darboux,[27] he provided fully rigorous proofs in [**1884b**].

Of most interest to us is his solution to the canonical Schröder equation

$$S \circ f = f'(\lambda) \cdot S \tag{1.9}$$

on a neighborhood D of λ, where λ is an attracting (but not superattracting) fixed point of f. He proved, first in [**1883**] and then more rigorously in [**1884b**], that

$$S(z) = \lim_{n \to \infty} \frac{f^n(z) - \lambda}{z - \lambda} \tag{1.10}$$

is an invertible, analytic solution to equation (1.9) on D. In the proof from [**1883**] he showed only that the limit converges pointwise on D; in his second proof, he used his extensions of Darboux's theorems to prove that the convergence is uniform, which allowed him to conclude that S is analytic. His proof also used the method of majorants, which we describe in a different context in Section 3.2.

Kœnigs derived a solution to the Abel equation by setting

$$A = \frac{\ln(S)}{f'(\lambda)}. \tag{1.11}$$

Routine calculations show that near λ, $A \circ f = A + 1$, but there is a logarithmic singularity of A at λ. This turned out not to be a serious problem. The linearization provided by the Schröder equation is an optimal model for iteration of f near an attracting fixed point λ satisfying $0 < |f'(\lambda)| < 1$, so the translation model suggested by the Abel equation is not particularly germane.

However, as suggested in our discussion in Section 1.4.2, there are times when iteration near a fixed point acts like a translation. While Kœnigs did not use the Abel equation to model this sort of behavior, its utility to do so was recognized by his student Leau in his study of rationally neutral fixed points in the late 1890s, as well as by Fatou over 30 years after Kœnigs' first paper.

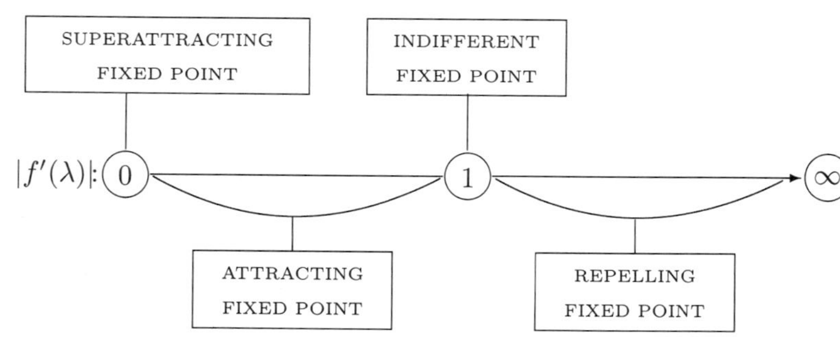

FIGURE 1.4. The local viewpoint, circa 1900: the classification of fixed points after Kœnigs. The horizontal line represents the modulus $|f'(\lambda)|$. Much of the post-Kœnigs 19th century French study of iteration focused on determining the properties of iteration near indifferent and superattracting points.

[27] Kœnigs was a student of Darboux, and at the onset of [**1884b**] he extended to complex functions two theorems of Darboux regarding the uniform convergence of series of real functions. For more information about Darboux and Kœnigs, in particular, the latter's extension of the former's theorems, see Alexander's paper [**1994b**].

1.7.2. Superattracting fixed points: Grévy and Böttcher. Kœnigs did not treat the superattracting case. That was left to his student Auguste Grévy who wrote two papers [**1894, 1896**] that generalized Kœnigs' work in two principal ways. One was to consider functional equations of the form

$$p_0 \cdot G + p_1 \cdot G \circ f + \cdots + p_n \cdot G \circ f^n \equiv 0, \tag{1.12}$$

on a neighborhood D of an attracting fixed point λ of f, where the p_k are given functions analytic on D and G is unknown. Setting $n = 1$, $p_0(z) \equiv f'(\lambda)$, and $p_1 \equiv -1$ yields the canonical Schröder equation. He initially treated the case where $0 < |f'(\lambda)| < 1$ but later extended his treatment to the superattracting case.

His second generalization of Kœnigs' study sought solutions to functional equations analogous to the Schröder equation in the neighborhood of a superattracting fixed point. One such functional equation is the Schröder equation $S \circ f = k \cdot S$, where, for $f(z) = a_k(z-\lambda)^k + \cdots$, with $a_k \neq 0$, he found a solution S with a logarithmic singularity at λ. Grévy's interest in functional equations was by no means restricted to iteration, and he did not seem to intend this equation as a means to model iteration, although if the fixed point were infinity, the mapping $z \mapsto kz$ would to some degree suggest the attracting behavior of a fixed point at infinity.[28]

However, this is not an optimal model. Schröder's heuristic (discussed on page 6), as well as elementary examples, such as $f(z) = z^k$, intimate that iteration near a superattracting fixed point acts like $z \mapsto z^k$. Indeed, the young Polish mathematician Łucyan Böttcher found a solution B to the so-called Böttcher functional equation,

$$B \circ f = B^k,$$

in the neighborhood of a superattracting fixed point λ of a function f satisfying $f(z) = a_k(z-\lambda)^k + \cdots$ in [**1904a, 1904b**], which were written in Russian. One can obtain B from Grévy's function S by setting $B = e^S$, as is easily seen. More will be said about Böttcher's body of work in Section 7.5.3.

1.7.3. Neutral fixed points: Kœnigs and Leau. Kœnigs' treatment of the neutral case was limited to elliptical linear fractional transformations, that is, linear functions L which satisfy

$$\frac{L(z)-a}{L(z)-b} = e^{i\theta}\frac{z-a}{z-b}.$$

He noted that if θ is commensurate with π, the iterative behavior was periodic, and if not, iterates of L would not converge at all. He then remarked, "I have excluded this case from my researches" [**1883**, p. 352]. Since this was the only instance where Kœnigs explicitly examined a neutral fixed point, it is reasonable to interpret this statement as an admission that he was unable to develop a general treatment of neutral fixed points—rational or otherwise.

Such an examination, at least in the rational case, was successfully carried out in 1897 by his student Leau. One of the difficulties in this case is that the Schröder functional equation does not have an analytic solution at a neutral fixed point.

[28]It turns out that a function very much like Grévy's G that also satisfies $S \circ f = k \cdot S$ but in a different context than he envisioned, called the Green's function, is quite useful in the study of iteration of polynomials on the basin of ∞. See for example, Branner [**1989**, p. 88] or Milnor [**2006a**, p. 100].

This can be seen as follows. Let f have a rationally neutral fixed point, which, for convenience we will assume is zero, and let

$$f(z) = a_1 z + a_2 z^2 + \cdots$$

be its Taylor expansion. Assume that an analytic solution

$$S(z) = c_1 z + c_2 z^2 + \cdots$$

to the canonical Schröder equation $S \circ f = a_1 \cdot S$ exists on a neighborhood of the origin.[29]

Solving $S \circ f = a_1 \cdot S$ directly for the c_n yields the recurrence relation

$$c_{n+1} = \frac{c_1[a_1^{n-1}(1-a_1)\cdots(1-a_1^{n-1})a_{n+1} + P_{n+1}(a_1,\ldots,a_n)]}{a_1^n(1-a_1)\cdots(1-a_1^n)},$$

where the P_n are polynomials of the a_k. Since a_1 is a root of unity, infinitely many coefficients of S will have zero in the denominator. Leau used a different method to show that analytic solutions to the Schröder equation do not exist. We choose this method since the above recurrence relation will appear again when we discuss irrationally neutral fixed points in subsequent chapters.

As noted above, a solution to the canonical Abel equation (1.4) exists, albeit one with a logarithmic singularity. However, there is another way to describe iteration near rationally neutral fixed points, the so-called Flower Theorem, articulated by Leau in [**1897**],[30] which yields a helpful visual depiction of iteration in the rationally neutral case, examples of which are given in Figure 7.3 in Section 7.2, where we will discuss the Flower Theorem in more detail.[31]

Iteration near irrationally indifferent fixed points presents much greater difficulties than it does in other cases. Like Kœnigs, Leau made no inroads into the irrational case:

> *...we will discard the case where* $\left|\frac{d\varphi}{dz}\right|_0 = 1$ *and the argument is irrational* [**1897**, p. 65].

1.8. The problem of the division of the plane

Kœnigs concluded his work [**1884b**] with remarks that suggest he was well aware that there was a significant gap in his understanding of the global behavior of iteration:

> *If one envisions the totality of points from the plane which are carried into the interior of* C_x[32] *and, as a result, to the point* x, *one can extend to this region our general theorems. But nothing is known of the general manner in which the region is bounded, and*

[29] As we have previously noted, routine calculations show that any solution S to a canonical Schröder equation maps the fixed point to the origin; thus, in this case, it fixes the origin.

[30] Another French mathematician, Ernest Lémeray, gave a less polished version of the Flower Theorem in [**1897a**], the same year as Leau. Lémeray's paper also contains an early example of the technique of graphical iteration which is used to depict the iteration of a real function near an attracting fixed point. See Figure 8.2 in Section 8.4 for a diagram illustrating this technique. Lémeray's work is discussed in Alexander [**1994a**, §5.2].

[31] The reader curious to know more of the mathematics of the Flower Theorem is invited to consult one of the standard complex dynamics references listed above in footnote 1. See Alexander [**1994a**, Chap. 5, 6] and Audin [**2009**, p. 93] for discussions of the history of the Flower Theorem.

[32] That is, a neighborhood of an attracting fixed point x.

one cannot affirm a priori that its mode of delimitation is not of a nature which restricts such an extension.

The importance of the division of the plane into regions according to the limit points to which they are carried is thus once again underscored. But in realizing that in general there are infinitely many circular limit groups,[33] *since the index* [of iteration] *can be arbitrarily large, one grasps the difficulty attached to this problem.* [**1884b**, p. 40]

The common interpretation of Kœnigs' remarks is that he was speculating that there could be infinitely many periodic attracting orbits. That may be so, but he could also have been suggesting that periodic points are not in general attracting, as Schröder showed in the case with Newton's method for the quadratic. If so, Kœnigs turned out to be correct. In any case, the fundamental theorem of algebra guarantees that for rational f, $f^n(z) = z$ has solutions for all n, so there are infinitely many periodic groups, and the task of classifying them would have been daunting, especially given the lack of set theoretic or topological tools available to him.

A little over a decade later, Leau echoed Kœnigs' pessimism at the conclusion of [**1897**]. His study of the domain of solutions of functional equations led him to also confront the intractability of the problem of "the division of the plane":

The solutions of functional equations are defined in a small domain [referred to below as the primitive domain]. *One can prolong the functions thus obtained to the regions from whence the points came if, after a finite number of iterations, they fall in the primitive domain. Naturally, we suppose that the solution of the equation is still defined in these new regions. ... The problem of the extension of these solutions reduces to the division of the plane into regions which are transformed into one another* [under iteration], *a problem which is without a doubt in general impractical, if one supposes that the function to be iterated is given in a series development.* [**1897**, p. 55]

What is particularly interesting is that he evidently did not find an analytic continuation, one of the most powerful tools for extending domains of solution, up to the task of finding maximal domains of solutions to functional equations. We will see Fatou's response to the issue of continuation in our discussion of Chapter VII of his monograph on rational functions, [**1919-1920**] in Section 7.3.

[33] Period p points.

CHAPTER 2

Introduction: Dynamics of a Complex History

2.1. Introduction

The development of complex dynamics is characterized by an ebb and flow that induces a partition of the study into four distinct phases. At the end of each of the first three phases lies a fallow period, in which interest in the subject seems to wane. But then a spark arises, and new ideas burst forth.

The first phase, discussed in the opening chapter, began in 1870 with Ernst Schröder's study and ended with a paper [**1897**] by Léopold Leau. Pierre Fatou's *Comptes Rendus* notice [**1906b**] initiated the next phase, which included works prepared in anticipation of the 1918 *Grand Prix des Sciences Mathématiques* and concluded with Carl Ludwig Siegel's paper [**1942**]. While characterized by the development of a robust global theory of rational dynamics,[1] it also witnessed the first attempts to understand the iteration of algebraic and transcendental functions.

The aftermath of World War II through the early 1970s constitutes the third phase. It is commonly thought that not much happened between Siegel's paper and the onset of the computer, but an examination of the record, about which we say more shortly, belies this. The advent of the personal computer helped trigger the fourth phase, in which the pioneering work of Adrien Douady, John H. Hubbard, and Benoit Mandelbrot in the late 1970s and early 1980s launched an energetic and far-reaching body of work that, as of the completion of our manuscript, shows no signs of abating.

Our book concerns the second phase and details its progress in Europe and the United States from 1906 to 1942. Alexander's book [**1994a**] treats the initial phase, as well as some of the details of Fatou and Julia's works. Audin's book [**2009**] discusses the events, people and mathematics surrounding the 1918 *Grand Prix* and follows certain reverberations through the 1960s. Histories of the other phases have yet to be written, although texts by Burckel [**1979**], Milnor [**2004**], and Steinmetz [**1993**], as well as Bergweiler's paper [**1993**], contain rich historical details.

[1] A note on terminology: The phrase *rational* (or *holomorphic*) *dynamics* refers to the iteration of functions which are analytic on the Riemann sphere $\overline{\mathbb{C}}$. Likewise, *transcendental dynamics* refers to the study of the iteration of a function defined throughout the plane with the point at infinity an essential singularity. *Complex dynamics* refers to the iteration of complex functions in general which includes rational dynamics, transcendental dynamics, meromorphic dynamics, as well as the iteration of algebraic functions. By *meromorphic dynamics* we mean transcendental functions with multiple poles (see Bergweiler [**1993**] for more on this subject). Unless otherwise specified, the functions we refer to are of a single complex variable.

PIERRE FATOU

FIGURE 2.1. Photo courtesy of Pierre Fatou's family members

2.2. Our story begins: Fatou's 1906 notice

As we observed in the last chapter, Schröder and Gabriel Kœnigs (1842–1917) independently mapped out a robust local theory of iteration in the last third of the 19th century. Beset by a seeming inability to extend the study beyond the neighborhood of a fixed point, interest in the study of iteration waned in the late 1890s and the early 1900s, and the study stalled.[2]

Progress came early in the new century with Pierre Fatou's *Comptes Rendus* notice [**1906b**] where he examined the iteration of two kinds of rational functions.[3]

---

[2]While there were no French studies of substance focusing on dynamical properties of iteration for several years following Leau's work, there was an interest in the functional equations related to iteration. See, for example, Carlo Bourlet's papers [**1897a, 1897b, 1899**], which examine iteration as well the related functional equations from the point of view functional operators. The Italian mathematician Salvatore Pincherle explored ideas similar to Bourlet's around the same time in [**1894-95, 1896a, 1896b, 1896c, 1897a**], and Pincherle's survey [**1912**, §7,§16] of functional equations discusses Bourlet's papers in the context of functional analysis. The lack of discussion of Bourlet's line of investigation in the work of Fatou and Julia suggests that his work did not have significant influence on the development of their ideas.

[3]There was one another mathematician who published an important paper on iteration in the first years of the 20th century, Lucyan Böttcher, whose work we touched upon in Section 1.7.2. His papers [**1904a, 1904b**] appeared in a Russian journal in 1904, but, perhaps due to the language barrier, there was virtually no mention of them in France until Fatou's monograph [**1919-1920**].

The first, which as Fatou observed is typified by the family $f_k(z) = z^k/(z^k+2)$ where k is a natural number greater than one, has a single attracting orbit consisting of a fixed point at the origin, and the second has two attracting orbits, one, again, a fixed point at the origin, and the other the point at infinity. The only example Fatou provided for the second kind of function was $g(z) = (z^2+z)/2$.

Fatou successfully solved the problem of the division of the plane referred to in Section 1.8 for both kinds of functions. The nature of the respective divisions is quite different. For functions f such as the family f_k, Fatou argued that all points in $\overline{\mathbb{C}}$ aside from those in a *totally disconnected perfect set*[4] J converge under iteration to the origin. Points from J remain in J under iteration by f; moreover, it is clear that solutions to $f^n(z) = z$ other than zero are in J as well—something Fatou pointed out in a footnote.

For g, both the origin and infinity are attracting fixed points, hence there are two domains of convergence. Fatou observed that a simple closed curve Γ divides these domains but is not analytic, by which he meant that points on Γ generally do not posses a well-defined tangent line. Both J and Γ are examples of the Julia set discussed in Section 1.6.2.

Comptes Rendus notices typically do not contain rigorous details of proofs, and Fatou's was no exception. Nonetheless, he indicated why the behavior he observed occurred and emphasized that the behavior observed in both g and f_k are not specific to his examples. Color Plates 2 and 3, respectively, depict the dynamics of f_2 and g.

Caption for Color Plate 2: (See p. 63.) We use a representative of the family $\{f_k\}$ referred to in Fatou's 1906 *Comptes Rendus* notice [**1906b**],

$$f_2(z) = \frac{z^2}{z^2+2},$$

to illustrate the iteration a function whose Julia set J is a Cantor set. The function f_2 admits fixed points $O = 0$, $A = \frac{1}{2} + i\frac{\sqrt{7}}{2}$, and $B = \frac{1}{2} - i\frac{\sqrt{7}}{2}$ (not indicated). The origin is superattracting, while A and B are repelling and belong to the Julia set J. All points on the complex plane except those in J give rise to orbits that converge to the origin.

These images are obtained via the escape-time method (see Appendix P) which assigns each point a particular color depending upon how many iterations are needed for its orbit to enter a predetermined region R surrounding an attracting point. In our case the region R is the dark blue ellipse centered at the origin. In each image, points whose color is closer to blue take fewer iterations to reach R, while those closer to red require more.

Top right: A blow-up of the region near A in the left image. Bottom right: A magnification of the region indicated by the box in the upper picture.

[4]Totally disconnected perfect sets are closed sets with no isolated points (every point of the set is a limit point of the set) whose maximal connected components are single points. The Cantor middle third set is perhaps the most well known totally disconnected perfect set, and for that reason such sets are often referred to as Cantor sets, a terminology we will adopt.

Fatou began his construction of the set J for a rational function f of the first kind by defining a circle C of radius r surrounding the fixed point where r satisfies two conditions. On the one hand, it has to be sufficiently large so that all the critical values of f are interior to C.[5] This ensures that all points on C and its exterior will have d preimages, where d is the degree of f. On the other hand, r must be small enough so that all points interior to C remain in its interior under iteration by f before converging to the origin. However, for the example of functions of the first kind that Fatou provided—the family $\{f_k\}$ given above—such a C does not exist because each f_k has the critical value $z = 1$, and any circle large enough to contain $z = 1$ on its interior no longer has the property that its interior is invariant under iteration by f_k.[6]

This does not present a serious problem since one can instead use a closed curve containing the critical values. We suspect that Fatou took it for granted that the reader would understand this. For example, the boundary of the dark blue region in Color Plate 2 is an ellipse with $z = 1$ in its interior for which the properties hold that Fatou claimed for C.

Fatou's analysis of the iteration of f relied on a family of sets E_n. His definition of these sets is a bit muddled since he defined E_n as the set of points $\{z\}$ such that $|z| > r, |f(z)| > r, \ldots, |f^n(z)| \geq r$ while $\left|f^{n+1}(z)\right| < r$, which is the set of points that remain outside of C for exactly n iterations but are interior to C starting with the $(n + 1)$st. This creates a family of disjoint sets since $z \in E_n$ satisfies $\left|f^{n+1}(z)\right| < r$ while all $z \in E_{n+1}$ satisfy the contrary condition $\left|f^{n+1}(z)\right| \geq r$. However, not only did he remark that the sets are contained within one another, but his argument that J is a Cantor set relies upon their nesting, [**1906b**, pp. 546–547].

In order to make the definition of E_n consistent with the details of Fatou's argument, we take the liberty of removing the condition that $\left|f^{n+1}(z)\right| < r$; hence, the E_n can be viewed simply as the set of points that remain exterior to C for at least n iterations. Since the points of E_{n+1} remain exterior to C for $n+1$ iterations, E_{n+1} is contained in E_n, as Fatou stipulated.

The set E_n has d^n components, where d is the degree of f, each of which is bounded by a simple closed curve denoted collectively as Γ_n that, because of the definition of E_n, satisfies $f(\Gamma_n) = \Gamma_{n-1}$. Fatou observed that if, beyond a certain index, $|f'(z)| > A > 1$ on Γ_n, then the circumference of the Γ_n go to zero as $n \to \infty$; hence, the intersection of the closure of the E_n (the set of points $\{z\}$ for which $f^n(z)$ is exterior to C for all n) gives rise to infinitely many nestings, each of which intersects down to a single point. For example, referring to the upper right image in Color Plate 2, we obtain one nesting if we take a route beginning with the upper light green component, followed by the leftmost yellow, orange, and red-components, and so forth, and another if we instead follow rightmost components.

[5]Recall from Chapter 1 that the critical values of a function are the branch points of the various branches of its inverse; hence for f_2, for example, two distinct inverses do not exist at the critical points. See Definition 1.11. For the functions f_k the critical values are $z = 0$ and $z = 1$.

[6]It is possible to choose a circle for C and still obtain a Julia set in the manner Fatou describes, but one gives up either the invariance of iteration on the interior of C or the property that the preimage of C consists of discrete curves. For example, if C is the unit circle, the preimage of C under f_2 consists of two branches of a hyperbola that meet at the point of infinity, forming a figure eight, which is replicated in subsequent preimages. For a depiction of the dynamics of f_2 when C is the unit circle, see Alexander [**1994a**, §7.5].

Since each component of a given E_n contains infinitely many other components of higher order whose diameters go to zero, within a given ϵ of each nesting will be infinitely many others, so each point in J will also be a limit of J. The result is a totally disconnected perfect set.

The analogy between J and the Cantor middle-thirds set is obvious. One way to view the formation of J is to first remove the points inside C, then the points which arrive in the interior of C after one iteration, and so forth.

Consistent with the nature of a *Comptes Rendus* notice, Fatou provided very little justification for his assertion that the lengths of Γ_n go to zero, which is the cornerstone of his argument that J is a Cantor set. A little over a decade later he gave a similar construction in his monograph [**1919-1920**, Chap. IV, p. 245 ff.],[7] where he showed that the Julia set for a particular class of rational functions is a totally disconnected perfect set.[8]

As was the case in his 1906 notice, Γ_n comprised d^n closed curves. Fatou showed that for a representative function f from the class he was considering, there was an N such that when $n \geq N$, $|f'(z)| > A > 1$ on Γ_n, which implies that the length of Γ_n (and with it, those of the curves that compose it) go to zero. His justification for this last statement was based on differentials and is essentially as follows.

Since $f(\Gamma_n) = \Gamma_{n-1}$, we can write $\gamma_{n-1} = f(\gamma_n)$ where γ_n is one of the curves that make up Γ_n and γ_{n-1} is its image. An application of the chain rule yields

$$d\gamma_n(t)\,|f'(\gamma_n(t))| = d\gamma_{n-1}(t),$$

where $d\gamma_n$ is an arc length element of γ_n. For $n \geq N$, $|f'(z)| > A > 1$ on Γ_n, hence

$$d\gamma_n < \frac{1}{A} d\gamma_{n-1},$$

which, because the union of the γ_n is Γ_n, implies

$$\ell(\Gamma_n) < \frac{1}{A}\ell(\Gamma_{n-1}).$$

It readily follows that

$$\lim_{n\to\infty} \ell(\Gamma_n) = 0.$$

Was this the argument he had in mind in 1906? We of course cannot be certain. But the fact that his justification from [**1919-1920**] followed the outline he had sketched in 1906 suggests that it may have been.

The dynamics involved in the iteration of the second kind of function, typified by the function $g(z) = (z^2 + z)/2$, are quite different from those of the f_k. As noted above, Fatou observed that for g, the boundary between the domains of convergence of $z = 0$ and the point at infinity is a simple closed curve Γ whose points do not in general posses a well-defined tangent line. The curve Γ and the domains of convergence are depicted in Color Plate 3.

Caption for Color Plate 3: (See p. 63.) At left: The Fatou and Julia sets for the function

$$g(z) = \frac{z^2 + z}{2},$$

[7]We will discuss this monograph briefly in the next section and in much greater detail in subsequent chapters.

[8]See Theorem 7.13 in Chapter 7.

also from Fatou's 1906 notice, which admits the fixed points $O = 0$ (superattracting), $A = 1$ (repelling), and the point at infinity (attracting). Points in the red region converge to zero under iteration; points on its boundary are in the Julia set J. Points shaded blue are in the basin of attraction for the point at infinity and are colored according to the number of iterations it takes for their moduli to reach a particular magnitude M.

At right: Two magnifications near $z = 1$ reflecting the fact that J retains the same general pattern of irregularity regardless of how much it is magnified. This property is often referred to as self-similarity or homogeneity. The upper image at right is a magnification of the region near A indicated in the larger image. The lower image is a magnification of the region indicated by the square in the top right image. The cusps indicate why tangents do not exist: they pervade Γ and become more evident upon subsequent magnifications.

Fatou's justification that Γ is not analytic is a little obscure. He noted first that Γ is symmetric with respect to the real axis, hence, if Γ is analytic, then there would be an analytic function φ mapping a point z on Γ to its complex conjugate, $\overline{z}$, which, due to Γ's symmetry, is also on Γ. Although he did not say why this is so, it follows from the Schwarz reflection principal and φ—whose domain can be extended to an annulus surrounding Γ—is sometimes referred to as a Schwarz function. Since g is a polynomial with real coefficients, one can easily show that $\overline{g(z)} = g(\overline{z})$, from which $\varphi \circ g = g \circ \varphi$ readily follows. Without going into details, Fatou noted the fact that g and φ satisfy the system of functional equations $\varphi \circ g = g \circ \varphi$ and $\varphi^2 = I$ leads to a contradiction;[9] hence he concluded that Γ is not analytic.

It is unclear what argument he had in mind, but an argument for a similar proposition from his *Comptes Rendus* notice [**1918a**] may offer some clues. In that article he showed, under similar conditions, that Γ is analytic only if it is a circle. His argument relied on the Schwarz reflection principle and asserted that if Γ is analytic, it is the image under a linear mapping of a circle surrounding the origin. Perhaps he had an inchoate version of this argument in mind. Indeed, if φ is linear, one can use the conditions $\varphi(-2) = -2$, $\varphi(1) = 1$ (which follow from the fact that $z = 1, -2$ are points on Γ) and $\varphi^2 = I$ to deduce an explicit equation for φ,

$$\varphi = \frac{4 - z}{1 + 2z}.$$

Since g and φ are defined on a set with a limit point, the relation $\varphi \circ g = g \circ \varphi$ holds beyond Γ. But one can easily choose a point not on Γ for which $\varphi \circ g = g \circ \varphi$ does not hold.

An example where Γ is analytic occurs with $g(z) = z^2$, where $\varphi(z) = 1/z$, Γ is the unit circle, and the system of functional equations $\varphi \circ g = g \circ \varphi$ and $\varphi^2 = I$ is not only satisfied on Γ, but throughout $\mathbb{C}$. The inference from Fatou's argument from 1918 is that if the dynamical system given by $z \mapsto z^2$ is perturbed in such a way that Γ is not a circle, its Julia set is no longer an analytic curve.

Because it appeared during one of those dips of interest that occur periodically within the study of iteration, Fatou's groundbreaking notice garnered no response

[9] I represents the identity function.

for quite some time. Even though it would be ten years before Fatou would again take up the study of the iteration of rational functions in print, Audin [**2009**, p. 17], citing a 1907 letter from Fatou to Maurice Fréchet, points out that Fatou continued to explore the subject subsequent to the publication of his 1906 notice.

2.3. Where we are headed

In 1917 Fatou, Gaston Julia, Samuel Lattès, and Joseph Fels Ritt published a flurry of papers on the global behavior of iterates of complex rational functions. The motivation for Lattès and Julia—and most likely Fatou—was the 1915 announcement by the French Académie des Sciences that its 1918 *Grand Prix des Sciences Mathématiques* competition would be devoted to the study of the iteration of complex functions. This flurry became a blizzard, and two lengthy monographs from Fatou and Julia ([**1919-1920**] and [**1918a**], respectively) brought a substantial global theory of rational dynamics into being.

There was still work to do, however. Despite their best efforts, neither Fatou nor Julia were able to determine the iterative behavior of rational functions near an irrationally neutral fixed point, and they left essential questions unanswered. An understanding of iteration near these points began to emerge in the late 1920s with the work of the young German mathematician Hubert Cremer [**1927**], and Carl Ludwig Siegel gave a conclusive answer in [**1942**] to one of the most significant questions left open by Fatou and Julia.

Moreover, the studies Fatou and Julia prepared in anticipation of the *Grand Prix* focused almost exclusively on the iteration of rational functions. Consequently, very little was known about the global behavior under iteration of other kinds of complex functions. In an attempt to remedy this, Fatou published two introductory papers, one on the iteration of algebraic functions [**1922a**] and another [**1926**] on transcendental dynamics. These papers found little response for quite some time. Due in part to the disruptions caused by World War II, Siegel's 1942 paper signaled an end to the phase of complex dynamics that began with Fatou's 1906 notice.

It would be another 40 years or so before the study of complex dynamics would again receive the sort of attention that it had at the time of 1918 *Grand Prix*. Beguiling images such as the Mandelbrot set and Newton's method for the cubic circulated widely in the early 1980s, even gaining the attention of audiences beyond the community of mathematicians.[10] Images which Fatou and Julia could only describe with metaphor could now be seen. Extraordinarily bountiful lines of inquiry opened, and advances in topology and analysis provided the tools mathematicians needed to analyze what they saw. The field had entered its mature phase.

Before we continue, we would like to dispel the commonly held notion that very little happened between 1942 and the studies of Adrien Douady, John H. Hubbard, and Benoit Mandelbrot of the late 1970s and early 1980s. Not only did Irvine Noel Baker develop a substantial research program on the iteration of entire, transcendental maps beginning in 1955 (Rippon and Stallard's tribute to Baker lists 79 papers by Baker related to the iteration of entire maps [**2008**, pp. xv–xix]

[10]The August 1985 popular science magazine *Scientific American* featured the Mandelbrot set on its cover, and complex dynamics was often portrayed as part of the emerging science of chaos, which caught the popular imagination in the 1980s. The best-selling general-audience book *Chaos* by James Gleick, published in 1987, devoted a chapter to complex dynamics.

over half of which appeared before 1980) but we found well over 20 other works on the iteration of complex functions written between 1942 and 1980.

Some of these works are a residue of genealogy: Cremer's student Hans Töpfer wrote two papers on iteration [**1939, 1949**], as did his own student Bruno Klingen [**1962, 1968**]. Likewise, Baker's students Prodipeswar Bhattacharya [**1969**] and Lennox Liverpool[11] [**1971**] devoted their doctoral dissertations to iteration. Other works written during this period include Hans Brolin [**1965**], Thomas Cherry [**1964**], Maurice Heins [**1941**], Pekka Myrberg [**1960**], Hans Rådström [**1953**], and Paul Charles Rosenbloom [**1948, 1952**].

Perhaps one reason for the perception that interest in iteration withered is that its principals did not continue to research it: Ritt, Julia, and Cremer moved on to study other things; Fatou died early, at age 51, in 1929; Siegel's interest in complex dynamics was passing, his 1942 paper being a bit of a one-off. Nor, with the exception of Cremer, did they develop students who carried on the study. But interest did not fade after 1942 or even go into hiatus. Rather it persisted, but just not as prominently as it once had and would be again.

In the next two chapters we devote considerable space to the study of differential equations. While the studies of iteration as practiced by Fatou and Julia are divergent from the use of iteration in celestial mechanics, as exemplified by Henri Poincaré, the two studies are not unrelated. First of all, in its *Grand Prix* announcement the Académie referred to Poincaré's use of iteration. We wish to explore that reference with a discussion that includes Poincaré's return map. To put the return map in its proper context requires an excursion into Hamiltonian systems and the concept of stability. Moreover, Lattès extended Poincaré's method of investigating differential equations and adapted them to his own study of the functional equations that naturally arise in iteration, and our examination of Poincaré's return map will help the reader to fully appreciate what Lattès did.

We ask those readers whose interest is not differential equations to have patience: we will return to the study of complex dynamics, but not before we embark on an examination of the small divisors problem in the context of celestial mechanics and perturbation theory in Chapter 4. We do this for two reasons: The first is that small divisors surface at several turns in our text. They are the principal impediment to understanding the dynamics of iteration near an irrationally neutral fixed point, and they vexed many of the mathematicians we discuss, from the American mathematician George Adam Pfeiffer, who was the first to encounter them in the context of complex dynamics around 1915, to Fatou, Julia, and Cremer. We hope that our portrayal of the difficulties that small divisors problems posed for these mathematicians will foster a deeper appreciation of their struggles and accomplishments. The second reason is that Siegel's demonstration of the existence of centers[12] in complex dynamics represents a rare—at least for the time—success with a small divisors problem. This event influenced subsequent developments in KAM theory, an influence we explore first in Chapter 4 and then, again, in more detail in Chapter 12.

We return, finally, to complex dynamics in Chapter 5 with an account of the early 20th century developments in the U.S., which largely took place prior to the

[11]Liverpool's own student, Kenneth Yugunda, wrote his thesis on iteration as well, in 1998, at the University of Jos in Nigeria.

[12]Centers are discussed in Section 1.6.2.

U.S. entry into World War I, and before the appearance of the works of Fatou and Julia prepared in anticipation of the 1918 *Grand Prix*. While informed by the work of Kœnigs and his students, particularly their work with functional equations, Edward Kasner's study of invariants in conformal geometry [**1913**] was also essential to the beginnings of the American study. The work of two of the Americans discussed in this chapter, Ritt and Pfeiffer, factors into subsequent European developments.

The 1918 *Grand Prix des Sciences Mathématiques* is obviously at the center of our story, and we devote the next several chapters to events revolving around the competition for the prize. We examine the responses of Fatou, Julia, Lattès, and others, as well as questions left unanswered. Paramount among the open questions is what is often referred to as the *center problem*, which seeks to determine whether the iteration of a nonlinear rational function near an irrationally neutral fixed point is ever conformally equivalent to a rotation of the unit disc.

A reason to suspect that it might be stems from an examination of the canonical Schröder functional equation discussed in Section 1.4.1,

$$S \circ f = f'(\lambda) \cdot S, \tag{2.1}$$

where λ is a fixed point of f. This equation is fundamental to the study of iteration, and almost everyone who studied iteration in our story considered it. As we noted, when λ is an attracting but not superattracting fixed point (i.e., the modulus $|f'(\lambda)|$ is strictly between zero and one), this equation can always be solved, and its effect is to linearize iteration via the mapping $z \mapsto f'(\lambda)z$. It is natural to ask, then, if solutions exist in the irrationally indifferent case. If so, iteration would be conformally equivalent to the rotation $z \mapsto e^{i2\pi\theta}z$ where $f'(\lambda) = e^{i2\pi\theta}$.

However, due in large measure to issues involving small divisors, proving the existence of such behavior is quite difficult, and though they tried, Fatou and Julia made little progress toward the solution of the center problem. Nonetheless they explored what might happen if centers did exist and gained a not inconsiderable insight into the dynamics of the problem. We describe their examination in Chapter 9. In Chapter 10, the last to treat the *Grand Prix* and its immediate aftermath, we survey the study of iteration in the 1920s.

In Chapters 11 and 12 we continue with the history of the center problem. The former chapter concerns the mathematics of Cremer who, in the late 1920s and 1930s, found conditions that imply solutions to equation (2.1) do not exist; the latter chapter discusses Siegel, who found conditions under which they do. The final chapter surveys mathematicians around the world who studied iteration of complex functions in the 1920s and 1930s.

Our book closes with a series of appendices by present-day researchers that treat technical details involved in the topics we investigate and provide additional historical perspectives by describing subsequent developments. We thus hope to link old discoveries to current investigations, thereby providing a fuller context to our discussion. Included in the appendices are brief accounts of the lives of Fatou and Julia, capsule biographies of other mathematicians involved in our story, and a discussion of the technical details of the computer graphics we include.

CHAPTER 3

Iteration and Differential Equations I: The Poincaré Connection

SESSION OF DECEMBER 27, 1915
GRAND PRIX DES SCIENCES MATHÉMATIQUES
(Prize budget: 3000 fr.)
Biennial prize with variable topic

The iteration of a substitution in one or more variables, that is to say, the construction of a system of successive points $P_1, P_2, \ldots, P_n, \ldots$, where each is deduced from the preceding by a given operation:

$$P_n = \phi(P_{n-1}) \qquad (n = 1, 2, \ldots, \infty)$$

(where ϕ depends rationally, for example, on the point P_{n-1}) and where the initial point P_0 is given, arises in many classical theories as well as in some of the most celebrated memoirs of Poincaré. Up to the present, the most well-known works have been devoted exclusively to the "local" point of view.

The Académie wishes to move the investigation to the entire domain of values which the variables can assume. In this spirit it has put forth the following question for the year 1918:

To perfect in a meaningful way the study of the successive powers of a given substitution, when the index of the power grows indefinitely.

One should consider the influence of the initial value P_0, the given substitution, and one may confine oneself to the most simple cases, such as the rational substitution of one variable. [**CR-1915**, p. 921]

3.1. Introduction: Poincaré and the roots of complex dynamics

Note the reference to Poincaré in the above announcement. We want to explore this. Why was it included? What "celebrated memoirs" did the Académie have in mind?

We assume that the announcement refers to either (or perhaps both) Poincaré's study of Kleinian groups in the early 1880s, or to the investigation of ordinary differential equations (ODEs) and celestial mechanics that occupied him for much of the last two decades of the 19th century, and included the paper that earned him the prize King Oscar II of Sweden offered in celebration of his 60th birthday [**1890b**] as well as the three-volume *Les méthodes nouvelles de la mécanique céleste* [**1892a, 1893, 1899b**].

There are indeed similarities between the iteration of complex functions and the automorphic functions that arise in the study of Kleinian groups, which Fatou pointed out in his 1906 notice when he observed that the set of points which did not converge to the attracting fixed points of the examples he gave are "analogous" to sets found in the study of automorphic functions [**1906b**, p. 548]. He and Julia pointed out other similarities between iteration and Kleinian groups; see, for example, Fatou [**1919-1920**, pp. 269, 314] and Julia [**1918a**, pp. 51, 106–107, 202, 213, 276–277].

Audin, however, argues in [**2009**, p. 15, fn.] that the Académie was not referring to Poincaré's study of Kleinian groups in its announcement. We agree. The iteration found in his study of Kleinian groups involves Möbius maps and is so elementary that neither Fatou nor Julia bothered with it, and therefore it seems an unlikely motivation for the Académie. Moreover, although examples of what would come to be called the Julia set are present in Fatou's 1906 notice, there was no sense of its centrality to the study of iteration at the time of the announcement, so it is equally unlikely that the Académie had this connection in mind. It turns out that there are links between complex dynamics and Kleinian groups that run quite deep, even though they were not understood at the time. Appendix D discusses these links.

We think it likely, then, that the Académie was referring to Poincaré's studies of ODEs, where iteration occurs with Poincaré's return map, one of the tools he developed to study the stability of periodic solutions that arise in the n-body problem, which we discuss in Section 3.3.

However, the nature of iteration in Poincaré's studies are quite different from extant works in complex dynamics. Poincaré generally iterated functions of multiple real variables and many of the techniques and theorems developed in the real case do not apply to complex functions, although under certain conditions it is useful to view the iteration of real functions as taking place on the complex plane, something we will touch upon briefly at the end of Section 3.5.

Fatou delivered an interesting verdict on whether Poincaré's use of iteration in celestial mechanics influenced his own study. He only mentioned Poincaré's use of iteration once in his monograph, at the beginning of its first installment [**1919-1920**, pp. 161–162], where he made an explicit reference to pages 193–195 of Poincaré's paper, *Sur les courbes définies par une équation différentielle* [**1886**], during his discussion of the works that came before his own. He observed that these studies—presumably including Poincaré's—were local in nature. Moreover, he noted that he was restricting his own work to the iteration of functions of a single variable while Poincaré (along with a few others) undertook the "more difficult" case of two or more variables.

This was Fatou's sole reference to Poincaré's use of iteration in the study of ordinary differential equations. Julia made not a one, and it is clear from his citations that he saw the genesis of his own study in the work of Kœnigs and, to a much lesser degree, his student Leau. Likewise, several references to Kœnigs and his students indicate that Fatou also saw his work as flowing out of the Kœnigs school. Had either Fatou or Julia felt Poincaré's use of iteration to be a significant influence, we assume they would have said so.

Perhaps we are digging too deeply here, and the reference to Poincaré was not intended to motivate or to suggest any specific connection between Poincaré's work and the iteration of complex rational functions, but we merely observe that iteration is a widely used tool in mathematics, and that among those who used it was the celebrated Poincaré.

Whatever the Académie's motivation, its mention of Poincaré raises several interesting questions: Did Poincaré's use of iteration somehow influence Kœnigs' study—or vice versa? Is there a connection between complex dynamics and Poincaré's use of iteration in his study of ODEs? Are there perhaps other connections between the two fields? If so, did any of the principals in complex dynamics explore these interrelations? Finally, did Fatou and Julia find anything in Poincaré's *œuvre* that would prove useful?

We will explore each of these questions in turn. Our examination of the connections between ODEs and complex dynamics will occupy significant portions of this chapter and the next, as well as Chapter 12, where we investigate connections between complex dynamics and the origins of KAM theory. We will address the others in this chapter, beginning with the question of common origins.

Two things might suggest that Kœnigs' study of iteration and Poincaré's use of iteration in ODEs have a common origin. One is that iteration began to appear in the works of the two mathematicians at roughly the same time. Kœnigs began to study iteration in [**1883**], while Poincaré's use of iteration dates back to 1881 in [**1881-82**]. The other is that both used a similar terminology to describe the process of iteration, words such as *antecedent* and *consequent* and later, with Julia and others, *center*.[1]

When he began his study in 1883, Kœnigs may have been aware of Poincaré's application of iteration in differential equations and Kleinian groups. However, Kœnigs gave no indication that he was influenced by Poincaré, and the only explicit example he examined from Poincaré's work came in the third of his three papers to concern iteration and functional equations:

> *I will conclude with the case of linear substitutions. This case is simple, but it provides an interesting connection. In his investigation of circles passing through the double* [fixed] *points of a substitution, Poincaré formed a pencil of orthogonal circles. So, if x and x' are the double points of the substitution* $[Z]$, *it is well-known that it can be put in the form*
>
> $$\frac{Z-x}{Z-x'} = k\frac{z-x}{z-x'}.$$
>
> *Therefore* $B(z) = (x-x')\frac{z-x}{z-x'}$. [**1885b**, p. 404]

The function B is the solution to the Schröder equation $B \circ Z = k \cdot B(z)$, where k and Z are as above. He observed that the curves of "equal modulus" of B, that is, the level curves $|B(z)| = c$ for a given constant c, are the orthogonal circles while the curves of "equal argument," which are the curves orthogonal to the level curves, form the set of circles passing through x and x'.

He then suggested,

> *The important role that these circles enjoy in the representation of linear substitutions leads one to think that the curves of equal*

[1] An *antecedent* of a point z under f is a point w with $f(w) = z$ and is equivalent to the concept of a preimage given in Definition 1.8. The *consequent* of a point z is its image under f, $z_1 = f(z)$ and is synonymous with iterate.

> *modulus and equal argument of the function $B(z)$ may prove to be very useful in the study of any substitutions whatever; I have attempted here to provide a first step in that direction.* [**1885b**, p. 404]

Thus, rather than signaling Poincaré's influence, Kœnigs' discussion instead suggests that the tools he has developed in his study of iteration might be helpful in many other areas, including the study of linear groups.

The other examples and applications Kœnigs gave generally involve numerical algorithms (for example Newton's method) or functional equations such as the Schröder and Abel equations (we examine an application he gave involving the permutable functional equation $f \circ g = g \circ f$ in Section 10.2). The papers Kœnigs cited were likewise generally works discussing iteration and its relation to functional equations (for example, Farkas [**1884**], Korkine [**1882**], and Schröder [**1870, 1871**]); as with Fatou and Julia, we believe that had Poincaré's work influenced Kœnigs, he would have said so.

We do not know why Kœnigs began to study iteration. Salvatore Pincherle, writing in the *Enzyklopädie der Mathematischen Wissenschaften* [**1906**], and again six years later in the French edition, the *Encyclopédie des Sciences Mathématiques Pures et Appliquées*, [**1912**], observed that the field of functional equations generally concerns equations with unknown functions that have not yet become a field of study in their own right. Perhaps Kœnigs chose to group the functional equations useful in the study of iteration with the intent of creating a new field. If so, he was quite successful.

In any event, there seems to be no evidence that Kœnigs' study of iteration did not develop independently from Poincaré. That they share a common terminology for elementary objects is, in the absence of any other suggestion of a common origin, indicative of very little aside from the fact they share the process of iteration.

We do not mean to suggest that Poincaré had no influence on the work of Fatou and Julia, only that his work with iteration did not seem to. Indeed, two of Poincaré's papers proved important to the development of complex dynamics. One was his 1907 paper *Sur l'uniformisation des fonctions analytiques* [**1908**], which we will refer to in Chapter 7. Appendix I also contains a brief discussion on the development of uniformization theory.

The other, his 1890 paper *Sur une classe nouvelle de transcendantes uniformes* [**1890a**], involved what in the 1910s began to be called the Poincaré functional equation,

$$f \circ \theta = \theta(s \cdot I), \tag{3.1}$$

where I is the identity, s a complex constant and f a given function. The Poincaré equation has come to be almost as fundamental to iteration as the Schröder equation. Yet, Poincaré did not state the functional equation in this form. Rather, it is a special case of a system of equations that he studied,

$$\theta_1(su) = R_1[\theta_1(u), \ldots, \theta_n(u)]$$

$$\cdots$$

$$\theta_n(su) = R_n[\theta_1(u), \ldots, \theta_n(u)],$$

with s a complex constant, the R_k rational functions from $\overline{\mathbb{C}}^n$ to $\overline{\mathbb{C}}$, and the θ_k unknown.

It is one of the ironies of complex dynamics that while the above system of functional equations gave birth to equation (3.1)—a tool that proved extraordinarily useful to some of the mathematicians we will study—it did not itself arise from the study of iteration. That noted, Poincaré did use iteration to solve the system in the case where the funcational equations define a Cremona substitution [**1890a**, pp. 557–563].

The fact that iteration proved useful in his solution prompted Poincaré to pose a question he remarked he would leave unanswered: are there regions on which Cremona substitutions converge to a common limit under iteration [**1890a**, p. 561]?

Georges Valiron was perhaps the first mathematician to realize the utility of the single variable special case of the above system of equations in complex function theory. His 1913 thesis concerned entire functions of finite order and to demonstrate that functions of order zero arise naturally from functional equations, Valiron gave an example involving the Poincaré equation [**1913**, pp. 203–206]. Valiron's thesis was one of the sources Fatou suggested readers of his monograph on rational function consult, should they wish to know more about the Poincaré equation [**1919-1920**, p. 287]. It is possible, then, that Valiron may well have been the inspiration for its use in the theory of iteration.

The Poincaré functional equation is the inverse of the Schröder functional equation introduced in Definition 1.4 since each expresses a particular functional relationship, but from a reciprocal point of view. To see this, let $f(z) = sz + \sum_{k=2}^{\infty} a_k z^k$, with $s \neq 0$, be a function with a fixed point at zero. In studying iteration near a fixed point, it is very helpful to linearize it. As we noted in Section 1.7, Kœnigs chose to linearize the attracting case, that is where $0 < |s| < 1$, by seeking an analytic function S satisfying the Schröder equation $S \circ f = f'(0) \cdot S$. Around 1918, perhaps inspired by Valiron's special case, Lattès and Ritt independently chose to linearize the repelling case, where $|s| > 1$, by seeking a function θ satisfying $f \circ \theta = \theta \circ (s \cdot I)$.

Thus, linearization near a fixed point can be achieved in two opposite but equivalent ways, via the Poincaré equation or the Schröder equation:

(1) Schröder: $\quad S \circ f = s \cdot S \iff S(f(z)) = sS(z)$,
(2) Poincaré: $\quad f \circ \theta = \theta \circ (s \cdot I) \iff f(\theta(z)) = \theta(sz)$.

Just as the Schröder equation offers a convenient way to model the convergence under iteration of points near an attracting fixed point, the Poincaré equation is very useful in depicting the dispersive nature of iteration around a repelling point. Using the Schröder equation, if $0 < |s| < 1$ and z is in a neighborhood D of zero within the domain of S, routine calculations show that $f^n(z)$ converges to zero.

On the other hand, suppose $|s| > 1$. It can be shown (see, for example, Steinmetz [**1993**, p. 71]) that a meromorphic solution θ exists to the Poincaré equation for f. It follows that $f^n(\theta(z)) = \theta(s^n z)$. Picard's Little Theorem implies that there are at most two values of $w \in \overline{\mathbb{C}}$ where there is no z with $\theta(z) = w$. These w are called *exceptional values*. Thus, $f^n(\theta(z)) = \theta(s^n z)$ implies that on a neighborhood N of a repelling point of f, the union of the $f^n(N)$ for $n \in \mathbb{N}$ cover $\overline{\mathbb{C}}$ with the exception of at most two points. In other words, iterates of points near a repelling fixed point are dispersed throughout the plane.

That Poincaré's use of iteration did not influence the genesis of Kœnigs' study or have much an effect on Fatou and Julia does not mean that there is no connection between complex dynamics and ODEs or that Poincaré did not influence others besides Fatou and Julia in the field of complex iteration. Lattès, who was one of the three mathematicians to submit an entry for consideration in the 1918 *Grand Prix*, tried to reconcile the two subjects and was clearly influenced by Poincaré and his return map.

In the next section we lay the foundation for the remainder of our chapter with a short examination of Augustine-Louis Cauchy's use of undetermined coefficients and majorant functions. Our discussion of Cauchy will be historical in intent, although it is not intended as a definitive historical analysis of what he did.

While the techniques pioneered by Cauchy were adopted by Poincaré, they also proved crucial in determining the solutions to functional equations in the work ofLattès, Pfeiffer, and Siegel. Kœnigs also employed majorant functions in his solution of the Schröder equation in the neighborhood of an attracting (but not superattracting) fixed point; see equation (1.10).

In subsequent sections we review the work of Poincaré and Tullio Levi-Civita. For Poincaré, we rely on his papers [**1885**] and [**1886**], a decision we will justify below. One of the foci of our discussion is the concept of stability. It is for this reason we include Levi-Civita, whose refinement of Poincaré's concept of stability for periodic orbits is both elegant and close in spirit to its use in complex dynamics. While there are discussions of Levi-Civita's work in the literature (see Barrow-Green [**1997**, pp. 181–182]), they are generally brief, and we hope ours proves useful.

We conclude this chapter with the work of Lattès. One fundamental element that emerged from Lattès' work is the primary role of functional equations in describing the invariant curves associated with the iteration of a given function. Exploiting this notion, Lattès gave an example attesting to the close relationship between differential equations and iterated maps (see the discussion following equation (3.40) below).

3.2. Methods for continuous systems

As soon as differential equations became a tool for mathematically describing the evolution in time of physical phenomena, it became evident that their reduction to pure quadrature problems in order to get an analytic, closed form solution via integration was not in general possible. So mathematicians sought power series solutions.

The use of series methods (in particular power and trigonometric series) with undetermined coefficients became a primary tool in the study of ordinary and partial differential equations in the 18th and early 19th centuries. But it was only after Augustine Cauchy and Karl Weierstrass that the fundamental question about whether the series thus obtained are convergent was fully appreciated.

These two mathematicians combined the method of majorants and of undetermined coefficients to obtain bounds on the terms of the power series in order to determine convergence. Their approaches and their scopes were different. Cauchy focused on existence and uniqueness results as well as estimating the degree to which a truncated series approximates the actual solution, while Weierstrass viewed a differential equation as a mathematical tool to define an analytic function via its solution.

In the following quotation extracted from the introduction of his short paper *Mémoire sur un théorème fondamental, dans le calcul intégral* [**1842**], Cauchy explained his point of view:

> *In the theory of algebraic equations, geometers have rightly considered of fundamental importance the question of whether a general equation admits a root. Analogously, in the theory of integral calculus, one of the most important questions is whether ordinary and partial differential equations can be integrated.... Despite the numerous works of geometers regarding the integral calculus, this important question has not been thoroughly solved in its full generality.... By virtue of a fundamental theorem,* [not only can one obtain convergence of the formal power series representing the solution] *but one may also find a bound for the error that arises when one stops the expansion after computing a certain number of terms.* [**1842**, p. 1020]

Although Cauchy, whose work took place in the 1830s and 1840s, preceded Weierstrass, the latter was not initially aware of Cauchy's results according to Cooke [**2001-W**, p. 1] and considered a less general problem by assuming that all the functions defining the ODE were of polynomial type. This allowed him to construct majorants that were polynomials and, therefore, easily integrated.

Another important difference between the two approaches was Cauchy's use of the Cauchy integral formula for the derivatives of a function $f(z)$ holomorphic (i.e., analytic) on a disc $B(z_0, R)$ of radius R about a point z_0,

$$f^{(n)}(z_0) = \frac{n!}{2\pi i} \oint \frac{f(\zeta)}{(\zeta - z_0)^{n+1}}\, d\zeta, \qquad n = 0, 1, \ldots,$$

which yields Cauchy's inequality

$$|f^{(n)}(z_0)| \leq \frac{n! M(\rho)}{\rho^n}, \qquad n = 0, 1, \ldots, \tag{3.2}$$

where $M(\rho)$ is the maximum of $|f(z)|$ on the circle $|z - z_0| = \rho$. The convergence of the Taylor series on the disc $B(z_0, R)$ is then easily deduced:

$$\left| \sum_{n=0}^{\infty} \frac{f^{(n)}(z_0)}{n!}(z - z_0)^n \right| \leq M(\rho) \sum_{n=0}^{\infty} \frac{|z - z_0|^n}{\rho^n} = \frac{M(\rho)}{1 - \frac{|z-z_0|}{\rho}} < +\infty, \tag{3.3}$$

where ρ is such that $0 \leq |z - z_0| < \rho < R$.

Given Cauchy's influence on the subsequent French study of differential equations, we will describe Cauchy's point of view in order to set the stage for our discussion of Poincaré and Lattès.

In [**1842**] Cauchy's purpose was to apply the technique of majorant functions to partial differential equations by considering the generalization of (3.2) to an arbitrary number of variables.[2] For a function of two variables $f(t, z)$ the inequality becomes

$$\left| \frac{\partial^{m+n} f}{\partial t^m \partial z^n}(t_0, z_0) \right| \leq \frac{m! n! M(r, \rho)}{r^m \rho^n}, \qquad m, n = 0, 1, \ldots, \tag{3.4}$$

with $M(r, \rho)$ an upper bound of $|f(t, z)|$ in the region

$$D = \{|t - t_0| \leq r,\ |z - z_0| \leq \rho\}.$$

[2]Cauchy called this method the "*calcul des limites*" in [**1842**].

Following the approach of Jean-Claude Briot and Charles A. Bouquet in [**1856**], we will consider an initial value problem of the form[3]

$$\begin{cases} y' = f(t,y), \\ y(t_0) = y_0, \end{cases} \tag{3.5}$$

where $f(t,y)$ is taken to be analytic on an open set containing D:

$$f(t,y) = \sum_{m=0}^{\infty}\sum_{n=0}^{\infty} c_{mn}(t-t_0)^m(y-y_0)^n, \qquad \text{for } (t,y) \in D, \tag{3.6}$$

with the coefficients of the Taylor expansion given by

$$c_{mn} = \frac{1}{m!n!}\frac{\partial^{m+n} f}{\partial t^m \partial z^n}(t_0, y_0).$$

Cauchy's approach to the existence and uniqueness of the solution to (3.5) consists of two steps.

(a) First, to find a *formal series solution*[4] via the method of undetermined coefficients for series,

$$y(t) = y_0 + \sum_{k=1}^{\infty} y_k(t-t_0)^k. \tag{3.7}$$

(b) Next, to use the method of majorants to compare the coefficients of this formal solution with those of a known convergent series in order to obtain the convergence and therefore the analyticity of (3.7).

For Cauchy the natural place to develop the theory of differential equations was the complex plane.

Substituting (3.6) and (3.7) into (3.5), Cauchy obtained

$$y_1 + \sum_{k=1}^{\infty}(k+1)y_{k+1}(t-t_0)^k = \sum_{m=0}^{\infty}\sum_{n=0}^{\infty} c_{mn}(t-t_0)^m\Big(\sum_{k=1}^{\infty} y_k(t-t_0)^k\Big)^n,$$

from which the unknowns y_k, $k = 1, 2, \ldots$, are recursively defined in terms of the coefficients c_{mn}. The first few terms are

$$\begin{aligned} y_1 &= c_{00}, \\ 2y_2 &= c_{10} + c_{01}y_1, \\ 3y_3 &= c_{20} + c_{11}y_1 + c_{02}y_1^2 + c_{01}y_2, \\ 4y_4 &= c_{30} + c_{21}y_1 + c_{11}y_2 + c_{01}y_3 + c_{12}y_1^2 + 2c_{02}y_1y_2 + c_{03}y_1^3; \end{aligned}$$

and, in general,

$$y_k = \frac{1}{k}\sum_{\substack{m,n=0 \\ m+n\le k}}^{k} c_{mn} \sum_{k_1+\cdots+k_n=k-m} y_{k_1}y_{k_2}\cdots y_{k_n}, \qquad k = 2, 3, \ldots, \tag{3.8}$$

with $k_i \geq 1$ and where we assume that for $n = 0$, $k_1 + \cdots + k_n = 0$ and $y_{k_1}\cdots y_{k_n} = 1$ (so some c_{i0} may not explicitly appear in the formula). The recursion relation (3.8) uniquely defines the coefficients of the formal series (3.7) thus providing a uniqueness result for the solution.

---

[3]Our presentation of the method of majorants borrows from the approach followed by Roger L. Cooke in his paper [**2001-W**] which summarizes Briot and Bouquet's exposition.

[4]A *formal series solution* (or a *formal solution*) satisfies a functional or differential equation but its convergence has not been established.

The convergence on an interval $|t - t_0| < T \le r$ is accomplished via step (b). Cauchy's approach stems from his observation that a Taylor series converges if and only if it is dominated by a geometric series, as demonstrated by (3.3) and its generalizations to functions with several variables. His method of majorants is based on comparing the formal solution of the original problem (3.5) to a convergent solution of a second initial value problem

$$\begin{cases} z' = g(t, z), \\ z(0) = 0, \end{cases} \tag{3.9}$$

where $g(t, z)$ (which will be constructed below) is analytic on an open domain containing D whose Taylor expansion

$$g(t, y) = \sum_{m=0}^{\infty} \sum_{n=0}^{\infty} g_{mn} t^m y^n, \qquad \text{for } (t, y) \in D, \tag{3.10}$$

consists of nonnegative coefficients g_{mn} that bound those of f:

$$|c_{mn}| \le g_{mn}, \qquad \text{for all } m, n = 0, 1, \dots\ . \tag{3.11}$$

Furthermore, equation (3.9) is assumed to posses an analytic solution $z(t) = \sum_{k=1}^{\infty} z_k t^k$ for $|t| < T$, whose convergence depends on that of g. The coefficients z_k satisfy a relation of the form (3.8) with the coefficients c_{mn} and y_{k_i} replaced by g_{mn} and z_{k_i}, respectively. A straightforward induction argument shows that all the z_k are nonnegative and, comparing the absolute value of both sides of (3.8) and exploiting the inequality (3.11), satisfy $|y_k| \le z_k$. It then follows that the series for $y(t)$ converges at the very least on the time interval $|t - t_0| < T$ as long as g does.

The final step is to construct g. This is trivially accomplished by Cauchy's inequalities (3.4) which suggests the obvious choice $g_{mn} = M(r, \rho) r^{-m} \rho^{-n}$ and therefore

$$g(t, z) = \frac{M(r, \rho)}{\left(1 - \frac{t}{r}\right)\left(1 - \frac{z}{\rho}\right)},$$

allows the explicit computation in closed form of the solution of (3.9) as

$$z(t) = \rho \left(1 - \left(1 + \frac{2rM(r, \rho)}{\rho} \log\left(1 - \frac{t}{r}\right)\right)^{\frac{1}{2}}\right),$$

which proves convergence of $y(t)$ at least on the interval $|t - t_0| < r(1 - e^{-\rho/(2Mr)})$.

An important by-product of Cauchy's method of majorants is an estimation of the error of the approximating solution $\tilde{y}(t)$ obtained by truncating the series (3.7) at a particular term. From the above discussion it easily follows that this error is bounded by the corresponding one in the series for $z(t)$. Thus, Cauchy's method also provides a useful tool to numerically approximate solutions of differential equations.

3.3. Poincaré's return map: An early use of iteration

The method of majorant functions proved to be so powerful that it was readily imported into subjects related to ordinary differential equations. Poincaré applied it to his study of dynamical systems; it was also successfully applied by Kœnigs and Lattès to functional equations.

As we will see below, Poincaré's use of majorants is tied to his return map, an iterative technique he developed to investigate the stability of periodic orbits, first in autonomous ODEs of dimension three, and later in the three-body problem.[5]

One way to view the return map is as a method of generating a sequence of points $\{(x_n, y_n)\}$ which can be viewed as the orbit[6] emanating from an initial value (x_0, y_0). The mapping which produces these points is summarized below as equation (3.15).

Poincaré's study of the three-body problem was motivated by the larger goal of ascertaining whether or not the solar system is stable. Due to the high degree of freedom of the associated Hamiltonian differential equation, the general n-body problem is quite complex and its analysis is notably simplified by considering the so-called *circular (or elliptical) restricted three-body problem.* This describes the behavior of three bodies on a plane, subject to the mutual Newtonian forces, one of which (the planetoid) has negligible mass with respect to the others which describe circular (elliptic) orbits around their common center of gravity.

For Poincaré the key to understanding the behavior of a dynamical system was to understand the stability of its periodic orbits, a point of view that became a predominant theme in the study of dynamical systems. By stability of an orbit (sometimes also referred to as a trajectory), Poincaré meant the following, a definition he ascribed to Siméon Denis Poisson:

> *We say that the trajectory of a moving point is stable if, describing any circle or sphere of radius r around the starting position, the moving point, after escaping the circle or sphere, enters it again infinitely often, no matter how small one chooses r.* [**1885**, p. 171]

In what follows, we trace Poincaré's construction of the return map, as well as his subsequent examination of the stability of the trajectories associated with it.

Our discussion of Poincaré's return map is based primarily upon his papers from the mid-1880s [**1881-82, 1885, 1886**] rather than on his more celebrated works from the early 1890s [**1890b, 1893**]. These latter works represent the climax of Poincaré's research activity regarding the dynamics of the three-body problem and include the well-known formalization of the *general problem of dynamics* that gave rise to the development of perturbation theory within celestial mechanics (which will discuss in Section 4.4). Considering that understanding the three-body problem was Poincaré's dominating motivation during his extended study of ODEs, the former works are rightly perceived as preparatory to his later developments.

Nevertheless, it was in the earlier papers that Poincaré introduced almost all the important tools which he later skillfully refined and adapted to handle the enormous mathematical difficulties emerging from his investigation of the stability of periodic orbits in Hamiltonian systems. It is also noteworthy that both Levi-Civita and Lattès, in works we will examine below, independently cite Poincaré's early works as the basis for their own studies.

---

[5]Details on the three-body problem and related questions may be found in Barrow-Green's excellent monograph [**1997**].

[6]A word of caution: In complex dynamics orbits are almost always discrete sets; in ODEs they are generally continuous. However, Poincaré's return map does produce discrete orbits. Unless it is clear we are talking about an orbit induced by the return map, the reader should assume that during our discussion of ODEs, orbits are continuous.

Moreover, in his introduction to his monograph on the iteration of rational functions [**1919-1920**, pp. 161–162], Fatou made explicit mention of the use of iteration in Poincaré's 1886 paper, as noted above. Fatou's reference to Poincaré's work is perhaps somewhat tangential since Poincaré had no direct connection with the theory Fatou developed in subsequent chapters of [**1919-1920**].

One purpose of Fatou's reference was to emphasize the particular difficulties associated with the iteration of functions of two or more variables, such as the one conducted by Poincaré, and to contrast it to his own study concerning single variable functions.

Another, as we have also noted, was to include Poincaré's work among the local studies of iteration that preceded his own global study. Strictly speaking, this is accurate. Poincaré's treatment of iteration is indeed local. However, this characterization is a bit misleading because Poincaré's intention was not the study of the iteration of his return map in and of itself; rather, it was a means to examine the stability of the solutions to ODEs. It should also be pointed out that Poincaré's study of ODEs was global in nature.

The specificity of Fatou's citation (he referred to Poincaré [**1886**, pp. 193–195]) attests to his detailed knowledge of Poincaré's use of iteration, and in particular, the return map. This argument is reinforced by the fact that Fatou was also an active researcher in celestial mechanics, especially in the late 1920s, when he carried out studies of perturbation theory in a spirit similar to Poincaré's (frequently referencing his works) but with more restrictive assumptions (see [**1921e, 1923c, 1928b, 1929b, 1931**]). By "restrictive" we mean they addressed particular problems but were not necessarily intended to open new horizons.

Finally, even though it had minimal impact upon the genesis of Fatou's monograph, the early phase of Poincaré's study of the stability of specific solutions of ODEs has the more relevant connections to the iteration of functions in the complex plane because there are several analogies between the two theories, which we will highlight during our discussion of Lattès in Section 3.7.

The initial concern of Poincaré's investigation of the ODEs related to the three-body problem was the dynamics of low dimension general ODEs. The discoveries he made were then finely tuned over time to better handle the Hamiltonian case.

His intuition that a qualitative study of the stability of solution curves in the entire phase space is preferred to merely regarding solutions as functions of time expressible as power series, as was the custom of predecessors, is quite clear from the inception of his investigations.

He began the series of papers [**1881-82, 1885**] by constructing the solution curves of the two-dimensional equation

$$\frac{\mathrm{d}x}{\mathrm{d}t} = X, \qquad \frac{\mathrm{d}y}{\mathrm{d}t} = Y. \tag{3.12}$$

His approach to the problem, as well as his desire to examine the geometry of the solution curves, suggested to him that the most convenient environment to describe and visualize the dynamics of the solution curves was a sphere tangent to the xy plane at the south pole. This is analogous to the view that the natural place to study the iteration of a complex rational function is on the Riemann sphere, $\overline{\mathbb{C}}$.[7]

[7]One difference is that Poincaré used a gnomonic, rather than stereographic, projection, where the intersection of the straight line joining a point (x, y) of the plane with the center of the sphere locates two antipodal points on the sphere. To create the Riemann sphere, stereographic projection

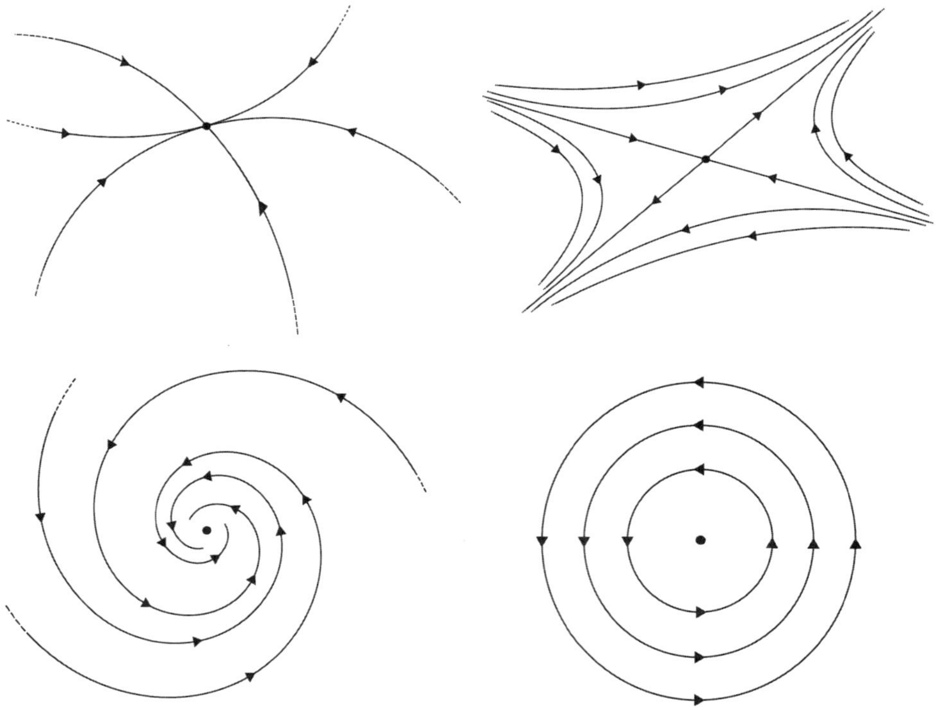

FIGURE 3.1. The nature of an equilibrium point of an ODE in the plane. Four outcomes for the neighboring trajectories near an equilibrium point of a two-dimensional autonomous dynamical system. Upper left, attracting node; upper right, saddle point; bottom left, attracting focus; and bottom right, center.

To study the solutions of equation (3.12), Poincaré coined the terms *nœdes* (nodes), *cols* (saddle points), *foyers* (foci), and *centres* to describe the manner they approach (or depart from) equilibrium points[8] (see Figure 3.1 and Barrow-Green [**1997**, pp. 30–33]). In investigating the existence of periodic orbits in the phase plane, he was led to the discovery of *limit cycles*, which are particular periodic orbits that the neighboring solution curves eventually approach.

It was here that his return map (sometimes referred to as the *Poincaré first return map*) made its appearance. He defined a generic curve C in the plane, transversal to the flow at each point, that is generally referred to as a *Poincaré section* and studied orbits originating from a point located on C under the assumption that it crosses C an infinite number of times, thus generating a sequence of points $\{(x_n, y_n)\}$, where (x_n, y_n) is the nth crossing point of the orbit. We can induce the return map $\mathcal{P}$ from this procedure, defined by $(x_n, y_n) \mapsto (x_{n+1}, y_{n+1})$. See Figure 3.2 for a depiction of the return map in two and three dimensions.

---

is used to induce a bijection between the compactified complex plane and the (Riemann) sphere whereby an arbitrary point P of the complex plane is mapped to a point P' on the sphere defined by intersection of the line joining P to the sphere's north pole.

[8]Sometimes an equilibrium point is called a singular point.

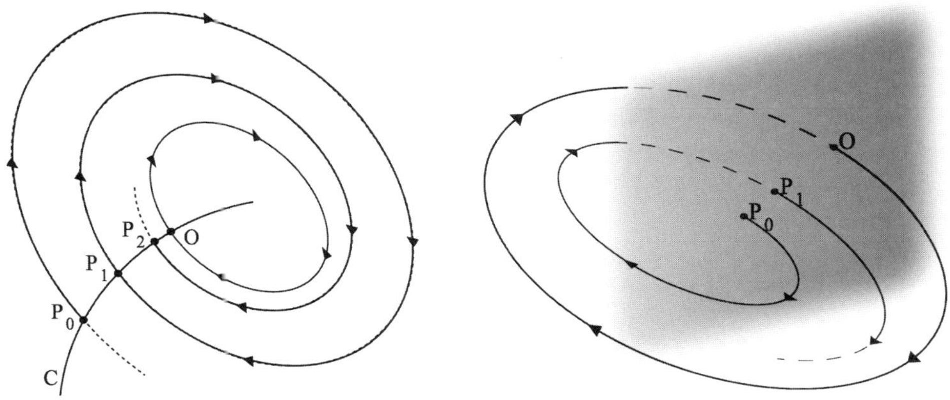

FIGURE 3.2. The Poincaré return map. An orbit originating from a point of a Poincaré section close to an attracting periodic solution in the plane (left picture) and in the space (right picture). The orbit emanating from O is periodic. The point P_1 represents the first return of P_0.

More appropriate for our purposes is Poincaré's study of the three-dimensional system

$$\frac{\mathrm{d}x}{\mathrm{d}t} = X, \qquad \frac{\mathrm{d}y}{\mathrm{d}t} = Y, \qquad \frac{\mathrm{d}z}{\mathrm{d}t} = Z. \tag{3.13}$$

In this case, the Poincaré sections are planes over which the iteration of his return map is to be investigated. His geometrical insight led him to treat periodic orbits as fixed points for the map on the plane, while he reduced the study of the asymptotic behavior of neighboring solution curves (which accounts for the stability character of the periodic orbit) to the computation of the successive iterations of points close to the fixed point.

There are crucial differences between his earlier (i.e., from the mid-1880s) and later, more mature studies on this topic. In the latter, Poincaré's main objective was to analyze changes in the orbit structure under perturbations of the vector field. He considered dynamical systems depending on a small parameter μ and, by exploiting a continuation argument, he hoped to transfer known properties of the solutions corresponding to the choice $\mu = 0$ to the case where $\mu \neq 0$ but was sufficiently small. This approach is consistent with the requirements of the restricted three-body problem, the parameter μ being the ratio between the planetoid and the masses of the primaries. This is exactly the question that KAM theory addresses. We will talk more about this in Sections 4.4 and 12.5.

To face this problem, Poincaré extended his earlier use of majorants since the solution is now expanded in powers of the time t and of the parameter μ. Moreover, he made extensive use of the theory of invariant integrals.[9] In particular, he combined the technique of the return map with the volume and area-preserving

[9] An *invariant integral* (or *first integral*) for system (3.13) is a nonconstant scalar function $I = I(x, y, z)$ satisfying the equation

$$\frac{\partial I}{\partial x}X(x,y,z) + \frac{\partial I}{\partial y}Y(x,y,z) + \frac{\partial I}{\partial z}Z(x,y,z) = 0, \qquad \forall x, y, z,$$

transformations which are characteristics of the symplectic structure of the flow associated with a Hamiltonian system, and he studied the iteration of points on an invariant curve depending on μ.

Returning to his study of ODEs of dimension three and the related return map in Chapter XIX of [**1886**], Poincaré assumed the existence of a closed curve for the autonomous system

$$\frac{dx}{dt} = X, \quad \frac{dy}{dt} = Y, \quad \frac{dz}{dt} = Z, \tag{3.14}$$

where X, Y, and Z are polynomials in the unknowns x, y, and z, respectively. His aim was to study the trajectories of (3.14) in a neighborhood of the closed curve. Starting from a reference point O on the curve, he introduced the curvilinear abscissa s to locate a generic point M on the curve, whose length he denoted l. He then considered a plane passing through the point M and orthogonal to the closed orbit, and he introduced an orthogonal coordinate system on the plane with M as the origin and Mx and My as axes. A generic point is then uniquely identified by the triple (x, y, s). Poincaré showed that such a coordinate system is particularly useful in representing trajectories emanating from points very close to the periodic orbit.

Indeed, he started by considering a point $P_0 = (x_0, y_0, 0)$ on the plane $s = 0$ close enough to O so that the corresponding orbit could intersect the generic orthogonal plane for all values $0 \leq s \leq l$ (see the right picture in Figure 3.2). He denoted by $P_1 = (x_1, y_1, 0)$ its intersection with the plane $s = l$ (which coincides with $s = 0$) and observed that since P_0 is close to O, x_1 and y_1 are holomorphic functions of x_0 and y_0 that vanish when $P_0 \equiv O$. He could then define a holomorphic mapping Φ on the plane $s = 0$ by $(x_0, y_0) \in D \mapsto (x_1, y_1)$, where D is a neighborhood of the origin, which assumed the form

$$\begin{cases} x_1 = \alpha x_0 + \beta y_0 + R(x_0, y_0), \\ y_1 = \gamma x_0 + \delta y_0 + R'(x_0, y_0), \end{cases} \tag{3.15}$$

with higher order terms R and R'.

The point (x_0, y_0) is in essence the projection of the trajectory (X, Y, Z) from the three-dimensional phase space. At time t_1 we obtain the "iterate" (x_1, y_1) via the return map $\mathcal{P}$ and, continuing in a similar manner, generate the sequence $\{(x_n, y_n)\}$.[10]

The same reasoning that led to the construction of the set $\{(x_n, y_n)\}$ allowed Poincaré to reduce the ODE (3.14) to a two-dimensional nonautonomous ODE of the form

$$\frac{dx}{dt} = X(t, x, y), \quad \frac{dy}{dt} = Y(t, x, y),$$

---

which implies that every solution of (3.13) satisfies $I(x(t), y(t), z(t)) = I(x_0, y_0, z_0) \equiv c$, where (x_0, y_0, z_0) represents the initial condition. Geometrically, a first integral of (3.13) may be thought of as a family of surfaces in the three-dimensional phase space $Oxyz$ where the initial condition determines the surface on which the dynamics takes place. As a consequence, the dimension of the dynamical system may be lowered by the number of known independent first integrals, which results in a reduced system which is easier to analyze. In the next chapter we will describe the role of first integrals in the context of the Hamiltonian systems that arise in celestial mechanics.

[10] We give an example of the return map depicting a quasi-periodic orbit on a two-torus in Section 4.4, equation (4.28).

whose solution $(x(t), y(t))$ describes the dynamics of the system on the moving plane $s = \frac{lt}{2\pi}$. The functions $X(t,x,y)$ and $Y(t,x,y)$ are analytic with respect to x and y, and the coefficients of their expansion in powers of x and y are trigonometric series of the variable t. The linearized system

$$\left\{\begin{aligned} \frac{dx}{dt} &= h(t)x + k(t)y, \\ \frac{dy}{dt} &= h'(t)x + k'(t)y, \end{aligned}\right. \tag{3.16}$$

admits the solution

$$\left\{\begin{aligned} x(t) &= A_1 e^{\lambda_1 t}\varphi_1(t) + A_2 e^{\lambda_2 t}\varphi_2(t), \\ y(t) &= A_1 e^{\lambda_1 t}\psi_1(t) + A_2 e^{\lambda_2 t}\psi_2(t), \end{aligned}\right. \tag{3.17}$$

where $\varphi_1(t)$, $\varphi_2(t)$ $\psi_1(t)$, $\psi_2(t)$ are again trigonometric series of t, where λ_1 and λ_2 are defined by $S_1 = e^{2\pi\lambda_1}$ and $S_2 = e^{2\pi\lambda_2}$, with S_1 and S_2 being the eigenvalues of the matrix

$$\begin{pmatrix} \alpha & \beta \\ \gamma & \delta \end{pmatrix}, \tag{3.18}$$

where the entries are given in equation (3.15).

Poincaré distinguished different cases according to the values assumed by S_1 and S_2 and, for each of these cases, deduced the stability character of the neighboring orbits. The case of interest for our present purposes is when S_1 and S_2 are both real and positive with $S_2 < 1 < S_1$;[11] this means that λ_1 and λ_2 are also real and of opposite sign. Poincaré asserted the existence of two curves lying on the plane $s = 0$ which pass through the origin O and satisfy the following two properties:

(a) The equations of these curves may be recast in the form

$$x = \varphi(y) \quad \text{and} \quad y = \psi(x),$$

where φ and ψ are analytic functions that vanish for $y = 0$ and $x = 0$, respectively.

(b) If (x_0, y_0) is chosen on one of these curves, the same will hold for its consequent (x_1, y_1) under the return map $\mathcal{P}$; that is, each curve is invariant for the map $(x_0, y_0) \mapsto (x_1, y_1)$.

As we will see, Poincaré's computation also shows that the curves ψ and φ are tangent to the axes Ox and Oy, respectively. More importantly, points close to the origin on ψ will eventually approach 0 while analogous points on φ will be repelled.

To prove the existence of such curves, Poincaré first performed a linear change of variables that puts the linear part of (3.15) in diagonal form:

$$\left\{\begin{aligned} y_1 &= S_1 y_0 + \Phi_1(x_0, y_0), \\ x_1 &= S_2 x_0 + \Phi_2(x_0, y_0). \end{aligned}\right. \tag{3.19}$$

He then considered the Taylor expansion of $\psi(x)$

$$y = \psi(x) = \alpha_2 x^2 + \alpha_3 x^3 + \cdots + \alpha_n x^n + \cdots$$

[11] In Section 4.2 we will discuss the case where S_1 and S_2 are complex conjugates of unit modulus.

and tried to determine the unknown coefficients α_i by imposing the invariancy condition $y_0 = \psi(x_0) \Rightarrow y_1 = \psi(x_1)$ or equivalently

$$S_1\psi(x_0) + \Phi_1(x_0, \psi(x_0)) = \psi(S_2x_0 + \Phi_2(x_0, \psi(x_0))).$$

The coefficients α_i, $i = 2, 3, \dots$ are then uniquely defined by exploiting the method of undetermined coefficients which involves the Taylor expansion of Φ_1 and Φ_2. Rather than determining the explicit expression of the α_i, Poincaré was mainly interested in showing the convergence of the corresponding series, and this was carried out by exploiting the method of majorants in a manner analogous to Cauchy's treatment of ODEs, discussed in the previous section.

Poincaré introduced two suitable constants M and β such that

$$-\Phi_1(x_0, y_0) \ll \frac{M\beta^2(x_0 + y_0)^2}{1 - \beta(x_0 + y_0)}, \qquad \Phi_2(x_0, y_0) \ll \frac{M\beta^2(x_0 + y_0)^2}{1 - \beta(x_0 + y_0)},$$

where the symbol "$\ll$" indicates that the coefficients of the expansion of $-\Phi_1(x_0, y_0)$ and $\Phi_2(x_0, y_0)$ in a neighborhood of O are bounded by those of the geometric series $1/(1 - \beta(x_0 + y_0))$ times the factor $M\beta^2(x_0 + y_0)^2$. Again, instead of (3.19), one studies the easier map

$$\begin{cases} y_1 = S_2 y_0 - \dfrac{M\beta^2(x_0 + y_0)^2}{1 - \beta(x_0 + y_0)}, \\ x_1 = S_2 x_0 + \dfrac{M\beta^2(x_0 + y_0)^2}{1 - \beta(x_0 + y_0)}, \end{cases} \tag{3.20}$$

for which an analytic function $y = \psi'(x) = \sum_{i=2}^{\infty} \alpha_i' x^i$ satisfying

$$S_2\psi'(x_0) - \frac{M\beta^2(x_0 + \psi'(x_0))^2}{1 - \beta(x_0 + \psi'(x_0))} = \psi'\left(S_2x_0 + \frac{M\beta^2(x_0 + \psi'(x_0))^2}{1 - \beta(x_0 + \psi'(x_0))}\right)$$

is easily shown to exist.

Perhaps the easiest way to see this is to introduce the change of variable $x+y = z$ in (3.20) which leads to the system[12]

$$\begin{cases} z_1 = S_2 z_0, \\ x_1 = S_2 x_0 + \dfrac{M\beta^2 z_0^2}{1 - \beta z_0}. \end{cases}$$

The invariancy condition for a curve in the form

$$x = \Theta(z) = z + \vartheta_2 z^2 + \vartheta_3 z^3 + \dots + \vartheta_n z^n + \cdots$$

is

$$\Theta(S_2 z) = S_2\Theta(z) + \frac{M\beta^2 z^2}{1 - \beta z},$$

or explicitly,

$$S_2 z + \vartheta_2 S_2^2 z^2 + \vartheta_3 S_2^3 z^3 + \dots = S_2 z + (\vartheta_2 S_2 + M\beta^2) z^2 + (\vartheta_3 S_2 + M\beta^3) z^3 + \cdots$$

yielding

$$\vartheta_i = -\frac{M\beta^i}{S_2 - S_2^n}.$$

[12]For more details, see the analogous computations in Lattès' thesis [**1906**].

The series for $\Theta(z)$ is therefore convergent for $|z| < 1/\beta$. By substituting $x + y = x+\psi'(x)$ back in place of z in the expansion of $\Theta(z)$, one sees that $\psi'(x)$ is implicitly defined by

$$\psi' = \frac{M\beta^2}{S_2 - S_2^2}(x+\psi')^2 + \frac{M\beta^3}{S_2 - S_2^3}(x+\psi')^3 + \cdots + \frac{M\beta^n}{S_2 - S_2^n}(x+\psi')^n + \cdots .$$

Finally, an induction argument shows that $\psi(x) \ll \psi'(x)$, which proves the convergence of the series for $\psi(x)$. The existence of the other curve $x = \varphi(y)$, tangent to Oy, is proven analogously.

The asymptotic behavior of the sequence $\{(x_n, y_n)\}$ gives obvious information about the stability of the related trajectories of the original system (3.13). In fact, this procedure allowed Poincaré to deduce that the origin O behaves like a saddle node since there exist one attracting and one repelling manifold (the two curves ψ and φ) intersecting at O.

As we will see in Section 3.5, Lattès grafted Levi-Civita's concept of stability to Poincaré's computational methodology to study curves left invariant by planar substitutions.

3.4. Stable or not stable: That is the question

Levi-Civita refined Poincaré's techniques in [**1901**] and assumed a more restrictive definition of stability in the sense that his definition applied to an equilibrium point of an ODE, here chosen as the origin of the phase space (that is, the null solution), while Poincaré's definition also referred to a periodic orbit (see page 42). This did not prevent Levi-Civita from treating periodic orbits since, as we will see, he introduced a change of variables that transforms the periodic orbit to an equilibrium point.

Levi-Civita's understanding of stability, in turn, differed from the one introduced by Aleksander M. Lyapunov a few years earlier in [**1907**].[13] Lyapunov's definition is preferred in the present-day literature, and is as follows:

> *The solution $x_i = 0$ will be stable if and only if, for every neighborhood E of the origin, no matter how small, there exists a second neighborhood $H \subset E$ such that, choosing in H the initial position of the moving point, it remains in E for any positive value of t.*

In contrast, Levi-Civita's definition of stability is

> *The solution $x_i = 0$ will be stable ... if and only if, for every neighborhood E of the origin, no matter how small, there exists a second neighborhood $H \subset E$ such that, choosing in H the initial position of the moving point, it remains in E for any positive or negative value of t.* [**1901**, p. 262]

Levi-Civita thus required the asymptotic behavior of neighboring solutions both in the future and in the past, while Lyapunov only required it for the future. For example, a saddle point is unstable in both definitions, while attracting nodes and

[13]We reference the French translation which appeared in 1907. Lyapunov's work appeared in a Russian language journal in 1892, and Levi-Civita evidently had some knowledge of its contents.

foci are stable in the former but unstable in the latter.[14] Centers are stable under both definitions.

Lattès implicitly used Levi-Civita's concept of stability in his thesis [**1906**], at least in his early chapters when he studied trajectories; in later chapters, where he focused on iteration, he did not use the terms "stable" and "unstable". Rather, he examined a number of cases and described the behavior of each (see the discussion following equation (3.35)). For example, he may characterize a particular fixed point as a node towards which only consequents converge (stable according to Lyapunov), or a node towards which only antecedents converge (unstable under both definitions).

Levi-Civita considered the nonautonomous m-dimensional system of differential equations

$$\frac{dx_i}{dt} = X_i(x_1, x_2, \ldots, x_m, t), \qquad i = 1, \ldots, m, \tag{3.21}$$

with X_i regular real functions of period T with respect to the time variable t. His aim was, like Poincaré's, to describe the dynamics around a periodic orbit $x_i = \varphi_i(t)$, $i = 1, \ldots, m$.

First of all, he observed that one can always assume all φ_i are identically null, after the change of variables $y_i = x_i - \varphi_i(t)$, which does not alter the nature of (3.21). Therefore, the existence of a periodic orbit can be expressed by assuming for $i = 1, \ldots, m$ that $X_i(0, \ldots, 0, t) = 0$ for every time t.

Levi-Civita's paper [**1901**], *Sopra alcuni criteri di instabilità*,[15] demonstrates the existence of a family of ∞^2 (nearly uniformly circular) periodic orbits for the restricted three-body system and then proves that most of them are unstable.

Like Poincaré, his approach is based upon the study of the iteration of a suitable transformation, but he extended it to m dimensions. In Poincaré the iteration was induced by a process; in Levi-Civita we can express it explicitly in terms of functions canonically induced by the continuous flow

$$x_i(t) = F_i(x_1^{(0)}, x_2^{(0)}, \ldots, x_m^{(0)}, t), \qquad i = 1, \ldots, m. \tag{3.22}$$

Here the $x_i^{(0)}$ refer to the initial conditions (the solution at time $t = 0$), while the F_i are the transition functions that yield the solution at time t and satisfy $F_i(0, \ldots, 0, t) = 0$ for any t.

Let $P_n = (x_1^{(n)}, x_2^{(n)}, \ldots, x_m^{(n)})$ be the state vector at time $t = nT$, and consider the functions $f_i(x_1, \ldots, x_m)$ defined as

$$f_i(x_1, \ldots, x_m) = F_i(x_1, \ldots, x_m, T),$$

[14] Using modern terminology, equilibrium points that are stable *only* under Lyapunov's definition are said to be *asymptotically stable*, while Levi-Civita's definition of stability, when applied to a Hamiltonian system, could be more appropriately labeled *marginal stability*. Levi-Civita remarked that Lyapunov distinguished between past and future stability while he did not, but he observed that the cases he worked differed from those of Lyapunov's, in particular because the Hamiltonian (autonomous) systems with isolated stable equilibria that arise in his study of celestial mechanics are not just asymptotically stable but are always marginally stable.

[15] Levi-Civita's title, which translates as *On some criteria of instability*, reflects the fact that stable equilibrium points and orbits only occur under very special circumstances. Indeed, his title calls to mind Poincaré's famous sentence, "Instability is thus the rule, and stability the exception" [**1885**, p. 172], and suggests that Levi-Civita was on the same wavelength as Poincaré in regarding stable solutions as rather rare.

that is, $f_i(x_1,\ldots,x_m)$ yields the solution of (3.21) at time T starting from the initial vector $P=(x_1,\ldots,x_m)$. We can view the generation of the P_n by iteration of the f_i as follows.

From the property of flows,

$$F_i(x_1^{(n)},x_2^{(n)},\ldots,x_m^{(n)},t-nT)=F_i(x_1^{(n-1)},x_2^{(n-1)},\ldots,x_m^{(n-1)},t-(n-1)T).$$

Setting $t=nT$, we get

$$x_i^{(n)}=f_i(x_1^{(n-1)},x_2^{(n-1)},\ldots,x_m^{(n-1)}),\quad i=1,\ldots,m,\quad n=\pm1,\pm2,\ldots, \tag{3.23}$$

that is, the state vector P_n at time $t=nT$ may be retrieved from the initial position P_0 by repeatedly iterating the functions f_i.

Exploiting a continuity argument, Levi-Civita easily showed that he could transfer questions about the stability of a periodic solution to (3.21) to corresponding questions regarding the fixed point $O=(0,\ldots,0)$ of the transformation (3.23): Levi-Civita's definition of stability for such maps is precisely the same as the one given in his quotation from the beginning of this section, the only difference being that the moving point is here represented by the sequence of points $\{P_n\}$.[16]

In the first part of the paper, Levi-Civita cataloged a number of instances where the origin is an unstable point. His first result concerns the case where at least one of the eigenvalues (or multipliers) of the matrix associated with the linear part of (3.23) has modulus different than one.

In the second chapter he considered the case where all the multipliers have unitary modulus but, for simplicity, confined his study to two-dimensions ($m=2$). After a suitable linear change of variables, Levi-Civita put the transformation in the form

$$\begin{cases} x_1=\omega_1 x+f_1'(x,y),\\ y_1=c_{21}x+\omega_2 y+f_2'(x,y),\end{cases} \tag{3.24}$$

where $f_1'(x,y)$ and $f_2'(x,y)$ are higher order terms in both x and y, and the multipliers ω_1 and ω_2 assume one of the following values:

$$\begin{array}{lll} (1) & \omega_1=\pm1, & \omega_2=\mp1,\\ (2) & \omega_1=\pm1, & \omega_2=\pm1,\\ (3) & \omega_1=e^{i\vartheta}, & \omega_2=e^{-i\vartheta}.\end{array}$$

In the first case, iterating (3.24) two times yields

$$\begin{cases} x_2=x+f(x,y),\\ y_2=y+g(x,y),\end{cases} \tag{3.25}$$

where again $f(x,y)$ and $g(x,y)$ are of order greater than one. One is led to the same formulae for the second case, under the assumption $c_{21}=0$. Alternatively, one can recast (3.24) in the form

$$\begin{cases} x_2=x+f(x,y),\\ y_2=x+y+g(x,y).\end{cases} \tag{3.26}$$

Finally, the third case corresponds to the system

$$\begin{cases} x_1=\cos\vartheta\,x-\sin\vartheta\,y+f(x,y),\\ y_1=\sin\vartheta\,x+\cos\vartheta\,y+g(x,y).\end{cases} \tag{3.27}$$

[16]For references to Lyapunov's stability analysis, see, e.g., Lakshmikantham, Leela, and Martynyuk [**1989**] or Rouche, Habets, and Laloy [**1977**] for the continuous time case (ODEs), and Lakshmikantham and Trigiante [**2002**] for the discrete-time case (difference equations).

Levi-Civita left aside the case of complex conjugate multipliers, observing that (3.27) reduces to (3.25) whenever ϑ is commensurate with 2π. In Section 4.3 we will see that the complex conjugate case with $\theta \neq \frac{2h\pi}{k}$, where h and k are coprime, is the most relevant to our study from the point of view of celestial mechanics and Hamiltonian systems as well as the most difficult to handle, due to the appearance of small divisors.

He then proved instability in the following situations:

(a) for (3.25), in the case where the quadratic parts of $f(x, y)$ and $g(x, y)$ have no common factor;

(b) for (3.26), in the case where the quadratic part of $f(x, y)$, say $\varphi(x, y) = a_{11}x^2 + 2a_{12}xy + a_{22}y^2$, is such that $a_{22} \neq 0$.[17]

Returning to (3.23), his next step was to link the transformations f_i to the data of the original ordinary differential system, namely the X_i's in (3.21). This was accomplished by considering the variational equations

$$\frac{dx_i}{dt} = \sum_{j=1}^{m} a_{ij}(t)x_j, \qquad i = 1, \ldots, m, \tag{3.28}$$

obtained by expanding X_i around the periodic solution (here assumed to be zero) and retaining the linear part. Let $\Xi(t) = \{\xi_{ij}(t)\}$ be the fundamental solution system of (3.28) defined by $\xi_{ij}(0) = 0$ for $i \neq j$ and $\xi_{ii}(0) = 1$. The coefficients $a_{ij}(t)$ are periodic with period T, and the transition matrix, which we will refer to as C, yields $\Xi(t+T)$ when multiplied by $\Xi(t)$ and must have constant coefficients. When $t = 0$, $C = \Xi(T)$.

Denoting the m multipliers of $\Xi(T)$ (i.e., the roots of $\det(\Xi(T) - \omega I)$, with I the identity matrix of dimension m) as $e^{\alpha_1 T}, e^{\alpha_2 T}, \ldots, e^{\alpha_m T}$, a fundamental solution of (3.28) takes the form

$$x_{ij}(t) = e^{\alpha_j t}\sigma_{ij}(t), \qquad i, j = 1, \ldots, m,$$

with $\sigma_{ij}(t)$ periodic or, in the case of nondistinct α_j, of the form $\tau_0(t)t^{\mu} + \tau_1(t)t^{\mu-1} + \cdots + \tau_{\mu}$, with $\tau_k(t)$ periodic and $\mu < m$. According to Poincaré, the coefficients α_i are called *characteristic exponents* of the given periodic solution.

Levi-Civita showed that the transformation (3.23) also shares the same multipliers $e^{\alpha_j T}$. In fact,

$$x_i^{(1)} = f_i(x_1^{(0)}, x_2^{(0)}, \ldots, x_m^{(0)}) = F_i(x_1^{(0)}, x_2^{(0)}, \ldots, x_m^{(0)}, T) = \sum_{j=1}^{m} \xi_{ij}(T)x_j^{(0)} + \cdots .$$

This allowed him to obtain results about the stability of the periodic solution of the continuous system (3.21) by studying the iteration defined by (3.23) at the origin. For example, observing that $|e^{\alpha_j T}| = 1$ if and only if a_j has null real part, he deduced the well-known result due to Lyapunov: a periodic solution is unstable if at least one of the its characteristic exponents has nonzero real part.

Thus, Levi-Civita reinforced the notion introduced by Poincaré that iteration can be a very useful tool for studying the stability of solutions to differential equations. This notion would be refined further by Lattès.

---

[17]He also proved instability for some special subcases corresponding to $a_{22} = 0$.

3.5. Lattès' thesis: Closer connections with complex iterates

Samuel Lattès wrote one of the three works submitted for the 1918 *Grand Prix des Sciences Mathématiques*. His entry concerned the iteration of rational functions of one and two complex variables. The report on the 1918 *Grand Prix* competition[18] cited his investigation of the iteration of a function of two complex variables as a particular strength of his paper. We discuss his entry to the *Grand Prix* in Section 7.5.

The present section focuses on Lattès' doctoral thesis,[19] [**1906**] where he generalized Kœnigs' study of single-variable iteration to higher dimensions in order to examine curves and surfaces left invariant under the iteration of functions of two and three variables. He showed that this task is in turn equivalent to solving functional equations which are canonically induced by the functions being iterated.

Lattès' thesis connects the work of Poincaré and Levi-Civita to complex dynamics and explicitly explores links between differential equations and the functional equations arising from the iteration of complex rational functions of a single variable. A number of his results are explicit generalizations of Kœnigs' results to higher dimensions. In many respects then Lattès' thesis stands at the crossroads of several avenues of exploration related to complex dynamics, including its relations to the study of ODEs as personified by Poincaré and Levi-Civita.

Lattés' functional equations were not motivated by specific examples, such as the local behavior of iteration, but rather as means to study objects, such as a curve or surface, left invariant under a mapping. Nor were his applications to differential equations intended to solve a specific problem, such as the three-body problem, or to analyze a specific equation; rather, they served to deepen understanding of functional equations:

> *Functional equations are, in the most general sense, equations defining unknown functions through relations between unknown functions and given ones.*
>
> *There exist categories of functional equations according to the nature of these relations as well. For example, a first class consists of differential equations.*[20] *Other equations, studied by various geometers, give rise to a second class: Those defining an unknown function through the values taken on by defined integrals depending on this function.*
>
> *One can conceive other species of functional equations as well....* [**1906**, p. 1]

In his introduction Lattès referred, on the one hand, to studies of functional equations by Abel, Kœnigs, Grévy, Leau, Picard, and the French mathematician Carlo Bourlet, and, on the other, to Poincaré and Levi-Civita's studies of the stability of equilibrium points of differential equations (see Figure 3.1):

[18]See Appendix A for our translation of the report and [**CR-1918-Rapport**] for the original.

[19]Lattès dedicated his doctoral dissertation to Jacques Hadamard who, along with Paul Painlevé and with Kœnigs (who served as *président*), was a member of Lattès' examination committee.

[20]This notion of a differential equation being at its essence a functional equation was shared by Pincherle, who expressed that view in his entries to both the *Enzyklopädie der Mathematischen Wissenschaften* [**1906**], and the French version, the *Encyclopédie des Sciences Mathématiques pures et appliquées* [**1912**].

> *The results to which I was led present a striking analogy to those obtained by Poincaré in the study of the integral curves of a differential equation in a neighborhood of an equilibrium point....* [**1906**, p. 5]

He also noted a specific outcome of his study:

> *I discuss the method used by Poincaré and extend it in a way he did not envision.* [**1906**, p. 3]

What he refers to is a method involving both functional equations and the Poincaré return map, one we will discuss below, to approximate trajectories near equilibrium points. This method suggests an explicit link between the theory of functional equations and Poincaré and Levi-Civita's respective studies of the stability of equilibrium points, which Lattès described as follows:

> *We will see the reason for the link between these two questions and show that the theory of differential equations may be regarded as being a limit of the theory of iteration and of functional equations which attach themselves to this theory ....* [**1906**, Ch. III, p. 63]

Lattès' work thus represents a bridge between the theory of differential and functional equations. Pincherle found Lattès' exploration quite interesting and included a discussion of his work in relation to Poincaré's in his entry on functional equations in the French mathematical encyclopedia [**1912**, p. 70ff].

Lattès was by no means the first to notice similarities between differential equations and the functional equations related to iteration. Appell applied Kœnigs' study of the Schröder functional equation to differential equations in [**1891**]. Grévy ([**1894**, pp. 307, 314] and [**1896**, p. 314]) and Leau ([**1897**, Chap. VIII]), students of Kœnigs whose work we discussed in Sections 1.7.2 and 1.7.3, also spoke of analogies between differential and functional equations.

Bourlet investigated the nature of these connections in his *Comptes Rendus* notice [**1897b**] using what he called the transmutation operator—a functional operator that included the operations of differentiation and integral. He defined "additive transmutation" as an operator $\mathcal{T}$ satisfying $\mathcal{T}(\phi + \psi) = \mathcal{T}(\phi) + \mathcal{T}(\psi)$ and, referring to results from Pincherle [**1897b**], showed that the only equations resulting from the additive transmutation operator were functional equations of the form Grévy discussed (see equation (1.12)), as well as certain differential equations:

> *Thus, the only operative equations whose theory is analogous with the theory of ordinary differential equations are the iterated functional equations. The striking parallels between the theories of Grévy, Leau and others who study similar functional equations and that of differential equations is therefore naturally explained.* [**1897b**, p. 1443]

3.6. Lattès' extension of the return map

The first two chapters of Lattès' thesis are devoted to the study of invariant curves or surfaces of maps of two or three real variables. In keeping with the general nature of his study, the invariant curves are not necessarily periodic orbits, although, as Lattès pointed out, they could be.

In Chapter I he considered the two-dimensional substitution

$$\left\{ \begin{array}{l} X = f(x,y), \\ Y = \varphi(x,y), \end{array} \right. \tag{3.29}$$

where $f(x,y)$ and $\varphi(x,y)$ are analytic on a domain containing the origin, where they vanish, making the origin a fixed point for the substitution. His aim was to search for invariant curves passing through the origin of the form $y = \psi(x)$ and $x = \psi'(y)$, which are analytic on a domain surrounding the origin. This in turn begs the question of finding the solution of the following functional equation in the unknown $\psi(x)$,

$$\psi(f(x,\psi(x))) = \varphi(x,\psi(x)), \tag{3.30}$$

which is generated as follows. If $P_0 = (x,\psi(x))$ is on the invariant curve, so is $P_1 = (X,Y)$, where $Y = \varphi(x,\psi(x)) = \psi(X) = \psi(f(x,\psi(x))$. An analogous equation can be generated for $\psi'(x)$.

Lattès pointed out that for appropriate choices of φ, (3.30) reduces to several specific functional equations:

(*a*) $\psi(f(x)) = \psi(x) + a$, the Abel equation, obtained by assuming f is independent of y, and $\varphi(x,y) = a + y$;

(*b*) $\psi(f(x)) = a\psi(x)$, the Schröder equation, obtained by assuming f is independent of y and $\varphi(x,y) = ay$;

(*c*) $\psi(\psi(x)) = \varphi(x)$, the Pfeiffer equation,[21] obtained by assuming φ is independent of y and $f(x,y) = y$;

(*d*) $\psi(f(x)) = \varphi(\psi(x))$, Fatou's version of the permutable functional equation (see equations (10.2) and (J.1)) obtained by assuming f is independent of y, and φ is independent of x.

Lattès rewrote the system (3.29) in a more convenient form via a linear change of variables that puts the linear part of the system in (3.29) (the Jacobian) in diagonal form (he continued to use the symbols f and φ):

$$\left\{ \begin{array}{l} X = Sx + f(x,y), \\ Y = S'y + \varphi(x,y), \end{array} \right. \tag{3.31}$$

where $f(x,y)$ and $\varphi(x,y)$ are analytic in a neighborhood of the origin and their series expansions begin with terms of second degree in x and y. S and S' are the eigenvalues of the Jacobian matrix of the vector field $[f(x,y),\varphi(x,y)]^T$, obtained as roots of the equation

$$\left| \begin{array}{cc} \frac{\partial f}{\partial x}|_{(0,0)} - S & \frac{\partial f}{\partial y}|_{(0,0)} \\ \frac{\partial \varphi}{\partial x}|_{(0,0)} & \frac{\partial \varphi}{\partial y}|_{(0,0)} - S \end{array} \right| = 0. \tag{3.32}$$

The map (3.31) is the same as (3.19). Lattès' purpose was to extend the result Poincaré obtained by assuming S and S' to be real with $S' > 1 > S > 0$ (see page 47). In order to deal with less restrictive conditions, Lattès instead assumed that S and S' are distinct, nonzero, and possibly complex (but with modulus different from one), satisfying

$$S' - S^n \neq 0 \tag{3.33}$$

---

[21] So named because it was studied in detail by Pfeiffer in 1918; see Pfeiffer [**1918**] as well as equation (5.19) in Section 5.7.

for all positive values of n (since $S' - S^n$ appear as denominators in the expression of the coefficients of the expansion of $\psi(x)$).

After proving the existence and uniqueness of two curves C and C' tangent to the Ox and Oy axes respectively,[22] Lattès applied his knowledge of invariant curves to put the substitution (3.31) in a more convenient form where C and C' become the two axes Ox and Oy,

$$\begin{cases} X = x(S + F(x,y)), \\ Y = y(S' + \phi(x,y)). \end{cases} \tag{3.34}$$

The functions $F(x, y)$ and $\phi(x, y)$ are holomorphic in a domain containing the origin and vanish for $x = y = 0$. The form (3.34) is particularly appropriate for studying the iteration of a point in a neighborhood of the origin,[23] which Lattés did, under the assumption that S and S' are real and distinct. Starting from a given point (x, y) close to the origin, he considered the sequences of iterates (x_1, y_1), (x_2, y_2), $\ldots$, (x_n, y_n), obtained by

$$\begin{cases} x_{n+1} = x_n(S + F(x_n, y_n)), \\ y_{n+1} = y_n(S' + \phi(x_n, y_n)). \end{cases} \tag{3.35}$$

These are analogous to the points Poincaré obtained via his return map (see equation (3.15)).

He then described three subcases of the asymptotic behavior of the sequence $\{(x_n, y_n)\}$:

(a) $|S'| < |S| < 1$. For small values of x and y, say $|x| < \varepsilon, |y| < \varepsilon$, the two functions $S + F(x,y)$ and $S' + \phi(x,y)$ have maximum modulus M less than one. Consequently,

$$|x_{n+1}| < M|x_n| \qquad \text{and} \qquad |y_{n+1}| < M|y_n|,$$

which yields

$$|x_n| < M^n|x| \qquad \text{and} \qquad |y_n| < M^n|y|,$$

thus the x_n and y_n tend to zero.

The existence of the two invariant curves (here transformed into the two coordinate axes) yields additional information about how $\{(x_n, y_n)\}$ approaches the origin O. From (3.35) one gets

$$\frac{y_{n+1}}{x_{n+1}} = \frac{y_n}{x_n}\left(\frac{S'}{S} + \theta(x_n, y_n)\right),$$

where $\theta(x, y)$ is a holomorphic function near the origin and $\theta(0,0) = 0$. From the above assumption on S and S' it then follows that

$$\left|\frac{y_{n+1}}{x_{n+1}}\right| \leq K\left|\frac{y_n}{x_n}\right|$$

with $K < 1$. Therefore $|y_{n+1}/x_{n+1}| \to 0$, which means that the sequence $\{(x_n, y_n)\}$ approaches the origin tangentially to the Ox axis (or, for the general substitution (3.31), tangentially to the curve C). The only exception is when the point (x, y) is chosen on the other invariant curve C' in

[22] The proof is completely analogous to Poincaré's, including the choice of majorant functions.

[23] Here Lattès' approach covers all cases examined by Poincaré in [**1886**] but uses a different technique than Poincaré did.

which case the iterates of (x, y) remain on C'. The origin O is therefore an asymptotically stable node.

(b) $|S'| > |S| > 1$. Lattès treated this reciprocal case by observing that the inverse transformation of (3.35) has the form

$$\begin{cases} x = X\left(\dfrac{1}{S} + \cdots\right), \\ y = Y\left(\dfrac{1}{S'} + \cdots\right). \end{cases}$$

One has $|1/S'| < |1/S| < 1$, which leads to the conclusions of the previous case for the antecedents of (x, y). The origin O is an unstable node.

(c) $|S'| < 1 < |S|$. This is the case Poincaré studied above, following equation (3.18) (see page 47). Near the origin one has

$$|S' + \phi(x, y)| < M < 1 \qquad \text{and} \qquad |S + F(x, y)| > m > 1,$$

which, assuming (x, y) is close enough to O, yields $|x_n| > m^n|x|$ and $|y_n| < M^n|y|$. The first n consequents approach the Ox axis (or in general the curve C) and depart from the Oy axis (the curve C').[24] The only exception is when (x, y) lies on one of the two invariant curves. The fixed point O is a saddle node.

Lattès next examined the case where S and S' are complex conjugates: $S = \rho e^{i\alpha}$, $S' = \rho e^{-i\alpha}$. In this case he showed that (3.34) can be conveniently cast in the form

$$\begin{cases} X + iY = (x + iy)(\rho e^{i\alpha} + H(x, y) + iK(x, y)), \\ X - iY = (x - iy)(\rho e^{-i\alpha} + H(x, y) - iK(x, y)), \end{cases} \tag{3.36}$$

where, as usual, $H(x, y)$ and $K(x, y)$ are two holomorphic functions that vanish at O. It is worth noting that the two relations above actually represent the same map, as is to be expected since a map $\mathbb{R}^2 \to \mathbb{R}^2$ may be equivalently expressed as a map $\mathbb{C} \to \mathbb{C}$. From (3.36) one deduces

$$X^2 + Y^2 = (x^2 + y^2)(\rho^2 + \Theta(x, y)), \qquad \Theta(0, 0) = 0.$$

This leads to two more subcases:

(d) $|S| = |S'| = \rho < 1$. In a neighborhood of O the function $\rho^2 + \Theta(x, y)$ is bounded in modulus by a constant $M < 1$. It then follows $x_n^2 + y_n^2 < M^n(x^2 + y^2)$ and the consequents (x_n, y_n) eventually approach the origin O. Furthermore

$$\arg(X + iY) = \arg(x + iy) + \arg(\rho e^{i\alpha} + H(x, y) + iK(x, y)).$$

Denoting $\omega_n = \arg(x_n + iy_n)$, it follows that $\omega_{n+1} = \omega_n + \alpha + \varepsilon_n$, with $\varepsilon_n \to 0$ as $n \to \infty$. One concludes that $\omega_{n+1} - \omega_n$ admits a limit equal to α, which means that for large n the consequents (x_n, y_n) rotate around the fixed point O (while approaching it) at a constant angle. Lattès showed that in this case the invariant curves are logarithmic spirals and the fixed point is a stable focus.

(e) $|S| = |S'| = \rho > 1$. It is sufficient to consider the inverse substitution to deduce the result from the previous case for the antecedents of the original map: O is an unstable focus.

---

[24] One can choose n arbitrarily large by suitably reducing the values of $|x|$ and $|y|$.

Although Lattès assumed (3.29) is a map $\mathbb{R}^2 \to \mathbb{R}^2$, with (3.36) he recast it as a map $\mathbb{C} \to \mathbb{C}$. Since every complex transformation

$$Z(= X + iY) = g(z)(= u(x, y) + iv(x, y))$$

may be put in the form (3.29) by defining $f \equiv u$ and $\varphi = v$, this allowed Lattès' results to be extended to complex maps. However, when confining the analysis to complex maps that are holomorphic in an open set containing the double point, the Cauchy-Riemann equations $\partial u/\partial x = \partial v/\partial y$ and $\partial u/\partial y = -\partial v/\partial x$ imply that the solutions of (3.32) are always complex conjugate pairs. Therefore when S, S' are not real and have modulus different than 1, the above two subcases account for the behavior of the iteration of g locally in a neighborhood of the fixed point.

3.7. Analogies between iteration and differential equations

Section 28 of Chapter III of Lattès' thesis has the intriguing title "*Analogies entre la théorie de l'itération et la théorie des courbes définies par une équation différentielle dans le domaine d'un point singulier*".[25] Once again, Lattès related his investigation to Poincaré's studies of the trajectories of a system of differential equations in a neighborhood of an equilibrium point or periodic orbit. This time his goal was to argue that certain differential equations can be viewed as the limit of a functional equation.

He began by considering the substitution

$$\left\{ \begin{array}{l} X = x + f(x, y)\delta t, \\ Y = y + \varphi(x, y)\delta t, \end{array} \right. \tag{3.37}$$

where δt is a parameter different than zero. The fixed points of (3.37) are obtained by setting both $f(x, y) = 0$ and $\varphi(x, y) = 0$. Lattès assumed that the origin O is a fixed point and that the eigenvalues λ_1 and λ_2 of the Jacobian of the vector field $[f(x, y), \varphi(x, y)]$ are distinct. A suitable linear change of variables then puts the substitution in the more convenient form

$$\left\{ \begin{array}{l} X = x + (\lambda_1 x + \cdots)\delta t, \\ Y = y + (\lambda_2 y + \cdots)\delta t, \end{array} \right. \tag{3.38}$$

whose linear part has distinct eigenvalues $S_1 = 1 + \lambda_1 \delta t$ and $S_2 = 1 + \lambda_2 \delta t$. The functional equation associated with the analytic invariant curve $y = \psi(x)$ is

$$\psi(x + f(x, \psi(x))\delta t) = \psi(x) + \varphi(x, \psi(x))\delta t. \tag{3.39}$$

Taking the expansions of both sides in powers of δt yields

$$\psi'(x) f(x, \psi(x))\delta t + h.o.t. = \varphi(x, \psi(x)\delta t + h.o.t.$$

from which, dividing first by δt and then letting $\delta t \to 0$, yields

$$\frac{dx}{f(x, y)} = \frac{dy}{\varphi(x, y)}, \tag{3.40}$$

where $\psi'(x)$ has been replaced by dy/dx. Lattès concluded:

[25] "Analogies between the theory of iteration and the theory of curves defined by a differential equation in the vicinity of a singular point."

The functional equation that defines an invariant curve for (3.38) *becomes in the limit* [as $\delta t \to 0$] *the differential equation* (3.40). *The invariant curves defined by equation* (3.39) *become the curves defined by this differential equation; the fixed points of the substitution* (3.38) *become the equilibrium points of the differential equation.* **1906**, p. 64]

Lattès constructed an arbitrary (but not necessarily analytic) invariant curve of the substitution (3.38) from an initial point P_0, its consequents $P_1, P_2, \ldots, P_n$, and its antecedents $P_{-1}, P_{-2}, \ldots, P_{-n}$ by constructing a continuous curve joining P_0 to P_1 and observing that its consequent (respectively, antecedent) will be a curve joining P_1 with P_2 (respectively, P_0 with P_{-1}). The curve obtained by forward and backward iteration of the initial arc will then be invariant by its very definition. Letting $\delta t \to 0$, every point approaches its consequent and, in the limit, this invariant curve becomes an integral curve of the differential equation (3.40).

Lattès noted that the assumption $S_1 \neq S_2^n$ (see equation (3.33)) implies that $(1 + \lambda_1 \delta t) \neq (1 + \lambda_2 \delta t)^n$ which eventually becomes $\lambda_1 \neq n\lambda_2$ (and equivalently for $S_2 \neq S_1^n$). Thus the ratio λ_1/λ_2 is neither an integer nor the reciprocal of an integer, which is usually considered in the theory of differential equations (here Lattès referred to Picard's work [**1908**]). As an application of the results obtained by studying the iteration for different choices of S_1 and S_2, Lattès derived, in the limit for $\delta t \to 0$, well-known results regarding the local stability character of an equilibrium point of two-dimensional differential equations:

(*a*) λ_1 *and* λ_2 *real and of the same sign*: The eigenvalues S_1 and S_2 are both greater or less than one. This becomes one of the cases (*a*) or (*b*) above, where all the consequents or the antecedents of a given point close to the origin approach this point. In the limit $\delta t \to 0$, all the integral curves will pass through O which therefore is a node.

(*b*) λ_1 *and* λ_2 *real and of opposite sign*: For small δt, S_1 and S_2 are one greater and the other less than one (case (*c*) above), and the differential equation admits only two invariant curves passing through the origin, which is a saddle point.

(*c*) λ_1 *and* λ_2 *complex conjugates*: In this case S_1 and S_2 are also complex conjugates (cases (d) and (e)), and the equilibrium point is a focus for the differential equation.[26]

3.8. A numerical method

An interesting application of this method, one that Lattès did not discuss, is a numerical method for solving differential equations. Since Lattès' work is rooted in techniques originally developed by Poincaré to construct trajectories, one can say that not only did Lattès show that Poincaré's methods can be extended to functional equations, but, implicitly at least, they can also be extended to create a

[26]Actually, in this case the stability character of the focus may be not preserved by the map (3.37), since it may turn an asymptotically stable equilibrium point into an unstable double point. The logarithm spiral shape of the curve is preserved excluding, possibly, exceptional values of the parameter δt.

numerical method: given a system of two differential equations in the form

$$\left\{\begin{aligned} \frac{dx}{dt} &= f(x,y), \\ \frac{dy}{dt} &= \varphi(x,y), \end{aligned}\right. \tag{3.41}$$

the mapping (3.38) is nothing but the explicit Euler method applied to the above system (δt stands for the integration step).

Starting from an initial point $P_0 = (x_0, y_0)$, it produces a numerical solution $P_1, P_2, \ldots, P_n$ that approximates the true solution with an error $O(\delta t)$ on finite time intervals. In other words, this connection between functional equations and differential equations allows one to use the functional equation (3.39) associated to the substitution (3.37) to solve the differential equation (3.41).

This leads to the association displayed in the following diagram (we have put $F(z) = (f(z), \varphi(z))^T$, $z = (x,y)^T$ and $y = \psi(x)$).

$$\begin{array}{ccc} z' = F(z) & \longrightarrow & \frac{dx}{f} = \frac{dy}{\varphi} \\ {\scriptstyle \delta t \to 0}\big\uparrow & & {\scriptstyle \delta t \to 0}\big\uparrow \\ z_{n+1} = z_n + \delta t\, F(z_n) & \longrightarrow & \psi(x + f(x,y)\delta t) = y + \varphi(x,y)\delta t \end{array}$$

The two equations on the left define the evolution in time of the continuous and discrete state vectors $z(t)$ and $z_n \simeq z(t_0 + n\delta t)$, respectively. The equations on the right define the solution curves on the phase plane (x,y). It is worth noting that, while the functional equation possesses exactly one analytic solution, for equation (3.40), under regularity assumptions for the functions f and φ, we obtain an analytic solution curve regardless of the choice of the initial point P_0.

3.9. Concluding remarks

Although Poincaré's use of iteration in his studies of ODEs—in particular his return map—is sometimes cited as an inspiration for Fatou and Julia's monographs on the iteration of rational functions (perhaps due to the Académie's reference to him in its 1915 announcement of the competition for its 1918 *Grand Prix des Sciences Mathématiques*), we found no evidence that this is the case.

Neither Fatou nor Julia cited him as an inspiration, and a far more likely scenario is that their works grew out of the long-standing French study of the iteration of rational functions inaugurated by Kœnigs in the mid-1880s and carried forth into 1890s by two of his students, as well as other French mathematicians.

Kœnigs' study emerged at around the same time as Poincaré's early studies of ODEs and linear groups, both of which employed iteration. Despite common terminology involved in Kœnigs and Poincaré's use of iteration, it seems their studies had an independent genesis. As is the case with Fatou and Julia, Kœnigs did not cite Poincaré's study as an inspiration, and his only reference to it was to point out an aspect of his own work that might prove useful to the study of linear groups.

An investigation of the use of iteration involved with Poincaré's return map reveals that it is of a different character than what was seen in the French study of the iteration of complex functions. Nonetheless, there are connections between the iteration of complex functions and Poincaré. One is a special case of a functional equation he introduced in his paper [**1890a**] that proved very useful in characterizing the iteration of a complex function near a repelling point that will also be discussed later in our work.

Others arise in the generalization of Kœnigs' study to functions of more than one variable. In 1907 Lattès extended the investigation of the functional equations that arise in the "Kœnigs school" to multivariable functions. Drawing as well on Levi-Civita's work in ODEs, Lattès pointed out connections between his own work and Poincaré's study of ODEs, and he modified Poincaré's return map for his own ends.

Although the hint by the Académie about possible relationships with Poincaré's work—which we believe was most likely a reference to the return map—was not acted upon by Fatou and Julia, we will see in the next chapter that there are clear mathematical analogies between the return map and the iteration of complex functions around irrationally neutral fixed points, often referred to as the center problem. The Académie could not have foreseen this connection since there was nothing known about iteration around irrationally indifferent points at the time of its announcement, aside from an emerging investigation in the U.S. that the French were evidently not aware of. This U.S. study is the subject of Chapter 5.

The center problem and the investigation of the stability of periodic orbits of canonical Hamiltonian systems each encounter a formidable common obstacle—the *small divisor problem.* This is our principal concern in the next chapter.

COLOR PLATE 1. A Herman ring for the function $f(z) = \frac{e^{i2\pi\theta}z^2(z-4)}{1-4z}$. See page 16.

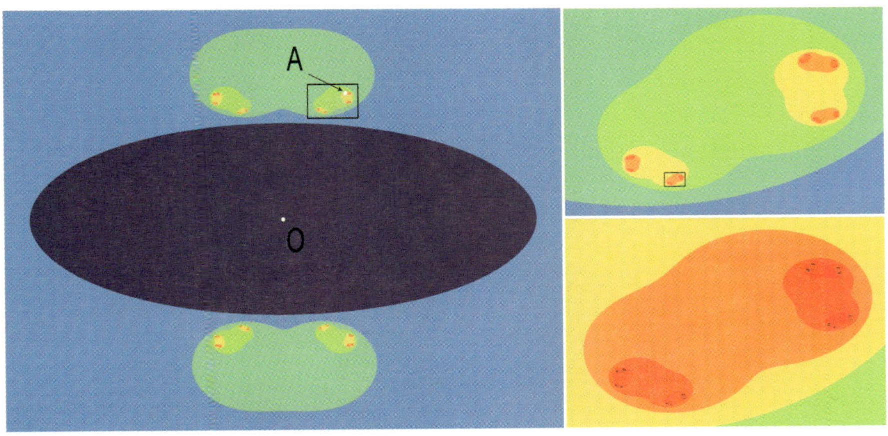

COLOR PLATE 2. The formation of the Julia set for $f_2(z) = \frac{z^2}{z^2+2}$. See page 25.

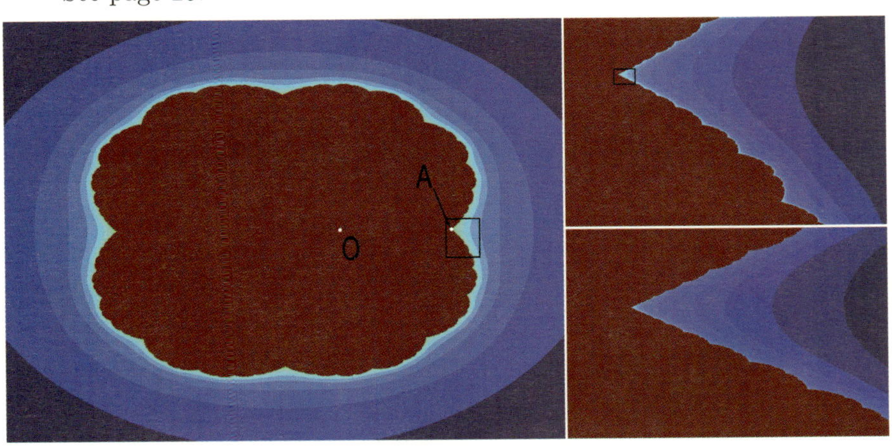

COLOR PLATE 3. The Fatou and Julia sets for $g(z) = \frac{z^2+z}{2}$. See page 27.

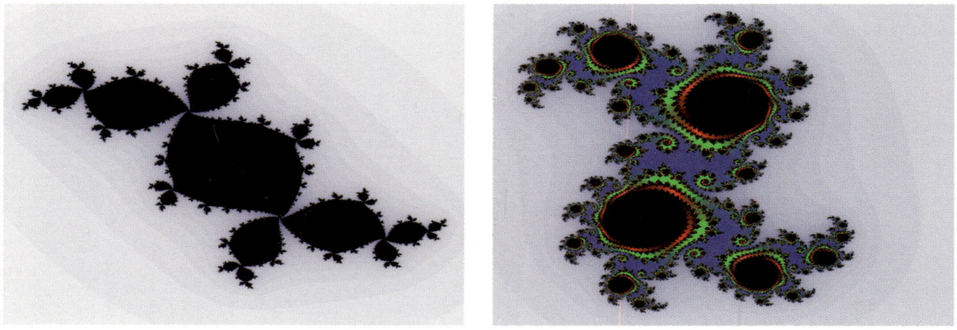

COLOR PLATE 4. Julia sets for $f(z) = z^2 + c$. See page 141.

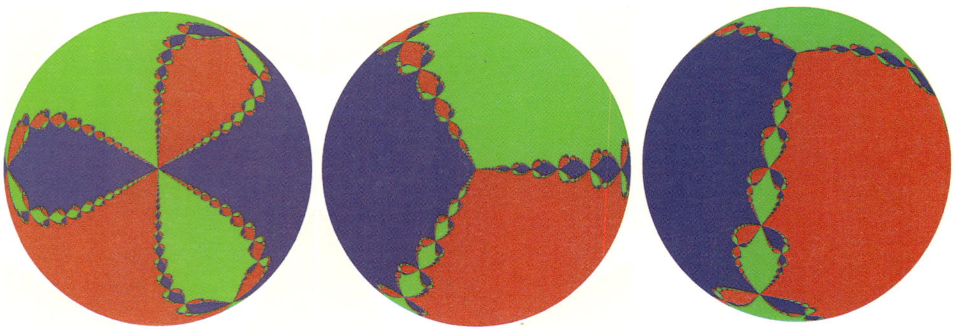

COLOR PLATE 5. Newton's method for the cubic. See page 143.

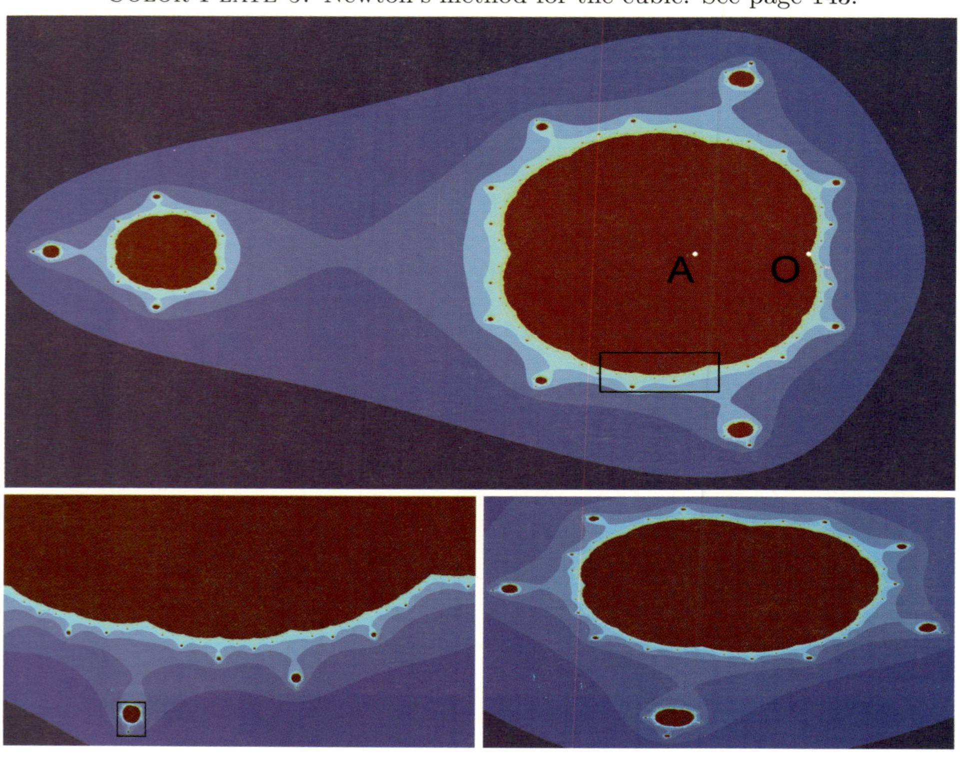

COLOR PLATE 6. Fatou's Archipelago, $R(z) = z^3 + 3z^2 + 2z$. See page 162.

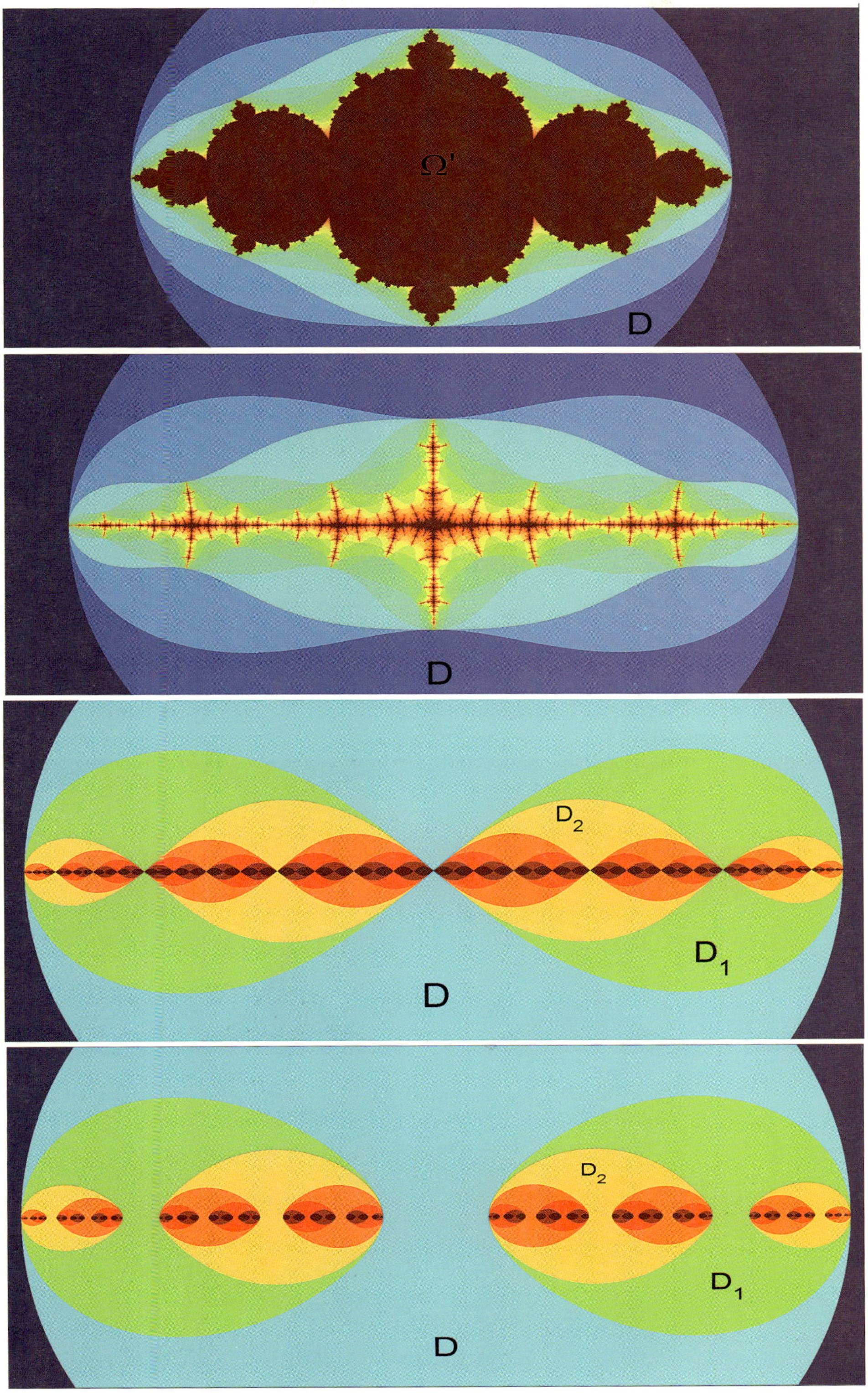

Color Plate 7. Pincherle's $\alpha_a = z^2 - a$ for $a = 0.7, 1.4, 2, 2.1$.
See page 195.

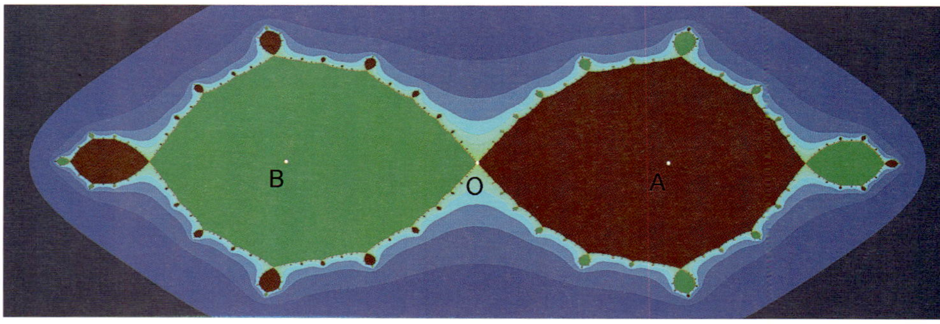

Color Plate 8. The Julia set for $z \mapsto \frac{-z^3+3z}{2}$. See page 144.

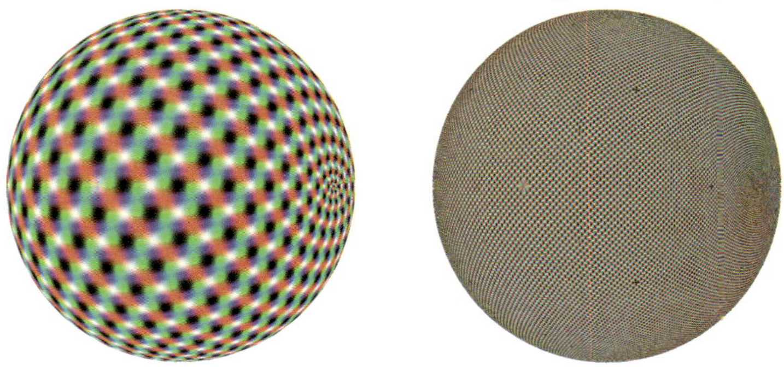

Color Plate 9. Repelling periodic points for Lattès' example. See page 170.

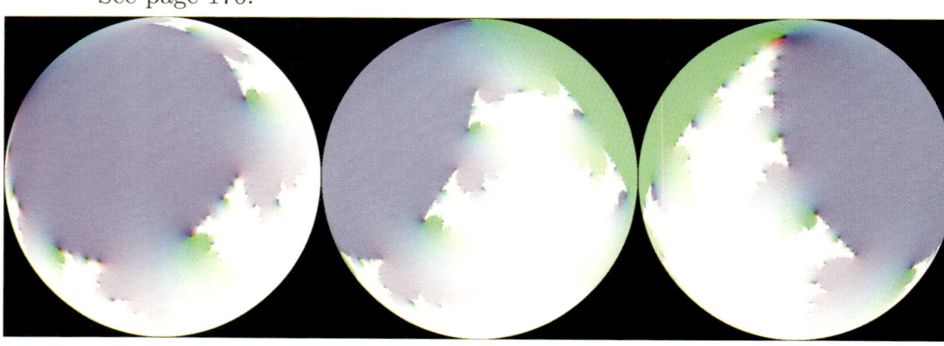

Color Plate 10. Views of the Baker domain for $E(z) = e^z + z$. See page 222.

Color Plate 11. Examples of domain coloring. See page 376.

CHAPTER 4

Iteration and Differential Equations II: Small Divisors

4.1. Introduction: Big problems from small divisors

If arbitrarily many denominators of the constant terms a_k in a power series $\sum a_k z^k$ approach zero in such away that the terms grow without bound, the radius of convergence of the series may be zero, causing the series to diverge. In the mid-1800s this phenomena of arbitrarily small denominators began to appear in the study of differential equations and became known as the problem of small divisors. The presence of small divisors, which occurred in other kinds of series as well, developed into a significant impediment to the understanding of the differential equations that arise in the study of our solar system, in particular those related to the n-body problem.

Small divisors are also intimately involved with many of the developments in complex dynamics, and we will encounter them at several turns. Not only does the problem of small divisors in complex dynamics mirror its appearance in differential equations, but the solution of the small divisors problem in complex dynamics played an inspirational role for one of the mathematicians who successfully attacked the problem of small divisors in the context of differential equations.

The n-body problem addresses one of the oldest scientific questions of humankind: What governs the motions of heavenly bodies? While humans have accurately charted the movements of the planets for millennia, it was not until Johannes Kepler developed his famous laws of planetary motion in the early 1600s that scientists had the tools to reliably analyze their motions.

Since the planets in our solar system have small masses compared to that of the sun, their orbits approximately conform to Kepler's laws.[1] In fact, if one ignores all the mutual planet-to-planet interactions, such a system is equivalent to a number of decoupled sun-planet, two-body systems, where each planet describes elliptical orbits around the sun which occupies one of the two foci.

The earliest approaches to the n-body problem[2] took this point of view, articulating what is often called the *direct problem of dynamics*: obtain all the possible configurations of the orbits of a mass point moving in a known force field.[3] The

[1] Unlike Newton's universal gravitational law, Kepler's laws do not take into account the weak gravitational interactions between the planets.

[2] The ultimate goal of the n-body problem is to take into account all gravitational interactions, both weak and strong, and to explain the dynamical interactions between n bodies. The most famous n-body problem is to explain the dynamics of the solar system. Our use of the term "celestial mechanics" refers to the study of the planetary n-body problem.

[3] In contrast, the *inverse problem of dynamics* consists of determining a stationary force causing a free body with given initial conditions to move in a given orbit. This problem was first explored in Newton's *Principia* [**1687**], where the determination of the gravitational potential

two-body problem, in particular, is an *inverse square force problem* because, in compliance with Newton's gravitational law, the force is proportional to the inverse of the square of the distance between the two planets. The first successful solutions to the two-body problem were published by Johann Bernoulli [**1710**] and Isaac Newton in the second edition of his *Principia* [**1713**].[4]

However, this huge simplification was not consistent with astronomical observations. The deviations of the motion of a planet from Kepler's laws due to attraction from other planets are called perturbations, and Newton's gravitational law showed that the gravitational forces between the planets, while far weaker than the pull of the sun, produces significant perturbations of their elliptical orbits over long periods of time. The model based on Newton's gravitational law thus improves upon Kepler's and gives a much better fit to the observations.

After Newton's law described the interactions among celestial bodies, celestial mechanics drove both mathematical and scientific research to new and unexpected frontiers. Devising advanced mathematical tools turned out to be of primary importance in making predictions about the effect of gravity on the orbits of the planets which were consistent with astronomical observations, and the solar system was recast as a planetary many-body problem formed by point masses interacting only through gravity, one of which, the sun, is much larger than the others.

As is now well known, excluding a few exceptional configurations, no exact solution in closed form for the differential equations of motion of the n-body problem is possible when $n \geq 3$. However, since all the masses involved are small compared to that of the sun, the motion of a particular planet in this multibody dynamical frame may be regarded as a slightly perturbed elliptical motion, and astronomers beginning with Newton tried to detect the small changes in the parameters of the ellipse resulting from small forces as time progresses.

The simplest extension of a completely solvable two-body problem is a three-body problem, such as the one formed by the earth, the moon, and the sun. This problem is referred to as the *main problem of lunar theory*, and it is the inclusion of the sun in the two-body earth-moon problem that causes a perturbation on the motion of the moon, the gravitational forces exerted by the remaining planets being negligible. The consequent anomaly in the monthly revolution of the moon around the earth, formerly called the *moon's second anomaly* and now referred to as *evection*, was known in ancient times and was first formalized by Claudius Ptolemy in the 2nd century. Newton's analysis of evection improved the concordance between observational measurements and computational results, and it anticipated the perturbation methods of Leonhard Euler and Pierre-Simon Laplace.

These early efforts to understand the dynamics of more complicated systems by regarding them as perturbations of simpler and completely solvable configurations constituted the foundation of perturbation theory and date back to 18th century.

---

based on Kepler's laws was established. Bertrand later showed [**1877**] that Newton's gravitational force may be deduced by only considering Kepler's first law, namely that the orbit of every planet is an ellipse with the sun at one of the two foci.

[4]Newton had left out this important investigation from his first edition of the *Principia* [**1687**], but nonetheless he gave some geometrical hints to its solution. The first researcher who applied the differential calculus to solve the inverse problem of dynamics was Jacob Hermann [**1710a, 1710b**], a student of Jacob I. Bernoulli. See Otto Volk [**1976**] and Michael Nauenberg [**2010**] for a historical perspective on the inverse square force problem.

They allowed astronomers to obtain satisfactory starting approximations by treating the planets as independent two-body problems, as described above, which then were refined by an iterative technique that at each step added a correction to the previously computed approximate solution. The correcting function is typically represented by a Fourier series, whose convergence will depend upon the size of the orbital eccentricities and inclinations. Usually, one or a very few iterations sufficed to retrieve very good approximations of the true solution which were entirely consistent with the astronomical observations of the day.

Using these perturbation techniques, a number of mathematical successes followed, among them the independent predictions of the existence of Neptune in 1846 by the English astronomer John Couch Adams and the French astronomer Urbain Le Verrier, each of whom analyzed perturbations of Uranus which were not accounted for by the known planets, as well as Laplace's explanation of the "great inequality of Jupiter and Saturn" around 1785.[5]

These kinds of successes led to Laplace's well-known quotation about the causal determinism induced by Newton's law that reflects the extent to which the neighboring universe was perceived as a precise, almost perfect, mathematical system:[6]

> *We ought then to regard the present state of the universe as the effect of its anterior state and as the cause of the one which is to follow. Given for one instant an intelligence which could comprehend all the forces by which nature is animated and the respective situation of the beings who compose it—an intelligence sufficiently vast to submit these data to analysis—it would embrace in the same formula the movements of the greatest bodies of the universe and those of the lightest atom; for it, nothing would be uncertain and the future, as the past, would be present to its eyes. The human mind offers, in the perfection which it has been able to give to astronomy, a feeble idea of this intelligence. Its discoveries in mechanics and geometry, added to that of universal gravity, have enabled it to comprehend in the same analytical expressions the past and future states of the system of the world.* [**1840**, p. 4]

However, in the early 1800s, following Joseph-Louis Lagrange and Laplace, the iteration procedure was carried out beyond the first order and serious drawbacks to this approach became evident. The convergence of the series of perturbative developments could not be established due to the systematic appearance of small divisors (which are also referred to as *small denominators*): the denominators appearing in the generic term of the perturbative expansion could become arbitrarily small, causing the corresponding correction to become larger than the first-order correction, putting in doubt the convergence of the series itself. Among the first to realize this problem was Charles Eugène Delaunay who encountered it while

---

[5]Dating back to Kepler, astronomers had observed unexplained perturbations in the trajectories of Jupiter and Saturn that Laplace was able to account for. See, for example, Curtis Wilson [**1985**] and Edgar Odell Lovett [**1895**] for the details of the great inequality of Jupiter and Saturn.

[6]The quotation is taken from Truscott and Emory's English translation of Laplace's *Essai philosophique sur les probabilités*, [**1840**]; also see, B. Hasselblatt and A. Katok [**2002**]. We would also like to observe that several writers, including Hasselblatt and Katok, have traced the flow of events from Newton to Poincaré along the same lines that we do. It strikes us as the natural way to describe things.

studying the perturbative expansion for the Earth-Moon-Sun three-body system, [**1846, 1860, 1867**].

Much later in the 19th century, Poincaré [**1886**, Chap. XIX] introduced the question of small divisors in an abstract setting, where the ODE is not necessarily linked to a particular problem in celestial mechanics. Prior to that, he had successfully carried out the study of stability of singular points (equilibria) of autonomous two-dimensional ODEs in [**1885**, Chap. X-XI] (see Section 3.3 of our text). In particular, he classified singular points by looking at the behavior of neighboring trajectories in the phase plane and fully characterized the existence of centers in differential equations, which are particular singular points whose neighboring trajectories are closed orbits (see Figure 3.1).[7]

Such a characterization goes a long way towards proving the existence of a first integral $F = F(x, y)$ of the ODE which is positive definite in a neighborhood of the equilibrium point.[8] This, in turn, is equivalent to requiring that an infinite sequence of coefficients, which depend on the specific problem under analysis, vanish. In fact, the search for the first integral F is carried out by determining the coefficients in its Taylor expansion according to an iterative procedure at each step of which a new term is determined. Under the above-mentioned "vanishing assumption" (see the next section for details), Poincaré had no problem showing convergence of the formal series for F he had obtained.

So complete was the success of the techniques exploited to investigate the stability properties of two-dimensional autonomous ODEs that Poincaré tried to export them to the three-dimensional (autonomous) context. Actually, Poincaré did much more: He introduced a skillful tool, his *return map* (see Figure 3.2), to reduce the dynamics of the system around a periodic orbit to a two-dimensional nonautonomous dynamical system defined in a neighborhood of an equilibrium point. He then tailored the argument from the two-dimensional autonomous case to investigate neighboring trajectories, and, as he did previously, he was able to classify the stability properties of the equilibrium point (which he then transferred to the closed orbit), for three cases of interest—foci, nodes, and saddle points (this latter case was thoroughly discussed in Section 3.3). Likewise, he approached the question of the existence of centers in differential equations, which is the most relevant one for the applications in celestial mechanics, by searching for a first integral $F = F(x, y, t)$, now depending on the time t.

But a new, subtle difficulty arose: The vanishing assumption of the infinite sequence of coefficients proved to be necessary, but no longer sufficient, to determine the stability character of the equilibrium point. In particular, although he could formally construct the Taylor series for F, Poincaré did not succeed in proving its convergence, due to the presence of small divisors in its coefficients. As we will see in Section 5.4 (and elsewhere) a similar problem arises in complex dynamics when the multiplier of the fixed point is one in modulus but not a root of unity: while a formal series solution for the appropriate Schröder functional equation could easily be found in the neighborhood of the fixed point, determining its convergence or divergence is quite difficult.

[7]Since, at this early stage, the system was supposed autonomous, this is the only possible phase portrait for which a center can occur.

[8]See footnote 9 from Chapter 3 for a definition of a first integral for a three-dimensional ODE. The definition for a two-dimensional system is analogous. Positive definiteness implies that the level curves surrounding the equilibrium point are closed, which in turn implies stability.

The issue of small divisors is a very serious problem since, as Poincaré observed, such a situation is recurrent in celestial mechanics and, when it occurs, it is persistent, meaning that it does not depend upon the specific technique used to solve the problem (the Poincaré quotation at the conclusion of Section 4.2 speaks to this issue).

Centers in complex dynamics can be viewed as a discrete counterpart to centers in ODEs. Indeed, centers of both continuous (e.g., Hamiltonian) and discrete (e.g., complex) dynamical systems may be roughly perceived as marginally stable equilibria, that is, they are stable but not asymptotically stable (see page 50): surrounding trajectories never approach the center although, depending on the specific initial condition, they may remain in a small neighborhood of the center itself.

Due to the qualitative similarities of orbits near centers in ODEs and complex dynamics, it is not at all surprising that other analogies arise. The problem of small divisors is arguably one of the deepest of these, and for many, many years it cast doubt upon the existence of centers in both fields.

Although its genesis was in the context of ODEs, it was in complex dynamics that the first comprehensive solution of a small divisors problem occurred, in Siegel's 1942 proof that centers exist, discussed in Chapter 12. Siegel's proof encouraged researchers in celestial mechanics, in particular Jürgen Moser, to devise new tools to face the more delicate problem of small divisors arising from the stability analysis of nearly integrable Hamiltonian systems.

A little over ten years after Siegel's paper, Andrei Kolmogorov [**1954**] introduced a completely new and innovative technique reminiscent of Newton's method to treat small divisors, although it is not clear whether Siegel influenced Kolmogorov. The techniques developed by Kolmogorov and modified in subsequent works by Arnold [**1963**] and Moser [**1966**] form the bedrock of what is nowadays called KAM theory.

Our intent in this chapter is provide context for the center problem in both fields by presenting a brief, episodic history of the small divisors problems. We will present examples where problems of small divisors constitute a concrete barrier against a full understanding of the stability of equilibrium points or orbits in both the general theory of ODEs and perturbation theory in particular. At the conclusion of the present chapter, we will adumbrate the role of KAM theory in overcoming these obstacles in perturbation theory. In subsequent chapters we will carry out a detailed analysis of the small divisors problem in complex dynamics. Finally, in Chapter 12, we will explicitly discuss the connection between Moser's articulation of KAM theory and Siegel's resolution of the small divisors problem in complex dynamics.

We close this section with a quotation from Poincaré, reflective of the doubt introduced by small divisors and of the historical imperative to overcome them, that describes a very different universe from Laplace's:

> *The three-body problem is of such importance to astronomy, and at the same time so difficult, that mathematicians have for a long time devoted every effort to its resolution. A complete and rigorous solution to it is manifestly impossible, so we must appeal to methods of approximation. ... At the beginning of this century the conquests of Lagrange and Laplace, and more recently the work of Le Verrier, resulted in a degree of perfection that up to now has*

> *satisfied our practical needs. I can add that they may continue to do for quite some time to come, despite some differences in detail; yet, it is certain that they will not always suffice, as is quite evident after a brief reflection.*
>
> *The ultimate goal of celestial mechanics is to solve the formidable question of ascertaining whether Newton's law alone may explain all astronomical phenomena; the only means to accomplish such a task is to make as many precise astronomical observations as possible, and then to compare them to computational results. These are approximate in nature, and therefore it turns out to be of no utility to ask for an accuracy higher than the one deduced from direct observation. ... Moreover, the degree of approximation which may be considered satisfactory nowadays, is doomed to be insufficient in a few centuries.* [**1892a**, pp. 1–2]

4.2. Poincaré encounters small divisors

As was seen in Section 3.3, Poincaré introduced an innovative tool in Chapter XIX ("*Étude des Courbes Fermées*") of [**1886**] to investigate the stability of closed orbits in a three-dimensional phase space, the return map, that allowed him to recast an autonomous system of dimension three into a nonautonomous system with only two variables whose dynamics take place on a moving plane orthogonal at each point to the closed orbit given by

$$\frac{dx}{dt} = X(t, x, y), \quad \frac{dy}{dt} = Y(t, x, y).$$

The functions $X(t, x, y)$ and $Y(t, x, y)$ are analytic with respect to both x and y, and the coefficients of their expansion in powers of x and y are trigonometric series of the variable t.

The great advantage of such a reduction is that, from the point of view of stability, it allows us to treat a closed orbit as just an equilibrium point, so, by considering the location in the complex plane of the two roots S_1 and S_2 of the characteristic equation

$$S^2 - (\alpha + \delta)S + (\alpha\delta - \beta\gamma) = 0$$

associated with the linear part of his return map, he was able to classify the stability of the closed orbit according to the behavior of the sequence produced by the return map on the Poincaré section (the plane $s = 0$).[9]

After examining the first three cases, namely, when S_1 and S_2 are complex conjugates but with modulus different from 1, when $S_1, S_2 > 1$, and finally $0 < S_2 < 1 < S_1$, he made clear the existing analogy between the present analysis in [**1886**] and the theory of equilibrium points he had developed previously in [**1885**]: The first case corresponds to that of foci, the second to that of nodes, and the third to saddle points, discussed in Section 3.3 (see Figure 3.1).

There was still a final case to examine, the most relevant one in view of the applications of the results to ODEs arising from celestial mechanics, when S_1 and S_2 are complex conjugate and of unit modulus.

[9]The generation of this sequence is discussed immediately below equation (3.15).

He articulated his reasons for postponing his treatment of the last case in the following quotation:

> *It remains to examine the exceptional case where S_1 and S_2 are complex conjugates and of modulus 1. This case corresponds to the one we studied in detail in Chapter XI* [of [**1885**], namely centers]. *We will see that analogies will occur up to a given point.... But, after entering into the details of our analysis, we will recognize some essential differences, and completely novel difficulties will come into play.* [**1886**, p. 195]

What are the difficulties Poincaré referred to? They involve the small divisors that occur when he tried to find an expression for the constant of motion F associated with this case. The construction is involved, but worthwhile, since it shows explicitly how small divisors arise.

Poincaré began his construction by expanding the analytic functions $X(x, y, t)$ and $Y(x, y, t)$, which describe the orbit in the case where the S_i are complex conjugates of unit modulus, into their Taylor series with respect to x and y,

$$\begin{aligned} \frac{dx}{dt} &= X_1 + X_2 + \cdots + X_n + \cdots &&= X, \\ \frac{dy}{dt} &= Y_1 + Y_2 + \cdots + Y_n + \cdots &&= Y, \end{aligned}$$

where X_i and Y_i are homogeneous polynomials of degree i in x and y whose coefficients are trigonometric series in t.

To construct the constant of motion F, he assumed it could be expressed as a series in x and y

$$F = F_2 + F_3 + \cdots + F_n + \cdots, \tag{4.1}$$

where again F_i contains terms of degree i in x and y whose coefficients, like those of the X_i and Y_i are trigonometric series in t (note that the linear part of F is null). In perfect accord with what he did earlier while studying the existence of centers for autonomous two-dimensional systems in [**1886**], he tried to determine the unknown terms F_i in such a way that F is a constant of motion for the system.

This is accomplished by imposing the condition that the function

$$\Phi = \frac{\partial F}{\partial x} X + \frac{\partial F}{\partial y} Y + \frac{\partial F}{\partial t}$$

vanishes identically.[10] In particular, he set all the terms in the expansion of Φ with degree less that an arbitrary integer p equal to zero; this is known as the conservation requirement.

Setting

$$\Phi_{ik} = \frac{\partial F_i}{\partial x} X_k + \frac{\partial F_i}{\partial y} Y_k, \tag{4.2}$$

[10]The scalar function F is constant along a solution $(x(t), y(t))$ of the ODE if and only if $\frac{d}{dt} F(x(t), y(t)) = 0$ for all times t, in which case one easily observes that

$$\frac{d}{dt} F(x(t), y(t)) = \frac{\partial F}{\partial x} X + \frac{\partial F}{\partial y} Y + \frac{\partial F}{\partial t} = \Phi.$$

the conservation requirement up to degree $p-1$ is equivalent to the system

$$
(4.3)\qquad \begin{cases}
\dfrac{\partial F_2}{\partial t} + \Phi_{21} & = \quad 0, \\
\dfrac{\partial F_3}{\partial t} + \Phi_{31} & = \quad -\Phi_{22}, \\
\dfrac{\partial F_4}{\partial t} + \Phi_{41} & = \quad -\Phi_{32} - \Phi_{23}, \\
\cdots \quad \cdots & \quad\quad \cdots \quad \cdots \\
\dfrac{\partial F_{p-1}}{\partial t} + \Phi_{p-1,1} & = \quad -\displaystyle\sum_{j=2}^{p-2} \Phi_{p-j,j}.
\end{cases}
$$

The structure of the system suggests that the topmost equation in (4.3) should be solved first since it will yield the function F_2, which will then determine the term at the right-hand side of the second equation, yielding the function F_3, and so forth.

To find F_2, Poincaré applied the solution derived at equation (3.17) to the linearized system (3.16)

$$\frac{dx}{dt} = X_1, \quad \frac{dy}{dt} = Y_1,$$

where now S_1 and S_2 have unit modulus. This application in turn yields

$$
\begin{cases}
x(t) = A\left(\cos(\lambda' t + \theta)\sigma(t) - \sin(\lambda' t + \theta)\sigma'(t)\right), \\
y(t) = A\left(\cos(\lambda' t + \theta)\gamma(t) - \sin(\lambda' t + \theta)\gamma'(t)\right),
\end{cases}
$$

where λ' is such that $e^{2i\pi\lambda'} = S_1$ and $\sigma(t)$, $\gamma(t)$ and $\sigma'(t)$, $\gamma'(t)$ are the real and imaginary parts of the functions φ_1 and ψ_1, respectively (see Section 3.3). In order to further simplify the shape of the linearized equation, Poincaré introduced the linear change of variables[11]

$$\begin{pmatrix} -\sigma'(t) & \sigma(t) \\ -\gamma'(t) & \gamma(t) \end{pmatrix} \begin{pmatrix} \xi \\ \eta \end{pmatrix} = \begin{pmatrix} x \\ y \end{pmatrix}$$

so that the solution now becomes

$$\xi(t) = A\sin(\lambda' t + \theta), \quad \eta(t) = A\cos(\lambda' t + \theta).$$

The linearized system then uncouples as

$$\frac{d\xi}{dt} = \lambda'\eta, \quad \frac{d\eta}{dt} = -\lambda'\xi.$$

That we can always perform such a change of coordinates allows, without loss of generality, the assumption

$$X_1 = \lambda' y, \quad Y_1 = -\lambda' x.$$

Substituting X_1 and Y_1 into (4.2) with $i = 2$ and $k = 1$, one easily finds that

$$F_2 = x^2 + y^2$$

solves the first equation in (4.3). The remaining equations in (4.3) then become

$$(4.4)\qquad \frac{\partial F_q}{\partial t} + \lambda'\left(y\frac{\partial F_q}{\partial x} - x\frac{\partial F_q}{\partial y}\right) = H_q, \qquad q = 3, \ldots,$$

[11]Notice that this change of variable is nonlinear with respect to the time t.

where

$$H_q = -\sum_{j=2}^{q-2} \Phi_{q-j,j}.$$

In the polar coordinates $x = \rho\cos\omega$, $y = \rho\sin\omega$, the two q-degree polynomials F_q and H_q become

$$F_q = \rho^q \varphi(\omega, t), \quad H_q = \rho^q \psi(\omega, t),$$

where $\varphi(\omega, t)$ and $\psi(\omega, t)$ are trigonometric series in ω and t. Substituting F_q and H_q back into (4.4) yields[12]

$$\frac{\partial\varphi}{\partial t} - \lambda' \frac{\partial\varphi}{\partial\omega} = \psi(\omega, t). \tag{4.5}$$

Since, by virtue of (4.1), the left-hand side of equation (4.5) has no constant term, a necessary condition for the equation to admit a solution is for the constant term C_0 in the trigonometric series

$$\psi(w, t) = C_0 + \sum_{m,n} A_{mn} \cos(mt + n\omega + \theta)$$

to vanish, yielding

$$\frac{\partial\varphi}{\partial t} - \lambda' \frac{\partial\varphi}{\partial\omega} = \sum_{m,n} A_{mn} \cos(mt + n\omega + \theta). \tag{4.6}$$

Poincaré assumed that λ' is irrational (otherwise the formal solution (4.8) he derived below could have terms with a zero in the denominator) and for all q the corresponding constants C_0 in the expansion of the functions ψ are null. Before integrating (4.5) to find the F_q, Poincaré stressed the importance of the case where the C_0 vanish:

> *The case where all the constants C_0 are actually zero seems rather exceptional, since its occurrence requires an infinite number of conditions to be satisfied; nevertheless its relevance resides not only in the difficulties it brings, but also because it arises in the study of the general equations of dynamics.* [**1886**, p. 199]

To emphasize the importance of this case, Poincaré showed that, in some situations, one can establish a priori the vanishing of all the infinite constants C_0. In particular, by exploiting Green's theorem, he easily showed that this is the case when there exists a positive analytic function $M(x, y, t)$ defined on an open region containing a closed orbit satisfying

$$\frac{\partial MX}{\partial x} + \frac{\partial MY}{\partial y} + \frac{\partial M}{\partial t} = 0. \tag{4.7}$$

As an example of such an M, he noted that well-known astronomers, such as Hugo Gyldén, Anders Lindstedt, and Octave Callandreau, had already studied the equation

$$\frac{d^2x}{dt^2} = \varphi_1 x + \varphi_2 x^2 + \cdots,$$

[12] One can easily verify that $\frac{\partial F_q}{\partial\omega} = y\frac{\partial F_q}{\partial x} - x\frac{\partial F_q}{\partial y}$.

where the φ_i are trigonometric series in t. Such a second order equation is equivalent to the system of first order equations

$$\frac{dx}{dt} = y \equiv X, \quad \frac{dy}{dt} = \varphi_1 x + \varphi_2 x^2 + \cdots \equiv Y,$$

which satisfies condition (4.7) with $M = 1$. Therefore, the equation

$$\frac{\partial F}{\partial x} X + \frac{\partial F}{\partial y} Y + \frac{\partial F}{\partial t} = 0$$

may be successfully integrated by successive approximations without incurring any secular terms. Poincaré thus explained, in a more formal and less restrictive context, the success of the method of Lindstedt.

After this aside, Poincaré returned to the construction of F: in the event that all the constants C_0 are null, a series

$$F = F_2 + F_3 + \cdots$$

may be formally constructed by a direct integration of equation (4.6) which yields

$$\varphi(\omega, t) = \sum_{m,n} A_{mn} \frac{\sin(mt + n\omega + \theta)}{m - n\lambda'}. \tag{4.8}$$

It is here that the analogy to the theory of centers developed in Chapter XI of Poincaré's [**1885**] evaporates, and the difficulty he referred to in the quotation at the beginning of this section finally arises: whereas previously Poincaré showed that the series F is always uniformly convergent, he can now no longer draw that conclusion due to the presence of the divisors $m - n\lambda'$ that clearly may become arbitrarily small.

Poincaré's conclusion to [**1886**] is so emblematic of the difficulty he encountered that quoting it is perhaps the best way to summarize his thoughts:

> *Even in the event all the C_0 are null, it cannot be deduced that the series*
>
> $$F = F_2 + F_3 + \cdots,$$
>
> *defined above, is convergent, nor that a point close to the closed orbit will remain always in a neighborhood of it, nor that after departing from this orbit it will eventually approach it anew.... From what has been seen above, one can easily understand the extent to which the difficulties that one encounters in celestial mechanics due to small divisors and the quasi-commensurability of mean motions are tied to the intrinsic nature of things and cannot be overcome. It is extremely probable that such difficulties will arise whatever the method used.* [**1886**, p. 217]

Poincaré's investigations led to the concept of deterministic chaos in the theory of dynamical systems, the reason being the pervasive nature of instability. Poincaré's famous sentence, *"Instability is thus the rule, and stability the exception"* [**1885**, p. 172], attests to this and is indicative of the fact that stable equilibrium points or orbits only occur under very special situations.

4.3. Small divisors in the work of Cigala and Levi-Civita

Although small divisors do not arise directly in the work of Levi-Civita, they are implicit in a case he began in [**1901**] that was continued by the Italian mathematician Angelo Raffaello Cigala in [**1904**]. In Section 3.4, we described Levi-Civita's use of iteration to study the dynamics of a nonautonomous ODE in a neighborhood of a periodic orbit that, after a suitable change of coordinates, he located at the origin. To analyze the nature of a periodic orbit, he looked at the iteration induced by

$$x_i^{(n)} = f_i(x_1^{(n-1)}, x_2^{(n-1)}, \ldots, x_d^{(n-1)}), \qquad i = 1, \ldots, d, \quad n = \pm 1, \pm 2, \ldots$$

(see (3.23)).

After showing instability in the case where one of the multipliers has a modulus different than one, he embarked upon the study of stability in the simpler $d = 2$ case where both multipliers have unitary modulus. However, he left aside the case where the two multipliers, $\omega_1 = e^{i\vartheta}$ and $\omega_2 = e^{-i\vartheta}$, are complex conjugates, which leads to the study of the iteration $(x, y) \mapsto (x_1, y_1)$ defined by the mapping

$$\left\{ \begin{array}{l} x_1 = \cos\vartheta\, x - \sin\vartheta\, y + f(x, y), \\ y_1 = \sin\vartheta\, x + \cos\vartheta\, y + g(x, y). \end{array} \right. \tag{4.9}$$

Levi-Civita was well aware of the difficulties in treating this case when ϑ is incommensurable with 2π and observed that he was unable to adapt his approach, which had proved successful in other cases, to this special situation. He simply concluded that he would have to return to this task later [**1901**, p. 224]. He did not, although under his direction, Cigala took up this case with his paper [**1904**].[13]

Cigala first introduced the linear complex change of variables

$$\left\{ \begin{array}{l} x' = x + iy, \\ y' = x - iy, \end{array} \right. \qquad \left\{ \begin{array}{l} x_1' = x_1 + iy_1, \\ y_1' = x_1 - iy_1, \end{array} \right.$$

after which the iteration becomes

$$\left\{ \begin{array}{l} x_1 = e^{i\vartheta} x + \displaystyle\sum_{m=2}^{\infty} P_m(x, y), \\ y_1 = e^{-i\vartheta} y + \displaystyle\sum_{m=2}^{\infty} Q_m(x, y), \end{array} \right. \tag{4.10}$$

where we have used x, y, x_1, y_1 instead of x', y', x_1', y_1'. For $m = 2, 3, \ldots$,

$$P_m(x, y) = \sum_{p=0}^{m} \binom{m}{p} c_{mp} x^{m-p} y^p,$$

$$Q_m(x, y) = \sum_{p=0}^{m} \binom{m}{p} \bar{c}_{mp} x^p y^{m-p}.$$

The coefficients c_{mp} and $\bar{c}_{mp}$ are complex conjugates. Cigala's approach to the problem is close to the standard approach, which consists of trying to define a

[13] Cigala acknowledged Levi-Civita's help with the preparation of his manuscript.

suitable nonlinear change of variables that conjugates the map (4.10) to its linear part:[14]

> *We want to show how, by means of suitable substitutions, it is possible to cancel out a number of terms in our transformation and how, according to the results obtained, we can deduce our criterion for instability. ... We could first plug in a substitution of degree two and see whether we can conclude anything* [about the instability of (4.10)]; *we could then consider a substitution of degree three, and so on, and so forth.* [**1904**, p. 70]

Therefore, Cigala considered a generic substitution of degree $n > 1$, assuming that the first nonlinear term in (4.10) has degree n:

$$\left\{\begin{aligned} x_1 &= e^{i\vartheta}x + \sum_{m=n}^{\infty} P_m(x,y), \\ y_1 &= e^{-i\vartheta}y + \sum_{m=n}^{\infty} Q_m(x,y). \end{aligned}\right. \tag{4.11}$$

This defines a mapping F_n.

A novelty in Cigala's approach is that he then considered a complex substitution, Ψ_n

$$\left\{\begin{aligned} \xi &= x + \varphi_n(x,y), \\ \eta &= y + \psi_n(x,y), \end{aligned}\right. \qquad \left\{\begin{aligned} \xi_1 &= x_1 + \varphi_n(x_1,y_1), \\ \eta_1 &= y_1 + \psi_n(x_1,y_1), \end{aligned}\right.$$

whose inverse, Ψ_n^{-1}, in a neighborhood of the origin is

$$\left\{\begin{aligned} x &= \xi - \varphi_n(\xi,\eta) + \cdots, \\ y &= \eta - \psi_n(\xi,\eta) + \cdots. \end{aligned}\right.$$

The conjugate homogeneous forms of degree n, φ_n, and ψ_n, read

$$\begin{aligned} \varphi_n(\xi,\eta) &= \sum_{p=0}^{n} \binom{n}{p} a_p \xi^{n-p}\eta^p, \\ \psi_n(\xi,\eta) &= \sum_{p=0}^{n} \binom{m}{p} \bar{a}_p \xi^p \eta^{n-p}, \end{aligned} \tag{4.12}$$

where the unknown coefficients $a_0, a_1, \ldots, a_n$ (and their conjugates $\bar{a}_p$) are to be determined.

Introducing this new substitution into (4.11) yields

$$\left\{\begin{aligned} \xi_1 &= e^{i\vartheta}\xi + U_n + \text{higher order terms}, \\ \eta_1 &= e^{-i\vartheta}\eta + V_n + \text{higher order terms}, \end{aligned}\right. \tag{4.13}$$

where U_n and V_n are homogeneous polynomials of degree n in ξ and η defined as

$$\left\{\begin{aligned} U_n &= -e^{i\vartheta}\varphi_n(\xi,\eta) + P_n(\xi,\eta) + \varphi_n(e^{i\vartheta}\xi, e^{-i\vartheta}\eta), \\ V_n &= -e^{-i\vartheta}\psi_n(\xi,\eta) + Q_n(\xi,\eta) + \psi_n(e^{i\vartheta}\xi, e^{-i\vartheta}\eta). \end{aligned}\right.$$

---

[14] As the paper's title "*Sopra un criterio di instabilità*", or "On a criteria of instability", suggests, Cigala was, in the spirit of Levi-Civita, principally interested in proving the nonexistence of such a substitution.

Consequently, the two conditions

$$e^{-i\vartheta}U_n = 0, \qquad e^{i\vartheta}V_n = 0,$$

become, in terms of the unknowns a_p and $\bar{a}_p$ from (4.12)

$$(e^{i\vartheta(n-2p-1)} - 1)a_p = -e^{-i\vartheta}c_{np}, \qquad p = 0, \ldots, n, \tag{4.14}$$

and its conjugate.

Cigala distinguished between two subcases:

(a) If n is even, all the terms of degree n may be eliminated in (4.13). In this case $e^{i\vartheta(n-2p-1)} \neq 1$, since $n - 2p - 1 \neq 0$ and ϑ is incommensurable with 2π. After this substitution, the transformation (4.11) takes the desired form

$$\begin{cases} x_1 = e^{i\vartheta}x + \displaystyle\sum_{m=n+1}^{\infty} P'_m(x, y), \\ y_1 = e^{-i\vartheta}x + \displaystyle\sum_{m=n+1}^{\infty} Q'_m(x, y). \end{cases} \tag{4.15}$$

(b) However, when n is odd, say $n = 2\nu + 1$, $\nu > 0$, it is not possible to make all terms of degree n vanish, except when $P_n(x, y)$ (and, consequently, $Q_n(x, y)$) satisfy a special vanishing requirement, namely that $c_{2\nu+1,\nu} = 0$. In fact, when n is odd, $n - 2p - 1$ is always different from zero except when $p = \nu$. Therefore, conditions (4.14) will yield all the a_p except a_ν, which is left as a free parameter. By virtue of this substitution, all the degree n terms, except perhaps one, are conjugated away and the new transformation is

$$\begin{cases} x_1 = e^{i\vartheta}x + \dbinom{2\nu+1}{\nu} c_{2\nu+1,\nu}x^{\nu+1}y^{\nu} + \text{ higher order terms}, \\ y_1 = e^{-i\vartheta}y + \dbinom{2\nu+1}{\nu} \bar{c}_{2\nu+1,\nu}x^{\nu}y^{\nu+1} + \text{ higher order terms}. \end{cases} \tag{4.16}$$

Cigala then showed that if the coefficient $c_{2\nu+1,\nu}$ is nonzero (which it is unless the special requirement referred to above holds), one can deduce instability of the periodic orbit.

We are more interested in the case where the special vanishing requirement holds, and $c_{2\nu+1,\nu}$ is in fact zero, which is a necessary condition for stability. In this instance, Cigala imposed the condition that all terms of degree $2(\nu + 1)$ (the even case) are zero. After that, one arrives at a new coefficient $c'_{2\nu+3,\nu+1}$ via the iteration induced by $(x, y) \mapsto (x_1, y_1)$ whose nonnullity must be checked.

Cigala realized that determining the instability of the transformation (4.16) may necessitate the computation of an infinite sequence of terms. What about the possibility of stating the stability of the transformation? Cigala opened the last section of his paper by remarking that

> *There is a case—unfortunately the most important for the application in dynamics—where one can establish a priori that all the coefficients $c_{2\nu+1,\nu}$ are surely null, and for which absolute uncertainty reigns regarding the question of the stability of the transformation, which, in fact, cannot be deduced from the behavior of a finite number of terms.* [**1904**, p. 77]

Here, Cigala was referring to the special situation for which the determinant of the Jacobian matrix associated with the transformation (4.9) is equal to 1. For Hamiltonian systems, such a property comes directly from the symplecticity of the associated flow.[15]

Formalizing Cigala's procedure for the case where all coefficients $c_{2\nu+1,\nu}$ in (4.16) are null, as is the case with Hamiltonian problems arising in celestial mechanics, yields an interesting link between his work and procedures typically used in perturbation theory—and in particular KAM theory—many years later to confront the problem of small divisors (see the next section and Section 12.5).

We have seen that the introduction of the *elementary* substitution[16] Ψ_n to both the input variables (x, y) and output variables (x_1, y_1) for the mapping F_n defined by (4.11) leads to the definition of a new mapping F_{n+1} defined by (4.15), when n is odd, and by (4.16) when $c_{2\nu+1,\nu} = 0$ and n is even. Cigala's procedure may be interpreted as imposing the condition that the following diagram commutes:

$$\begin{array}{ccc} (x,y) & \xrightarrow{F_n} & (x_1,y_1) \\ \Psi_n \big\downarrow & & \Psi_n \big\downarrow \\ (\xi,\eta) & \xrightarrow{F_{n+1}} & (\xi_1,\eta_1), \end{array}$$

that is,

$$F_{n+1} = \Psi_n \circ F_n \circ \Psi_n^{-1}. \tag{4.17}$$

From (4.17) we obtain the formalized version of his procedure,

$$F_n = \Phi_n^{-1} \circ F_2 \circ \Phi_n, \qquad \text{with } \Phi_n = \Psi_2^{-1} \circ \Psi_3^{-1} \circ \cdots \circ \Psi_n^{-1}, \tag{4.18}$$

where F_2 is the original mapping (4.10).

Many years later, Kolmogorov and Moser each devised an analogous substitution, albeit in a different context, to successfully overcome the problem of small divisors (compare (4.18) with (12.21) and with the map ψ_n defined in (12.23)).

Although Cigala did not explicitly point it out, the quotation above suggests that he encountered a small divisors problem: Since ϑ is incommensurable with 2π, the quantity $e^{i\vartheta(n-2p-1)} - 1$ has zero as a limit point, and, following from equation (4.14), infinitely many a_p will be arbitrarily large. Thus, even if the infinite sequence of the substitutions Ψ_n at (4.18) could be determined, nothing a priori can be said about the convergence of their composition, that is, the limit of the sequence Φ_n in (4.18).

In summary, there are two interesting issues with Cigala's approach: The first is the use of complex elementary substitutions Ψ_n. The other is that the substitution Φ, defined by the limit as $n \to \infty$ of Φ_n which would allow the conjugation of the nonlinear transformation (4.10) to its linear part, is a composition of an infinite number of elementary substitutions, each of which may be understood as a perturbation to the identity map. As was pointed out above, this is similar in spirit to the infinite composition, to be discussed in Section 12.5, that is instrumental in the approach furnished by KAM theory. In the next section we give a sketch of

[15] In two dimensions these two properties are equivalent.

[16] We call the substitution Ψ_n "elementary" since it represents a correction of order n of the identity map I, that is, as is evident from (4.12), $\Psi_n(x,y) = I + R_n(x,y)$, where $R_n(x,y) \equiv [\varphi_n(x,y), \psi_n(x,y)]^T$ is an infinitesimal of order n as $(x,y) \to (0,0)$.

the difficulties arising from the appearance of small divisors in perturbation theory associated with the early attempts to approach the problem.

4.4. Small divisors in perturbation theory

Several years after encountering small divisors in his study of equilibria in the context of autonomous systems of dimension three in [**1886**], Poincaré confronted them in the more general context of perturbations of a Hamiltonian system in Volume II of *Les méthodes nouvelles de la mécanique céleste* [**1893**].

After providing a detailed description of both the restricted and general three-body problem in early sections of Volume I [**1892a**], Poincaré formalized what he termed the "*Problème général de la Dynamique*", a phrase still in use, and, emphasizing its importance, devoted all of §9 to its definition. We quote it in its entirety:

> *We are thus led to propose the following problem: To study the canonical equations*
>
> $$\frac{dx_i}{dt} = \frac{\partial F}{\partial y_i}, \quad \frac{dy_i}{dt} = -\frac{\partial F}{\partial x_i}, \tag{4.19}$$
>
> *assuming that the function F may be developed in terms of increasing powers of a very small parameter μ, as follows*
>
> $$F = F_0 + \mu F_1 + \mu^2 F_2 + \cdots, \tag{4.20}$$
>
> *and, in addition, assuming that F_0 depends on x while it is independent of y, and that $F_1, F_2, \ldots$ are periodic functions of period 2π with respect to y.* [**1892a**, pp. 32–33]

The quantity μ is a small parameter whose presence reflects the gravitational influence of external celestial bodies on a completely integrable system (which corresponds to the $\mu = 0$ case). The phrase "completely integrable" means that the equation of motion can be solved analytically. Indeed, choosing $\mu = 0$ and using vector notation, the equations of motion (4.19) reduce to

$$\frac{dx}{dt} = 0, \quad \frac{dy}{dt} = \omega(x) \equiv -\frac{\partial F_0(x)}{\partial x}, \tag{4.21}$$

since F_0 only depends on x. Consequently, equations (4.21) may be solved by a simple quadrature yielding

$$x(t) = x^{(0)}, \quad y(t) = y^{(0)} + \omega(x^{(0)})t, \tag{4.22}$$

where $x^{(0)}$ and $y^{(0)}$ are the initial conditions of the system.

That the functions F_i in (4.20) are 2π-periodic with respect to the variables y_i may be stated equivalently by assuming that the dynamics of the state vector (x, y) in (4.19) takes place on $\mathbb{R}^n \times \mathbb{T}^n$, where $\mathbb{T}^n$ is the standard n-dimensional torus

$$\mathbb{T}^n = \mathbb{R}^n / 2\pi\mathbb{Z}^n \equiv \{(y_1 \bmod 2\pi, y_2 \bmod 2\pi, \ldots, y_n \bmod 2\pi), \ y_i \in \mathbb{R}\}.$$

Consequently, all points y whose coordinates differ from an integer multiple of 2π are identified by a single point on the torus or, alternately, addition involving the vector $y(t)$ in (4.22) is considered modulo 2π.

The vector ω is the *frequency vector*, while its components ω_i are the frequencies. The frequencies ω_i are said to be rationally independent (or nonresonant) if

$$\langle \omega, k \rangle \equiv \sum_{\nu=1}^{n} m_\nu \omega_\nu \neq 0, \qquad \text{for all vectors } m \in \mathbb{Z}^n \setminus \{0\}; \tag{4.23}$$

otherwise, they are called rationally dependent (or resonant).

It is well known that if the frequencies ω_ν, $\nu = 1, \ldots, n$, are rationally independent, the corresponding solutions of (4.21) will be quasi-periodic, and, therefore, each of them will be dense on the torus $\mathbb{T}^n$. The relevant question then is to ascertain the extent to which the condition of quasi-periodicity persists under slight perturbations, that is, for $\mu \neq 0$ but small. When the qualitative behavior of the trajectories is unaffected by small perturbations of the parameter, the system is said to be *structurally stable*. Unlike the classical stability that we described previously[17] which considers perturbations of initial conditions in a neighborhood of a fixed point or a periodic orbit, structural stability deals with perturbations of the system itself.

From a topological viewpoint, structural stability means that the qualitative behavior of the trajectories in the phase space is not destroyed when slightly changing the value of μ, which allows us to conclude that stability properties valid for the case $\mu = 0$ can be transferred to the more complicated system corresponding to a nonnull but small value of μ. However, there is a trade-off: when $\mu \neq 0$, the solution can no longer be derived in closed form.

When discussing Poincaré's studies, the completely integrable system (the case $\mu = 0$, see (4.21)) typically refers to the two-body problem, while the case $\mu \neq 0$ but small implies the presence of a third body that exerts very weak forces on the the two bodies. The Hamiltonian function F can then be expanded in power series of the parameter μ, as indicated in equation (4.20).

The appearance of small divisors coupled with the considerable technical difficulties involved in treating them prevented Poincaré from carrying out an exhaustive study of the three-body problem, and it continued to bedevil researchers for many years to come.[18]

For quasi-periodicity to persist under perturbation, not only must the frequencies ω_i be rationally independent under the criterion of (4.23), but they must also be "far" from rationally dependent in the sense that the frequency vector $\omega \in \mathbb{R}^n$ satisfies the following condition:

There exist positive constants $c_0, \tau > 0$, such that

$$\left| \sum_{\nu=1}^{n} m_\nu \omega_\nu \right|^{-1} \leq c_0 \left(\sum_{\nu=1}^{n} |m_\nu| \right)^{\tau}, \qquad \text{for all vectors } m \in \mathbb{Z}^n \setminus \{0\}. \tag{4.24}$$

Such a condition is said to be Diophantine, and we will refer to it as the vector Diophantine condition in contrast to the following Diophantine condition for scalars that we will also discuss in subsequent chapters.

If for $\alpha \in \mathbb{R}$ there exist real constants $\epsilon > 0$ and $\kappa > 2$ such that

$$\left| \alpha - \frac{p}{q} \right| > \frac{\epsilon}{q^\kappa}, \quad \text{for all } \frac{p}{q} \in \mathbb{Z} \times \mathbb{N}, \tag{4.25}$$

---

[17]See Section 3.4.

[18]The monographs by June Barrow-Green [**1997**] and Florin Diacu and Philip Holmes [**1996**] detail the historical and mathematical richness of the n-body problems

then α is said to be Diophantine.

The scalar version has been around at least since the time of Liouville (who we will discuss briefly in Chapter 11), and one can view the vector version as a generalization of the scalar notion in the following sense. Suppose equation (4.25) holds for α. Since increasing κ only decreases the right-hand side of the inequality, we can, without loss of generality, assume $\kappa > 1$. The inequality at (4.24) holds for $n = 2$, $\omega_1 = 1$, and $\omega_2 = \alpha$, as can be easily verified by setting $(p, q) = (-m_1, m_2)$, $\epsilon = c_0^{-1}$, and $\kappa = \tau + 1$.

The connection between the application of the Diophantine conditions in both ODEs and complex dynamics is quite strong, and it involves an obvious but interesting analogy between the use of the Poincaré return map on the 2-torus $\mathbb{T}^2 = \mathbb{T}^1 \times \mathbb{T}^1$, and the iteration of a complex analytic function f of the form

$$f(z) = e^{2\pi i \alpha} z + g(z), \quad \text{with } \alpha \in (0, 1) \text{ and } g(z) = \sum_{k=2}^{\infty} a_k z^k. \tag{4.26}$$

Since the state variables x_i in (4.22) remain constant, we can skip them for the time being[19] and look at the dynamics of the y_i on the 2-torus where

$$\frac{dy_1}{dt} = \omega_1, \qquad \frac{dy_2}{dt} = \omega_2. \tag{4.27}$$

Consider an initial point $P_0 = (0, y_2^{(0)}) \in \mathbb{T}^1 \times \mathbb{T}^1$ at time $t_0 = 0$. Assuming $\omega_1 \neq 0$ assures us that at (positive) time $t_1 = \frac{2\pi}{\omega_1}$, $y_1 = 2\pi$ and consequently the trajectory returns on the "circle" $y_1 = 0$ (see Figure 4.1) at a certain point $P_1 = (0, y_2^{(1)})$. This procedure defines the Poincaré return map

$$\mathcal{P} : y_2^{(0)} \in \mathbb{T}^1 \mapsto y_2^{(1)} \in \mathbb{T}^1, \quad \text{with } y_2^{(1)} = y_2(t_1) = \left(y_2^{(0)} + \frac{\omega_2}{\omega_1} 2\pi \right) \bmod 2\pi. \tag{4.28}$$

It is clear that the iteration defined by $\mathcal{P}$ is equivalent to a rigid rotation of $\mathbb{T}^1$ through angle $2\pi\beta$, where $\beta = \frac{\omega_2}{\omega_1}$. That the two frequencies ω_1 and ω_2 are rationally independent may be stated equivalently by the assumption that β is irrational: If β is rational and equals $\frac{a}{b}$, then $b\omega_1 - a\omega_2 = 0$, in contradiction of (4.23). Since β is irrational, each infinite sequence generated by $\mathcal{P}$ is dense on $\mathbb{T}^1$, which, in light of the remarks immediately following (4.23), implies that the solution of (4.27) is quasi-periodic on $\mathbb{T}^2$.

The analogous situation in complex dynamics occurs with the iteration of the function f (4.26) around the origin when it is either linear, i.e., $g(z) \equiv 0$, or conjugate to its linear part via the Schröder functional equation, $S \circ f = 2\pi\alpha \cdot S$. In either case, the orbit of a point z_0 under f near zero lies on an invariant closed curve conformally equivalent to the complex circle $\{z : z = |z_0|\}$ and is quasi-periodic if and only if α is irrational. A fundamental issue in complex dynamics until 1942 was whether it was possible for f to be conjugate to its linear part when it is nonlinear. An answer of "yes" would guarantee the existence of centers, that is, fixed points around which the dynamics are equivalent to a rigid rotation, analogous to the centers whose existence Poincaré investigated by means of his return map.

---

[19]However, as is clearly pointed out by Moser in [**1966**], the variables x_i can no longer be neglected during the study of the stability of a perturbed Hamiltonian system of the form given at (4.19) and (4.20). See the *nondegeneracy condition* introduced in footnote 29 on page 88 of this chapter.

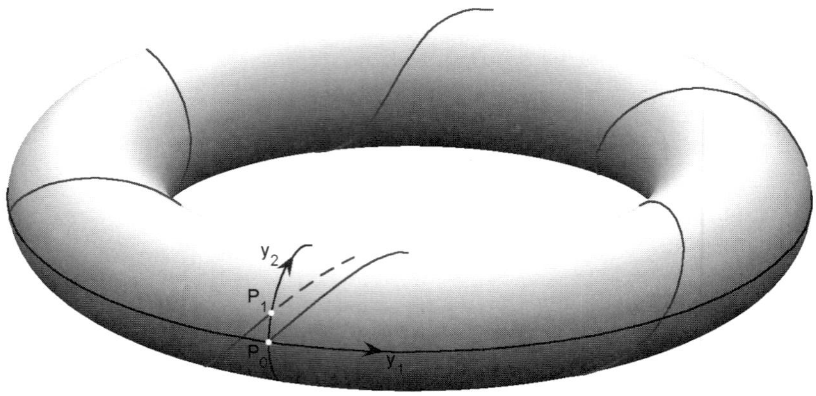

FIGURE 4.1. The Poincaré return map on a 2-torus. The dynamics of system (4.27), with y_1 and y_2 taken modulo 2π, can be conveniently visualized on a 2-torus $\mathbb{T}^2$ with an orthogonal coordinate system made up of two circles (1-tori $\mathbb{T}^1$) cutting $\mathbb{T}^2$ transversally and longitudinally. The circle $y_1 = 0$, referred to as a Poincaré section, will be transversal to the flow provided $\omega_1 \neq 0$. This allows the construction of the Poincaré return map defined in (4.28). An orbit originating at a point $P_0 = (0, y_2^{(0)})$, here assumed to be the origin of the axes, propagates on the 2-torus and intersects the circle $y_1 = 0$ at times $t_k = \frac{2k\pi}{\omega_1}$. The sequence of points P_k on the section generated by (4.28) will form a periodic or quasi-periodic orbit according to whether the frequencies ω_1 and ω_2 are rationally dependent or rationally independent.

The formidable question of the existence of centers in complex dynamics is a thread that will run throughout the remainder of our monograph, beginning with the work of Pfeiffer who started to address the problem in 1915 (Chapter 5), through Julia and Fatou's attempts to settle the question of their existence in the works they prepared in anticipation of the 1918 *Grand Prix* (Chapter 9) as well as Cremer's exploration of the question in the late 1920s and 1930s (Chapter 11), and finally to Siegel's 1942 paper showing that centers do indeed exist (Chapter 12) if the rotation number[20] α from (4.26) satisfies the scalar Diophantine condition (4.25).[21] Thus, in both ODEs and complex dynamics a Diophantine condition must be satisfied for quasi-periodic orbits to exist.

Returning to the general problem of dynamics discussed by Poincaré, the perturbed Hamiltonian system (4.19) is thus "close" to the unperturbed one (4.21) and therefore is said to be *nearly integrable*. As was seen above, the perturbation parameter μ is meant to account for the presence of weak forces that are introduced

[20]For a rational complex function with a center, the rotation number α can be easily gleaned from the multiplier of an irrationally neutral fixed point λ when it is expressed in the form $f'(\lambda) = e^{2\pi i\alpha}$.

[21]At the end of Chapter 12 we will compare Siegel's result to Kolgomorov and Moser's solution of the small divisors problem in perturbation theory.

in a simpler, fully understood system whose solutions may be computed in closed form.

The solution of the general problem is generally sought in form of a series (either a power series or a trigonometric one) which, under suitable assumptions, can be formally defined. Even with the relatively simple three-body problem, the dynamics become extremely complicated: in [**1892a**] Poincaré showed that it does not admit, in general, a sufficient number of first integrals[22] to allow the solution of the problem in closed form.

We see an example of Poincaré seeking a series solution to an ODE only to encounter a small divisors problem in [**1893**] when he generalized Lindstedt's approach from [**1883**] to the perturbed harmonic oscillator

$$\frac{d^2x}{dt^2} + n^2x = \mu\Phi(x,t), \tag{4.29}$$

where μ is small and $\Phi(x,t)$ may be expanded in powers of x with coefficients which are periodic functions of t.

Poincaré's generalization of Lindstedt's approach (see Barrow-Green [**1997**, pp. 155–158]) for a detailed analysis) led him to search for a formal solution to (4.19)–(4.20) in the form of $2n$ power series in the parameter μ:

$$\begin{cases} x_i(t) = x_i^0 + \mu x_i^1 + \mu^2 x_i^2 + \cdots, & i = 1,\dots,n, \\ y_i(t) = \varphi_i + \mu y_i^1 + \mu^2 y_i^2 + \cdots, & i = 1,\dots,n, \end{cases}$$

where the coefficients x_i^k and y_i^k are periodic functions with argument $\varphi_i = \omega_i t + \bar{\varphi}_i$, the frequencies ω_i being, in turn, expanded as series

$$\omega_i = \omega_i^0 + \mu\omega_i^1 + \mu^2\omega_i^2 + \cdots.$$

It follows from setting $\mu = 0$ in the equations for $x_i(t)$ and $y_i(t)$ that x_i^0 and $y_i^0(t) = \omega_i^0 t + \bar{\varphi}_i$ is the solution of the unperturbed problem (that is, system (4.19)–(4.20); compare also to (4.22)).

When $n = 2$, he showed that the coefficients $x_i^k(t)$ admit the formal series representation

$$x_i^k(t) = A_0 + \sum B_{m_1 m_2} \frac{\sin(m_1\varphi_1 + m_2\varphi_2) + h_{m_1 m_2}}{m_1\omega_1^0 + m_2\omega_2^0}, \tag{4.30}$$

from which it is clear that, depending on the nature of the frequencies ω_i^0, the denominators can become arbitrarily small so that the convergence of the series cannot be easily determined.

In order to place the issue of small divisors that Poincaré confronted in [**1893**] in a more general context using modern perturbation theory, we will use the introductory description from Alessandra Celletti's paper [**2009**], although we modify her notation slightly in order to make it consistent with our own. One reason for our choice of Celletti's paper is that she provides a lucid discussion of the classical approach to the problems caused by the presence of a perturbation parameter such as μ. Another is that the approach she discusses is similar in spirit to Cigala's since both employed a stepwise process that generates a sequence of new problems where the perturbation term becomes higher with each step.

[22] See footnote 8 on page 70 for a definition of first integral.

Of course, Cigala was studying a different problem. He tried to conjugate a nonlinear ODE to its linear part, under the assumption that the solutions of latter were stable. His stability analysis was carried out in a neighborhood of an equilibrium point or periodic orbit. Perturbation theory, at least in its early stages, principally focused on Hamiltonian systems which depend on a perturbation parameter. One searches for a canonical map that conjugates the perturbed system to the unperturbed one that could be nonlinear as well, as long as its stability properties are thoroughly understood.

Nonetheless, as we will see, the technique described in Celletti's paper of making the perturbation term $O(u^2)$, then $O(u^3)$, and so forth, is analogous to Cigala's recursive construction of F_n, defined at (4.11), which conjugates away the second-degree term of (4.10), then the third-degree term, etc.

Consider a nearly integrable Hamiltonian system analogous to system (4.19) with n degrees of freedom defined by an analytic Hamiltonian function in the form

$$F(x,y) = F_0(x) + \mu f(x,y), \tag{4.31}$$

where f is 2π-periodic with respect to the components y_i of the y-vector and acts as a perturbation term due to the presence of the small parameter μ. The unperturbed system (4.21) corresponds to $\mu = 0$.

The main task in perturbation theory is to determine a canonical map[23] which conjugates the system to a completely integrable system. Usually this is done by composing an infinite number of "elementary" canonical maps that may be understood as perturbations of the identity function, reminiscent of our formalization of Cigala's procedure in equation (4.18). At the very first step, an elementary canonical change of variables comes into play, which makes the new perturbed Hamiltonian μ^2-close (analogous to a first order approximation with a power series) to the unperturbed problem; likewise, the subsequent steps will remove higher and higher powers of the perturbation parameter μ.

We will detail the first step of the procedure. Consider the canonical change of variables $G : (x,y) \to (x',y')$, implicitly defined by a generic (unknown) generating function Φ:[24]

$$\begin{cases} x = x' + \mu \dfrac{\partial \Phi}{\partial y}(x',y), \\ y' = y + \mu \dfrac{\partial \Phi}{\partial x'}(x',y). \end{cases} \tag{4.32}$$

---

[23]By definition, a differentiable map $G : D \subset \mathbb{R}^{2n} \to \mathbb{R}^{2n}$ is a canonical or symplectic map if its Jacobian $G'(x,y)$ is a symplectic matrix, that is, it satisfies the relation $G'(x,y)^T J G'(x,y) = J$, for any $x, y \in D$, where

$$J = \begin{pmatrix} 0 & I \\ -I & 0 \end{pmatrix};$$

I is the identity matrix of dimension n. The conjugacy map is required to be symplectic since canonical Hamiltonian systems

$$\begin{pmatrix} \dot{x} \\ \dot{y} \end{pmatrix} = J \begin{pmatrix} \partial F/\partial x \\ \partial F/\partial y \end{pmatrix}$$

remain invariant after a canonical change of coordinates.

[24]After the work of William Hamilton and Carl Gustav Jacob Jacobi in the mid-1800s it was known that a change of variables in the form (4.32) is canonical (or symplectic); that is, it does not alter the canonical structure of a Hamiltonian system. Poincaré discusses this in §4 and 5 of [**1892a**] and references Jacobi.

We want the new Hamiltonian function, after this change of coordinates, to take the form

$$F'(x', y') \equiv F \circ G(x, y) = F_0'(x') + \mu^2 f'(x', y'), \tag{4.33}$$

where F_0' and f' are the new unperturbed and perturbing terms, respectively. To this end, the perturbing function f is decomposed as

$$f(x, y) = f_0(x) + \tilde{f}(x, y),$$

where $f_0(x)$ is computed by averaging $f(x, y)$ with respect to the angle y.

Plugging (4.32) into (4.31) and considering the Taylor expansion in a neighborhood of $\mu = 0$, yields

$$\begin{aligned} F'(x', y') &= F_0(x' + \mu \frac{\partial \Phi}{\partial y}(x', y)) + \mu f(x' + \mu \frac{\partial \Phi}{\partial y}(x', y), y) \\ &= F_0(x') + \left(-\omega(x') \cdot \frac{\partial \Phi}{\partial y}(x', y) + f_0(x') + \tilde{f}(x', y)\right) \mu + O(\mu^2), \end{aligned}$$

where $\omega(x') = -\partial F_0/\partial x'$ is the frequency vector (see (4.21); the dot notation "·" denotes the Euclidean scalar product between vectors). Therefore, F' will take the desired form (4.33) if

$$-\omega(x') \cdot \frac{\partial \Phi}{\partial y}(x', y) + \tilde{f}(x', y) = 0, \tag{4.34}$$

so that the new integrable term is

$$F_0'(x') = F_0(x') + \mu f_0(x').$$

It remains to be seen whether a solution for the unknown generating function Φ may be retrieved from (4.34). This is done by, first, expanding both $\tilde{f}$ and Φ in Fourier series ($m = (m_1, \ldots, m_n)$ represents an integer vector),

$$\begin{aligned} \tilde{f}(x', y) &= \sum_{m \neq 0} \hat{f}_m(x') e^{im \cdot y}, \\ \Phi(x', y) &= \sum_{m \neq 0} \hat{\Phi}_m(x') e^{im \cdot y}, \end{aligned}$$

and then inserting these expansions in equation (4.34), which yields

$$i \sum_{m \neq 0} \omega(x') \cdot m \, \hat{\Phi}_m(x') e^{im \cdot y} = \sum_{m \neq 0} \hat{f}_m(x') e^{im \cdot y},$$

from which the unknown coefficients $\hat{\Phi}_m(x')$ may be determined:

$$\hat{\Phi}_m(x') = \frac{\hat{f}_m(x')}{i \, \omega(x') \cdot m} = -i \frac{\hat{f}_m(x')}{m_1 \omega_1 + m_2 \omega_2 + \cdots + m_n \omega_n}. \tag{4.35}$$

Clearly, the coefficients are well defined if and only if

$$\omega(x') \cdot m = 0 \implies \hat{f}_m(x') = 0,$$

which is certainly satisfied when the frequencies $\omega_1, \ldots, \omega_n$ are rationally independent. This, however, only means that the corresponding series for Φ may be formally derived; the question of convergence is a different issue. Even though we assume the frequencies are rationally independent, the denominator in (4.35) may become arbitrarily small in absolute value, leading to a small divisor problem. This explains, from a contemporary point of view, the difficulties Poincaré confronted in the general problem of dynamics, equations; see (4.19)–(4.20).

As we can see, this approach is not enough to prevent the phenomenon of small divisors. A faster procedure was needed, and Kolmogorov, Moser, and Arnold developed one.[25]

It was not until 60 years after Poincaré's statement of the general problem of dynamics in the quotation at the beginning of this section that Kolmogorov delivered the first positive response to the problem of small divisors in perturbation theory in the Russian paper [**1954**], as well as in an address in Amsterdam at the International Congress of Mathematicians, published three years later as [**1957**]:[26] Invariant tori carrying a quasi-periodic flow with Diophantine frequencies persist under small perturbations.[27] We will summarize the meaning of Kolmogorov's statement below, after introducing Moser's version of Kolmogorov's results.

Kolmogorov's innovative idea relies on a completely different approach to the small divisors problem than the one presented above. He developed a Newton's method–like iteration procedure to define a sequence of elementary canonical maps whose compositions converge at a very fast rate to a map that, under suitable assumptions, conjugates the perturbed system to its unperturbed counterpart. This proves the structural stability of the latter, that is, it proves that the qualitative behavior of a given dynamical system, in our case the unperturbed one ($\mu = 0$), is unaffected by small perturbations on the system itself ($\mu \neq 0$).

Kolmogorov, however, did not fill out the details of his approach ([**1954**] was only four pages) although he provided a clear, general outline. Nine years later Arnold [**1963**] published a detailed proof, which Moser referenced in [**1966**].[28]

We give Moser's statement of Arnold's results, which appeared as Theorem 2 in Moser's paper (we introduce small changes in notation to conform to our own):

> *Under the above hypotheses*[29] *and for sufficiently small* $|\mu|$ *there exists a real analytic canonical transformation*
>
> $$\begin{aligned} x &= g_1(x', y', \mu), \\ y &= g_2(x', y', \mu), \end{aligned}$$
>
> *such that* $g_2 - y'$ *and* g_1 *have period* 2π *in* $y'_1, \ldots, y'_n$*, that* $g_1 = x'$*,* $g_2 = y'$ *for* $\mu = 0$ *and such that* [the Hamiltonian system defined by (4.31)] *is transformed into a system which reduces to*
>
> $$\frac{dx'}{dt} = 0, \qquad \frac{dy'}{dt} = \omega \left[\equiv - \left. \frac{\partial F_0}{\partial x} \right|_{x = x^{(0)}} \right] \tag{4.36}$$

[25]For other accounts of the early development of KAM theory see Barrow-Green [**1997**, §10.3] and Hasselblatt and Katok [**2002**, §4]. For an overview of the mathematics of KAM theory, see Arnold, Kozlov, and Neishtadt [**2006**] or Rafael de la Llave [**2001**].

[26]Kolmogorov also published a related paper [**1953**] that Moser mentions in his review [**1959**] of Kolmogorov [**1957**], but it does not seem to have had the impact of either his 1954 paper or his address.

[27]Invariant in this context means that any trajectory originating from a point on the torus remains on the torus for any value of the time.

[28]Moser did not reprove Arnold's result, rather he referred the reader to Arnold's paper.

[29]Moser's hypotheses are that the Hessian $\partial^2 F_0/\partial x^2$ is nonsingular when evaluated at $x^{(0)}$ (nondegeneracy condition) and the frequencies $\omega_1, \ldots, \omega_n$ satisfy the Diophantine condition (4.24).

ANDREI NIKOLAJEWITSCH KOLMOGOROV

FIGURE 4.2. Kolmogorov is second from the left. The picture was taken in Helinski in 1965; courtesy of the Archives of the Mathematisches Forschungsinstitut Oberwolfach.

for $x' = x^{(0)}$. *Hence*

$$x = g_1(x^{(0)}, \omega t + y^{(0)}, \mu), \qquad y = g_2(x^{(0)}, \omega t + y^{(0)}, \mu)$$

represent a family of quasi-periodic solutions of the given unmodified system [i.e., (4.31)], [**1966**, p. 516].

To appreciate the topological implications of the above result, first recall that (4.36) is the dynamical system defined by the Hamiltonian function (4.31), namely $F(x,y) = F_0(x) - \mu f(x,y)$, after setting the perturbation parameter μ equal to zero; that is, it is the unperturbed, completely integrable case.

As was observed at the beginning of the present section, if the components of the frequency vector $\omega = -\frac{\partial F_0}{\partial x}(x^{(0)})$ are rationally independent, its solutions $x(t) = x^{(0)}$, $y(t) = \omega t + y^{(0)}$ densely cover the n-torus $\{x^{(0)}\} \times \mathbb{T}^n$, hence are quasi-periodic. The question then arises whether this stability character persists if the perturbation parameter μ is different from zero but sufficiently small. In other words, do quasi-periodic solutions in the unperturbed system remain so under a small perturbation? Prior to Kolmogorov the potential of small divisors, for example, those arising in the coefficients given at (4.35), prevented mathematicians from demonstrating that they do.

However, Moser's result implies that if, in addition, ω is Diophantine, and the Hessian matrix of F_0 is nonsingular at $x^{(0)}$, a $\mu_0 > 0$ exists such that for $|\mu| < \mu_0$, the unperturbed torus $\{x^{(0)}\} \times \mathbb{T}^n$ can be analytically deformed into an invariant torus, namely

$$G_\mu(\{x^{(0)}\} \times \mathbb{T}^n), \qquad \text{with} \quad G_\mu(x', y') \equiv \begin{pmatrix} g_1(x', y', \mu) \\ g_2(x', y', \mu) \end{pmatrix},$$

over which the flow associated with the perturbed system (4.31) is analytically conjugate to the linear flow associated with (4.36); thus, the solution remains quasi-periodic and the stability character is preserved.

The corpus of techniques and theorems that stem from the works of Moser, Kolmorogov, and Arnold is usually referred to as KAM theory. In order to better delineate the connections between Siegel's demonstration of the existence of centers in complex dynamics and Moser's description of Kolmogorov's method, we will defer further discussion of KAM theory until after we discuss Siegel's proof in Chapter 12.

4.5. Concluding remarks

As the studies of the ODEs that occur in celestial dynamics evolved over the course of the 19th century, a serious obstacle arose: when determining the coefficients of a series—be they power series or trigonometric—their denominators can become very small, making it quite difficult to determine if the series converges or diverges. Delaunay, who ran into small divisors in the mid-1800s, was one of the first to encounter them, and by the end of the century Poincaré realized that they arise with startling frequency and, moreover, persist, meaning that they do not depend on the technique one used to approach the ODE.

Poincaré's early encounters with small divisors go back to his study of the stability of periodic orbits of three-dimensional autonomous ODEs in the mid-1880s. His return map allowed him reduce the analysis of stability around an equilibrium point to a two-dimensional nonautonomous ODE. This in turn led him to search for the existence of closed trajectories in the phase plane surrounding the equilibrium point. He hoped to formally define these trajectories as a power series expansion, but he found himself unable to ascertain their convergence due to the presence of small divisors.

As we will see in the next chapter, a small divisor problem also appeared in the study of complex dynamics in the early part of the 20th century, and the occurrence of small divisors in both studies is not unrelated. In both two-dimensional ODEs and complex dynamics, they occur in the study of centers, that is, equilibrium points around which orbits fall along circular orbits.

In complex dynamics, a center is a fixed point λ of a function f where $f'(\lambda)$ is unimodular but not a root of unity around which iteration can be linearized by solving the canonical Schröder equation, $S \circ f = f'(\lambda) \cdot f$. The existence of centers in complex dynamics was hypothetical for quite some time because a small divisors problem occurs when one attempts to determine the coefficients of S. The question of their existence would not be settled until 1942.

In ODEs we encounter an analogous situation when orbits in a neighborhood of an equilibrium point undergo bounded oscillations that persist indefinitely. In the autonomous case, this means that the linear part associated with the system has pairs of conjugate eigenvalues with nonpositive linear part, some of which are purely imaginary.

The use of the return map to conduct the stability analysis of an ODE in a neighborhood of an equilibrium point or periodic orbit leads to a discrete dynamical system where the pairs of purely imaginary eigenvalues mentioned above are transformed into complex conjugates with unitary modulus. Levi-Civita did not study this case, but in the early 1900s, Cigala, who was mentored by Levi-Civita, did, and his investigation implied that an attempt to linearize the system would lead to a small divisors problem. Although Cigala was unsuccessful in controlling the effect of small divisors, his technique, which involved an infinite composition of functions, deserves study since it prefigures an analogous, more powerful, technique that would eventually overcome the problem.

In the 1890s Poincaré again encountered small divisors in his study of Hamiltonian systems arising from celestial mechanics. His approach this time was to consider a completely integrable (and therefore stable) Hamiltonian system and then perturb it. His hope was that the system is structurally stable, that is, that under suitable conditions the stability of the unperturbed state would persist under a small perturbation, but Poincaré was unsuccessful.

This approach, however, would eventually bear fruit in a short paper by Kolmogorov [**1954**], albeit with a very different set of techniques that involve an infinite composition of functions. While Kolmogorov neglected to include many of the technical details of his approach in [**1954**], he provided a clear outline that Arnold and Moser used as a guide in the early 1960s. Together, the work of Kolmogorov, Arnold, and Moser formed the bedrock of KAM theory.

A few years prior to Kolmogorov the center problem in complex dynamics was solved by Siegel. His solution, to be discussed in Chapter 12, shared a very important detail with Kolmogorov's: the recognition of the centrality of a Diophantine condition. While there is no evidence that Kolmogorov was influenced by Siegel's approach, Moser certainly was, as we will see in Chapter 12.

Part 2

The Core (1906–1920)

CHAPTER 5

Early Overseas Results: The United States

5.1. Introduction

With the turn of the 20th century, the emerging U.S. mathematical community was producing home-grown research mathematicians. The founding of the American Mathematical Society (AMS) in 1888 (initially as the New York Mathematical Society) reflected these changes, as did the establishment of the AMS family of journals, first the *Bulletin of the American Mathematical Society* and then the *Transactions of the American Mathematical Society*, whose goal was to join the *American Journal of Mathematics* as an outlet for the American research mathematician.

While some of the papers in the new research journals had little impact, perhaps deservedly so, substantial works were also published, although not all of them were recognized by the European community in a timely fashion.

Emblematic of the emergence of an American research community were a group of mathematicians, two from Columbia University and one from Princeton, who collectively published over 15 articles directly related to the study of the iteration of complex functions, heretofore an exclusively European subject, in American journals between 1915 and 1925. The mathematicians from Columbia, George Adam Pfeiffer and Joseph Fels Ritt, were students of Edward Kasner at Columbia.[1] The mathematician from Princeton was Albert A. Bennett.

While Kasner published no papers on iteration, a remark he made in 1912 is often cited in the literature as a conjecture: When the modulus of the multiplier of a fixed point is one but not a root of unity, iteration is always conformally equivalent to a rotation of the disc. In other words, centers always exist for irrationally indifferent fixed points (see, for example, Milnor's book [**2006a**, p. 126]). This remark—and its interpretation—will be discussed in Section 5.3. Pfeiffer and Ritt each made significant contributions to complex dynamics which are referenced in contemporary research.

Ritt, whose work on iteration is probably the best known among the three, published several articles on the iteration of rational functions, and a notice from his early research [**1918a**] appeared alongside the work of Fatou, Julia, and Lattès in *Comptes Rendus*.

Pfeiffer's work on the Schröder equation grew out of his exploration of conformal geometry. Although he did not apply his findings to iteration, that he

[1] Ritt obtained his Ph.D. from Columbia in 1917 under Kasner's direction. Pfeiffer received his Ph.D. from Columbia in 1914 and took courses from Kasner to whom his work is closely related. He frequently cited Kasner in his doctoral thesis, which appeared as [**1915a**] in the *American Journal of Mathematics*. Aside from his wartime service, Ritt spent his entire professional career at Columbia; Pfeiffer taught briefly at Princeton and Harvard before World War I, and after the war he taught at Columbia until his death. See their respective biographies in Appendix O.

explicitly cited the 19th century principals—Schröder, Kœnigs, Leau, Grévy, and others—leaves little doubt that he was well aware of the import of his findings to the field.

The presence of George David Birkhoff hovers behind this group, as well: as we will see below, Pfeiffer credits a breakthrough in his study of the Schröder functional equation to a suggestion from Birkhoff.

Bennett treated the iteration of real and complex functions in both the single and multivariable cases. But his approach to the iteration of complex functions was largely that of a literature survey and had no lasting impact on the field.

The work of Kasner, Pfeiffer, and Bennett will be discussed in this chapter; the work of Ritt will be discussed in Sections 7.5, 10.2, 13.2, and J.3.

5.2. Pfeiffer and irrationally neutral fixed points

Up until 1915 almost nothing was known about iteration of a nonlinear rational function f near an irrationally indifferent fixed point. The obvious thing to do would be to see if iteration could be linearized via the Schröder equation

$$S \circ f = f'(\lambda) \cdot S \quad \text{where } |f'(\lambda)| = 1. \tag{5.1}$$

If so, then λ is a center, and iteration is conformally equivalent to a rotation of a disc (see Figure 1.2). Prior to Pfeiffer, no mathematician had made any progress in understanding iteration near an irrationally neutral fixed point; indeed, as we noted on page 19, Kœnigs in essence threw up his hands in frustration. None of his students tackled the problem either, although Leau characterized iteration for rationally neutral fixed points in his Flower Theorem.

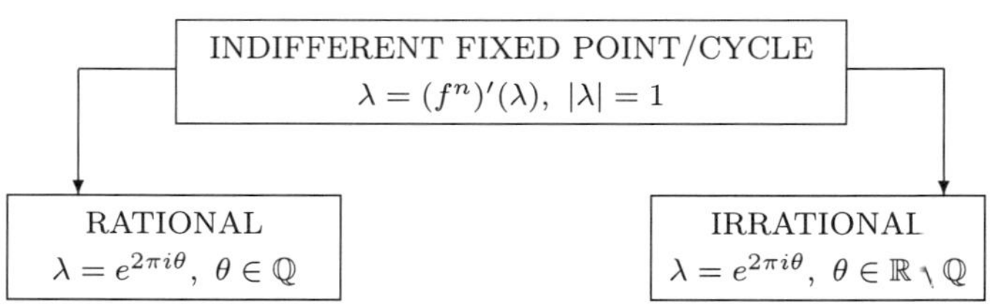

FIGURE 5.1. Iteration around a neutral fixed point. By the beginning of the 20th century, iteration near rational fixed points was well understood while virtually nothing was known regarding the irrational case.

In October 1915, two months before the Académie des Sciences announced it would offer its 1918 *Grand Prix* for the study of iteration, Pfeiffer asserted in a footnote midway through a 36-page paper—without proof—the existence of a class of functions for which the above Schröder equation had no convergent solutions, suggesting perhaps, that, as was the case with rationally neutral fixed points, linearization was not possible in the irrational case [**1915a**, p. 407]. A stand-alone announcement of the result appeared in a note the next year in the *Bulletin of the AMS* [**1916b**]. The full proof appeared a year later in the *Transactions of the AMS* [**1917**].

Pfeiffer's initial interest in the Schröder equation for the irrationally indifferent case lies not in its relations to iteration but rather to a problem of conformal geometry that Kasner examined in his paper [**1913**].

5.3. Kasner's mapping of conformal angles

Our interest in Kasner is two-fold. One is to discuss the remark he made in [**1913**] regarding irrational multipliers that we alluded to in the opening section of this chapter; the other is to provide context for Pfeiffer's investigation of the Schröder functional equation.

EDWARD KASNER

FIGURE 5.2. Photo of Edward Kasner; courtesy of University Archives, Columbia University in the City of New York

Kasner's paper [**1913**] was delivered at the 1912 International Congress of Mathematicians. In it he discussed the following problem involving higher order invariants:[2]

> *If two angles are equal in magnitude, that is have the same θ, will they necessarily be conformally equivalent? This is certainly a fundamental question in conformal geometry. We shall show the answer is not always in the affirmative. Curvilinear angles exist, which while equal in magnitude, are not conformally equivalent.*
>
> ...
>
> *This fact indicates the existence of other invariants besides θ. The principal problem of the present paper is to find these absolute*

[2]A higher order invariant is a quantity that is a function of the higher order derivatives of the curves forming the angle which remains the same after a conformal transformation. This contrasts with the angle θ, which is a linear invariant.

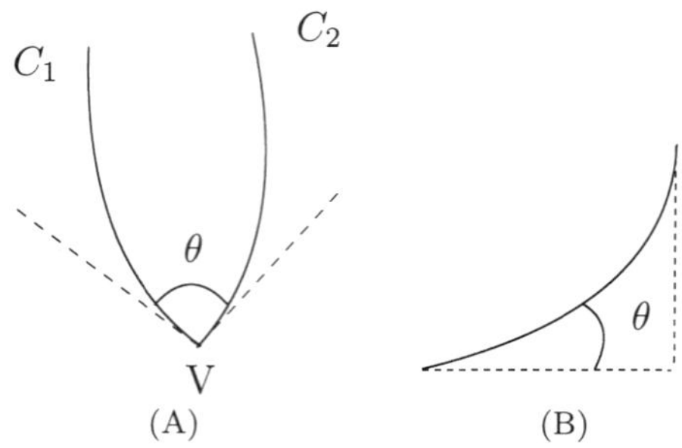

FIGURE 5.3. Curvilinear angles: (A) the usual configuration; (B) the reduced form assumed by Kasner.

invariants of higher order. ... Only invariants of finite order are discussed. The existence of invariants of infinite order, that is, those involving all *the coefficients of a power series representing the sides of the angle, is not settled. It is shown* [in the present paper] *that if θ is a rational part of π (in which case we shall term the angle rational), there exist higher invariants. If the angle is irrational, it is shown that no higher invariants exist.* [**1913**, pp. 81–82]

In the process of investigating irrational angles, he used a configuration that he called "reduced form" (pictured in Figure 5.3 B), in which one side of the angle is the real axis, and the other an analytic arc γ of the form

$$\gamma(x) = a_1 x + a_2 x^2 + \cdots, \tag{5.2}$$

where x is real, $a_1 = \tan(\theta)$, and the angle θ between γ and the real axis is irrational.

To prove that no higher order invariant exists, he first assumed the existence of another reduced form angle,

$$\Gamma(X) = A_1 X + A_2 X^2 + \cdots, \tag{5.3}$$

whose magnitude equals that of γ; that is, $a_1 = A_1$.

Both γ and Γ can viewed as planar curves $(x, \gamma(x))$ and $(X, \Gamma(X))$, which in turn can be transformed into complex variables. Kasner showed that a conformal complex function

$$\Phi(z) = \sum_{k=1}^{\infty} c_k z^k, \tag{5.4}$$

with $c_k \in \mathbb{R}$, could be formally defined that maps $(x, \gamma(x))$ to $(X\ \Gamma(X))$:

$$X + iY = \Phi(x + iy) = c_1(x + iy) + c_2(x + iy)^2 + \cdots,$$

which yields

$$X = c_1 x + c_2(x^2 - y^2) + \cdots \qquad \text{and} \qquad Y = c_1 y + 2c_2 xy + \cdots.$$

Since Φ maps γ to Γ, Kasner substituted $y = \gamma(x)$ into the above equations, and, because X and $Y = \Gamma(X)$ satisfy (5.3), he used the method of undetermined coefficients to obtain the c_k in terms of preceding coefficients $c_i, i = 1, \ldots, k-1$. In particular, Kasner observed that the kth equation is linear in c_k.

For example, routine calculations yield $c_1 a_1 = c_1 A_1$, which, since $a_1 = A_1$, leaves c_1 as a free parameter. Solving for c_2, we get

$$(5.5)\quad c_1 a_2 + 2c_2 a_1 = c_2 A_1(1-a_1^2) + c_1^2 A_2 \Rightarrow (2a_1 - A_1(1-a_1^2))c_2 = (A_2 c_1 - a_2)c_1.$$

These examples will help us to see why there are no higher order invariants: if the coefficient of c_k is zero, then one can readily deduce the existence of an invariant, since all the remaining quantities could be expressed as a function of just the a_i and A_i, and would therefore be present in any conformal mapping of the same form as Φ. If, for example, it turned out that $2a_1 - A_1(1-a_1^2) = 0$ in equation (5.5), then the right-hand side would be an invariant. However this is not the case, since $2a_1 - A_1(1-a_1^2) = \tan\theta(1+\tan^2(\theta))$.

Kasner observed a similar occurrence with each of the other c_k, and remarked,

> *The coefficient of c_k may be put in the form*[3]
>
> $$\sin(k\theta) - \tan\theta\cos(k\theta).$$
>
> *This cannot vanish for any integer $k > 1$; for if it did θ would be a rational part of π. Hence for arbitrary values of $a_1, a_2, \ldots, a_n$ and $A_1, A_2, \ldots, A_n$ values of $c_1, c_2, \ldots, c_n$ can be determined: in fact in ∞^1 ways since c_1 is arbitrary.*
>
> *For arbitrary curves* (5.2) *and* (5.3) *the series* (5.4) *can be formally calculated, the coefficients being rational functions of the parameter c_1. Probably, for some value of c_1, the corresponding series* (5.4) *will be convergent. This, however, the author has not succeeded in proving. If there are actually cases where* (5.4) *is divergent for all values of c_1 this would indicate the existence of invariant relations involving all the coefficients of the curves, that is, invariants of infinite order.* [**1913**, p. 83]

The above quotation is the one that is interpreted as Kasner's conjecture that convergent solutions always exist for the Schröder equation $S \circ f = f'(\lambda) \cdot S$ when λ is an irrationally indifferent fixed point of f.

Such an interpretation may not be entirely accurate. He was talking neither about iteration nor the Schröder equation, and it is up to the reader to determine the strength of Kasner's conviction that such a c_1 always exists. Is it a conjecture? Or just a speculation? He also seemed to be holding open the possibility that Φ is divergent for all choices of c_1, perhaps because it would suggest something interesting about the problem of infinite invariants, which, as he noted in the first quotation, "is not settled." Moreover, Pfeiffer, whose work rises directly from Kasner's study of invariants, did not even refer to Kasner's remark.

Although he did not do so in his 1913 paper, Kasner evidently discussed the relation between the problems of the conformal mapping of angles and functional equations in his courses. At the beginning of [**1917**] Pfeiffer observed,

> *The equivalence of these two problems, the geometrical and the functional one, was pointed out by professor Kasner in a course of*

[3]Computing the coefficients c_k is an easy matter since only linear terms from $\gamma(x)$ and $\Gamma(x)$ are involved. We carried out this computation and discovered that the coefficient c_k actually equals $(1+\tan^2(\theta))^{k/2}(\sin(k\theta) - \tan(\theta)\cos(k\theta))$. The factor $(1+\tan^2(\theta))^{k/2}$ is always positive, and its presence has no effect on Kasner's argument.

> *lectures at Columbia University.*[4] *The connection between the two problems has not been worked out in detail as here presented as far as the writer knows.* [**1917**, p. 191, fn.]

Kasner followed his examination of irrational angles with a discussion of the higher order invariants of rational curvilinear angles, which in turn led him to examine the invariants of horn angles (which have only a single higher order invariant). But these examinations are not relevant to our current study. What matters here is that Kasner played a role in the genesis of Pfeiffer's examination of the Schröder equation in the irrationally indifferent case. He also opened up the possibility, at least to those who were able to see the implications of his speculations, that there was perhaps more than one kind of dynamic in the neighborhood of an irrationally neutral fixed point, depending on whether the Taylor series of the mapping function (5.4) converged or diverged.

5.4. Pfeiffer: The dawn of a new perspective

Although Pfeiffer did not explicitly relate his investigation of the Schröder equation

$$S \circ f = f'(\lambda) \cdot S, \tag{5.6}$$

to iteration, it is quite clear he understood the connection.

Very soon into his investigation, he observed, "This equation has been treated by various writers" [**1917**, p. 185] and proceeded to list almost all the European mathematicians—major and minor alike—who treated the Schröder equation in relation to iteration.

He continued:

> *... but it appears that under the hypotheses made in this paper*[5] *the existence of divergent solutions has never been proved, although the existence of formal solutions is obvious and divergence of these solutions is probably the general case.*[6] [**1917**, p. 185][7]

His treatment of the Schröder equation as a stand-alone problem consisting of the solution of a functional equation involving complex analytic functions therefore suggests a clear awareness of how it fits into the study of the iteration. It is only after showing the existence of divergent formal solutions to (5.6) that he explicitly relates it to conformal mapping.

Further evidence of his familiarity with the 19th century French school of iteration—evidence that also deepens the mystery of why he did not explicitly connect his work to iteration theory—is given in 1915 by what he thought was

---

[4]These lectures were evidently given in a seminar on differential geometry Kasner taught. The 1913 Division of Mathematical and Physical Science Bulletin of Barnard College, then Columbia's sister-college, described the seminar: "This is open, on consultation with the instructor, to students who have a working knowledge of ordinary differential equations and the differential geometry of plane and twisted curves. It will deal with selected topics relating to systems of curves, conformal transformations, and the geometry of optics."

[5]Namely, that $f('\lambda)$ is an irrationally indifferent fixed point.

[6]It is interesting that the speculation has gone from Kasner (perhaps) thinking that solutions always exist in the irrationally indifferent case to Pfeiffer suggesting they never do.

[7]Moreover, around the time [**1917**] was published, his colleague Ritt was involved in the study of iteration of complex rational maps, so it is conceivable that through Ritt he would have been aware of the current state of the field and known that the 1918 *Grand Prix* was devoted to the study of iteration.

an improved version of Kœnigs' theorem that convergent solutions exist to the Schröder equation on a neighborhood of an attracting fixed point [**1915a**]. While examining the mapping of curvilinear angles in a slightly different context, Pfeiffer showed that the Schröder equation occurs naturally, and solved it, but in a different manner from Kœnigs. After concluding his proof he remarked:

> *This result is contained in the work by G. Kœnigs* [**1884b**]. *However, the above proof is much more direct and more simple.* [**1915a**, p. 407, fn.]

Whether one agrees with Pfeiffer probably depends upon whether or not one prefers a proof based on the uniform convergence of series, as in Kœnigs' proof, or a direct computation, such as Pfeiffer's.

Pfeiffer constructed a function f with an irrationally indifferent (neutral) fixed point whose associated Schröder equation $S \circ f = f'(\lambda) \cdot S$ has no convergent solution S in a step-wise manner, which we will describe, that forces any power series expression for S to have a radius of convergence of zero. The only assumptions he imposed on f at the onset were that zero is an irrationally indifferent fixed point, and f is analytic on an unspecified neighborhood surrounding the origin. All the coefficients of f, including a_1, were initially left undefined, but would be determined during his construction.

He assumed f has the form

$$f(z) = a_1 z + \sum_{k=2}^{\infty} a_{n_k+1} z^{n_k+1}, \tag{5.7}$$

where only the nonzero coefficients are displayed, as well as the existence of a function

$$S(z) = \sum_{k=1}^{\infty} c_k z^k$$

which formally satisfies $S \circ f = a_1 \cdot f$.

This allowed him to find a recurrence relation for the coefficients of S in terms of c_1 and the coefficients of f, similar to what Kasner did with the function Φ at (5.4), except that the coefficients of Kasner's curve γ are known.

Pfeiffer's choice for the a_{n_k+1} was governed by the need to make f converge, but it turns out this is not much of a constraint since his construction easily allows for the assumption that $|a_{n_k+1}| \leq 1$; his construction of a_1, by far the most complicated part of his algorithm, was designed to make the root test limit for S go to infinity regardless of the choice of c_1.

5.4.1. Pfeiffer's construction: Defining the recurrence relation. Pfeiffer's construction is verbose and involved. His choice of notation was not robust, and he did not always justify his assertions. We have therefore streamlined his exposition using notation that Pfeiffer did not use and by providing possible explanations for his claims. However, we remain faithful to the order and details of his procedure. In other words, while the notation and explanatory pauses are our own, the conception, construction, and order of attack are his.

Pfeiffer obtained a recurrence relation for the coefficients of S by comparing coefficients in the expression

$$S(f(z)) = \sum_{k=1}^{\infty} c_k \left(a_1 z + \sum_{j=2}^{\infty} a_j z^j \right)^k = a_1 \left(\sum_{k=1}^{\infty} c_k z^k \right) = a_1 S(z),$$

which yields

(5.8)

$$c_{n+1}(a_1, \dots, a_{n+1}, c_1) = \frac{c_1[a_1^{n-1}(1-a_1)\cdots(1-a_1^{n-1})a_{n+1} + P_{n+1}(a_1, \dots, a_n)]}{a_1^n(1-a_1)\cdots(1-a_1^n)}$$

$$= c_1 R_{n+1}(a_1, \dots, a_{n+1}) \cdot \frac{1}{1-a_1^n},$$

where

$$R_{n+1}(a_1, \dots, a_{n+1}) = \frac{a_1^{n-1}\prod_{k=1}^{n-1}(1-a_1^k)a_{n+1} + P_{n+1}(a_1, \dots, a_n)}{a_1^n \prod_{k=1}^{n-1}(1-a_1^k)} \tag{5.9}$$

and $P_{n+1}(a_1, \dots, a_n)$ is a polynomial in the a_k. Since the a_k are undetermined at this point, Pfeiffer viewed P_{n+1} and R_{n+1} as functions in the unknown variables a_k. Note that this implies that these functions—in particular P—have a finite number of roots. We will refer to this fact below. Moreover, Pfeiffer's construction will force most of the a_j to be zero.

5.4.2. Pfeiffer's construction: Determining a_1, step 1. Pfeiffer's procedure was to construct a sequence of roots of unity $\{t_k\}$ that converges to an irrationally indifferent multiplier a_1 which in turn forces the radius of convergence of S to be zero. The order of the roots of unity go to infinity very quickly; if r is the order of the root of unity t_k, then $r+1 \gg r$.

Pfeiffer noted that his proof was due in part to an unspecified suggestion from Birkhoff (see Pfeiffer's papers [**1915a**, p. 407 fn.] and [**1917**, p. 186]). A glance at the recurrence relation (5.8) may provide a small clue. Since $|a_1| = 1$, the denominators in the relation can be very close to zero, so Pfeiffer was faced with a small divisors problem of the sort described in the previous chapter. Birkhoff was certainly quite familiar with such problems, and perhaps he had insight about how to treat small divisors that was helpful to Pfeiffer.

Pfeiffer's first step was to choose a primitive mth of unity with $m > 1$, denoted b_1, which will be used to determine t_1. Let $n_1 = m$, choose $a_{n_1+1} \neq 0$ and set $a_k = 0$ for $1 < k \leq n_1 (= m)$; hence, f is of the form $f(z) = a_1 z + a_{n_1+1} z^{n_1+1} + \cdots$. Assume $|a_{n_1+1}| \leq 1$ which will ultimately allow f to converge on the unit disc.[8]

Since b_1 is a primitive n_1th root of unity and a_{n_1+1} is now determined, it follows that if b_1 is substituted for a_1 in the expression for R_{n+1} at equation (5.9) we get

$$R_{n_1+1}(b_1) = \frac{b_1^{n_1-1}\prod_{k=1}^{n_1-1}(1-b_1^k)a_{n_1+1} + P_{n_1+1}(b_1, 0, \dots, 0)}{b_1^{n_1}\prod_{k=1}^{n_1-1}(1-b_1^k)},$$

where neither product equals zero since $k < n_1$.

[8]Pfeiffer did not specify conditions on the a_{n_k+1}, only that they are chosen to allow f to converge on a disc surrounding the origin.

To choose the first term of the sequence, t_1, let $N_\epsilon(b_1)$ be a neighborhood of the primitive n_1th root b_1. Restrict this neighborhood to the unit circle C_1 via the intersection $N_\epsilon(b_1) \cap C_1$, shrinking ϵ if need be so there are no kth roots of unity for $k < n_1$. This last condition implies that, as was the case with b_1, there is no $t \in N_\epsilon(b_1) \cap C_1$ satisfying $1 - t^k = 0$ for $k < n_1$, hence the products in the denominator and numerator of $R_{n_1+1}(t)$ are nonzero for all $t \in N_\epsilon(b_1) \cap C_1$.

Without explicitly saying why, Pfeiffer assumed $R_{n_1+1}(t)$ is nonzero on $N_\epsilon(b_1) \cap C_1$. His reasoning might have been that since the numerator of R_{n_1+1} has a finite number of roots on $N_\epsilon(b_1)$,[9] we can, if necessary, further shrink ϵ or consider a connected, open subset of $N_\epsilon(b_1)$ on which the numerator has no roots.

Thus, there is an ϵ_1 such that $|R_{n_1+1}(t)|$ is bounded away from zero on $N_{\epsilon_1}(b_1)$, so there exists a positive quantity μ_1 with

$$|R_{n_1+1}(t)| > \mu_1 \tag{5.10}$$

for all t in $N_{\epsilon_1}(b_1)$.

Finally, since $b_1^{n-1} - 1 = 0$ and the roots of unity of order greater than n_1 are dense on $N_\epsilon(b_1)$, an appropriate choice of t will make $\frac{1}{|t^{n_1}-1|}$ arbitrarily large. Thus, for a given $M_1 > 1$, we can find an ϵ_1 such that for all t in $N_{\epsilon_1}(b_1)$,

$$\left| c_1 \frac{R_{n_1+1}(t)}{1 - t^{n_1}} \right| > c_1 M_1^{n_1+1}.$$

Pick one such t and denote it t_1. Placing the exponent $n_1 + 1$ on M_1 ensures that when the root test is applied to the formal series for S, its radius of convergence will be zero.

It follows from equation (5.8) that the choice of t_1 yields a first approximation of c_{n_1+1}, namely

$$c_{n_1+1}(t_1, \ldots, 0, c_1) = c_1 R_{n_1+1}(t_1) \cdot \frac{1}{1 - t_1^{n_1}}.$$

This approximation will be modified with each successive t_k.

5.4.3. Pfeiffer's construction: Determining a_1, step 2. The key to Pfeiffer's construction is that, as in the first step, he can bound $|R_{n_k+1}(t)|$ away from zero and by shrinking ϵ make

$$\frac{1}{1 - t^{n_k}}, \tag{5.11}$$

and hence the product

$$c_1 \frac{R_{n_k+1}(t)}{1 - t^{n_k}},$$

as large in modulus as desired.

To determine the next nonzero term in f, cut the radius of $N_{\epsilon_1}(b_1)$ in half and choose a pth $> n_1$th root of unity (with $p > n_1$) from $N_{\epsilon_1/2}(b_1)$. Denote this pth root of unity b_2, and let $n_2 = p$. Set all the coefficients a_k strictly between a_{n_1+1} and a_{n_2+1} equal to zero, and let $a_{n_2+1} \neq 0$. To ensure the convergence of f, we can assume $|a_{n_2+1}| \leq 1$.

Using the procedure from the first step, find an ϵ such that there are no kth roots of unity on $N_\epsilon(b_2)$ with $k < n_2$ so we can likewise bound $R_{n_2+1}(t)$ away from

[9] For the sake of notational simplicity we will assume that all ϵ neighborhoods used in the construction are restricted to the unit circle.

zero on $N_\epsilon(b_2)$. Shrinking $N_\epsilon(b_2)$ if necessary, we can choose $M_2 > 1 + M_1$ and an $\epsilon_2 < \frac{\epsilon_1}{2}$ so that

$$\left|\frac{R_{n_2+1}(t)}{1 - t^{n_2}}\right| > M_2^{n_2+1}$$

for all $t \in N_{\epsilon_2}(b_2)$. Choose one, and denote it t_2. Finally, we can choose our ϵ_2 so that $N_{\epsilon_2}(b_2)$ is properly contained in $N_{\epsilon_1}(b_1)$.

5.4.4. Pfeiffer's construction: Determining a_1, final step. Pfeiffer thus created several interlocking infinite sequences: the coefficients of f, $\{a_{n_k+1}\}$; the roots of unity, $\{b_k\}$; the neighborhoods, $N_{\epsilon_k}(b_k)$; their radii, $\{\epsilon_k\}$; the lower bounds, $\{M_k\}$; and, finally, the points $\{t_k\}$ from each $N_{\epsilon_k}(b_k)$ for which

$$\left|\frac{R_{n_k+1}(t_k)}{1 - t_k^{n_k}}\right| > M_k^{n_k+1}. \tag{5.12}$$

These sequences satisfy

(1) $M_k > 1 + M_{k-1}$ (with $M_1 > 1$) so $M_k \to \infty$,
(2) $\epsilon_k < \epsilon_k/2$ so the $\epsilon_k \to 0$,
(3) $N_{\epsilon_{k-1}}(b_k) \subset N_{\epsilon_k}(b_{k-1})$,
(4) $N_{\epsilon_k}(b_k)$ contains no jth primitive roots of unity with $j < n_k$,
(5) the degree of the roots of unity goes to infinity, and
(6) $|a_{n_k+1}| \leq 1$, hence the radius of convergence of f is at least one.

Since the $N_{\epsilon_k}(b_k)$ are properly nested, their intersection is nonempty; moreover, since the radii of the $N_{\epsilon_k}(b_k) \to 0$, the intersection is a single point a_1 on the unit circle. This point a_1 is the limit of the t_k as can be easily verified; in addition, a_1 is not a root of unity. Suppose that it were. Then for some j, $a_1^j = 1$. If true, then by item (4) above, if $j < n_k$, then $a_1 \notin N_{\epsilon_k}$. Since the n_k go to infinity with k, this means that a_1 is not in the intersection of the $N_{\epsilon_k}(b_k)$.

The lower bound $M_k^{n_k+1}$ at equation (5.12) still holds for a_1 because our only requirement on t_k was that it be in N_{ϵ_k}, which is true of a_1 as well.

Finally, since it was used at each step in the process—and will be used again below—we will state the principal articulated in equation (5.11) as a lemma, something Pfeiffer did not do:

LEMMA 5.1. *At any stage in the construction, ϵ_k can be chosen so that*

(1) *$|R_{n_k+1}(t)|$ is bounded away from zero on $N_{\epsilon_k}(t)$;*
(2) *$\frac{1}{1-t^{n_k}}$ and $c_1\frac{R_{n_k+1}(t)}{1-t^{n_k}}$ can be made arbitrarily large in modulus.*

5.4.5. Pfeiffer's construction: A Schröder equation with no solution. We can now define f. Let a_1 be the limit of the sequence $\{t_k\}$, and chose the $\{a_{n_k+1}\}$ so that their moduli is at most one. The analyticity of f on $N_1(0)$ follows.

Any formal solution S to $S \circ f = a_1 S$ has a radius of convergence of zero. Why? From the recurrence relation at equation (5.8),

$$c_{n_k+1}(a_1, \ldots, a_{n_k+1}, c_1) = c_1 R_{n_k}(a_1)\frac{1}{1 - a_1^{n_k}},$$

hence

$$|c_{n_k+1}| = \left|c_1 R_{n_k}(a_1)\frac{1}{1 - a_1^{n_k}}\right| > |c_1| M_k^{n_k+1}.$$

Apply the root test to the formal series for S:

$$\begin{aligned} \lim_{k\to\infty} \sqrt[n_k+1]{|c_{n_k+1}|} &\geq \lim_{k\to\infty} \sqrt[n_k+1]{|c_1|\, M_k^{n_k+1}} \\ &= \lim_{k\to\infty} \sqrt[n_k+1]{|c_1|} M_k \\ &= \infty \end{aligned} \tag{5.13}$$

since $M_k \to \infty$. Thus, the radius of convergence of the series for S is zero as long as $c_1 \neq 0$, hence all formal solutions S to $S \circ f = a_1 \cdot S$ diverge.

Pfeiffer did not construct an explicit example of such a function. One would be difficult given that not only did he not provide an explicit algorithm for finding a_1, but given a multiplier $f'(\lambda)$ of an irrationally indifferent fixed point, it would be difficult to determine if it could be produced by Pfeiffer's procedure. Nor did Pfeiffer explore the possibility that there might be irrational a_1 for which convergent solutions to the Schröder equation exist, although the quotation near the top of this section suggests that perhaps this was because he did not believe it possible.

Despite the fact that Pfeiffer's breakthrough represented the first progress towards the solution of the Schröder equation in the irrationally indifferent case, neither the announcement of the result in [**1915a**] nor its proof in [**1917**] garnered any immediate response from the European community.

Julia examined the solution of equation (5.6) in his entry to 1918 *Grand Prix* [**1918a**], and Fatou explored irrationally neutral fixed points extensively in his monograph [**1919-1920**].[10] Neither gave any indication that they were aware of Pfeiffer's work. Eventually, however, Pfeiffer's accomplishment became known in Europe and was cited by Cremer in [**1927**] and Siegel in [**1942**]. Pfeiffer's result is frequently referenced in present-day surveys of complex dynamics.

One explanation for the desultory recognition of Pfeiffer's breakthrough might be a European indifference to the American mathematical community. It might also be due to the precarious nature of—and low priority given to—shipping mathematical works across the Atlantic during World War I. That Pfeiffer's discoveries were not made in the context of iteration theory might also have contributed.[11] Indeed, one is left to wonder why Pfeiffer never related his solution directly to the study of iteration since it was clear he understood its import.

5.5. Pfeiffer and Diophantine numbers

Both Cremer and Siegel's studies of iteration have a much different flavor than the works of Fatou and Julia, in part because their results link to number theory, Siegel's in a deeper way than Cremer's. Indeed, in many respects they have more in common with Pfeiffer's approach. It is not a surprise then, that hints of what they did can be found in Pfeiffer's construction of a_1.

Before we explore this notion, a definition (also stated in Section 4.4):

DEFINITION 5.2. A number t is Diophantine of order less than or equal to $\kappa > 2$ if there exists an $\epsilon > 0$ such that for all rational $\frac{p}{q} \in \mathbb{Z} \times \mathbb{N}$,

$$\left| t - \frac{p}{q} \right| > \frac{\epsilon}{q^\kappa}.$$

[10] See Chapter 9. Julia, in fact, came to believe solutions did not exist in the irrationally neutral case [**1919a**], although he later realized he had made an error in his assumptions. We discuss this in Section 9.3.

[11] More is said about this in Section 9.6.

The following theorem, offered without proof,[12] states an equivalent condition for $|t| = 1$ to be Diophantine.

THEOREM 5.3. *If there exists an $\epsilon > 0$ (not necessarily the same ϵ as in the above definition) and $\kappa > 0$ such that for all $q \in \mathbb{N}$*

$$|1 - t^q| > \frac{\epsilon}{q^{\kappa-1}}, \tag{5.14}$$

or, equivalently,

$$\frac{1}{|1 - t^q|} < \frac{q^{\kappa-1}}{\epsilon}, \tag{5.15}$$

then t is Diophantine of order at most κ.

Irrational numbers which are not Diophantine are said to be Liouville numbers, or well approximated by rationals. Diophantine numbers, on the other hand, are sometimes said to be poorly approximated by rationals.

In Chapter 12 we discuss how Siegel proved in [**1942**] that if f is analytic with expansion $f(z) = a_1 z + a_2 z^2 + \cdots$ with a_1 Diophantine on a neighborhood of an irrationally neutral fixed point (assumed to be zero), then a convergent solution S to the Schröder equation $S \circ f = a_1 \cdot S$ exists. From this it follows that iteration near the fixed point acts like the irrational rotation $z \mapsto a_1 z$, in which case the fixed point is a center.

As we will see in Chapter 11, Cremer pointed out that the multipliers he used to produce divergent formal solutions to the Schröder equation in 1927 are well approximated by rationals. Although it is explicit in Cremer's case, the observant eye might intuit that Pfeiffer's a_1 is a Liouville number: looking at equation (5.15) from Theorem 5.3, if a_1 were Diophantine, there would exist κ and ϵ such that for all q

$$\frac{1}{|1 - a_1^q|} < \frac{q^{\kappa-1}}{\epsilon}.$$

If true, this would seem to be at odds with the notion that Pfeiffer exploited in his construction that he had complete freedom to make

$$\frac{1}{|1 - a_1^{n_k}|}$$

arbitrarily large.

If we alter the construction we used to produce a_1, it is clear that a_1 is not Diophantine. In our new construction of a_1 we will increase the size of $1/|1 - t^{n_k}|$ in two ways. First, we will increase the size of the lower bound M_k. We originally chose M_k to satisfy $M_k > M_{k-1} + 1$ in order to meet Pfeiffer's requirement that the M_k approach infinity. Now, we will insist that

$$M_k > M_{k-1} + (n_k - n_{k-1}) \tag{5.16}$$

with $M_1 > n_1$. It follows then that $M_k > n_k$.

Next, since Lemma 5.1 allows us to make $1/|1 - t^{n_k}|$ as large as we want, we will assume that

$$\frac{1}{|1 - t^{n_k}|} > M_k^{n_k+1} \tag{5.17}$$

[12] A proof is outlined in Milnor [**2006a**, p. 129].

for all t on $N_{\epsilon_k}(b_k)$, including a_1. Construct a_1 according to equations (5.16) and (5.17) and assume it is Diophantine. Theorem 5.3 implies there exist κ and ϵ_0 such that for all $q \in \mathbb{N}$

$$\frac{1}{|1 - a_1^q|} < \frac{q^{\kappa-1}}{\epsilon_0}.$$

In particular, the above inequality holds when q equals any of the exponents n_k that arise out of the construction of a_1.

Since the sequence $\{n_k\}$ is infinite and κ and ϵ_0 are fixed, it follows that n_k can be chosen so that $n_k^{n_k+2-\kappa} > 1/\epsilon_0$ which implies that

$$n_k^{n_k+1} > \frac{n_k^{\kappa-1}}{\epsilon_0}.$$

Because $M_k > n_k > 1$,

$$M_k^{n_k+1} > \frac{n_k^{\kappa-1}}{\epsilon_0}.$$

Equation (5.17) yields

$$\frac{1}{|1 - a_1^{n_k}|} > M_k^{n_k+1} > \frac{n_k^{\kappa-1}}{\epsilon_0} > \frac{1}{|1 - a_1^{n_k}|},$$

a contradiction. Thus, our assumption that a_1 was Diophantine fails.

Our point is that when reviewing the literature, someone as conversant in number theory as Siegel would easily see that Pfeiffer's a_1 are, like Cremer's numbers, well approximated by rationals. It would make sense, then, to look for convergent solutions to the Schröder equation among those fixed points with Diophantine multipliers.

5.6. A brief glance at Cremer's 1927 paper

Cremer, citing Pfeiffer's work a decade later, proved a related result for rational complex functions. He showed (see Theorem 11.1) that if $|\beta| = 1$ but is not a root of unity, if d is a natural number strictly greater than one, and if

$$\liminf_{n\to\infty} \sqrt[d^n]{|\beta^n - 1|} = 0, \tag{5.18}$$

then any rational function f of degree d with $f(\lambda) = \lambda$ and $f'(\lambda) = \beta$ does not have a center at λ, hence the associated Schröder functional equation diverges.

Pfeiffer's a_1 can be chosen so it satisfies Cremer's result, assuming f is a rational function (although as noted, the only condition Pfeiffer imposed was that f is analytic on a neighborhood of the origin). Rather than assuming that the terms between the a_{n_k+1} are zero as Pfeiffer did, no assumption is made about them, other than conditions analogous to those at equations (5.12) and (5.17) apply.

Again, we will alter the size of $1/|1 - a_1^{n_k}|$ on $N_{\epsilon_k}(b_k)$, choosing it so

$$\frac{1}{|a_1^{n_k} - 1|} > M_k^{d^{n_k}} \Rightarrow |a_1^{n_k} - 1| < \frac{1}{M_k^{d^{n_k}}}.$$

Our only demand on the M_k is that they go to infinity. Take the (d^{n_k})th root of both sides of the inequality following the arrow to obtain

$$\sqrt[d^{n_k}]{|a_1^{n_k} - 1|} < \frac{1}{M_k}.$$

Since the right-hand side goes to zero as n_k goes to infinity, Cremer's result follows.

5.7. Pfeiffer and the functional equation $g^2 = f$

Pfeiffer's examination of the Schröder equation in the irrationally indifferent case is clearly his most significant result for our purposes. However, he examined other aspects of the Schröder equation, as well as a related functional equation, $g^2 = f$, sometimes called the Pfeiffer functional equation, in passing in [**1915a**] and in greater detail in [**1918**]. We will briefly discuss the 1918 paper since, as he notes, it has some parallels to his investigation of the Schröder equation described above.

A second geometric problem that Pfeiffer considered in both [**1915a**] and [**1918**] was the bisection of an angle formed by a straight line and an arc that reduces to the solution of the functional equation $g \circ g = g^2 = f$,[13] where f is given.

In the case where $f(z) = a_1 z + a_2 z^2 + \cdots$ with $|a_1| = 1$ but not a root of unity, Pfeiffer again showed that the existence of solutions to a functional equation depends upon the nature of a_1.

His examination proceeded along analogous lines to that of the Schröder functional equation described in previous sections. Comparing coefficients, he obtained an expression very similar to equation (5.8), except that the numerator consists of terms of the form $1 + a_1^k$ rather than $1 - a_1^k$, and he used similar methods to construct the multiplier a_1.

He showed that if a_1 satisfies a condition equivalent to $\liminf\ a_1^n + 1| = 0$, then no convergent solutions to $g^2 = f$ exist. By varying his proof, he next determined conditions for the existence of a unique solution to $g^2 = f$ and concluded with an example of an equation for which two solutions exist.

He delineated the connections between $g^2 = f$ and the associated Schröder equation $S \circ f = a_1 \cdot S$. First, he showed that if $g^2 = f$ has fewer than two convergent solutions, the Schröder equation has no convergent solutions. If two solutions exist to $g^2 = f$, the associated Schröder equation may or may not have a solution.

Thus, while solutions to the associated Schröder equation can be used to obtain solutions to $g^2 = f$, solutions g to $g^2 = f$ do not necessarily lead to solutions of the Schröder equation, since functions f exist for which there are solutions to $g^2 = f$, yet none for $S \circ f = a_1 S$.

To illustrate this point, Pfeiffer constructed an example of a function f with a fixed point at zero for which at least one g exists with $g^2 = f$ but no S exists satisfying $S \circ f = f'(0) \cdot S$.

The first step was to construct a function $\tilde{f}(z) = \alpha z + \cdots$, where $|\alpha| = 1$ but α is not a root of unity, whose Schröder equation $S \circ \tilde{f} = \alpha \circ S$ has no convergent solution S. Next, he defined $f = \tilde{f}^2 = \alpha^2 z + \cdots$. Setting $g = \tilde{f}$ it follows immediately that the Pfeiffer equation

$$g^2 = f \tag{5.19}$$

has at least one solution. Pfeiffer then deduced, by contradiction, that the Schröder equation $S \circ f = \alpha^2 \cdot S$ has no solution.

---

[13]This equation also appears in our discussions of Lattès' analysis of functional equations, Section 3.6, page 55, and Pincherle's study of iteration, Section 8.4, page 185.

TABLE 1. **Pfeiffer's equations.** A comparison of Pfeiffer's results for solutions to functional equations near an irrationally indifferent fixed point. Here, PFE denotes the Pfeiffer functional equation $g^2 = f$. Elsewhere, PFE denotes the Poincaré functional equation, $f \circ \theta = \theta(s \cdot I)$ introduced at equation (3.1). SFE denotes the Schröder functional equation.

Convergent solutions to SFE	**Convergent solutions to PFE**	**Linearization near the fixed point?**
none	none	no
none	one	no
maybe	two	maybe

5.8. On Bennett's work

Beginning in 1915, Albert Bennett published three articles on the theory of iteration in one or more variables [**1915, 1916a, 1916b**]. A letter from Oswald Veblen to Birkhoff might shed some light on why Bennett took up the subject. While traveling in Europe, Bennett paid a visit to Veblen in the spring 1914. Veblen wrote to Birkhoff that he "urged" Bennett to use the European visit as an opportunity to develop new mathematical directions. Although this visit took place over a year before the Académie's announcement of the 1918 *Grand Prix* competition, perhaps the visit to Paris and Veblen's urgings played a role in the process that led Bennett to study iteration.[14]

The paper most relevant to us is [**1915**], which in many respects is a survey of the basics of iteration with few new results. While his paper is clear and thorough in its treatment, and displays considerable computational skill, his survey is broader than it is deep, and on the whole, he did not seem interested in treating some of the more difficult unsolved questions such as the division of the plane into regions of convergence.

However, there are items of interest in his paper, such as his summary of the possibilities for solutions S to

$$S \circ f = f'(0) \cdot S$$

near the origin, where$|f'(0)| = 1$ but is not a root of unity. Unlike Pfeiffer, he explicitly followed the implications that the existence of convergent solutions of the Schröder equation would have on our knowledge of iteration around an irrationally neutral fixed point. The freshness of Bennett's assertiveness perhaps makes up for what it lacks in applicability.

> ... *$S(z)$ may be either convergent or divergent. If $f(z)$ is convergent certain conditions must be satisfied in order that $S(z)$ be convergent. There are not formal conditions on the first p coefficients, $p = 1, 2, \ldots$, since the problem is one having significance only for convergent series $f(z)$, and convergence is not dependent upon the values of any finite number of coefficients.* ...

[14]Our source for this letter is an article not yet published by Barrow-Green [**appear**], concerning three letters from Veblen to Birkhoff.

> *A necessary and sufficient condition for* [$S(z)$ to be convergent]*: It must be possible to find a circle about the origin in the complex plane such that if any point P be taken within this circle, the totality of the negative and positive iterates of P constitute a set of points lying everywhere densely upon a simple closed curve of finite length, and with a continuously turning tangent, encircling the origin,....* [**1915**, pp. 41–42]

Unfortunately, Bennett gave no examples of either convergent or nonconvergent solutions. Moreover, after stating his argument that a solution to the Schröder equation exists if and only neighborhoods of the fixed point can be mapped conformally to a disc upon which iteration acts like an irrational rotation, he gave no indication how one might actually solve the Schröder equation unless, perhaps, by finding the invariant curve upon which iterates lie densely and mapping it to a circle, one deduces $S(z)$. But finding such a curve would prove very difficult, except in isolated special cases.

Despite this, Bennett's quotation is not without interest. Julia gave a similar characterization of conditions for linearization, and Bennett's is perhaps the first explicit description in complex dynamics of what a linearization around an irrationally neutral fixed point might be like.

Its publication evidently predates Pfeiffer's 1915 announcement of his result about the existence of divergent solutions to the Schröder equation.[15] and Bennett's assertion that the first p coefficients of $f(z)$ have no effect on the convergence of S suggests he was not aware of Pfeiffer's work, which showed just how important the first coefficient was.

Bennett's discussion of stability is reminiscent of Levi-Civita's definition, examined in section 3.4, and Bennett in fact referred the reader to Levi-Civita's paper [**1901**]. This reference suggests that others besides Lattès were interested in the connection between the iteration of complex functions and differential equations that we discussed in Chapter 3. Clearly, though, Bennett's concern seems more informal and much less rigorous than that demonstrated in Lattès' investigation (see Section 3.5).

Bennett did not follow up upon these early researches in iteration, perhaps due to the interruption of World War I, although he distinguished himself in other areas of mathematics, as well as in his involvement with the Mathematical Association of America.

5.9. Concluding remarks

An interest in iteration began to take hold in United States prior to its entry into World War I. It was centered at Columbia University, and two students of Kasner, Ritt and Pfeiffer, achieved significant results. That works by Ritt and Pfeiffer, as well as by Bennett, whose Ph.D. was from Princeton, appeared in American journals prior to the Académie's announcement regarding its 1918 *Grand Prix* attests to the competition not being the motivation for the genesis of American interest.

Although Kasner was not directly involved with the study of the iteration of complex functions, his work in the theory of conformal invariants seems to have sparked Pfeiffer's interest. Certain functional equations arise from Kasner's work,

[15] Bennett's paper [**1915**] appeared in the September 1915 edition of the *Annals of Mathematics*; Pfeiffer's [**1915a**], in the October issue of *The American Journal of Mathematics*.

and one of them, the Schröder functional equation, caught the eye of Pfeiffer, who, in 1917, constructed a function with an irrationally neutral fixed point whose associated Schröder equation did not have a convergent solution. In other words, he showed there were irrationally neutral fixed points for which centers did not exist.

This was a significant result. Kœnigs in the 1880s and his student Leau in the late 1890s both failed in their attempts to understand the behavior of iteration near such a point, so prior to Pfeiffer's paper it was not known whether centers existed, or whether there were functions, like the one he had constructed, for which they did not.

Thus Pfeiffer answered a long-standing question. His results, however, were slow to gain recognition in Europe. Neither Fatou nor Julia, whose examination of irrationally neutral fixed points in papers published between 1918 and 1920 resulted in incisive (in Julia's case, the results were mistaken) results regarding the existence of centers, gave any indication that they were aware of Pfeiffer's accomplishment, and it was not until 1927, ten years after Pfeiffer's paper appeared, that Pfeiffer's ideas started to percolate, via the work of the German mathematician Cremer.

Birkhoff's presence also hovers about the American interest in iteration. Pfeiffer evidently ran into difficulties with his construction that were related to the existence of small divisors and cited Birkhoff's help in overcoming them.

Perhaps prompted by the 1918 *Grand Prix* competition, Ritt contributed a brief notice on the study of iteration of complex functions to *Comptes Rendus* in 1918 and published several other related articles in the 1920s that will be discussed in subsequent chapters.

CHAPTER 6

The Road to the *Grand Prix des Sciences Mathématiques*

6.1. The 1918 edition

The announcement of the 1918 *Grand Prix des Sciences Mathématiques*, quoted in full at the beginning of Chapter 3, certainly accounts for the appearance of Fatou and Julia's studies of iteration. Yet, although the competitions held by the Académie des Sciences often focused on existing open problems, one is left to wonder why a contest devoted to the iteration of complex functions was announced in 1915, when the study had laid dormant for quite some time. The last work of note to focus on the iteration of rational functions of a single variable was Fatou's notice [**1906b**], which appeared nine years earlier, and we have to go back nine years prior to that to find another significant French paper on the subject, Leau's paper [**1897**].

The only other French mathematician to develop a body of work involving the iteration of complex functions in the decade before the announcement was Lattès, whose doctoral thesis [**1906**] we discussed in the final section of Chapter 3. While it drew upon results from the "Kœnigs school", Lattès' reason for doing so was to study curves and surfaces left invariant by the functional equations that arise in the iteration of multivariate functions and to link these equations to the study of differential equations. Lattès wrote several related papers in the intervening years, including [**1908**, **1911a**, **1911c**], none of which directly address the mathematical topics described in the announcement.

It also seems that there was not unanimity on the part of the Académie to make iteration the subject of the 1918 *Grand Prix*. According to Audin, progress in number theory at that time evidently led some members of the committee—Gaston Darboux, Camille Jordan and Émile Picard—to consider offering the prize for the study of Fermat's Last Theorem [**2009**, pp. 14–15].

Nevertheless, it turned out to be a very fortuitous time to stage a competition to develop a global theory of iteration since the recently developed theory of normal families of functions (which Montel had begun to articulate in inchoate form as early as 1904, although he would not use the term "*familles normale*" until [**1911**, p. 996]) would hold the key to Fatou and Julia's respective development of a robust global theory of iteration.[1]

[1]Appendix B discusses the development of the theory of normal families as well as several related theorems useful to the study of iteration including Montel's normality criterion referred to there as "*Critère fondamental*," and connections to Picard theory (see page 286). See Chapter 8 of Alexander [**1994a**] for a brief historical account of the role of normal families in the development of the theory of iteration of rational functions.

If there was an inkling "in the air" that normal families might unlock the secrets of iteration, it was not one that Montel acted upon, although he later expressed great admiration for the accomplishments of Fatou and Julia in his 1927 book *Normal Families*:

> *One of the most interesting applications of the theory of normal families concerns the iteration of rational functions, the study of which advanced very far in the beautiful works of Fatou and Julia.* [**1927**, p. vi]

As we noted above, competition topics often concerned open questions, and there were many of these in the theory of iteration. We have already quoted Leau and Kœnigs' descriptions of questions they could not answer in Section 1.8, and the questions they posed remained open in 1915.[2] Moreover, while Fatou was successful in classifying the global iterative properties of two types of functions in his notice [**1906b**] (see Section 2.2), he had not publicly followed up on his results prior to the announcement, although, according to a letter he wrote to Maurice Fréchet, he evidently continued to make progress with his research on the iteration of functions, as we noted in Chapter 2.[3] Audin suggests that in view of the closeness between Fréchet and Jacques Hadamard, a member of the Académie,[4] the former might well have communicated this knowledge to the latter [**2009**, p. 17], which may offer insight into the choice of iteration as a topic for the *Grand Prix*.

There were also others who might have communicated to the Académie their interest in seeing the study of iteration revived. By 1915 Kœnigs was a prominent French mathematician. In 1892 he was nominated to fill Pierre Ossian Bonnet's seat in the Académie des Sciences, though it was ultimately awarded to Appell, and he would eventually gain entry into the Mechanics Section in 1918. Although he had not published any new works on iteration since his papers of the 1880s, he influenced its development through the 1890s via his students Léopold Leau and Auguste Grévy, and Kœnigs also served as *président* of Lattès' dissertation committee; we assume he maintained his interest in iteration.

While there is no indication that Kœnigs had the ears of the Académie's mathematicians on the subject of iteration, he was on familiar terms with several of them. He served with Appell (who was *président*) and Picard on the committee for Leau's doctoral dissertation [**1897**], which concerned several aspects of iteration theory, most notably the first version of the Flower Theorem; Painlevé and Hadamard served with Kœnigs on Lattès' doctoral dissertation committee. Moreover, Appell's paper [**1891**] referenced Kœnigs' work several times, and Picard in [**1900**] and Hadamard in [**1901**] discussed attracting fixed points for real functions

[2]The competition helped resolve a number of them: the report on the 1918 *Grand Prix* noted that "*questions posed by Kœnigs*" were finally answered (see Appendix A, p. 283).

[3]It could be that either his work at the Observatoire de Paris or health issues, which plagued him throughout his life, prevented him from bringing this work to the point of publication prior to 1917. See Appendix M or Audin [**2009**, Chap. V] for more about Fatou's life.

[4]In 1918 the Académie des Sciences was divided into two divisions, *Sciences Mathématiques* and *Sciences Physiques*, each headed by a *Secrétaire Perpétuel*. The *Sciences Mathématiques* division consisted of five sections, *Géométrie*, *Mécanique*, *Astronomie*, *Géographie et Navigation*, and *Physique Générale*. What the contemporary reader would think of as pure mathematics was generally the province of the *géométrie* section.

In 1915 the members of the *géométrie* section of the Académie were Paul Appell, Hadamard, Georges Humbert, Jordan, Paul Painlevé, and Picard, with Darboux the *Secrétaire Perpétuel* of the division of *Sciences Mathématiques*.

in papers which referenced Poincaré. It would make sense, then, that the Académie would consider the extension of Kœnigs' local theory of iteration as a suitable open question. (But again, why for the 1918 competition and not earlier?)

The competition for the 1918 *Grand Prix* sparked a number of responses: Fatou published the first on May 21, 1917 in *Comptes Rendus* [**1917a**], and between December 17 and the following April, Fatou, Julia, and Lattès each submitted several additional *Comptes Rendus* notices mapping out approaches to a global theory of iteration for rational complex functions (see Figure 6.3 for a timeline of the submissions). In their reliance on normal families, Julia and Fatou's responses were very similar in both scope and content. Although they took a completely different approach from Fatou and Julia, the responses of Ritt and Lattès shared a similar perspective. However, the theory they outlined is nowhere near as broad or robust as that of Fatou and Julia. Perhaps this was a result of the route they took, which did not utilize normal families but instead was based on the Poincaré functional equation $f \circ \theta = \theta(f'(\lambda) \circ I)$, that was discussed in the opening section of Chapter 3. We will discuss the mathematical content of the various responses in the next chapter.

While it may not be correct to say tempers flared, words were indeed exchanged. The similarity of Fatou and Julia's entries would prompt Julia to instigate a judgement of priority (discussed in Section 6.4), and Lattès pointed out in *Comptes Rendus* that he had published his results prior to Ritt; Julia also used *Comptes Rendus* as a forum to downplay the importance of Lattès' remarkable example of a function whose Julia set is the entire Riemann sphere.

In addition to these contretemps, there would be mystery: When the 1918 *Grand Prix* commission[5] awarded its prize to Julia, it noted that there were three entries, one by Lattès (which was subsequently misplaced), one by Julia (which appeared in the *Journal des Mathématiques Pures et Appliquées* as [**1918a**]), and a third entry whose author the prize report did not name, only remarking that it had been returned. However, before we discuss these things, a few words about World War I are in order since it served as a backdrop for the priority dispute between Julia and Fatou.

6.2. One

Julia, who joined the Army of the Saint-Cyriens in August 1914 at the age of 21,[6] was among the many young French academics involved in the war. He was trained as a sublieutenant in the 9th company of the 34th regiment of the French infantry, which went off to battle in late 1914. His regiment was soon sent to the northeastern front to fight the Germans.

---

[5]The prize commission's members were Appell, Hadamard, Jordan, Painlevé, as well as Joseph Boussinesq and Lèon Lecornu from the *Mécanique* section. Humbert and Picard were the *rapporteurs*, or reporters, and Humbert would eventually write the internal report on the entries for the prize [**1918**].

[6]*Saint-Cyriens* refers to the soldiers of the École Spéciale Militaire de Saint-Cyr, a renowned, French military school founded by Napoleon Bonaparte in 1802 at Fontainebleau (near Paris) to educate new officers. It moved a few times in its early days before finding a home in Saint-Cyr-sur-Loire (near Tours) in 1808. It was disbanded briefly after Waterloo, but reopened in 1818 in Saint-Cyr where it remained until the occupation in 1940. Many of the original buildings were destroyed by Allied bombing during World War II. After the war the French consolidated their officer training in Coëtquidan, in Bretagne.

GASTON JULIA

ÉMILE BOREL

FIGURE 6.1. Photos courtesy of Académie des sciences—Institut de France.

Julia's mathematical activity did not cease with his enlistment. Hand-written letters from Julia to Borel during his service time attest to the fact that when the situation allowed, he worked on the redaction of Borel's *Leçons sur les fonctions monogènes*, [**1917**]. In one letter[7] Julia referred to both Poincaré and the GermanPaul Koebe during a discussion of uniformization theory, a concept that became enmeshed in the bedrock of Julia's monograph of rational functions [**1918a**].

On Monday, January 25th, 1915, an intense German attack took place near a fortified quarry on the *Chemin des Dames* ("the Ladies' path"),[3] which was the nickname of a road along a high ridge running west to east, northeast of the town of Soissons, in the region of Aisne.

---

[7]Filed as M 145 in the Archives of the Académie of Sciences, Fonds Émile Borel, 44 J.

[8]This was fiercely contested ground. Napoleon fought here, and earlier in the war French and British troops along the road were overrun, although they managed to hold on to a position surrounding a fortified quarry called the *Caverne du Dragon*. With the battle of January 25, 1915, the Germans captured the quarry and secured the entire ridge. A massive French offensive in late April 1917 involving over 1,000,000 troops, referred to as *la bataille du Chemin des Dames* or *la seconde bataille de l'Aisne*, attempted to retake this ground. While it achieved small gains, it was perceived as a failure since it fell far short of its goals. The French suffered staggering losses, and their casualties ran well over 100,000. In early May, troops from the French 2nd Division refused orders to continue the attack, and a mutiny ensued. The initial failures of the campaign led to the removal of the commanding general of the offensive,Robert Georges Nivelle (1856–1924). After the unrest subsided, the campaign continued, initially at a more modest pace, and by late 1917 the Germans had been repulsed from the ridge. See Keegan [**2003**].

As has been described in several places, Julia fought bravely—according to Garnier [**1978**] he was the only officer of his company to survive—and was awarded the Legion of Honor. But he paid a severe price: He was terribly wounded in the face and left disfigured, losing his nose, a wound he would customarily cover with a leather patch. These wounds obliged Julia to a long convalescence[9] during which time he met his wife, Marianne Chausson, daughter of the composer Ernest Chausson. Not surprisingly, mathematical activity was a challenge for most of 1915, as the following quotation from Borel indicates:

> *Gaston Julia cannot attend the courses of 1915, since he joined the Army and was hospitalized to heal his wounds.* [**1926**, p. XVIII]

By 1916[10] he had recovered sufficiently to reengage deeply in mathematics, although he was still undergoing treatment for violent headaches with powerful medications. He read the works of Charles Hermite and worked intently on material that would become his thesis. He also resumed his collaboration with Borel. As Borel noted in his preface,

> *These lessons were to appear at the end of 1914. Gaston Julia, a student from the École Normale Supérieure, who very much wanted to work on their redaction, was grievously wounded in January 1915. I want to thank him for the care he took in completing the redaction and the correction of the drafts; through great sufferings he courageously persisted and pursued his own remarkable works at the same time. I very much appreciate the valuable help of this young scientist on whom we can count to continue the French mathematical traditions.* [**1917**, p. v]

The final stages of work on this book are also described in Julia's letter of May 19, 1917.[11]

Julia was gaining a reputation as a talented researcher; during his recovery Humbert and Picard both visited him, and he became close to both men. They later served on his thesis dissertation committee, and when he was able, he took a course in number theory from Humbert. Audin argues that not only did many members of the elder generation of mathematicians, including Picard and Borel, come to view Julia as a surrogate son in reaction to the losses they suffered during the war, but Julia came to look at Picard as a father figure (see Audin [**2009**, pp. 27–28]). These visits evidently buoyed Julia's spirits, and he later credited Humbert as helping him regain his zest for mathematics:

> *He assiduously came to visit me at the end of his course, encouraging me to fight through the medication, and helping me to recover my passion for mathematics.* [**A8**]

Garnier noted in his necrology of Julia that

> *... the discussions with his teachers, Georges Humbert and Émile Picard, oriented his existence.* [**1978**, p. 127]

In early 1917 Julia was researching the iteration of rational functions, and it seems that subject became part of Julia's conversations with these men:

---

[9]See Rosa [**2005a** for more details.

[10]See Julia's letter to Daum, written July 5, 1917, and included in [**1968-1970**], volume 6, p. 2.

[11]Letter M 153 in the Archives of the Académie des Sciences, Fonds Émile Borel, 44 J.

> *With Painlevé and Jordan, he* [Humbert] *pressed me to tell them about my recent works of analysis, on the problem of the iteration of rational fractions, which I had been studying. I had the joy of giving Jordan one of the first examples, not yet explicitly constructed, of a Jordan curve which naturally arises as a boundary....* [**A8**, p. 89]

It is not too much to speculate that perhaps they also encouraged him to enter the competition as well, despite the fact that rational functions were a new field for him.

6.3. Doctor in mathematics

Among Julia's concerns during the last months in 1917 was the completion and presentation of his doctoral thesis [**1917a**]. It was an ambitious paper, entitled *Study on the nonquadratic binary forms with undetermined real or complex coefficients, or with conjugated undetermined coefficients.*[12] Notices related to his thesis began appearing in *Comptes Rendus* in 1916–17 during his convalescence (see [**1916a, 1917c, 1917f**]).

Julia completed his dissertation in November 1917, dedicating it to those lost in the war,

> *To the memory of my comrades from the École Normale killed by the enemy.*

His dissertation topic was initially inspired by Hermite's study of binary quadratic forms,

$$Q(x, y) = ax^2 + 2bxy + cy^2,$$

but he studied a more extensive version

$$\varphi = t_1^2(x - \alpha_1 y)^2 + t_2^2(x - \alpha_2 y)^2 + \cdots + 2u_1^2(x - \beta_1 y)(x - \beta_1' y) + \cdots . \tag{6.1}$$

Rather than go into the detail of his thesis, we will quote from the examining committee's report, authored by its *président*, Humbert (Picard and Lebesgue were the other members):

> *Summarizing the first two parts of his remarkable work, the author develops, with all its consequences, a geometric interpretation of the continuous reduction methods, first, Hermite's, and then one he himself suggested in the complex field; this interpretation is of a rare simplicity and renders reduction properties intuitive. It ... is applicable to the quadratic case exclusively, which was studied by Stephen Smith*[13] *and Bianchi;*[14] [Julia's thesis] *significantly improves the arithmetics of forms of higher degree, whose applications to cubic and biquadratic forms increase its importance.*
>
> *The third part, equally inspired, leads to many new topics, and to new and unexpected results; the author shows a clever and skillful handling of the combined theories of arithmetic, analysis, and geometry, in a fortunate fusion ... .* [**1917b**]

[12] *Étude sur les formes binaires non quadratiques a indéterminées réelles ou complexes, ou a indéterminées conjuguées.*

[13] The British mathematician Henry John Stephen Smith, whose paper [**1875**] contained the first known Cantor-like set, presented in the context of Riemann integration.

[14] The italian mathematician Luigi Bianchi.

Julia entered his dissertation in the competition for the 1917 *Prix Bordin*[15] whose topic ("to improve, on several important points, the arithmetic theory of nonquadratic forms") had already been proposed for the 1913 *Prix Bordin*, but garnered no entries (see [**CR-1913, CR-1914**]).

According to the 1917 *Prix Bordin* report, there were two entries submitted but only Julia's was accepted.[16] On December 10th, Julia was declared the winner. Humbert's report on Julia's thesis, quoted above, was published in *Comptes Rendus* immediately following the award notification [**1917b**, pp. 820–823]. In addition, the memoir was published in the *Mémoires de l'Académie* [**1917a**].

Two days after the award, on December 12th, Julia received a letter[17] from Picard:

> *Sir, the work you presented as your thesis received an award from the Académie the day before yesterday. The next* Comptes Rendus *will include the report by Humbert. I shall repeat to you how we delighted in your remarkable work, whose second and third parts show a real spirit of invention. We appreciate the moral qualities which you have been showing these last three years. You have the good sense to dedicate your work to the memory of your comrades of the École Normale, who fell on the field of honor. You have been luckier than several young professors, whose heroic death will remain the glory of the École Normale. But you were cruelly wounded and, between many operations you found the energy to devote yourself to mathematical research. After the tragic time we are living in passes, France will need more than ever men of talent and of character. We count on you, sir, for the future. After this first work, others you have already in preparation will come. For now, I am happy to tell you that the Faculty awarded you the rank of doctor with a most honorable mention, and we join in our deepest and most warm congratulations.*
>
> *Émile Picard*

6.4. The controversy between Fatou and Julia

> *The thesis finished, I am doing something quite different, which I will publish only when it I have dug deeply enough into it.*

The above quotation is from a letter Julia wrote to Léon Daum excerpted in Garnier's necrology of Julia [**1978**, p. 128]. Not surprisingly this new project was his study of the iteration of rational functions. The deadline for entries into the 1918 *Grand Prix* was December 31, 1917, and both Julia and Fatou were busy.

While Julia flushed out the details of his new project, Fatou readied a note on the iteration of rational functions that was published in *Comptes Rendus* on May 21 [**1917a**]. It was not until December 17 that he published a follow-up, [**1917b**].

---

[15] The prize commission members were Appell, Boussinesq, Hadamard, Jordan, Lecornu, Painlevé, Picard, and Paul Vieille (from the *Mécanique* section) with Humbert the *rapporteur*.

[16] Entries in a competition sponsored by the Académie that were not deemed worthy of serious consideration were commonly returned to the entrant, who was not referred to by name in the prize report appearing in *Comptes Rendus*.

[17] Kept in volume 5542 of the series AJ16 at Archives Nationales, Paris.

In the lacunae between Fatou's two notices, Julia took part in a venerable tradition by depositing—but not publishing—a series of four "*plis cachetés*" (sealed letters) describing the fruits of his new project with the Académie. This procedure, which Julia used to establish priority, officially dates back to 1735, when a registry for their receipt was established, but the practice dates back longer to at least the late 1600s. Pierre Berthon in [**1986**] described Guillaume l'Hôpital opening a sealed letter from an unnamed mathematician on April 13, 1697, that had been deposited on March 16 of that year which purported to contain a solution to the brachistochrone problem (the curve of quickest descent), posed by Jakob Bernoulli.

Many of the earliest *plis cachetés* involved inventors of mechanical devices who wanted to establish priority as both a hedge against counterfeiters and an attempt to gain exclusive rights to their invention from the King of France [**1986**, p. 71].

What Julia did in submitting the letters was by no means unusual. According to Berthon, between 1837 and 1983 an average of 95 such letters were submitted to the Académie each year; moreover, during World War I this number increased a bit due to the scientific and medical discoveries prompted by the war [**1986**, p. 73].

The procedures regarding the depositing of the letters changed over time and we quote the rules in place at the time of the competition from the *Annuaire de l'Académie des 1918*:[18]

> *The Académie des Sciences accepts the deposit of sealed envelopes in its archives as an alternative to publication in order to give a specific date to the discoveries they are supposed to contain. This deposit does not provide or imply the legal prerogatives of a patent and is not intended to supplant one.* [**A13**, p. 333]

Since mathematical capital is generally measured in prestige rather than in monetary terms—aside from financial prizes involved in winning a competition—the sealed letters are, despite the Académie's wording, in some sense the mathematical equivalent of a patent.

Julia made it clear that his decision to deposit the sealed letters was prompted by the publication of Fatou's May 21 notice:

> *A bit of background is necessary to give an account of the contents of my letters. If I have to date my decision to deposit these letters, it was May* 21st *1917, when Fatou announced in these* Comptes Rendus *a number of results that I had already attained.* [**1917h**, p. 1098]

Fatou's May 21 note [**1917a**] represented considerable progress towards the prize, but by no means did it signal a winning hand, in part because his approach was not yet fully formed. In particular, he had yet to apply the theory of normal families. Rather, his approach was based on the role of a circle left invariant under iteration called *fundamental circle*. These circles are in some sense a special case because they do not occur with all rational maps. Yet they occur with enough frequency that they were an essential component to Fatou's program and can be used to illustrate general properties common to all rational functions, such as a Julia set, although Fatou did not do so in his May 21 notice.[19]

[18]The rules were revised again in 1990. For further information about these procedures see Berthon [**1986**] and Brian and Demeulenaere-Douyère [**1996**].

[19]We discuss the role of the fundamental circle in Fatou's work in Section 7.3.

If Fatou had not yet discovered the importance of normal families, it appears that neither had Julia. Recounting the events surrounding the sealed letters, Julia noted

> *Up to this point* [the appearance of Fatou's May 21st note] *I had ignored the work of Montel. My attention was drawn to it by his* [Montel's] *note of June 4th. I then studied an offprint that Montel was kind enough*[20] *to send to me.* [**1917h**, p. 1099]

Julia did not say exactly what caught his eye, but there is one passage from Montel's note that not only appears to be tailored to the study of the convergence of a sequences of iterates, but also bears a strong visual similarity to depictions of iteration:

> *Let* $f_1(z), f_2(z), \ldots, f_n(z), \ldots$ *be an infinite sequence of analytic functions bounded in their entirety interior to a connected domain* (Δ). *If this sequence converges uniformly on an arbitrarily small arc* (α) *inside the domain, it will also converge uniformly throughout the interior of that domain ....* [**1917b**, p. 880]

Fatou did not employ the strategy of the *plis cachetés* but rather choose the public route of publishing his results directly in *Comptes Rendus*. Julia's behavior in the competition was thus very different than Fatou's, and this difference would have significant consequences.

Fatou presented another notice to the Académie [**1917b**] during the December 17 session, and the results were striking. The theory of iteration of rational functions was now fully grown under his guidance. His notice lists more than ten major results. We do not know how Fatou came to apply normal families or whether he was similarly struck by Montel's June notice, but by the end of 1917 he had embraced the theory of normal families. The problem was, Julia had as well.

In making sense of what follows, three things are important: The first is that Julia felt it essential that he establish priority. The second is that Julia did not allege that Fatou somehow stole his results as has sometimes been suggested when people informally discuss these events; what was at stake was priority, not an accusation of plagiarism, and Julia did not overtly accuse Fatou of anything untoward. The third thing is that Julia was not naïve, despite his youth. He well understood the rules of the game, sealing his notes until the time to reveal them was right.

Julia was about to instigate something that was relatively rare. While the submitting of sealed letters was not unusual, full-blown priority disputes were. According to Berthon, most *plis cachetés* languished unopened, making the registry a kind of dead-letter file. Those letters that were unsealed were often done so for a variety of reasons that had nothing to do with establishing priority: their authors sometimes specified that their letters be opened upon their death, or on a particular date; others were opened by family members and some by historians. Berthon's article, written in 1986, was prompted in part by the Académie's decision to go through its vast score of sealed letters [**1986**, p. 71ff.].

[20] If Montel's courtesy indicated a deeper bond between the two men, it dissipated over time. Audin describes a priority dispute that broke out in the 1960s between Montel and Julia over the term "*points irrégulier*", which Montel used to describe points, as early as 1914, around which a particular family failed to be normal, and "*points de Julia*", which Julia used to describe the same phenomena, a usage which led to the term "lines (or curves) of Julia" (which are discussed in Appendix E). See Audin [**2009**, pp. 196–209] for a detailed discussion of the dispute.

Events unfolded one step at a time, according to the dictates of the procedure. Invoking the rules,[21] Julia submitted the following letter to the president of the Académie at the December 24 meeting:[22]

December 24, 1917

Mr. President of the Académie des Sciences,

I am honored to ask you to open the sealed envelopes I deposited to the Secretary of the Institute on the dates, June 4th, August 27th, September 17th, and December 10th, 1917, which have been registered under the numbers 8401, 8431, 8438, 8466.

You will please accept, Mr. President, the expression of my respectful devotion.

Gaston Julia

The next step took place later that same day when Julia submitted the notice [**1917h**], *Sur les substitutions rationnelles*, wherein he formally asked for a judgement of priority. His notice was not entered into the record until the next meeting, which is not uncommon. This also happened to the notice that followed [**1917h**] in *Comptes Rendus*, and it could be that both were submitted too late to be included in the December 24 session.

Julia's notice began with the words

> *I read with interest the article of Fatou, published in the* Compte Rendu *of December 17th 1917. I had previously entered the results contained therein in four sealed envelopes, deposited with the Secretary of the Institute, registered as n° 8401, 8431, 8438, 8466, and dated, respectively, June 4th 1917, August 27th 1917, September 17th 1917, December 10th 1917.*[23]
>
> *The Académie shall see, after opening these four envelopes, that they contain the results Fatou gives, with notations and examples, followed with brief indications of the methods, one of which, by curious coincidence, relies on precisely the same results from*

[21]Details on the procedure, from the 1918 *Annuaire*: "The author of a *plis cacheté* can ask that it be opened. Such a request shall be written and signed. The envelope is opened during a session of the Bureau [the group consisting of the *président* and *vice-président* of the Académie, and its two *secrétaires perpétuels*] who, having realized its contents can submit it, if need be, to a competent commission and suggest that the Académie insert all or part of it into its publications, or into it is archives" [**A13**, pp. 335–336]. In recent years, these procedures have been slightly modified.

[22]Archived in the *Dossier de Séance* of December 24, 1917.

[23]Our examination of the archives of the Académie revealed only that the four *plis cachetés* are entitled, respectively, *Sur les transformations ponctuelles*, *Sur les substitutions rationnelles*, *Sur les substitutions rationnelles* (two letters had the same title) and *Sur l'itération des fractions rationnelles*. The title *Sur les transformations ponctuelles* does not match any of Julia's *Comptes Rendus* notices, and, conversely, the title of [**1918c**], *Sur des problèmes concernant l'itération des fractions rationnelles*, does not match that of any of the letters. The letters themselves were not available, so it is unknown whether any of the notices are verbatim copies of one of the *plis cachetés*.

PAUL APPELL

GASTON DARBOUX

JACQUES HADAMARD

GEORGES HUMBERT

CAMILLE JORDAN

ÉMILE PICARD

FIGURE 6.2. Members of the *Géométrie* section of the Académie des Sciences around the time of the 1918 *Grand Prix des Sciences Mathématiques*. Darboux was the *Secrétaire Perpétuel* of the *Sciences Mathématiques* division until he died in 1917 and was succeeded by Picard. Paul Painlevé, who was also member, is not pictured. Photos courtesy of Académie des sciences—Institut de France.

Montel's theory of normal sequences of analytic functions that Fatou uses. The Académie can assess whose methods and whose results deserve priority. [**1917h**, p. 1098].

The phrase "curious coincidence" stands out, especially since it concerns Fatou's use of Montel's theory of normal families. Is it not clear exactly what about Fatou's use of normal families Julia found "curious."[24] To an English speaker it

[24]For those who wish to parse Julia's wording closely, here is the original French: "*L'Académie pourra, en ouvrant ces quarte plis, se rendre compte que les résultats que donne M. Fatou, aux notations et aux exemples près, s'y trouvent avec l'indication brève des méthodes suivies, dont l'une, coincidence curieuse, utilise précisément les résultats de M. Montel sur les suites normales de fonctions analytiques qu'utilise M. Fatou. L'Académie estimera, à la fois quant aux méthodes et quant aux résultats, à qui doit être attribuée la priorité.*"

sounds quite caustic. Perhaps it is intended as sardonic humor? Or does it contain a suggestion that there was indeed something "curious" about Fatou's use of normal families: perhaps Fatou only thought to use normal families because he found out—perhaps through Montel[25]—that Julia was? It is impossible to tell. Sometimes the inclusion (or omission) of a word can make a huge difference in how a statement is apprehended, and one wonders what effect Julia's use of the word "curious" has had on those who have read his letter to Académie.

The Académie responded very quickly to Julia's claim, resolving the matter in the the crucial week between the submission of Julia's request on December 24 and the December 31 deadline for the 1918 *Grand Prix*. Humbert was chosen to adjudicate the dispute, and his judgement came at the December 31 session, which we quote in its entirety:

> *The Académie has charged me to examine four sealed envelopes submitted by Julia, opened in the session of December 24, as well as a note where the same author raises a question of priority, related to certain statements that, it can be assured, he had consigned to those envelopes.*
>
> *In these statements, which he does not explicitly cite in his notice, he refers to Fatou's entire notice (Fatou* [**1917b**]*), excepting a numerical example and at least a portion of one of the results it contains.*
>
> *The comparison, between the texts of these two notes and the four envelopes, reveals that Julia's assertion is true: his envelopes include, among other things, all statements whose authorship he claims; they appear, either in equivalent terms, and also sometimes in a more general or extended form (Humbert* [**1917a**]*).*

According to the rules governing the process, Humbert is doing exactly what he is expected to do: He opened the *plis cachetés*, as Julia requested, and he reported that they contained exactly what Julia said they did. Since the Académie received the *plis cachetés* before it received Fatou's December 17 note, Julia had established priority.

Fatou did not submit an entry, although one can reasonably conclude that his 280-page monograph on the iteration of rational functions [**1919-1920**], published in three parts in the *Bulletin de la Société Mathématique de France* beginning in 1919, was at one time intended for the *Grand Prix*. We are left to speculate why he choose not to enter since he did not allude to it in subsequent writings, and we could find nothing in the public record about his reaction. It is not mentioned in necrologies by Léon Bloch [**1931**] and Jean Chazy [**1932**], nor in the private recollections of his nephew Robert Fatou [**F1**]. The only comment we have found from someone close to Fatou that even obliquely alludes to the priority decision are words spoken at Fatou's funeral in 1929 by his colleague Henri Mineur at the Observatoire de Paris where Fatou worked his entire professional life:

[25] Both Fatou and Montel received their doctorates from the University of Paris in 1907, and both were students of Lebesgue who served as chairman of Fatou's thesis committee and was a second reader on Montel's thesis—Borel was the first. Audin points out that Montel and Fatou were friends [**2009**, p. 169].

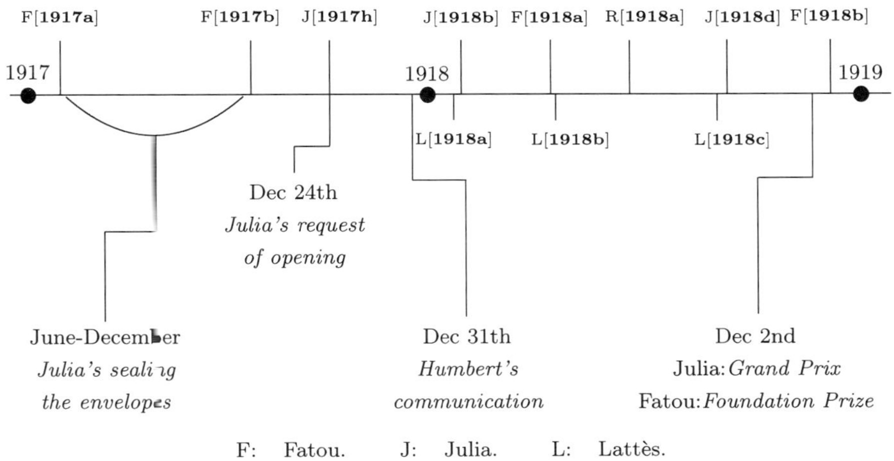

FIGURE 6.3. The flow of events. The first notice, from Fatou, appeared in the May 21 *Comptes Rendus*, although it was submitted on May 1; the last, from Julia, was submitted on April 8 and appeared in the April 15 *Comptes Rendus*.

> *To one of his most remarkable discoveries, i.e., to the iteration of rational functions, the name of Fatou is inalterably tied: he was the first to dare to attack the problem, the first one to solve it.* [**A11**]

Mineur was surely aware of the events surrounding the *Grand Prix*. We do not know, though, quite what he meant by this assertion of priority, or even if it was intended as a comment on the dispute.

Perhaps Fatou wished to avoid additional controversy. Such a decision would be consistent with descriptions of his personality.[26] Perhaps he felt his chance to win the *Grand Prix* evaporate with the closing paragraph of the Académie's verdict, written by Humbert who, as one of the two *rapporteurs* (the other was the ardent nationalist Picard, who, like Humbert, was close to Julia) would also judge the *Grand Prix*. One can certainly imagine a reaction of resignation from Fatou, perhaps compounded by the fact that Julia was a war hero, while his own chronic health issues rendered him unable to serve in the military.

A year later, when the Académie surprised no one in awarding the *Grand Prix* to Julia, it noted the following:

> *For the competition, three memoirs were deposited to the Secretary; the Commission kept only the memoirs by LATTÈS, professor at the University of Toulouse, and JULIA, Lieutenant of* 34th *Infantry Regiment, winner of the* Prix Bordin *in 1917.* [**CR-1918**, p. 812][27]

Just who the third entrant was a bit of a mystery.

---

[26]See Appendix M for a brief biographical sketch. See Audin [**2009**, Chapter V] for an in depth treatment of his life.

[27]The complete report on the *Grand Prix* that appeared in *Comptes Rendus* is included as Appendix A.

6.5. The third man

Fatou would normally be the prime suspect for the third entrant, but we know he did not enter the competition. Moreover, entries were generally returned only if they did not adequately address the topic of the prize, a description which does not come close to characterizing Fatou's monograph.

The third man was Salvatore Pincherle. Pincherle published several works on iteration which we discuss in our Chapter 8 on the study of iteration in Italy. At the time of the competition, he was considered an expert in the study of functional equations, as his authorship of chapters on functional equations in the *Encyklopädie der Mathematischen Wissenschaften* [**1906**] and the *Encyclopédie des Sciences Mathématiques Pures et Appliquées* [**1912**] attests. Both of these works include treatments of the iteration of complex functions as represented by Kœnigs and others.

Pincherle's body of work on iteration is largely independent of the French study and does not rely on normal families. One focus is on functional equations; the other concerns the iteration of real polynomials, principally $p(x) = x^2 - a$, and Pincherle published several articles on this topic prior to the 1918 *Grand Prix*, for example, [**1914, 1917a, 1918b**]. He expanded his study of $p(x) = x^2 - c$ to the complex case in his entry for the prize, *Sur l'itération de la substitution* $x^2 - a$ [**1917b**] as well as [**1920a**]. The scope of Pincherle's work is certainly less than that of Julia's, but perhaps not as much as the title might lead one to think; in many respects he came as close to the heart of as many fundamental ideas—albeit different ones—as Lattès did. We will talk more about Pincherle's entry in Sections 8.5 and 8.6.

6.6. Other conversations

In a contest such as the *Grand Prix*, authors are competitors and at times view each other that way. We have already seen that this is how Julia viewed Fatou. In this section we examine Fatou and Julia's reaction to Lattès, as well as Lattès' reaction to another potential competitor, Joseph Fels Ritt of Columbia University.

Although Lattès' entry was subsequently lost (see Section 7.5), we assume that the three short notices he submitted to *Comptes Rendus* which appeared early in 1918 were abstracted from it. The first of these, [**1918a**], appeared on January 7, 1918, and included a description of the orbits of arbitrary points for what has come to be called a *Lattès function*, a function whose Julia set is the entire sphere

$$l(z) = \frac{(z^2+1)^2}{4z(z^2-1)}. \tag{6.2}$$

Lattès was not fully aware of the scope of his discovery since the value of this example to him was not that its Julia set equaled the sphere (he did not even conceive of things in these terms). Rather, it was illustrative of a method by which he could solve what he saw as being the fundamental problem of iteration: the characterization of the orbit of an arbitrary point. Our discussion of Lattès' entry in Section 7.5 includes an examination of this example.

Prior to Lattès' note, the only Julia sets that had been found in the French study were either totally discontinuous perfect sets, analogous to the Cantor set, or continuous curves without an interior. So, while it was known that if a Julia set has an interior, it must equal the sphere, it was not known to Fatou and Julia if such a set actually existed until Lattès' note.[28] Julia noted in the introduction to his entry, which had already been submitted to Académie by the time Lattès' notice appeared, that, although he suspected a function exists whose Julia set had an interior, he had been unable to find one [**1918a**, p. 49].

If Lattès had not fully realized the import of his example, Julia certainly did. Two weeks after Lattès' note, Julia published a notice in *Comptes Rendus* which he began by explaining Lattès' note in the context of the theory he had already developed:

> *The oddities that Lattès noted seem less surprising; it is a very special case.* [**1918c**, p. 153]

However, in a footnote inserted into his *Grand Prix* entry upon its publication, Julia took a slightly different tone and referred to it as "interesting example" [**1918a**, p. 106, fn.] We will have more to say about these oddities in Section 7.5.

Fatou was not overly enthusiastic about Lattès' results either, as the following quotation from his monograph, published sometime after Lattès' notice suggests:

> *It is a very special case, but the iteration of this rational function is not so interesting from the viewpoint of the theory of functions.* [**1919-1920**, p. 41][29]

Three weeks later, on January 28, Lattès published a second notice [**1918b**], where, returning to ideas in his dissertation (see Section 3.5, he examined the iteration of complex functions of two variables, something neither Fatou nor Julia had yet looked at, although they would in the 1920s (see Chapter 10). On March 25, Lattès followed up upon results from his first two notices and, perhaps concerned with establishing priority for the approach he took, commented on Ritt's notice that had appeared three weeks earlier:

> *In a recent notice*[30] *J. F. Ritt announced several results on the iteration of rational functions: certain of these results were obtained by Ritt with the aid of a Poincaré function with a fixed point such that* $|s| > 1$ *(a repelling point, in the terminology we will use from now on).*[31] *I have myself shown in a note published earlier*[32] *the appeal the Poincaré functions hold for the study of iteration; among the propositions found there is one that Ritt also gave: one can find, for any point z whatsoever, a sequence of antecedents having as their limit any of repelling points.* [**1918c**, p. 486]

Julia presented his final notice concerning the prize on April 15, and the public conversation among the participants ceased.

---

[28]For a rational function of degree two or more, Julia sets are nonempty, perfect sets. We will see in Section 7.5.3 that others before Lattès found functions whose Julia sets have an interior, but either they did not understand the dynamics of their examples, or their findings were unknown to the wider mathematical community.

[29]It should be noted that some present day researchers are intrigued by Lattès' example and have investigated classes of functions with such properties. See Milnor's paper [**2004**].

[30]Ritt's notice [**1918a**].

[31]Ritt evidently introduced this terminology, as well as "attracting".

[32]Lattès' notice [**1918a**], which appeared on January 7th.

6.7. Concluding remarks: The awarding of the prizes

Samuel Lattès died on July 5 from typhoid fever at the age of 45. Sanitary conditions through France had deteriorated during the war (see Weber **1986**, p. 58ff.]), which may have contributed to his death.

When the results of the *Grand Prix* were announced on December 2, 1918, Lattès' contributions were acknowledged with a "most honorable mention". The prize commission's report particularly commended his work on the iteration of complex functions of two variables, observing

> *...the author proved to be a skilful and wise spirit, and he could have taken his discoveries further if death had not taken him earlier this year, still young and in full possession of his talent ....* [**CR-1918**, p. 812]

The Académie also posthumously awarded Lattès a prize from the Gegner Foundation which came with 2000 francs. The commissioners included Painlevé, Appell, and Picard, and the citation, written by Picard, is as follows:

> *Samuel Lattès, professor at the Faculty of Science in Toulouse is well known for his interesting work on substitutions and iteration studied from the point of view of the general theory of functions. His recent death is deeply regretted by all who had still expected so much from him. The commission wants to reward a work prematurely interrupted.* [**CR-1918**, p. 880]

These funds were no doubt intended for his wife, Jeanne Lattès, herself an accomplished mathematician and scientist.[33] From his dossier at the *Centre Historique des Archives Nationales* in Paris [**A9**], we know that soon after his death, his wife sent letters to the institutes where Lattès worked, perhaps to inquire about pensions.[34]

At least one of Lattès' colleagues assisted her. On July 16, soon after her husband's death, the Dean of the University of Toulouse wrote a letter on her behalf:

> *I have the honor of asking for your benevolent attention, Mr. Minister, to the situation of Mrs. Lattès, whose husband, professor of Mechanics at the Faculty, died on July 5th. Mrs. Lattès has one young baby to provide for* [she was four years old] *and is without personal fortune. I do not know also whether she is entitled to the pension, which would in any case be very low.*[35]
>
> *She had been a brilliant student of the University of Montpellier.*
>
> *For these various reasons she is thoroughly worthy of the benevolence of M. Minister, and I will be grateful to you for soliciting immediate aid and support for her, as you are able.*

---

[33]See Audin [**2009**, p. 76] for a brief account of her life.

[34]Letters were sent to the École Normale Supérieure, where he was a student, as well as to his places of employ: the Nice lycée, the Carnot at Dijon, the Technical Lycée Garçons d'Alger, the Faculty of Sciences at the University of Montpellier, the Lycée Garçons de Montpellier, and the Collège Mignet (Aix-en-Provence).

[35]Found in Lattès' dossier at the University Paul Sabatier, Toulouse.

At least in financial terms, neither Fatou nor Julia went home discontented: Julia was awarded 3000 francs for winning the prize, and Fatou 2000 francs, from the Henri Becquerel Foundation. This was a considerable sum of money. For Fatou both the award, which was for his work in trigonometric series and the iteration of functions, and the favorable mention in the prize commission's report, must have seemed a bit of well-deserved recognition for his accomplishments.

CHAPTER 7

Works Written for the *Grand Prix*

7.1. Introduction

In awarding Julia the *Grand Prix des Sciences Mathématiques*, the Académie was effusive in its praise:

> *The memoir by Julia bears the mark of a mathematical spirit of the highest order, one with the vigor to understand problems in all their generality and to follow their consequences no matter where they lead; he displays as well a profound knowledge of the results and methods of modern analysis, together with a remarkable aptitude in applying them. He made decisive progress in his investigation of iteration and showed once again how the most subtle properties from both the theory of sets and the study of Jordan curves can be naturally introduced into the study of analysis, and especially rational functions.* [**CR-1918-Rapport**, p. 814][1]

While the beginning of this passage is a prime example of the sort of purple rhetoric that often comes with the awarding of prizes, there is nothing inflated about the last sentence—and it applies equally to the work Fatou produced in anticipation of the prize.[2] The originality of their approaches was necessitated by the difficulties innate with repeated iteration: the limit functions that emerge are far too complicated to render explicitly, as are the boundary curves that line their domains of convergence, and so Fatou and Julia's best hope was to approach these objects qualitatively, via analysis, topology, the geometric properties of curves, and set theory. As Julia remarked near the beginning of his entry,

> *From my point of view, this is a most qualitative study that I undertake.* [**1918a**, p. 47]

In this chapter we describe the works that Julia, Fatou, and Lattès prepared in the run-up to the 1918 *Grand Prix*. Our descriptions are not intended to be exhaustive. We invite the reader who wants to know more to consult the original works we discuss since our summaries cannot do them complete justice. Our goal is to present highlights from their studies, and to convey the sense that while these works may not be at the forefront of the revolutions in mathematics that came with the new century, they were among its very first offspring.

In what follows we will examine in turn the works Julia, Fatou, and Lattès prepared in anticipation of the prize. In Chapter 8 we will discuss Pincherle's entry.

---

[1]Appendix A contains a translation of the entire report.

[2]We are working under the assumption that a principal motivation for Fatou's long monograph on rational functions [**1919-1920**] was the *Grand Prix*, and that prior to the eruption of the priority dispute, he had intended to enter the competition.

It is important to keep in mind the fact that while Julia and Fatou's works differ in matters of content and style, there are remarkable similarities, and these similarities mark the works far more than their differences (and we will get to those). So when we say, "Fatou showed P" or "Julia proved Q" the reader can assume that more than likely we could have just as easily said "Julia showed P" or "Fatou proved Q"; indeed, that should be considered the default. The same cannot be said of Lattès, whose work is rooted in different principles and covers different topics. Although focused on a single family of curves, we will see that Pincherle's work shares similar spirit, if not scope, with Fatou and Julia. Our intent below is to describe the papers of Fatou and Julia, not judge them against one another. Although one of us signaled a sight preference for Julia in a previous book, that preference has evaporated over time, and he now considers such judgements the province of readers.

Perhaps nowhere are these similarities more evident than in Fatou and Julia's use of Montel's theory of normal families[3] (the reader will recall that this similarity played a vital role in Julia's decision to instigate the priority dispute against Fatou). The theory of normal families is perfectly suited to the study of iteration, and once the insight to use it arises, there seems to be a natural way to apply it. One of the primary objects in the study of the iteration of a complex function f is the *iterative family* $\mathcal{I} = \{f^n\}$, and the theory of normal families partitions the sphere into domains where $\mathcal{I}$ is normal (the Fatou set) and nonnormal (the Julia set), which both Fatou and Julia did. The theory of normal families tells us much about the structure and properties of these domains, and Julia and Fatou came to understand them in similar ways.

Julia and Fatou also understood that changing the parameters in certain families of functions affected the Julia and Fatou sets which anticipated the kind of thinking that lies behind the Mandelbrot set; we will discuss this in Section 7.4. Each man also knew the crucial role of branch points of the inverses of f, and each proved that the number of attractive orbits for a rational function is limited by the degree d of the function to at most $2d - 2$.

The center problem[4] is one area where their studies differ. Neither was able to determine whether centers existed, nor was either able to find an example where they did not. They did show, however, that if a center exists, it is in the domain of normality, but, unlike the case with periodic attractive orbits, the limit functions of the iterative family $\mathcal{I}$ (i.e., the limits of subsequences from $\mathcal{I}$) are nonconstant on neighborhoods of λ.

At this point their studies of the center problem diverge. In 1919 Julia mistakenly thought he had a proof—one he withdrew many years later—that centers did not exist. This belief may have cut short his investigation of centers. Fatou, who remained agnostic towards their existence, carried his investigation a bit further than did Julia. As a result Fatou probably presented a deeper picture of the situation. Since the center problem was the most important explicit problem they left hanging, we devote a separate chapter to their investigations of it, Chapter 9.

---

[3]A family of functions $\mathcal{F}$ is normal on a domain D if every sequence of functions from $\mathcal{F}$ has a subsequence f_n which converges uniformly on compact subsets of D, or it converges uniformly to ∞ on D.

[4]If λ is a center, a solution S exists to the Schröder equation $S \circ f = f'(\lambda) \cdot S$, and iteration of f on a neighborhood of λ is conformally equivalent to a rotation of a disc surrounding the origin; see Section 1.6.2.

There are other differences as well, examples of things one did that the other did not: Fatou developed an inchoate version of the technique of symbolic dynamics [**1919-1920**, pp. 257–258, II][5] (and later in 1922 in a short article on the iteration of algebraic functions [**1922a**, pp. 192–193]) to code the forward and/or backward orbit of a point under iteration. Julia's innovations include an extension of Schwarz's Lemma.

There is one other important difference. While Fatou made liberal use of functional equations—the title of his work was *Sur les équations fonctionnelles*—except for the center problem, Julia purposefully and explicitly excluded them from his monograph. As he noted near the end of his introduction:

> *There remain, as has already been indicated, many other questions to treat I could apply the results obtained herein to the well-known functional equations from the work of Kœnigs and his successors. That study does not fall within the purview of this memoir, which is the study of iteration in and of itself: I may ultimately return to the questions thus left suspended.* [**1918a**, p. 128]

He ultimately did. Many of his works from the mid-1920s include the phrase "*équations fonctionnelles*" in their titles.

7.2. Julia: *Mémoire sur l'itération des fonctions rationnelles*

Julia's work [**1918a**] is structured very much like a textbook. Its exposition is prefaced with two sections, the *Introduction*, which summarizes many of the main results, and *Preliminaries*, which outlines the mathematical prerequisites for his monograph proper, which is sectioned into four chapters.

7.2.1. Setting the stage: *Introduction* and *Preliminaries*. The *Introduction* includes Julia's views on the evolution of the study of iteration and summarizes his most important results. He described his goal not only as fulfilling the Académie's request that the theory be extended from its local focus near a fixed (or periodic) point, but also in bringing a high degree of generality and abstraction to the subject. One sentence into his monograph he remarked,

> *Aside from the two notes by Fatou in* Comptes Rendus de l'Académie Sciences *from October 15, 1906, and May 21, 1917, there have only been local studies of the subjects. The results that he gave are most interesting but are examples which utilize particular properties of the function studied. One sees that these examples are the most simple one can imagine. I have tried, in this memoir to give those properties which, for any given rational function $f(z)$ whatsoever, are independent of such and such a particular property.* [**1918a**, p. 47]

It should be noted that Julia's monograph is, aside from the inclusion of a footnote or two and an addendum at the end where he corrects an error, what he submitted for the 1918 *Grand Prix* in late December 1917. Because of the timeline for submission—the deadline was December 31, 1917—when Julia wrote

[5] Fatou's monograph was published in three parts and covers seven chapters with a brief introduction. We chose to denote it as a single work because it was conceived as one. Unfortunately, there is overlap in the pagination of the first and third installments, so well we will add the chapter after the page reference to avoid ambiguity.

his entry he evidently did not realize just how far Fatou's study had advanced since his May 21 notice [**1917a**]. Fatou published two notices early the following year [**1918a, 1918b**], and the results outlined therein, along with Fatou's December 17 notice [**1917b**] are most certainly not based on the particular properties of simple examples. Rather, they demonstrate a depth of reasoning and a level of generality equal to that of Julia's work. Since Julia nowhere refers to Fatou's December 17 notice, which is the one that ignited the priority dispute discussed in Section 6.4, one can infer that Julia wrote his monograph with no knowledge of the extent and depth of Fatou's study.[6]

Julia was also evidently unaware that the Polish mathematician Łucyan Böttcher published several works written in German, Polish, and Russian, beginning with his dissertation [**1898**] written in German, that contain passages which treat iteration from a global point of view (see Section 7.5). Julia was not alone in being unaware of these works, since they were evidently not well known in France.

The point of the above quotation is not whether Julia is right in his characterization of Fatou's papers—we tend to think not since it is clear that Fatou understood, even in 1906, that his examples were not so much special cases, as they were particular functions that embodied general properties, but we encourage the reader to make his or her own judgement. Rather it is an attempt to capture Julia's sense of mission. Not only did he want to bring a global perspective to the study of the iteration, but he wanted to ground it on very general principles.

To carry his mission forward, Julia required new theoretical underpinnings which he found not only in set theory and the study of Jordan curves, but also in Picard theory and Montel's theory of normal families, which he used to great effect.

The set $E = \{z : f^p(z) = z \text{ and } |d/dz\ f^p(z)| > 1\}$, the set of repelling periodic orbits, was central to his study. As he noted in the *Introduction*:

> *I resolved to study a priori, in the general case, the set E of the points $z = f_p(z)$, ($p = 1, 2, \ldots, \infty$), for which $|f_p'(z)| > 1$. This is a denumerable set. The work of Montel on normal sequences immediately yields simple properties for all points P from E.* In any circle with center P the functions $f_1(z), f_2(z), \ldots, f_n(z), \ldots$, (the indefinite iterates of $f(z)$) take on all complex values except for two at most[7] ... *The comparison of these results to the classic theorems of Picard about essential singularities has led me to think that the set E, along with, naturally, its derived set E' constitutes the set of singularities for which I have searched.* [**1918a**, p. 48]

He observed that E' is a perfect set, and then described its possible configurations:

> *I have studied the* structure of E' without any *restrictive a priori hypotheses. E' can be discontinuous,*[8] *a linear continuity or a continuous surface. I easily showed the first two possibilities through examples, and for the third, I show that E' cannot contain any region of the plane without containing the entire plane. In comparing the elements in common between the present question and those of properly or improperly discontinuous* automorphic groups *of the*

[6]Indeed, that Fatou's study had progressed to the point where it became the equal of his own must have come as quite a surprise.

[7]Julia liberally used italicized text for emphasis.

[8]In contemporary parlance, a *discontinuous set* is one which is *totally disconnected.*

> *plane, I am led to believe that this last possibility will eventually present itself, but I have not succeeded in finding such an example.* [**1918a**, pp. 48–49]

Finally, he described the fractal property of E': "*the structure of E' is the same in all its parts*" [**1918a**, p. 49], that is, that E' is homogeneous, meaning it is has the same topological structure, or, in more contemporary parlance, is self-similar.

Montel's theory of normal families provided both Fatou and Julia with the means to conceptualize iteration of a rational function f. The set of points on whose neighborhood the iterative family $\mathcal{I}$ is normal, generally speaking, comprises regions on which f converges to an attracting point λ or comes arbitrarily close to a finite set of attracting periodic points. For rational functions there are two other possibilities for the domain of normality: f can have a center, whose existence neither Julia nor Fatou could confirm; or it can have a Herman ring (named for Michel Herman who discovered them around 1980; see Section 1.6.2), where iteration is conformally equivalent to a rotation of an annulus—a possibility neither even speculated about. A Herman ring and the component of the Fatou set containing a center are known as *rotations domains.*

The limit functions Julia referred to in his quotation are the functions to which a subsequence of $\mathcal{I}$ converges. Although he defined the set E' as the derived set of the set of repelling periodic orbits, it is also the set of points around which $\mathcal{I}$ fails to be normal[9] (and we are not sure when that phrase caught on), as he proved later, and it can be viewed as the set of essential singularities of the limit functions of $\mathcal{I}$. Fatou took the other path: he defined the set E' as the domain of nonnormality, and then showed it was also the closure of the union of the repelling and rationally neutral periodic points. This difference on how one defines the Julia set E' will matter when it comes to extending its definition to transcendental functions in the mid-1920s. We will come back to this in Section 10.4.

In what follows, we will simply say that the iterative family is normal on the Fatou set (or specific points therein) and nonnormal on the Julia set (or specific points within), with the understanding that we mean normal (or nonnormal) on neighborhoods of such points.

The link between the theory of normal families and the Picard theorems that Julia raised in the penultimate quotation above, harkens back to a paper Julia published in 1913, concerning lines of essential singularities [**1913**],[10] which he discussed further down in his introduction:

> *Therefore the set E' is the set of essential singularities for every limit function for the functions extracted from the sequence of*

[9]The terms "domain of normality" and "Fatou set" are synonymous, as are "Julia set" and and "domain of nonnormality". In fact, the precise definition of the Fatou set is the domain of normality for $\mathcal{I}$. Likewise, the Julia set is defined as the domain of nonnormality. As we noted in the opening chapter, our use of the terms "Fatou set" and "Julia set" is anachronistic.

The earliest use of the term "Julia set" we found was when Valiron referred to "functions for which Julia sets (*les ensembles de M. Julia*) do not exist," in the *Comptes Rendus* notice, [**1925c**, p. 1895]. However, the context was not the iteration of functions, but in Julia's application of normal families to the theorems of Picard in [**1919-21**] (see Appendix E for more details), where domains of nonnormality arise. We are not sure if this is the source for the use of the term in the study of iteration. The term "Fatou set" came into prominence after Paul Blanchard's survey [**1984**].

[10]Appendix C analyzes Julia's paper [**1913**].

> *the* $f_n(z)$. *This closely resembles the theorem I proved related to Picard's theorem on the exceptional values*[11] *of the sequence* $f_n(z)$ *around every point of* E'. [**1918a**, p. 49]

But it also beckons the future: the first major publication to follow his monograph on the iteration of rational functions was a three-part work between 1919 and 1921 [**1919-21**], which among other things investigates how an entire or meromorphic function behaves on an angle whose vertex is an essential singularity. The results are often referred to as curves, lines, or sequences of Julia, and Julia's study of them is discussed in Appendix E.

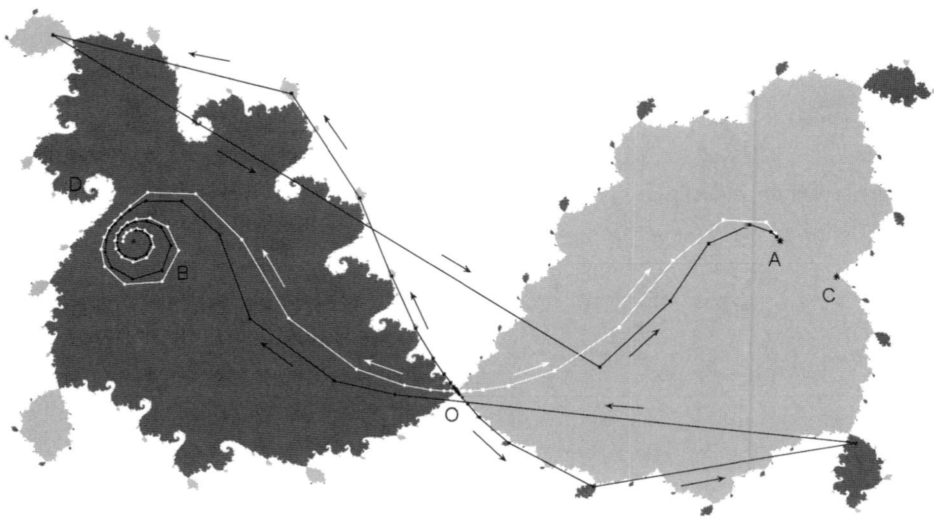

FIGURE 7.1. Iteration near a point in the Julia set. We show four orbits emanating from a neighborhood N of a point 0 in the Julia set of a fifth degree polynomial. The two in white converge directly to the attracting fixed points A and B, while the two others in black travel a more circuitous route. The multiplier of A is 0.5 and that of B is $0.95e^{i\pi/6}$, which accounts for the spiral nature of the orbits converging to B as well as the relative slowness of their convergence. The white region is part of the basin of infinity, hence there are also points in N that converge to the fixed point at infinity. The remaining fixed points, C and D, are repelling, hence in the Julia set.

In addition, the repelling points provide a bit of propulsion to iteration. Except when the Julia set is the entire Riemann sphere, neighborhoods of repelling points contain points which are in components of the Fatou set and are therefore eventually attracted to an attracting orbit[12] or rotation domain.

---

[11]An exceptional value or point of a family of functions $\mathcal{F}$ on an open set D is a point a such that for all $f \in \mathcal{F}$ there are no z in D with $f(z) = a$. If a family is not normal on any neighborhood N of a point z, then it will have at most two exceptional values on all such N. The theorem referred to in the quotation is Theorem 7.6 below.

[12]We use the phrases *attracting orbit*, *attracting cycle*, or *attractor* to refer generically to an attracting fixed point or periodic orbit.

Thus, one can view the iteration of a rational function as pushing points in the Fatou set away from repelling points until they fall into the pull of an attractor, a center, or a Herman ring.[13] This behavior is depicted in Figure 7.1.

For Julia this view of repelling points is a departure from the past where repelling points were viewed locally as attracting points of the inverse of f:

> *But one can find nothing in the literature which refers, for example, to* $z_1 = f(z)$ *when* $|f'(\zeta)| > 1$. *The authors, who have treated this case from the local viewpoint immediately pass to the inverse map* $z = \psi(z_1)$, *for which* $\zeta = \psi(\zeta)$ *and* $|\psi'(\zeta)| < 1$ *which they immediately reduce to the previous case (of* [an attracting] *limit point or limit cycle), but for* $\psi(z)$, not for $f(z)$; *one thus has indications of the* antecedents *of* z [under f] *near* ζ, *but not on its* consequents.[14] [**1918a**, pp. 82–83]

In other words, one of the deficiencies of the local perspective is that the view that repelling points are attracting (or periodic) points of an inverse does not give a true picture of the dynamics of iteration near repelling points. While those who studied repelling points from a local point of view understood points were pushed away under iteration from repelling points, they did not realize the degree to which this occurred. Again, since Julia submitted this work at the end of 1917, he did not realize that Ritt and Lattès shared his view of repelling points.

Julia followed his introduction with a section entitled *Preliminaries*, where he not only delineated his mathematical prerequisites, focusing on recent advances in topology, set theory, and uniformization, but also highlighted the new mathematical elements he would use to move the iteration theory of the Kœnigs school forward.

This section commences with a short description of Kœnigs' researches. Although Julia devoted only one paragraph to Kœnigs' work (perhaps suggestive of his sweeping out the old), it is worth noting that Kœnigs is a presence throughout the work. He cites him from time to time, sometimes in the context of problems he left open,[15] and adopted some of Kœnigs' notation and terminology, such as "circular limit groups" for a set of periodic points. But he ignored large portions of Kœnigs' work since much of it revolved around functional equations. The only person from the 19th century he cited with any regularity was Kœnigs. He referenced Leau's work on the rationally neutral case but did not refer to Grévy, Lémeray, or others from that tradition.

---

[13]Actually, neighborhoods of points from the Julia set have a stronger property: given any attracting cycle S or rotation domain D, in any neighborhood of a point from the Julia set are points which eventually are attracted to S or D. See Julia, [**1918a**, Theorem 7.6]. Julia's quotation is informed with this perspective.

This property can be viewed from another direction: given an attracting or periodic point λ of a function f, the limit set of the set of the preimages of a point arbitrarily close to λ is the Julia set. This property can in fact be extended to arbitrary points in the plane (with at most two exceptions — again, refer to [**1918a**, Theorem 7.6]), a notion Ritt and Lattès both articulated in a slightly different fashion in their respective *Comptes Rendus* notices from early 1918, when they each observed that the limit set of the preimages of an arbitrary point (excluding the exceptional points to which we just referred) contains all repelling points of f.

[14]Fatou and Julia used the words "antecedent" and "consequent" for, respectively, preimage and image.

[15]Including, for example, the determination of the number of attracting limit cycles and the problem of the division of the plane into convergence regions; see §34 of Julia's monograph [**1918a**].

Julia began his discussion of the new mathematics that he would bring to the study of iteration with Montel's theory of normal families, stating a particular interest in families of meromorphic functions, and he gave a brief statement of what he termed "a fundamental theorem", which is also known as Montel's normality criterion:

THEOREM 7.1 (Montel's Normality Criterion). *Meromorphic functions $f(z)$ in a domain D which never take distinct values a, b and c form a normal family in D.* [**1918a**, p. 56].

Julia next reviewed several elementary definitions from set theory and topology, including the concepts of an open set and the interior and exterior of a simple, closed Jordan curve. He defined a set to be "*d'un seul tenant*" (literally, "of one piece", but meaning something analogous to "path connected") if, given a positive ϵ and any two points from the set,

> *...one can find in the set a succession of points beginning and ending with the two points, such that the distance of each to the one following it is less than ϵ.* [**1918a**, p. 58]

The concept of uniformization was central to Julia's investigation. He quoted Poincaré's result from [**1908**] that any simply connected domain is conformally equivalent to an open disc, the entire complex plane (excluding the point at infinity) or the sphere. He also cited Carathéodory's papers [**1912, 1913a, 1913b**] on the conformal representation of any simply connected domain D to the interior of the unit disc $\mathbb{D}$, although he remarked that he found Lindelöf's investigation much simpler [**1918a**, p. 133]. Intertwined in his discussion are several references to the properties of Jordan curves.

One of the more interesting discussions in *Preliminaries* is a result sometimes referred to as Julia's extension of Schwarz's Lemma, which has sparked research interest in its own right (see Appendices F, G, H).

After considering the various Riemann surfaces created by the image of a mapping f on a simply connected region D, which without loss of generality can be taken to be the open unit disc $\mathbb{D}$, Julia proved Schwarz's Lemma, which we state in Julia's own words:[16]

THEOREM 7.2 (Schwarz's Lemma). *If $f(z)$ is holomorphic for $|z| < 1$ and if, throughout the domain $|z| < 1$, one has*

$$|f(z)| \leq 1 \quad \text{with} \quad f(0) = 0,$$

then for all points z, $0 < |z| < 1$, the inequality

$$|f(z)| < |z|$$

holds except in the case where $f(z)$ is the linear function $ze^{i\theta}$, in which case

$$|f(z)| = |z|.$$

[16]We give an English version of Julia's translation of Schwarz's statement of the lemma from Volume II of Schwarz's collected works [**1890**, pp. 109–111]. According to Reinhold Remmert [**1990**, pp. 272–273] a version of the lemma was included in Schwarz's formalized notes from a course he gave in 1869–1870 at the Eidgenössische Polytechnische Schule in Zürich. He used this early version of the lemma in a proof of the Riemann mapping theorem. Evidently, Schwarz's Lemma attracted little attention until Carathéodory realized its importance in 1912 and suggested it be called "Schwarz's Lemma" [**1912**, p. 110]. See Boas [**2010**] for an extended treatment of Schwarz's Lemma, including a discussion of Julia's contributions.

[**1918a**, p. 67]

Julia was particularly interested in the case where $|f(z)| < |z|$, which occurs when 0 is an attractive fixed point. He also stated a corollary:

COROLLARY 7.3. *Let $|f(z)| \leq 1$ for $z \in \mathbb{D}$. Suppose also that there is a nonzero λ in $\mathbb{D}$ with $f(\lambda) = \lambda$. Then*

$$\left|\frac{f(z)-\lambda}{f(z)-\lambda'}\right| \leq \left|\frac{z-\lambda}{z-\lambda'}\right|,$$

where λ' is symmetric to λ with respect to the unit circle. Equality occurs if and only if

$$\frac{f(z)-\lambda}{f(z)-\lambda'} = e^{i\theta}\frac{z-\lambda}{s-\lambda'}.$$

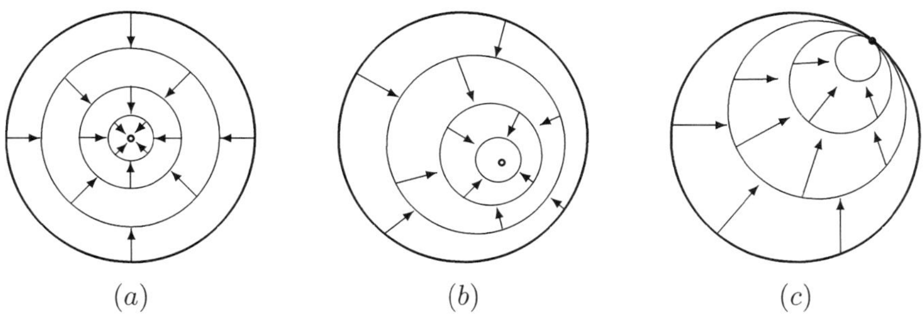

FIGURE 7.2. Schwarz's Lemma (a), the corollary (b), and Julia's extension (c).

Julia followed this with what he called a "New extension of the Schwarz's Lemma" [**1918a**, p. 72]. His statement of the extension, which we provide in translation, was discursive rather than mathematically concise. He mapped $\mathbb{D}$ to the upper half-plane $\mathbb{H}$, assumed that f is invariant on H, and then gave his statement of the theorem:

THEOREM 7.4 (Julia's Extension of Schwarz's Lemma). *Let Γ be a circle tangent to a point τ on the x-axis (denoted "Ox", which remains invariant under f), let $\tau_1 = f(\tau)$, and for a given z let $z_1 = f(z)$. Then*

> ... *the domain Δ, described by z_1 when z describes the interior of the circumference of a circle Γ tangent to Ox at τ, of any radius R whatsoever, is* interior *to the circle Γ_1 of radius $Rf'(0)$ tangent to Ox at τ_1, having no point in common with the circumference of Γ_1 other than τ_1 itself.* [**1918a**, p. 76]

To indicate how this would be useful, he offered the following "application of this proposition to the question of iteration" [**1918a**, pp. 75–76].

EXAMPLE 7.5. Let D be an open disc, let $C = \partial D$, and let f be analytic on $D \cup C$ with $f(D) \subseteq D$ and $f(C) = C$. Suppose as well that f has a fixed point λ on C with $0 < |f'(\lambda)| \leq 1$. Let Γ be a circle tangent to C at λ but otherwise interior to D. Finally, let $z \in \text{Int}(\Gamma)$. The sequence $\{f^n(z)\}$ converges to λ.

He observed that this application "demonstrates quite simply the results announced by Fatou" in his May 1917 note, [**1917a**] for functions with a fundamental circle [**1918a**, pp. 150–151]. Two years later, Julia devoted a short paper to his extension of Schwarz's Lemma [**1920**].

7.2.2. The set of repelling points: *Part One*. Julia discussed the notation and elementary properties involved in the iteration of a rational function of degree $d > 1$[17] in the initial section of *Part One*, after which he investigated the properties of the Julia set by first establishing the existence of a set which is a precursor to the Julia set, $E = \{z : f^n(z) = z,\ n \in \mathbb{N}, \text{with } |d/dz\ f^n(z)| > 1\}$, that we discussed above. He asserted that if the degree d of a rational function f is greater than one, then the set E is never empty. He based his claim on a formula sometimes referred to as the rational fixed point formula,[18]

$$\sum_{k=1}^{d+1} \frac{1}{f'(\lambda_k) - 1} = -1, \tag{7.1}$$

where the $d+1$ fixed points λ_k of f are assumed finite and distinct. One can show, as Julia and Fatou both did[19] that under these conditions, f must have at least one fixed point whose multiplier is strictly greater than one, hence the set E is nonempty.

However, there are rational functions of degree $d > 1$ where the largest multiplier of a fixed point is one, a fact which escaped Julia's attention when he originally wrote his monograph, so his argument fails to be completely general. Julia realized his error after his submission to the competition but not in time to revise his entry before its publication, so he attached an addendum to his monograph [**1918a**, §118], correcting his error and patching the holes that resulted. Since he did not rely on the set E after his first chapter, the only items that warrant discussion are theorems in *Chapter One* where E is assumed nonempty on the basis of equation (7.1). But the nonemptiness of E still holds because, as Julia observed in his addendum, even if the largest multiplier of a fixed point is one, there is at least one periodic point whose multiplier is greater than one. Moreover, he showed in *Part Four* that if a fixed point has multiplier 1, the iterative family is not normal around the fixed point.

Perhaps the most important theorem from *Part One* is the following, which Julia referred to as the "fundamental theorem" about E [**1918a**, pp. 88–93]:

THEOREM 7.6 (Fundamental Theorem). *Let f be a rational function. Let D_0 be a neighborhood of a point x from E, and define $D_n = f^n(D_0)$.*

There are only three possibilities for the $\{D_n\}$.

(1) There are exactly two distinct points—called exceptional values—in the extended plane $\overline{\mathbb{C}}$ that remain exterior to the union of all the D_n. We can map them

---

[17] In what follows d will always be assumed to be greater than one. Neither Fatou nor Julia treated the case where $d = 1$ since the dynamics were not only well known through the study of Kleinian groups, but are quite different from the higher degree case.

[18] We will examine this formula again during our discussion of Fatou's monograph Section 7.3. Julia expressed the formula in a slightly different but equivalent form.

[19] See Audin [**2009**, p. 92] for an argument similar to Julia's. Fatou did not provide his justification but likely proved it differently than Julia, as is discussed following equation (7.2) below.

to a and to ∞, respectively, by a suitable linear conjugation which reduces $z_1 = f(z)$ to one of these two forms:

$$Z_1 - a = (Z - a)^k \qquad \text{or} \qquad Z_1 - a = \frac{1}{(Z-a)^k}.$$

For a sufficiently large n, any bounded region of $\mathbb{C}$ which does not include a is interior to D_n and all subsequent $f^k(D_n)$.

(2) Only one point in the extended plane is exterior to all the D_n. If this point is mapped to infinity, f reduces to a polynomial (this conclusion is also true if f is a transcendental entire map).[20] *Under this hypothesis, any bounded region in the plane is, for a sufficiently large n, inside D_n.*

(3) Every point of the plane is inside the D_n and, since $\overline{\mathbb{C}}$ is a closed set, there will be an index n such that D_n contains $\overline{\mathbb{C}}$; the same will hold for all iterates of D_n. This is the general case for an arbitrary rational map f.

It is here that Julia first applies Montel's theory of normal families. He began his proof by assuming that the iterative family has three exceptional values on the union of the D_n, which by Montel's normality criterion (Theorem 7.1 above) is tantamount to the normality of the D_n, and he then showed that the iterative family is not normal on the D_n, hence can have at most two exceptional values.

His fundamental theorem, coupled with the converse of Montel's normality criterion, provided him with an alternative definition of the Julia set E' [**1918a**, p. 97]:

THEOREM 7.7. *A necessary and sufficient condition for a point $z \in \overline{\mathbb{C}}$ to be in E' is that the iterative family have at most two exceptional values on neighborhoods of z.*

Since this implies that E' is the domain of nonnormality, it is equivalent to the standard definition of the Julia set as the domain of nonnormality.

Julia's investigation of E' included the verification of the possibilities for E' that he listed in his *Introduction* (see page 134). In so doing, he thoroughly delineated its properties, spatial distribution, and structure. Of particular importance is its homogeneous nature which provides a topological coherence for E since a homogeneous set cannot be discontinuous in one region and continuous in another.

He ended *Part One* by describing the role of E' in the study of iteration, noting that when E' is continuous, it forms the boundary between such regions, and when it is totally disconnected, all points in $\overline{\mathbb{C}}$ complement E' converge under iteration to a unique attracting fixed point of f.

We close our discussion of *Part One* with an image illustrating two possibilities for E'; see Color Plate 4.

Caption for Color Plate 4: (See p. 64.) This image displays examples of Julia sets for the function $f(z) = z^2 + c$ at $c = -0.14035087 + 0.716981132\,i$ (left) and $c = 0.37596491 + 0.208679245\,i$ (right) at about 50 iterations. The Julia set at left is connected and is the boundary of the dark region.

On the right, we depict the formation of a disconnected Julia set similar to what occurs with Color Plate 2. As the index of

[20] Julia made occasional reference to transcendental maps in his monograph, but never devoted a stand-alone paper to them.

iteration increases, the domains ringed with green and red bands will shrink down towards single points. These formations will not only increase without bound, but they will become arbitrarily close to one another in the process. Thus, the Julia set will be a totally disconnected, perfect set.

7.2.3. Domains of attraction: *Part Two.*

The second chapter of Julia's entry is devoted to the study of attracting limit cycles and includes a fundamental result highlighting the important role of the postcritical set[21] [**1918a**, pp. 129–130]:

THEOREM 7.8. *The immediate basin D of an attracting fixed point ζ always includes at least one critical point of a branch of an inverse of f, in particular, the branch which fixes λ.*

Julia's proof proceeds by assuming that the derivative of the attracting fixed point λ is nonzero (otherwise, the fixed point would be a critical point of the inverse) and that there are no critical values in the immediate domain of attraction of λ. Since D is simply connected (something he showed earlier [**1918a**, p. 128]), it follows from the *Preliminaries* that it is conformally equivalent to $\mathbb{D}$, hence there is a one-to-one, onto map S from D to $\mathbb{D}$, with $S'(\lambda) = 1$.[22]

Julia then showed that $Z = S \circ f \circ S^{-1}$ is a map of $\mathbb{D}$ into $\mathbb{D}$ fixing the origin, hence, quoting a result from Poincaré's 1884 paper on groups of linear equations, [**1884**], it is a rotation of $\mathbb{D}$. In particular Z has a derivative of 1 at the origin. A simple application of the chain rule implies that $f'(\lambda) = 1$. But this contradicts the hypothesis that λ is an attracting fixed point.

One of the most important applications of Theorem 7.8 is its role in determining an upper bound for the number of attracting orbits of a rational function [**1918a**, § 34]). In 1884 Kœnigs wondered whether there was a limit on the number of attracting orbits [**1884b**, p. 40]. Directly quoting Kœnigs' remarks on the division of the plane problem (which we included in Chapter 1, page 21) Julia asserted that the number of attracting orbits was at most $2d - 2$, where d is the degree of the function [**1918a**, p. 137].

Julia then leaves a tantalizing hint of things to come before pausing to consider several examples:

> *We will discuss an application of Newton's method below; when the equation we are applying it to surpasses degree two, something quite novel and interesting arises.* [**1918a**, p. 138]

The examples Julia then considered [**1918a**, §35] involve relatively simple cases where there are either one or two attracting fixed points. In each case the domain of attraction for a fixed point consists of a single component.

Over the next several sections he turned his attention to more complicated circumstances and showed the following:

(1) If the immediate domain of attraction of a fixed point is also the complete domain of attraction (recall that such a domain is called a *completely*

[21] Recall from Definition 1.11 that a critical value of f (that is, a point $w = f(z)$, where $z \in D$ and $f'(z) = 0$) is sometimes called a critical point of the inverse. The *postcritical set* is the union of the forward orbits of the critical values; see Definition 1.12. Fatou denoted this set E_c and its derived set as E'_c.

[22] Solutions to the Schröder equation are generally normalized so the derivative at the fixed point is one.

invariant domain of attraction) and is simply connected, it contains $d-1$ critical values [**1918a**, §37].

(2) If f has more than two attracting fixed points, at most one them has an invariant domain of attraction, hence there must be a domain of attraction with more than one connected component [**1918a**, §38]. He showed it is possible that a domain of attraction can have infinitely many components: "This is completely new, and it seems to have escaped everyone who occupied himself with the study of iteration until now" [**1918a**, p. 148].

(3) If the immediate domain of attraction for a fixed point λ is not the total domain of attraction, the complete domain of attraction has infinitely many components. Therefore, if there are more than two attracting fixed points, at least two of them will have infinitely many connected components [**1918a**, §40–44]: "The fact that I announced on page 148 [see the previous point] is thus general" [**1918a**, p. 155].

He concluded the chapter with a detailed series of examples. As he hinted above, there would be a surprise awaiting the reader during his discussion of Newton's method. Of course, by the time Julia got there, the reader had probably guessed that the surprise would be that the convergence regions for Newton's method consist of infinitely many components.

Keep in mind, however, that when Julia was writing this in 1917, he was solving a long-standing problem. Both Arthur Cayley, most notably in [**1879a**], and Ernst Schröder [**1870, 1871**] sought to understand why Newton's method converged to roots of polynomials. Each of them completely explained the convergence in the degree two case and showed the domain of attraction for a quadratic was completely invariant (see Section 1.4.3). However, neither made any progress when the degree of the polynomial was greater than two. Julia's investigation explained why: the Newton's method function for the cubic has three attracting fixed points; hence, the domains of attraction for at least two of them consist of infinitely many components. Commenting about the convergence of Newton's method for $z^3 - a^3 = 0$, Julia observed,

> *Thus, each of the domains of convergence for* a, $a\omega$, *or* $a\omega^2$ [$\omega^3 = 1$] *is composed of* an infinitely of pieces .... *The total domain of each yields the other two through two successive rotations of* 120° *around the origin* ....
>
> *The linear continuity of* E' *which forms the boundary set of the preceding total domains (each* total *domain evidently has for its boundary the complete set* E', *thus any point of* E' *is a limit of the successive antecedents of an arbitrary point in each of the domains) conserves itself under a rotation of* 120° *around the origin. By a simple homographic transformation* ($z = aZ$), *it reduces to the case where* $a^3 = 1$. [**1918a**, p. 181]

The convergence properties of Newton's method for $z^3-1=0$ are depicted in Color Plate 5.

> **Caption for Color Plate 5:** (See p. 64.) Newton's method for the cubic. We show the distribution of the domains of attraction over the Riemann sphere for
>
> $$N_3(z) = \frac{2z^3+1}{3z^2},$$

the Newton's method function for the cubic $f(z) = z^3 - 1$. The blue region depicts points that converge to $z = 1$ under N_3. Points colored green converge to $z = e^{i2\pi/3}$, and those in red to $z = e^{i4\pi/3}$. On the left the unit disc is pictured. In the middle, we see the region $|z| \geq 1$, and on the right is a view of the Riemann sphere from the point at infinity down into the unit circle.

Another example involved a schematic of the Julia set for the function $z \mapsto (-z^3 + 3z)/2$ whose Julia set E' is a Jordan curve that crosses itself infinitely often. He provides a fascinating description of the Julia set:

> *It is rather difficult to give an exact representation of what this continuous E' is like. But it is not impossible to give an idea of it with the aid of a constructive process that I will now present that shows by example it is not impossible that the simply connected domain $\mathcal{R}_\infty$* [the domain of attraction for the attracting fixed point at ∞] *is bounded by a linear continuity E' whose multiple* [crossing] *points are everywhere dense on it, a continuity which divides the plane into an infinity of regions, each of which touches $\mathcal{R}_\infty$; the frontier of each tiny region is, moreover, a simple curve.* [**1918a**, pp. 169–170]

His aid is a schematic consisting of a nexus of infinitely many equilateral triangles, each one representing a component of the Fatou set, based on the work of Helgen von Koch, [**1906**]. Julia's diagram can be found in [**1918a**, p. 170].[23] Fatou also invoked von Koch's work, and we will discuss von Koch in more detail below.

> **Caption for Color Plate 8:** (See p. 66.) The domains of attraction for the superattracting fixed points $A = 1$ and $B = -1$ of $z \mapsto (-z^3 + 3z)/2$ are depicted. Components of the Fatou set are colored according to whether points within converge under iteration to A or B. The blue region is the domain of convergence for the point at infinity.

7.2.4. Julia Sets: *Part Three*. The focus of this chapter is on the nature of continuous Julia sets, and its methodology continues that of the end of *Part Two*, where Julia used examples to induce general properties.

> *With the example $z_1 = z \mapsto (-z^3 + 3z)/2$ we have already dealt with a continuity that is very complicated. One can show, however, that in very general circumstances the complexity of these continua can be assumed to be a conglomeration of linear continua whose nature is a much simpler, for example, Jordan arcs,* [i.e.,] *closed continuous curves without crossing points. I did not always succeed in delineating the characteristic properties of all continua which present themselves in the iteration of rational maps; I have, nevertheless, under very general hypotheses, succeeded in showing that they could be reduced to Jordan arcs.* [**1918a**, p. 187]

[23]There are reproductions of Julia's diagram in Audin [**2009**, p. 72] and Alexander [**1994a**, p. 130].

Although he did not establish many general theorems in this section, he suggested that his findings can be applied in much wider situations than the examples he considered.

> *I commence with a simple example which will make the method that I have followed understandable. Naturally, I always consider rational fractions where I know a priori that there are at least two distinct limit points where the convergence is regular; I will be thus assured that the set E' comprises a linear continuity that divides the plane into regions, and it is the nature of this continuum that is the goal of my research.* [**1918a**, p. 187]

The first example he gave, $f(z) = z^2 + z/2$, was considered by Fatou in his *Comptes Rendus* notice [**1906b**] (see Section 2.2 for our discussion of Fatou's analysis of f). The second was a bit more abstract: It could be "any fraction whatsoever of degree two" [**1918a**, p. 213], which, like f, contains two attracting fixed points, λ_1 and λ_2 with invariant domains of attractions, D_1 and D_2 separated by a closed Jordan curve.

Julia deduced the following result via a deliberate and careful examination of examples, which, unlike Julia we summarize as a theorem [**1918a**, pp. 215–216].

THEOREM 7.9. *Let f be a rational function of degree d with two attracting fixed points, λ_1 and λ_2. Suppose as well the following:*

(1) *There is a simple closed curve C with $\lambda_1 \in \mathrm{Int}(C)$. The set $\{f^{-1}(z) : z \in \mathrm{Int}(C)\}$, where f^{-1} is the total inverse (see Definition 1.8), defines a region bounded by a simple closed curve C_{-1}.*
(2) *Suppose there are analogous simple closed curves Γ and Γ_{-1} for λ_2 exterior to C.*
(3) *Finally, suppose that on the doubly connected region R between Γ and C (or between some antecedent of each) there is an index n with $|d/dz f^n(z)| > M > 1$ for all $z \in R$.*

The sphere is then partitioned into two simply connected invariant regions D_1 and D_2 separated by a Julia set which is a simple closed Jordan curve obtained by taking the limit of the preimages of C (or Γ) under f of Γ or C. The regions D_k are the respective domains of attraction for the λ_k.

As Julia pointed out [**1918a**, p. 216], this theorem explains a result involving fundamental circles from Fatou's May 1917 *Comptes Rendus* notice [**1917a**]. We will come back to this point during our discussion of Fatou's investigation of maps with a fundamental circle.

One of the more intriguing investigations in *Part Three* involved the concept of a "general" rational function $\Phi(z)$, obtained from a known function f by varying its coefficients. The coefficients of a general function "do not satisfy any particular relation" [**1918a**, p. 219]. A function with real coefficients would be an example of a function whose coefficients satisfy a particular relation. However, he noted that small variations in the coefficients can alter these relations:

> ... *for example, each coefficient of $\Phi(z)$ will be assumed to be in the interior of a small circle whose center is the fixed value of* a particular coefficient from $f(z)$. ... *One can see that as these parameters each vary within the small circles of their domain,* it is

> always possible to chose them in such a fashion that none of the preceding relations are satisfied. [**1918a**, p. 219]

Near the beginning of *Part Three*, during a lengthy investigation spurred by Fatou's function f [**1918a**, p. 188], Julia pointed out that, as is the case with f, if the argument of the derivative of an inverse of f^n at a point z in E' satisfying $f^n(z) = z$ is incommensurate with π then the curve E' does not have a tangent at that point. Periodic points are dense in E' so if the derivatives of "enough" of them have arguments incommensurate with π, then E' will be without tangents [**1918a**, p. 202].

After defining the concept of a general function, he asserted that the non-existence of tangents at points in the Julia set that he discussed earlier in the chapter would extend to a general rational function:

> *Because it is said to be general, we can require of* $\Phi(z)$ *that none of the roots of* $z = \Phi'_n(z)$ *for* $n = 0, 1, 2, \ldots, \infty$ *have an argument commensurate with* 2π*, which would mean that none of the parameters would satisfy any relation of equality among a denumerably infinite group of relations* $(\mathcal{R})$*. In any event, repeating an argument heretofore made, one sees that for a general fraction* $\Phi(z)$*, at no point* P *from* E *can the curve* E' *have a determined tangent.* [**1918a**, pp. 219–220]

Julia also pointed out what seemed to him to be an analogy between points in E' where the tangent does not exist to the limit set of a Kleinian group [**1918a**, §87–89]). He alluded to this relationship already at the end of §18 in *Part One*, as well as in the *Introduction* of this work, where he remarked that the analogy he perceived between automorphic groups and the iteration of rational maps convinced him of the existence of rational maps whose Julia set covers the sphere, even though he could not find one (see page 134).

7.2.5. Flowers: *Part Four*. The final chapter of Julia's monograph on rational functions is devoted to the study of neutral fixed points, near which iterates of f are both attracted and repelled. Leau summarized this behavior in his dissertation [**1897**] with an early version what has come to be known as the Flower Theorem. The year before Ernest Lémeray, another French mathematician, described an inchoate version of the Flower Theorem in *Comptes Rendus* [**1896**], and the following year he gave a fuller account, which included a schematic diagram illustrating the Flower Theorem [**1897a**, p. 313]. Leau's conception was more robust, but Lémeray's work was also well known and was cited by Leau.[24]

Julia discussed both the irrational and rational cases, and we will defer our discussion of the former until Section 9.2. His investigation of rationally indifferent fixed points used Leau's as a point of departure, but introduced the role of normal families.

For example, he showed that rationally neutral fixed points belong to the Julia set. He assumed for the sake of convenience that the origin is a fixed point whose multiplier equals one, hence is in the form $f(z) = z + a_{n+1}z^{n+1} + \cdots$, and it follows that $f^k(z) = z + ka_{n+1}z^{n+1} + \cdots$. If the origin is not in the Julia set, then there is a neighborhood N of the origin such that a subsequence f^{k_i} of the iterative family

[24] For more details about the history of the Flower Theorem, see Alexander [**1994a**, Chap 5, 6] and Audin [**2009**, p. 93].

$\mathcal{I}$ converges uniformly on compact subsets of N to a limit function Γ satisfying $\Gamma(0) = 0$ and $\Gamma'(0) = 1$. But since the $(n+1)$st degree coefficient of f^{k_i} goes to infinity with k_i, no infinite sequence f^{k_i} can converge to a finite limit function such as Γ. The case when the multiplier is a qth root of unity proceeds similarly, and uses the fact that f^q has a multiplier of one.

The Flower Theorem gives a rich visualization to iteration in the rationally neutral case, and its name derives from the fact that the domains of convergence take on the shape of a flower petal. Julia in fact compared these configurations to a loop ("*boucle*") of the rosette of $r^n = a^n \cos(n\theta)$ [**1918a**, p. 300]; Fatou on the other hand, characterized the appearance of the configuration by the intersection of the adjoining petals, which he saw as star shaped ("*étoile*") [**1919-1920**, p. 209, Chapter II].

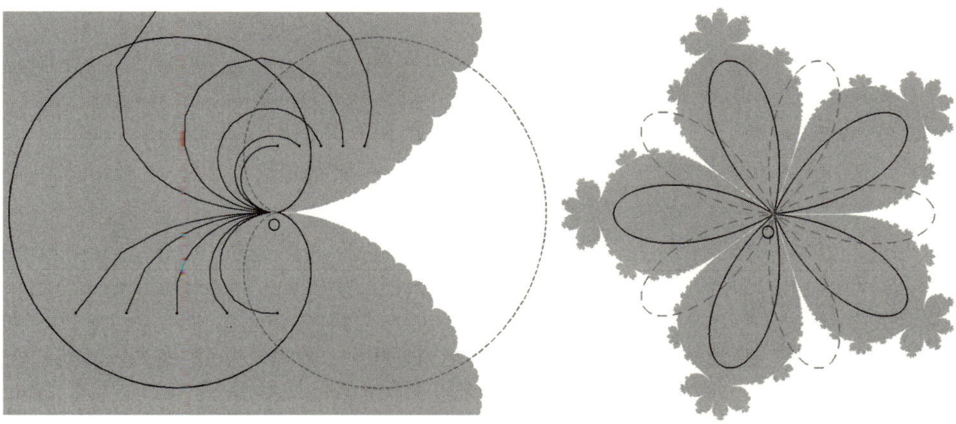

FIGURE 7.3. Depictions of Fatou-Julia-Leau flowers. On the left, $f(z) = z + z^2$; on the right, $f(z) = z + z^6$. The grey regions in each represent the domains of convergence for the neutral fixed point O; the white region is the basin of infinity. The attracting petals (see Theorem 7.9) are bordered by solid lines, and the repelling petals by the broken lines. We depict ten orbits that converge to the origin under iteration in the image corresponding to $f(z) = z + z^2$.

We will state a contemporary version below, adapted from Milnor [**2006a**, p. 112], which represents a distillation of Julia's ideas—Julia, in fact, did not present a concise statement of the Flower Theorem. His analysis was focused almost entirely on the case where $f'(\lambda) = 1$, and while he gave several examples of functions with rationally neutral fixed points of multiplier one, he gave no explicit analysis of the case where $f'(\lambda)$ is a primitive qth root of unity, noting that one only need apply his analysis to f^q [**1918a**, p. 297]. While this is true, Julia's remarks did not give the full richness of the root of unity case. Before stating the Flower Theorem, a definition is needed:

DEFINITION 7.10 (Multiplicity of a fixed point). Let λ be a fixed point of a function whose Taylor expansion around λ is $f(z) = \lambda + a_1(z - \lambda) +$

$a_{n+1}(z-\lambda)^{n+1}+\cdots$. The *multiplicity* or *order* of a fixed point λ is the multiplicity of λ viewed as a root of $f(z)=z$. Since $f(z)=z$ reduces to

$$0=(\lambda-z)+a_1(z-\lambda)+a_{n+1}(z-\lambda)^{n+1}+\cdots$$

near λ, its multiplicity is greater than one only when the multiplier $f'(\lambda)=a_1$ is one, in which case the multiplicity equals $n+1$.

THEOREM 7.11 (Flower Theorem). *Let λ be a fixed point of f of multiplicity $n+1>1$. Within a neighborhood of λ, there are are $2n$ simply connected domains, P_k, such that when k is odd, points in P_k converge to λ under iteration by f; P_k is referred to as an* attracting petal. *Conversely, when k is even, the orbit of any point in P_k eventually leaves it, and P_k is a* repelling petal. *The P_k can be chosen such that,*

(1) *for k odd, $f(P_k)\subset P_k$;*
(2) *the union of the P_k with λ forms a neighborhood of λ; and*
(3) *when $n>1$, each P_k intersects its immediate neighbors $P_{k\pm 1 \bmod 2n}$ but is disjoint from the other P_k. When $n=1$, $P_1\cap P_2$ is nonempty.*

If λ is a rationally neutral fixed point whose multiplier is a qth root of unity, then λ is a fixed point of f^q, and its order is a multiple of $2q$.

Referring to the depiction of the iteration of $f(z)=z+z^2$ in Figure 7.3, the region bounded by cardioid on the left is the *attracting petal* P_1, while the cardioid on the right is the *repelling petal* P_2. We show several orbits which converge to the fixed point O, which is at the shared cusp of the cardioids. Within the attracting petal, the Abel equation $A\circ f=A+1$ is satisfied, and iteration is equivalent to translation.

Note that orbits originating in the gray region of the repelling petal, in particular those in $P_1\cap P_2$, converge under iteration to the origin. Those in the white region of P_2 converge to the point at infinity, so the orbit of any point in the repelling petal eventually leaves. That some points surrounding the origin converge under iteration to the origin while others converge to infinity, provides an intuitive reason why rationally neutral fixed points are in the Julia set of f.

Again mapping the fixed point to the origin, if $n+1$ is the degree of the first nonzero higher order term of $f(z)=z+a_{n+1}z^{n+1}+\cdots$, the flower will consist of $2n$ overlapping petals, as indicated in the right image of Figure 7.3, where the multiplicity of the fixed point for $f(z)=z+z^6$ is 6. However, when the multiplier is a primitive qth root of unity, the number of petals is a multiple of $2q$.[25]

7.3. Fatou: *Sur les équations fonctionnelles*

Fatou's monograph came in three installments, the first one published in 1919, the other two in 1920, and it is clearly a unified work [**1919-1920**]. If Julia's monograph is structured as a textbook, Fatou's is a akin to a novel, one that sprawls—in a very positive sense—and at times revisits events. The effect is to draw the reader more deeply into a mathematical idea with each passing. An example of this is his treatment of the center problem, which we discuss in detail beginning

[25]The reader curious to know more of the mathematics of the Flower Theorem is invited to consult Milnor [**2006a**] or one of the other standard complex dynamics references listed in footnote 1 of Chapter 1.

in section 9.4. Rather than wrapping it into a series of consecutive sections within a larger chapter, Fatou considers it at three different places, §14, §28–31, and §54.

7.3.1. Why functional equations? Fatou's *Introduction*. We have noted how Fatou, unlike Julia, positioned his study of iteration squarely within the French tradition that connects iteration to the study of functional equations. This is evident from the outset of [**1919-1920**] which begins with a brief summary of the French study of iteration from Kœnigs' initial studies involving functional equations, to the work of his students Grévy and Leau, and, finally to the "very difficult" extensions to two or three variables by Picard, Poincaré,[26] Lattès, and others.

In Fatou's opinion the local nature of these studies limited them. While they demonstrated the existence of solutions to functional equations useful in the study of iteration near fixed points, they were unable find the full domains of solutions (see the quotations from Kœnigs and Leau in Section 1.8), and because of this, could not investigate the nature of the singularities of these solutions. Fatou asserted that by exploiting the link between the domains of convergence for functional equations and the convergence of the iterates of the given function f, he had overcome these difficulties, at least for the one variable case. As he noted, he restricted his monograph to the study of rational functions of a single complex variable.

In contrast to Julia's monograph, which, aside from the center problem, viewed the study of functional equations as somewhat distinct from the study of iteration, functional equations were not an add-on to the study of iteration for Fatou, but rather an integral part of it. The implication is that by divining the properties of the iterates of f, one simultaneously discovers properties of the solutions to these related functional equations.

Fatou's *Introduction* also depicts an evolution of the global theory of iteration, from his own notice [**1906b**] to the great leaps forward that came with the application of powerful theorems from analysis: the Picard theorems, Edmund Landau and Friedrich Schottky's extensions of Picard theory, and especially Montel's theory of normal families.

Before giving a chapter-by-chapter preview of his monograph, Fatou noted that he would restrict his study to rational functions of one complex variable, and then tipped his hat to the work of others who independently obtained similar results: Lattès, Ritt, and "above all ... to the remarkable researches" of Julia [**1919-1920**, p. 164, Intro.]

7.3.2. The rational fixed point theorem: *Chapter I*. In the first three chapters of his monograph, Fatou approached his investigation from what he termed a "purely algebraic point of view" [**1919-1920**, p. 164, Intro.] Although most of the algebra from this chapter is indeed elementary, he did invoke the theory of groups, forming the group G of the n-fold compositions of iterates of f and f^{-1}. Two points z and z' are equivalent under the action of this group if there exist $n, n' > 0$ with $f^n(z) = f^{n'}(z')$. He observed,

[26] Fatou explicitly referred to Poincaré's *Sur les courbes définies par une équation différentielle*. See page 34 for details.

> *We see that this group can be properly or improperly discontinuous;*[27] *its study is intimately linked to all the questions examined in this memoir.* [**1919-1920**, p. 166, I].

However, his explicit references to the theory of groups were rather infrequent.

An example of the algebra that typifies the mathematics in this chapter is his development of what he termed a "fundamental" relation among the multipliers of the fixed points. He began with the assumption that if f is a rational function of degree d with $d+1$ distinct finite fixed points, $\lambda_k, k = 1, \dots, d+1$ with multiplier s_k.[28] He showed that under these conditions,

$$\sum_{k=1}^{d+1} \frac{1}{s_k - 1} + 1 = 0, \tag{7.2}$$

and that if there are fixed points whose multiplier equals one, then

$$\sum_{k=1}^{m} \frac{1}{s_k - 1} + \sum_{k=m+1}^{n} a_k + 1 = 0. \tag{7.3}$$

Since rationally neutral fixed points have multiplicity greater than one, $n < d+1$. Both formulae can be readily obtained by calculating the residues of $1/(f(z) - z)$ around the fixed points, and the a_i are the residues around the neutral fixed points. Fatou, however, used neither the word nor the concept, and the argument he sketched relied on the partial fraction decomposition of $1/(f(z) - z)$. It is possible he used residues to determine the coefficients of the decomposition.[29]

As we noted above, Julia also used formula (7.2), which is sometimes called the rational fixed point formula, and his argument was very similar to Fatou's. However, Julia only presented the version where the fixed points were distinct, which may have led to his erroneous assumption that any rational function of degree greater than one had a repelling fixed point.

Invoking a notion similar to Julia's concept of a "general" rational function, described above on page 145, Fatou remarked,

> *If one views the coefficients of the fraction $f(z)$ of degree d as independent variables, the s_k are algebraic functions of the coefficients that always satisfy the fundamental relation* [i.e., equation (7.2)]. *If these coefficients tend to numerical values such that one of the s_k becomes equal to $+1$, and consequently $\frac{1}{s-1}$ becomes infinite,*

---

[27]According to Forsyth, who wrote a two volume work on the theory of functions (we used the third edition as a source [**1918**], roughly contemporaneous with Fatou's monograph), a group, or in more contemporary terms, its action, is discontinuous on a set D if G contains no infinitesimal transformations, that is, there is no element $g \neq I$ of G such that $g(z)$ is infinitesimally close to z for all $z \in D$. If there are points in D where an infinitesimal transformation exists, a discontinuous group is *improperly discontinuous* at that point, a term that is generally out of use nowadays; otherwise, its action is *properly discontinuous* [**1918**, pp. 717–719]. These seem to be the meanings that Fatou intends. Around the fixed point there are infinitesimal transformations, for example, f^n for $n > 0$.

[28]That there are $d+1$ fixed points counted with multiplicity is easily seen by solving for the fixed points of $f = p/q$, where deg $f = d$, by finding the roots of $p(z)/q(z) = z$. It follows from Definition 7.10 above that a fixed point has a multiplicity strictly greater than one if and only if its multiplier equals one. If the point at infinity is a fixed point, as is the case if deg $p >$ deg q, the coordinate change $z \mapsto 1/z$ maps it to zero.

[29]See Milnor's book [**2006a**, §12] for details on the mathematics of this formula.

> *there will be, by virtue of the fundamental relationship, another multiplier which tends towards* $+1$. [**1919-1920**, p. 169, I]

It is for this reason that Fatou developed (7.3).

Fatou teased out quite a bit of information about the multipliers of the fixed points from these two summations. For example, he showed that if the $d+1$ fixed points are distinct, then one of them must be repelling. To see this, express an arbitrary multiplier as $s_k = u + iv$. It follows that

$$\frac{1}{s_k - 1} = \frac{u-1}{u^2+v^2-2u+1} + i\frac{-v}{(u-1)^2+v^2} = \alpha + i\beta.$$

Fatou asserted, without demonstration, that $|s_k|$ is one, less than one or greater than one if α, respectively, equals, is less than or is greater than $-\frac{1}{2}$.

This is easily seen. Suppose first that $\alpha > -\frac{1}{2}$. Then

$$\begin{aligned} \frac{u-1}{u^2+v^2-2u+1} &> -\frac{1}{2} \\ \Rightarrow 2u-2 &< -(u^2+v^2)+2u-1 \\ \Rightarrow 1 &< (u^2+v^2). \end{aligned}$$

Since $|s_k| = \sqrt{u^2+v^2}$ it follows that $|s_k| > 1$. Although there are other ways to obtain this result, given Fatou's statement about the algebraic nature of this chapter, his proclivity towards direct computation and the fact that he took the time to break $\frac{1}{s_k-1}$ into its real and imaginary parts, it is fairly likely that he had a similar argument in mind.

That there is an s_k for which the corresponding α is greater than $-\frac{1}{2}$ follows from equation (7.2). The sum of the real parts of the $\frac{1}{s_k}$ equals -1. Since the degree of f is at least two, there are at least three fixed points, so it is impossible that the real parts of each $\frac{1}{s_k}$ is less than $-\frac{1}{2}$. From the computation directly above, it follows that a fixed point with multiplier greater than one exists.

After generalizing (7.3) to periodic points of f of order greater than one, he showed that there exist infinitely many periodic orbits of period 2 or more whose multipliers are at least one in modulus.

Fatou admitted that he could have used normal families to show the existence of infinitely many repelling orbits in a more precise fashion (and would later on), but seemed to have a fondness for equations (7.2) and (7.3):

> *...the elementary manner in which they were obtained merits attention, and the method may well lead to other results.* [**1919-1920**, p. 164, Intro.]

In fact, we will see shortly that he used (7.2) in his investigation of maps admitting a fundamental circle.

Like Julia, Fatou also established the importance of the critical values of f. There are at most $2d-2$ such points, and he used this fact to establish a bound on completely invariant domains. In a lengthy argument that went somewhat beyond elementary algebra [**1919-1920**, pp. 183–185, I], he proved that if an invariant domain D is simply connected, it must contain $d-1$ of the critical points of f^{-1}. From this it follows immediately that there can be at most two completely simply connected invariant domains.[30] Fatou extended the notion of a critical value to

[30] It is not until [**1919-1920**, p. 61, IV] that he shows that the domain of attraction of an arbitrary fixed point contains a critical value of f.

f^p, as well as that of an invariant domain: If a function is periodic of order p, and f permutes among only p domains, then the union of those p domains forms an invariant domain of the function f^p.

There are different ways of establishing the groundwork for things to come. One way is Julia's. Say what will come, as he did in his *Introduction*, and, then lay down as one does a foundation to a house all the basic principles on which he will draw.

Fatou chose a different way. In his *Introduction* he chose to speak generally and to hint at results he would prove rather than state them outright. In his first chapter he used elementary techniques—there was very little he used that was not available to the proponents of the Kœnigs school—to sketch out the scope of his study and to introduce, as it were, characters that he would bring back later in prominent roles.

Thus, Fatou accomplished much more in *Chapter I* than just establish the basic principles and terminology of iteration. Not only did he introduce several new ideas—critical values, invariant domains—but he foreshadowed many of the fundamental ideas that are to come later in his monograph. For example, by establishing the existence of repelling points long before defining the concept of a Julia set, he made it easier to show that it contains infinitely many repelling periodic points when the time is right. The notion of a completely invariant domain—and the fact that there can be no more than two—is a prelude to the division of the complex sphere into regions based on their convergence properties that will come in *Chapter IV* and *Chapter V*.

7.3.3. Local theory from a global viewpoint: *Chapter II*. Here, Fatou classified the local behavior of fixed points. It is interesting that Fatou did not begin with the global theory. One reason may be that the local theory is less restrictive since it only requires that f be analytic in a disc. Thus, for *Chapter II*, he dropped the assumption that f is rational which enabled him to develop a flexible local theory that could in subsequent works be applied to a wide range of functions. Indeed, in latter works he considered the iteration of algebraic and transcendental mappings, as well as functions of two variables (see, respectively, [**1922a**, **1926**, **1924**]).

He began his classification of fixed points with the case where $s = f'(\lambda)$ is strictly between zero and one in modulus, and he gave a stripped down version of Kœnigs' proof of the existence of an analytic solution S to the Schröder equation, $S \circ f = s \cdot S$, discussed in Section 1.7.1. He immediately followed with a proof of the existence of a solution to the Böttcher equation, $B \circ f = (B)^k$, in the superattractive case, that is, when $f'(\lambda) = 0$ and $k > 1$ is the order of the first nonzero derivative of the Taylor series of f expanded around the fixed point λ. The proof seems to be his own, although he attributed the first solution to Böttcher (see Section 7.5 for more on Böttcher), but without giving a specific reference. As we noted earlier, Böttcher's proof was in Russian so it would not be surprising if Fatou had not read it.

In the same spirit as Kœnigs' derivation of the solution to the Schröder equation (see (1.10) and Alexander [**1994a**, §3.5]), Fatou's solution of the Böttcher equation is derived from the the dynamics around a superattracting fixed point rather than algebraically from the functional equation. Assuming that $\lambda = 0$ and $f(z) =$

$z^k + az^{k+1} + \cdots$, with $k > 1$, he found a disc D surrounding the fixed point that is invariant under iteration by f, whose existence is guaranteed by Schröder's fixed point theorem (see Theorem 1.1), which Kœnigs made rigorous.

Fatou asserted that he could understand the behavior of the sequence $\{z_n = f^n(z)\}$ by studying the sequence $\left\{u_n = (z_n)^{\frac{1}{k^n}} = (f^n(z))^{\frac{1}{k^n}}\right\}$, where the (k^n)th root is understood to be the one whose Taylor expansion around the origin begins with z.

Fatou claimed that the sequence of holomorphic functions $\{(f^n(z))^{\frac{1}{k^n}}\}$ converges uniformly to a holomorphic function $B(z)$ on D. To show this, he introduced the auxiliary function

$$H(z) = \frac{1}{z}(f(z))^{\frac{1}{k}} = 1 + \frac{1}{k}az + \cdots, \tag{7.4}$$

whose relation to the u_n is given by

$$\frac{u_{n+1}}{u_n} = \frac{(z_{n+1})^{\frac{1}{k^{n+1}}}}{(z_n)^{\frac{1}{k^n}}} = \left(\frac{1}{z_n}(f(z_n))^{\frac{1}{k}}\right)^{\frac{1}{k^n}} = (H(z_n))^{\frac{1}{k^n}}.$$

Since $u_0 = z_0 = z$, it follows that

$$u_n = z\prod_{i=0}^{n-1}\frac{u_{i+1}}{u_i} = z\prod_{i=0}^{n-1}H(z_i)^{\frac{1}{k^i}}.$$

Consequently, the convergence of the u_n depends on that of the infinite product in H or, equivalently, on that of the series $\sum_{i=0}^{\infty}\frac{1}{k^i}\log(H(z_i))$, where log denotes the principal branch of the logarithmic function. Considering that $z_n \to 0$ whenever $z \in D$, the following bounds hold true beginning with a sufficiently large index N:

$$|H(z_n) - 1| < \frac{1}{2} \quad \text{and} \quad |H(z_n) - 1| < c|z_n|,$$

where c is a suitably chosen positive constant. After observing that, for $|w| < \frac{1}{2}$,

$$|\log(1 + w)| < |w| + |w|^2 + |w|^3 + \cdots = \frac{|w|}{1 - |w|} < 2|w|,$$

Fatou exploited the two bounds above to deduce that

$$|\log(H(z_n))| < 2c|z_n|,$$

which transfers the question of the uniform convergence of the original series

$$\sum_{i=0}^{\infty}\frac{1}{k^i}\log(H(z_i))$$

to that of the series $2c\sum_{i=0}^{\infty}\frac{|z_i|}{k^i}$, which is convergent whenever $z \in D$.

Fatou then easily demonstrated that on D,

$$B(z) = \lim_{n\to\infty}(f^n(z))^{\frac{1}{k^n}}$$

satisfies the Böttcher functional equation:

$$B(f(z)) = \lim_{n\to\infty}(f^{n+1}(z))^{\frac{1}{k^n}} = \lim_{n\to\infty}\left((f^{n+1}(z))^{\frac{1}{k^{n+1}}}\right)^k = B(z)^k.$$

As with his solution to the Schröder equation, the domain of his solution is local, but later, in *Chapter VII*, he widened the domains of the solutions to both equations.

He viewed the solutions of both the Schröder and Böttcher equations as a means to establish a conformal equivalence between an unspecified neighborhood D of the fixed point and a neighborhood N of the origin, which implies that iteration is conformally equivalent to either the map $w \mapsto s \cdot w$ or $w \mapsto w^k$, depending on whether the fixed point is attracting or superattracting. This permitted Fatou to easily classify the curves left invariant on N under iteration by f by first finding those left invariant under $w \mapsto s \cdot w$ or $w \mapsto w^k$, and then mapping the curves back to D.

In the attracting case, if s is real, then every line passing through the origin is invariant, while if s has a nonnull imaginary part, the invariant curves become logarithmic spirals of the form $z(t) = s^t z_0$, where z_0 is an arbitrary point on N and $t \in [0, +\infty)$. These two families of curves are mapped back to D, via the inverse transformation $S^{-1}(z)$ and become, respectively, analytic curves passing through the fixed point and spiral-shaped curves that approach the fixed point asymptotically.

The case where s is complex has a precise analogue with Lattès' study of invariant curves of two dimensional substitutions in a neighborhood of a fixed point, which we examined in Section 3.6. In particular, it corresponds to subcase (d) on page 57, where Lattès assumed that the eigenvalues associated with the linear part of the substitution are complex conjugates with modulus less than one. We recall that, under this assumption, Lattès found it convenient to interpret the dynamics in terms of a complex map rather than a map on $\mathbb{R}^2$, and this allowed him to deduce that the invariant curves were spiral shaped.

In the superattracting case, the invariant curves of the map $w \mapsto w^k$ come in two forms: lines pass through the origin with argument $\frac{n\pi}{k-1}$, $n = 0, \ldots, k-2$ and logarithmic spirals are described by the polar equations $\rho = e^{c(\omega + \frac{2n\pi}{k-1})}$, are $n = 0, \ldots, k-2$, where c is an arbitrary constant. These invariant curves are respectively transformed by the function B^{-1} into $k-1$ analytical curves on D that form angles of equal size at the fixed point, or $k-1$ sets of spiral-shaped curves approaching the fixed point asymptotically.

Fatou then turned his attention to functional equations for the rationally indifferent case in [**1919-1920**, §8–14]. Like Julia and Leau, he began with the case $f'(\lambda) = 1$ but his proof proceeds quite differently. He mapped the fixed point to infinity, and among other things, showed that the iteration of a rationally indifferent point near infinity is equivalent to translation by a, which suggests the Abel equation $A \circ f = A + a$. His approach to the rationally neutral case is discussed in greater detail in Alexander [**1994a**, §6.3] and Audin [**2009**, pp. 92–94].

He closed the chapter with a discussion of the local behavior of iteration around a repelling point λ. He showed, first, that no subset of the iterative family $\mathcal{I}$ converges uniformly on neighborhoods of a repelling fixed point (without invoking either Montel or the concept of normality), and then he induced a solution to the Poincaré functional equation $f \circ \theta = \theta \circ (f'(\lambda) \cdot I)$. Fatou introduced irrationally neutral fixed points at this point, noting that it was not at all clear that they could be linearized. He would continue the discussion of such points later, as we describe in Sections 9.4 and 9.5.

7.3.4. The fundamental circle: *Chapter III*. Fatou devoted the third chapter to the dynamics of rational maps with a fundamental circle (that is, a circle Γ that is invariant under f whose interior and exterior are also fixed by f) and extended his analysis to functions whose domains of attraction were bounded by an analogous simple closed curve. The unit circle for $z \mapsto z^2$ is a fairly trivial example of a fundamental circle.

Fatou asserted that there are four distinct possibilities for such maps, which we have summarized as the following theorem—it should be noted that Fatou normally did not distinguish results in the form of a theorem, and we do so here and elsewhere only to present them in a convenient form.

THEOREM 7.12 (Fatou's Fundamental Circle Classification). *Let f be a rational map of degree d with a fundamental circle Γ that leaves both* Int(Γ) *and* Ext(Γ) *invariant. One of the following must hold* [**1919-1920**, pp. 227–229, III]*:*

(1) *f has two attracting fixed points one exterior to Γ and one in its interior;*
(2) *f has a unique attracting fixed point on Γ;*
(3) *f has a neutral fixed point with multiplier one on Γ which is a double root of $f(z) = z$; and*
(4) *f has a neutral fixed point with multiplier one on Γ which is a triple root of $f(z) = z$.*

These results were also presented in a somewhat different arrangement in his first notice prepared in anticipation of the *Grand Prix* [**1917a**].

He referred to the first case as *species one* maps, the second as being of *species two*, and the others, *singular cases.* To get a flavor of his investigation of fundamental circles—and to show an application of the rational fixed point formula at equation (7.2) above—we will describe his proof of this classification theorem.

First, assume the fixed points are distinct. To normalize mappings with a fundamental circle, he conjugated f to a map n of the same degree which transforms Γ to the real axis, its interior to the upper half-plane, and the exterior to the lower half-plane. Such a map can be written

$$n(z) = cz + h - \sum_{k=1}^{d} \frac{A_k}{z - \alpha_k},$$

where $c \geq 0$, $A_k > 0$, and the α_k are real with $\alpha_k < \alpha_{k+1}$. He assumed in the general case that $c = 0$, otherwise there would be a fixed point at ∞ which can be removed by a suitable change of variables.

To find the fixed points of n, he examined the dynamics of $g(z) = n(z) - z$ and observed that

$$\lim_{z \to \alpha_k^-} g(z) = \infty \text{ and } \lim_{z \to \alpha_k^+} g(z) = -\infty,$$

where the poles α_k are approached along the real axis. The poles induce $d-1$ finite intervals (α_k, α_{k+1}) on which g goes from $-\infty$ to ∞. This implies that g must be zero at least once on each interval, and since g is generally increasing, for at least one of these zeros, the derivative of g is positive. The zeros of g are fixed points of n; denote the fixed point with positive derivative in the kth subinterval λ_k. Since

$$0 < g'(\lambda_k) = n'(\lambda_k) - 1,$$

it follows that $n'(\lambda_k) > 1$ so the $d - 1$ λ_k are repelling fixed points of n. Let $n'(\lambda_k) = s_k$.

There are two other fixed points of n, β, and β'. Denote their multipliers s and s'. Since all the fixed points are distinct, we can apply the rational fixed point formula, equation (7.2),

$$\sum_{k=1}^{d+1} \frac{1}{s_k - 1} + 1 = 0,$$

which yields

$$\frac{1}{s-1} + \frac{1}{s'-1} + \sum_{k=1}^{d-1} \frac{1}{s_k - 1} + 1 = 0.$$

Thus, s and s' (and Fatou observed β and β') are either real or complex conjugates.

If they are complex conjugates, the sum of the first two terms must be less than -1, since the other multipliers s_k are strictly greater than 1, so

$$-1 > \frac{1}{s-1} + \frac{1}{s'-1} = u + iv + u - iv = 2u \Rightarrow u < -\frac{1}{2}.$$

In our discussion accompanying the presentation of the fixed point formula, we noted that if the real part of $\frac{1}{s-1} < -\frac{1}{2}$, then $|s| < 1$. Thus, the two fixed points β and β' are attracting. Moreover,

> *We remark here that we have an application for what we said in Chapter I (§6) on the subject of invariant domains, namely that the exterior and interior of the fundamental circle constitute two completely invariant domains each including half of the critical points of the inverse function.* [**1919-1920**, p. 252, III]

If the fixed points β and β' are, on the other hand, real and distinct, they both must be in one the intervals (α_k, α_{k+1}) along with the repelling point λ_k, or in one of the extremal intervals, $(-\infty, \alpha_1)$ or (α_d, ∞) (or, if one prefers to view the real line as a circle, (α_d, α_1)). If only one fixed point, say β, is in an interval, then g would only have two zeros there. Since g goes from $-\infty$ to ∞, this can only happen if one of the fixed points is a double root. But this would mean it was a double fixed point of n as well, contrary to our assumption that the fixed points are distinct. It follows, then, that the derivative of g is positive at one of β or β', as well as at λ_k hence the mapping n has a unique attracting fixed point on the real axis.

The only other possibility is for β and β' to become a double root, or to join with a λ_k and form a triple root of g. In either of the two cases, $s = 1 = s'$ (and s_k as well, if λ_k is a triple root).

In some senses, maps with a fundamental circle are a special case that did not require the thorough analysis he allotted to them. Fatou indicated as much in his introduction when, discussing his treatment of maps with a fundamental circle, he remarked,

> *One could have shortened the treatment using the general theory from subsequent chapters.* [**1919-1920**, pp. 164–165, Intro.]

Nonetheless, as he noted at the beginning of §23,

> *We will show now that the methods used to study substitutions having a fundamental circle can be applied to grand number of cases.* [**1919-1920**, p. 252, III]

An example of the sort of application he had in mind involved his examination of a generic species two mapping f with a fundamental circle, that is, one with a unique attracting fixed point λ. He showed that the set of points P which do not converge to λ under the iteration of f form a totally disconnected perfect set, which we refer to as a Cantor set.[31] His method, reminiscent of the argument he outlined in [**1906b**] to show that the Julia set of $z \mapsto z^k/(z^k+2)$, $k = 2, 3, \ldots$, is a Cantor set (see Section 2.2), constructs P directly by finding a particular simple closed curve C (not the fundamental circle) such that the set

$$\bigcap_{n \in \mathbb{N}} f^{-n}\left[C \cup \operatorname{Int}(C)\right],$$

where f^{-1} is the total inverse, is a totally disconnected set P.

He then applied these same techniques to show that identical behavior also occurs in maps that do not possess a fundamental circle (for example $z \mapsto z^2 + 5$ and the family $z \mapsto z/(z^k + 2)$, for $k = 3, 4 \ldots$) to prove the following, which we state as a theorem, something Fatou did not do:[32]

THEOREM 7.13 (Unique Attracting Fixed Point Theorem). *Let f be a function with a single attracting fixed point λ. Suppose there is a simple closed curve C such that*

(1) *C partitions the plane into two regions, one of which contains λ and is designated* Int(C),
(2) *all the critical values of f are in* Int(C),
(3) *the interior of C is in the basin of attraction of the fixed point, and*
(4) *there is an N such that if $k \geq N$, then $|f'(z)| > r > 1$ for all z on the kth preimage of C.*

Then all points on the sphere converge to the fixed point under iteration, save for a Cantor set P.

For mappings of the first species (i.e., there is an attracting fixed point on either side of the the fundamental circle Γ) the set P of points which do not converge to either fixed point under iteration equals Γ and forms the boundary between two completely invariant domains, each of which contains an attracting fixed point. This behavior, too, is quite typical of functions that have exactly two attracting orbits.

Fatou showed that for maps of both species, P shares similar properties: Repelling periodic points are dense in P, and the iterative family takes on all values in the sphere on neighborhoods of points from these sets, except for at most two exceptional values (compare with Julia's Fundamental Theorem 7.6). This implies, as well, that for mappings of both species, the preimages of any point in the plane—aside from the exceptional values—have in their limit set all points in P.

Again, this is done in the context of the special case of the fundamental circle. No mention is made of Montel's theory of normal families, and Fatou made no explicit mention that analogous sets P occur with all rational functions of degree two

[31] In *Chapter IV* Fatou showed that the set P is the domain of nonnormality for the iterative family $\mathcal{I}$, nowadays referred to as the Julia set. Our use of the notation P to denote this set is not only consistent with Fatou's notation (P was chosen to highlight the fact that this set is perfect), but is also suggestive of Fatou's notion of the set P in *Chapter III* as a proto-Julia set.

[32] Fatou's rather leisurely discussion of these results extends over both §21 and §23 and is summarized on [**1919-1920**, pp. 256–257, III].

or more. Fatou was, of course, drawing the reader in with his comment, referenced above, that he could shorten his treatment of fundamental circles had he so desired. He immediately followed that remark by adding, "but it seemed preferable to us to include this elementary presentation, since the methods will be useful in solving other problems we will encounter" [**1919-1920**, pp. 164–165, Intro.]

Theorem 7.13 is in many respects Fatou's unique attracting fixed point analog of Julia's theorem regarding curves bounding the invariant domains of two attracting fixed points, Theorem 7.9, that we examined during our discussion of Julia's monograph. We mentioned at the time that one of Julia's applications of his theorem was to explain a result from Fatou's May *Comptes Rendus* note [**1917a**], where the curve bounding two such domains is a fundamental circle, showing that the behavior Fatou observed is easily explained by his own Theorem 7.9, and the fundamental circle is the simple closed Jordan curve his theorem guarantees. Thus, Julia observed, Fatou's notice involved rather special circumstance.[33]

While this may be true, Theorem 7.13 as well as several other arguments from Chapter III show that by the time he had written his monograph, Fatou well understood that the behavior he observed with fundamental circles were indeed specific embodiments of far more general principles. Recall that when Julia prepared his monograph, he only had the benefit of Fatou's May 1917 notice, so he could be forgiven for not realizing how deep Fatou's understanding of the mathematics associated with iteration had become.

No doubt Fatou would agree that Julia's theorem is an effective generalization of the concept of a fundamental circle, analogous to his own Theorem 7.13. After all, his choice to investigate fundamental circles in his monograph—and perhaps also in his May 1917 note [**1917a**]—was not because he could not address iteration in a more general fashion, but rather because he saw an elementary investigation of fundamental circles as an excellent means to introduce properties that were prevalent throughout the study of the iteration of rational functions.

Fatou's study of the Cantor sets that arise from Theorem 7.13 led him to devise a method to code the preimages of C that nowadays is called *symbolic dynamics*. Designating the d branches of the inverse as $f_0^{-1}, f_1^{-1}, \ldots, f_{d-1}^1$, he used sequences of base d numbers to track the preimages of C. For example, the curve corresponding to (δ_1, δ_2) is the preimage of C under $f_{\delta_2}^{-1} \circ f_{\delta_1}^{-1}$ and is denoted $C_{\{\delta_1,\delta_2\}}$; an infinite sequence $\{\delta_k\}$, however, would refer to the infinite intersection

$$\bigcap_{\{\delta_k\}} \overline{\mathrm{Ext}(C_{\delta_1,\ldots,\delta_k})},$$

where the overbar denotes the closure of a set, and Ext indicates the component of the complement of C that does not contain the fixed point. Fatou showed that the diameters of C_{δ_k} go to zero,[34] hence the intersection corresponding to a infinite sequence $\{\delta_k\}$ is a single point.

The fact that each infinite intersection represents a different point gave Fatou a means to represent elements in the set P:

> *In this manner there is a biunivalent correspondence between the points of P and the infinite sequence of digits. The sequence of digits corresponds, moreover, to a number between* 0 *and* 1*, such*

[33]The remarks of Julia to which we refer are interspersed between pages 138 and 145 of [**1918a**].

[34]See page 27 for an analogous argument that Fatou sketched in his notice [**1906b**].

> *that the set of digits after the decimal point are represented in the base d system of numeration, the correspondence being one-to-one except for a denumerable infinity of rational numbers that can be written in two ways.*[35] [**1919-1920**, p. 257, III]

From this correspondence between P and the set of distinct base d sequences on $(0, 1)$, several fundamental properties arise:

(1) periodic points correspond to periodic sequences;
(2) sequences which are "infinitely close" to one another correspond to points which are likewise "infinitely close" [**1919-1920**, p. 258, III];
(3) all points in P are limits of periodic points; and
(4) in any neighborhood of a point z from P there is a point z' such that the derived set of $\{f^n(z') : n = 0, 1, 2, \ldots\}$ is P.

He also induced a one-to-one mapping ϕ from the set sequences of the $\{\delta_k\}$ to P that is nowadays called a shift map and is defined by $\phi(t) = z$ and $\phi(dt) \bmod 1 = f(z)$. This mapping shifts the sequences one base d place to the left and can be used to represent iteration on P. Fatou gave a concrete example of symbolic dynamics in describing the set P for $f(z) = z^3$ [**1919-1920**, p. 242, III].

In his discussion of Cantor sets, Fatou used the point of view that the coefficients of rational functions are really variables in some parameter space—a point of view he shared with Julia—to argue that Cantor sets were by no means pathological, but were instead the norm:

> *There is another important remark that should be made, and that is, that the existence of a boundary of the domain of convergence that is totally disconnected is not a singular occurrence, that is to say, this circumstance can be produced without any particular relationship among the coefficients of the function $f(z)$; it suffices that the coefficients vary in a suitable domain.* [**1919-1920**, p. 258, III]

He would argue later (as did Julia) that curves without derivatives were also the norm, and one wonders if he took a bit of pride in establishing that objects that some might consider pathological, or even monstrous, occur quite naturally.

Given that Fatou applied measure theory to complex functions in his dissertation [**1906a**], it is not surprising that he investigated what he termed the Borel-Lebesgue measure of the sets P. Referring to item (4) above from Theorem 7.13, he noted that if $r > d$, the linear measure of P is zero, and that if $r > \sqrt{d}$, the "*superficielle*" (i.e., two dimensional) measure is also zero [**1919-1920**, p. 257, III]. Perhaps worried that such a set could be a "fat" Cantor set (that is, one with positive measure), he remarked that

> *It would be interesting to know if P is a set of null linear measure, as is the case in the study of single-valued functions which admit P as the set of their essential singularities. It suffices that the length of the curves C_{-n}* [which are the curves referred to as C_n above] *tend to zero with $\frac{1}{n}$. This always happens if $r > d$ (the degree of $f(z)$), which is not a necessary condition. I am not able to determine if P can have nonnull measure.* [**1919-1920**, p. 248, III]

[35] There are infinitely many numbers between 0 and 1 that have multiple representations. For example, the ternary decimal sequences $(0, 2, 2, \ldots)$ and $(1, 0, 0, \ldots)$ both correspond to $\frac{1}{3}$.

As he noted earlier in the chapter in a slightly different context regarding a totally disconnected perfect set,

> *These examples suffice to show that the study of the set E'*[36] *is a problem of an arithmetic nature to which one could apply the methods of Borel-Lebesgue to determine the measure of the sets, something which requires further research.* [**1919-1920**, p. 243, III]

One can view *Chapter III* as the end of the first stage of evolution of Fatou's investigation of the iteration of rational maps. It is the last chapter which relies primarily on elementary methods. By the end of *Chapter III* one gets the sense that Fatou has consciously taken the Kœnigs' approach as far as it can go. That would certainly explain his choices to use what he termed "elementary algebra" in *Chapter I*, to focus on iteration near fixed points in *Chapter II* and to use fundamental circles in *Chapter III*. Aside from Ritt and Lattès' application of the Poincaré functional equation, which we will discuss later in this chapter, there is virtually no place left to go with a direct approach after *Chapter III*. The recent innovations in French mathematics that Fatou would apply in subsequent chapters of his monograph, with the notable exceptions of set theory and measure theory that he used to analyze the set P, are largely absent from the first three chapters. Instead, he used tools that were available to the proponents of the Kœnigs school, including critical values, the fixed point formulae, functional equations, and the notion of a fundamental circle. But, as he showed, one can only go so far with these tools. It was time to move to the next stage.

TABLE 1. The evolution of Fatou's theory

1906	$z \mapsto \frac{z^2+z}{2}$ and $\frac{z^m}{z^m+2}$
early 1917	maps with a fundamental circle
late 1917–20	the family of rational maps
late 1920–26	permutable, transcendental, algebraic, and two-variable maps

7.3.5. Finally, normal families: *Chapter IV*. Finally bringing Montel's theory of normal families into his investigation, Fatou's second installment of his monograph, *Chapters IV* and *V*, represents the heart of his investigation. Almost immediately, a cavalcade of now classic results issued forth, and within the first twenty pages the bedrock of the modern study was established. After defining the domains of normality (Fatou set) and nonnormality (Julia set), he

(1) shows that the domain of nonnormality is the closure of the periodic points with multiplier in modulus one or greater;
(2) describes the complete invariance and possible topological configurations of the Fatou and Julia sets, as well as the latter's self-similarity;

[36]The set E' to which he refers is an actual Cantor middle-thirds set on the unit circle that is invariant under the map $z \mapsto z^3$ [**1919-1920**, p. 242, III].

(3) demonstrates that the Julia set contains a nonempty interior only if it comprises the entire sphere; and
(4) proves that the Fatou set has 0, 1, 2, or infinitely many components.

He followed these results with an investigation of the limit functions of the iterative family $\{f^n\}$ on the Fatou set, showing that when there was an attracting or neutral periodic orbit, the limit functions are constant and their values correspond to the limit points of the orbit.

As we will see, Fatou did not discount the possibility that there are regions on the Fatou set where the limit functions are nonconstant. He referred to such regions as *singular domains* and investigated them enthusiastically, if inconclusively, since he never was able to determine if they actually exist.[37] Indeed, Fatou and Julia's respective investigations of singular domains are quite different, and Julia did not explore them in quite the same manner as Fatou. More will be said about these matters in Chapter 9.

Fatou determined that the number of singular domains, should they exist, is at most $4d-4$, where d is the degree of f [**1919-1920**, p. 69, IV] and showed as well that there are no more than $2d-2$ attracting and rationally neutral orbits combined.[38] Fatou's proof of the bound on attracting and rationally neutral cycles turned on the fact that the immediate domain of attraction for an attracting or rationally neutral periodic orbit contains a critical value of f, and there are $2d-2$ critical values counted with multiplicity [**1919-1920**, pp. 60–66, IV].[39]

7.3.6. Exploring the Fatou set: *Chapter V*. Here, Fatou explored the connectivity of the Fatou set, as well as the crucial role of the post-critical set. For example, Fatou proved that if the immediate domain of attraction of a fixed point contains all the critical values of f, the corresponding Julia set (which he referred to as $\mathcal{F}$, a notation we will adopt during our discussion of Fatou) is a Cantor set, Fatou speculated that the converse was true as well:

> *The existence of critical points of $R_{-1}(z)$, belonging to $\mathcal{F}$ does not seem to be compatible with the total discontinuity of $\mathcal{F}$, but I was not able to prove it rigorously.* [**1919-1920**, p. 84, V]

He was incorrect. In 1965 Hans Brolin found a counterexample,

$$f(z) = 18(z^3 - 2z^2 + z),$$

where the domain of nonnormality is a Cantor set containing a critical value of f [**1965**, p. 136].

---

[37]Such domains do exist, and they are the rotation domains we discussed in Section 1.6.2. The limit functions correspond to the rotations $z \mapsto \theta z$, where $\theta \in [0, 2\pi)$. Actually, it would be more accurate to say that domains that Fatou would recognize as singular domains exist, since he never quite said how he thought iteration might behave on one.

[38]In [**1987**] Shishikura trimmed Fatou's estimate on singular domains considerably, proving that, collectively, the number of singular domains and attracting and rationally neutral cycles is at most $2d-2$.

[39]Since the attracting fixed petals are pairwise disjoint (see Theorem 7.11), the immediate domain of attraction for a rationally neutral fixed point may compose several components. For example, each of the n attracting petals for the fixed point at the origin of the mapping $z \mapsto z + z^{n+1}$ are in distinct components of the Fatou set. See Beardon [**1991**, p. 194]. In general the existence of a rationally neutral cycle can cut down on the number of attracting cycles. This underscores the point of view that a rationally neutral fixed point can bifurcate into attracting cycles.

Caption for Color Plate 6: (See p. 64.) The attracting domain for the fixed point $A = \frac{-3+\sqrt{5}}{2}$ of $R(z) = z(z+1)(z+2) = z^3 + 3z^2 + 2z$ (Fatou's Example IV from [**1919-1920**, p. 88, IV]) has infinitely many disjoint regions, shown in red. Points in the blue region converge under iteration to ∞. On the top, an overview of the basins of attraction for A and the point at infinity. The origin is a repelling fixed point and therefore lies in the Julia set. The image at bottom left is a magnification of the region marked by the rectangle in the larger image. On bottom right is a magnification of the region indicated near the bottom of the left image.

Fatou closed his second installment with a catalog of nine examples illustrating the principles he had established over *Chapters IV* and *V*. He described one such example, $z \mapsto z^3 + 3z^2 + 2z$, that we call "the archipelago":

> *One sees easily that the partial domains do not have a common boundary; they are limited by curves without double points. In this example, the set $\mathcal{F}$ consists of infinitely many distinct continuities, as does the previous example.*[40] *In addition, the set E'_c, which here consists of only the two attracting points, does not, as it did in* Chapter III, *partition the plane into two simply connected regions.* [**1919-1920**, pp. 88–89, V]

7.3.7. No tangents: *Chapter VI*. The third and final installment of *Sur les équations fonctionnelles* begins with a lengthy exploration of the conditions under which points in the Julia set do not have tangents. Like Julia, he showed that the existence of tangents is the exception rather than the rule. In particular, he proved a result that he stated as follows [**1919-1920**, p. 240, VI]:

THEOREM 7.14. *Let λ be an attracting fixed point of f whose immediate domain of attraction D is simply connected. Let $J_\lambda = \partial D$. If the closure of the post-critical set $E \cup E'$ has no points in common with J_λ, then no point of J_λ has a tangent unless D is a fundamental circle of f or a rational function of the form $f(w)$ where w is an algebraic function of degree two.*

Fatou closed his discussion of Julia sets that do not possess tangents by drawing a sharp contrast between such curves and Koch's construction of what nowadays is often referred as the Koch snowflake [**1906**], depicted in Figure 7.4:

> *These curves also have striking analogies with the curves studied by Helgen von Koch; but the curves studied by that author are defined in a constructive manner in which certain properties were assigned in advance. We, on the other hand, analyzed the mechanism by which our curves were formed, with a lengthy and patient investigation, in particular, the analytic properties of certain functional equations.* [**1919-1920**, p. 242, VI]

He seemed to be suggesting that while Koch's curves are similar to some Julia sets, they are in some senses artificial because they arise from an arbitrary construction without any connection to a particular function. The curves which arise from the study of iteration are anything but artificially constructed since they are the natural

---

[40] $R(z) = \dfrac{z^4+1}{4\sqrt{2}} + z.$

boundary between regions of convergence under iteration. Such curves also form the boundary of the domains of solutions to functional equations, as we will see below.

Julia also invoked Koch curves, as we noted, but to a different end since his purpose was to provide a schematic that wold help the reader visualize Julia sets with infinitely many crossing points.

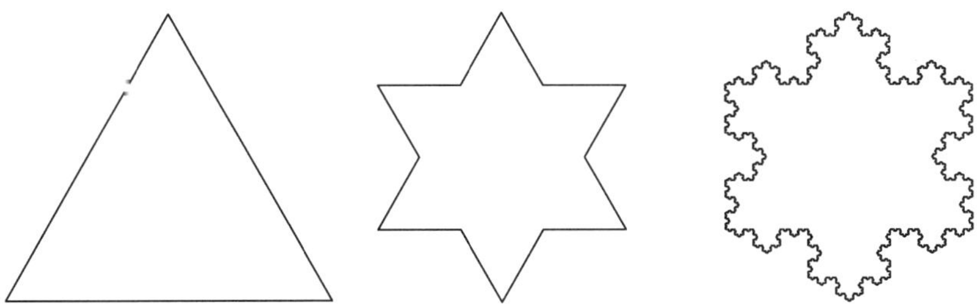

FIGURE 7.4. Genesis of the Koch snowflake. The process of replacing the middle-third of each side with two sides of an equilateral triangle scaled down by a factor of three is repeated indefinitely. The result is a curve without a tangent line at any point.

7.3.8. A return to functional equations: *Chapter VII*. Fatou devoted his concluding chapter to the subject suggested by the title of his monograph, the study of functional equations. Although he referred to functional equations elsewhere in his monograph (for example in *Chapter II*, when he reviewed the literature concerning the local solutions to the functional equations related to iteration, and in *Chapter VI*, during his examination of continuous Julia sets that did not possess a tangent), it was not until *Chapter VII* that he fully articulated his approach to the functional equations that are linked to the study of iteration.

There are three principal themes to his approach. One is to extend the domain of definition of solutions beyond the fixed point. Another is to relate functional equations to material from previous chapters which he did at several turns in *Chapter VII*: In §72–76 he returned to the Abel equation that he used in his study of neutral fixed points in *Chapter II* in much greater detail, and from a global perspective; he tied the study of functional equations to fundamental circles in §61–65 and again in §75; he revisited issues regarding the nonexistence of tangents from the point of view of the solutions of the function equations; finally, he used several examples he had considered previously in his monograph to illustrate various properties of the solutions of functional equations. Earlier we compared Fatou's monograph to a sprawling novel; if so, then in *Chapter VII* he resolved loose plot lines by linking functional equations to matters he studied earlier.

The third, and perhaps most important element of his study of functional equations, was his view that the solutions to the Schröder, Abel, and Poincaré equations are important functions in their own right. He concluded his monograph with a persuasive argument justifying that claim:

> ... *certain of these functions enjoy simple properties yet present singularities of a different nature than any encountered up until now. It seems to me that they merit an important position in*

> *analysis, beside elliptical functions and automorphic functions with whom they have an obvious kinship.* [**1919-1920**, p. 314, VII]

The singularities—at least for the Schröder equations and some classes of Abel equations—are the Julia set. One of their distinguishing peculiarities is their domain of definition, and, as we will see below, Fatou seemed caught between perspectives on how to view the entire domain of the solution S to the canonical Schröder equation $S \circ f = f'(\lambda) \cdot S$.

Until his monograph, the Schröder equation had only been studied in neighborhoods of an attracting fixed point. The uniformization theory developed by Koebe and Poincaré (see Appendix I) showed that these neighborhoods could be conformally mapped to the unit disc $\mathbb{D}$ by S. Fatou's goal was to study the entire domain of S.

However, developing a global study of solutions to functional equations presented a number of challenges. One arises with Fatou's attempt to define the domain of definition of a solution of the canonical Schröder equation beyond the immediate domain of attraction D_λ. Recall that D_λ may not be the complete domain of attraction for λ. If this is the case, then it has infinitely many preimages D_k with $f(D_1) = D_\lambda$ and $f(D_{k-1}) = D_k$. In other words, if the domain of attraction is not simply connected, it will have infinitely many components.

We can expect this to be the general case: The number of components of a nonempty Fatou set is one, two, or infinitely many. Since an arbitrary function is likely to have more than two attracting fixed points (for example, the Newton's method function for a cubic), the domain of attraction for a given fixed point is likely to have infinitely many components.

Given a solution to $S \circ f = f'(\lambda) \cdot S$ around a fixed point λ, the fundamental questions are, first, how far can we continue S within its immediate domain of attraction D_λ, and second, does the domain of S extend beyond D_λ, and if so how far?[41] Fatou wrestled with these issues almost immediately in *Chapter VII*:

> *The fundamental solution of the Schröder functional equation is a holomorphic function interior to D_λ. ...It follows, evidently, that the boundary points form a perfect set, continuous or discontinuous,*[42] *which are essential singularities of $S(z)$. The domain of existence of this function thus coincides ... with that portion of the domain of convergence that is* d'un seul tenant[43] *with λ. But suppose that the domain of convergence is not* d'un seul tenant*?* [**1919-1920**, p. 259, VII]

He then induced the following extension of S to an immediate preimage D_1 of D_λ:

$$S_1(z) = \frac{1}{f'(\lambda)} S(f(z)),$$

from which it follows that S_1 maps z to same place S maps a preimage of $f(z)$ contained in D_λ. One could extend this definition to all preimages of D_λ and obtain a different function on each preimage.

---

[41]See page 294 in Appendix C for a discussion of this issue.

[42]Fatou showed that this set is the Julia set.

[43]The phrase *d'un seul tenant* refers to path connectedness. In this usage, t means to be in the same connected component. See page 138 for the definition Julia gave.

> [If so], *this process of analytic continuation generalizes that of Weierstrass, and $S_1(z)$ will have essential properties of $S(z)$ as the continuation of $S(z)$ inside D_1. But it is necessary to note that the domains D_λ and D_1 may not have any common boundary points, or, at least in simple cases, only a finite number. ...*
>
> *So if we limit ourselves to the point of view of Weierstrass, the function S_1 cannot be prolonged beyond D_λ. Kœnigs' limit thus defines an infinity of analytic functions having distinct and limited domains of existence.* [**1919-1920**, pp. 259–260, VII]

Fatou thus seemed a little unsure about how to conceptualize S on the complete domain of convergence for λ. Is there a unifying expression for it that he was unable to find? Or is S an infinite collection of functions?

After this discussion, Fatou turned his attention to the properties of solutions to the functional equations, and studied in succession the Schröder, Poincaré, and Abel equations. He showed that if λ is an attracting fixed point, the boundary of D_λ, in particular that part of the Julia set bounding D_λ is a set of essential singularities for S. He was not able to do this for the Abel equation, except in special cases:

> *I cannot affirm in an absolutely general manner that the all the points on the boundary* [of the domain of definition for the solution to the Abel equation] *are the essential singularities of this function.* [**1919-1920**, p. 299, VII].

During his study of the Poincaré equation, Fatou pointed out, in a complimentary fashion, how much Lattès was able to achieve using the Poincaré equation, noting in particular, that Lattès obtained many of the same results he established in *Chapter IV* about exception values [**1919-1920**, pp. 281–282, VII]. He also suggested that readers wishing to know more about the Poincaré equation should consult Georges Valiron's 1913 thesis [**1913**] as well as the *Comptes Rendus* notices of Lattès [**1919-1920**, pp. 286–287, VII]. Fatou did not specify to which of Lattès' notices he referred, but we presume it was to Lattès [**1918a, 1918c**].

Valiron's thesis concerns entire functions of finite order, and to demonstrate that functions of order zero arise naturally from functional equations, he considered the Poincaré equation in the single variable case [**1913**, pp. 203–206]. As we noted in Chapter 3 this represents perhaps the first reduction of the multivariable functional equation that Poincaré treated in his 1890 paper [**1890a**] to complex functions of a single variable and it may have been the conduit that led to the introduction of the Poincaré equation to complex dynamics.

The remainder of Fatou's *Chapter VIII* deals with questions involving inverse maps and their critical points (§62, 71), local distribution of zeros for solutions to functional equations (§64, 65), local dynamics about a rationally indifferent fixed point (§72–74), and the solutions of other functional equations (§62, 69, 70). One of the tools he used was the group G, which we discussed at the beginning Section 7.3.2.

7.4. The first men into (parameter) space

In the opening pages of this chapter, we alluded to the fact that both Fatou and Julia anticipated ideas which lie behind the creation of the Mandelbrot set. The Julia set J_c of $z \mapsto z^2 + c$ varies with parameter c, and the Mandelbrot set M, is a way of encoding how J_c varies with c. It is defined as the set of c such that the Julia set J_c of $z \mapsto z^2 + c$ is connected, and if $c \notin M$, it is not difficult to show that J_c is a Cantor set. The exploration of M has proved extraordinarily fruitful, and was pioneered in the early 1980s by Adrien Douady, John L Hubbard, and Benoît Mandelbrot.[44] Notions similar to M have been extended to other kinds of functions.

We have already noted some examples where Julia and Fatou observed that certain properties of f persisted when its coefficients were perturbed. Julia's argument that points in a continuous Julia set generally did not have tangents was based on such reasoning (see page 146), as was Fatou's claim that Cantor-like Julia sets were not at all unusual (see page 159). Similar arguments were given at other points in their monographs as well, and we will cite three of them.

In discussing functions with two attracting fixed points whose Julia set J is a simple, closed, Jordan curve, Julia observed that J remains so if the coefficients of the functions vary slightly:

> *What has been said about $z_1 = (z+z^2)/2$ can evidently be extended to* any second degree rational map with two regular limit points, *that is, two points ζ_1 and ζ_2 where $z = \varphi(z)$ and $|\varphi(z)| < 1$. Such a rational map has complex coefficients that could vary between certain limits (for example, every coefficient can be represented by a point of the complex plane and required to remain in certain regions of that plane), but these limits will be more restrictive if we require, for example, that $z_1 = \varphi(z)$ conserve the interior of a fundamental circle.* [**1918a**, p. 140]

Fatou also noticed that small variations of coefficients may not affect Julia set topology. More than once, Fatou relied on such an exploration to highlight the dynamical properties of f under iteration. In the following quotation, $\mathcal{F}$ is the Julia set of a rational function R, and R_h is its iterate R^h.

> *If $\mathcal{F}$ is bounded and does not include any limit of the consequents of the critical points of $R_{-1}(z)$, $R'(z) > K > 1$ may not necessarily hold at all points of $\mathcal{F}$; but it always holds if R is replaced by R_h, h being a conveniently chosen whole number. It is easily shown that rational substitutions enjoying this property still do so when the change in coefficients of $R(z)$ is sufficiently small. It is probable, even though I did not investigate this question thoroughly, that this property holds in all general substitutions, that is, where the coefficients do not satisfy any special condition.* [**1919-1920**, 73, IV]

Finally, in investigating $z \mapsto z^2+5$, Fatou observed that its Julia set is a Cantor set, and then remarked,

> *One obtains similar results in numerous polynomials, for example, those of the form $z^m + a$, if $|a|$ is sufficiently large.* [**1919-1920**, p. 254, III]

---

[44] Accounts of the discovery of M are found in Peitgen, Richter, and Saupe [**2004**, pp. xii–xx] and Mandelbrot [**2004**, C23]. Excellent introductory accounts of the Mandelbrot set include Blanchard [**1984**], Devaney [**1989**], and Devaney and Keen [**1989**].

A few pages later he made this assertion more precise. Given a degree d polynomial $p(z)$ whose constant term is the parameter t, he outlined a procedure which ensures that t is large enough so that the Julia set is a Cantor set.

7.5. Lattès: The lost manuscript

Unfortunately, the manuscript Lattès submitted to the Académie for the *Grand Prix* has been lost. Our investigations at the archives of the Académie des Sciences revealed that Lattès' manuscript was borrowed from the Académie's archives on February 3, 1919, by Georges Humbert. He also borrowed Julia's manuscript about three weeks earlier. We could find no record of Lattès' entry being returned. Inquiries at the École Polytechnique, where some of Humbert's papers are kept, did not turn up the manuscript.

In his 1921 necrology of Lattès, Adolphe Buhl remarked that there were plans to publish Lattès' entry:

> *Thanks to the attentions of Émile Picard, the memoir of Samuel Lattès will be published in the* Annales de l'École Normale, *without a doubt in either 1921 or 1922.* [**1921**, p. 13]

This did not occur, and we do not know why. Perhaps it had already been lost.

What we do know of the manuscript is gleaned from the Académie's official report on the prize [**CR-1918**, pp. 811–814] (translated in Appendix A), Humbert's internal report about the Julia and Lattès' entries [**1918**], which the official report drew upon, and, finally, Lattès' three notices summarizing the highlights of his entry published in *Comptes Rendus* [**1918a, 1918b, 1918c**].

Lattès' interest in the French study of iteration dates back at least to his thesis [**1906**] in which he generalized Kœnigs' study of single-variable iteration and related functional equations to higher dimensions in order to study curves and surfaces left invariant by both functional equations and related differential equations (see Section 3.5). He was an active researcher, publishing about 15 papers between 1906 and 1916. His interests included differential equations [**1906, 1910b**], representations of functions by series [**1910a, 1914**], and iteration [**1911b, 1911c**].

Lattès' entry involved the iteration of functions of both one and two complex variables. The prize commission of the Académie spoke admiringly of his study of iteration of functions of two variables:

> *The portion of the memoir about substitutions of two variables is without a doubt its most interesting and novel part; it contains elegant results, notably about Cremona transformations, which although a little fragmentary, open up a field of research that could become very rich; ....* [**CR-1918**, p. 812]

Unlike Fatou and Julia, Lattès' investigation was not rooted in the theory of normal families. His only reference to the *Comptes Rendus* notice [**1918c**, p. 487] was a brief acknowledgment of its role that was evidently prompted by Fatou and Julia's use of it in their own notices. Consequently he did not decompose the sphere into domains of normality and nonnormality, as did Fatou and Julia (and Pincherle, too, although not in the context of normal families); in particular, he did not seem to have been in possession of the notion of a Julia set (as Pincherle surely was). Nevertheless, he had a robust understanding of the dynamical properties of several examples and found their Julia sets, although in a different, and perhaps more limited, context, as we will see. As Humbert noted in his internal report,

> *The author has not seen the role of the set that Julia called E', although he encountered this set in his examples.* [**1918**, p. 21]

Lattès saw the fundamental problem of iteration in both the one and two variable case as determining the orbit of an arbitrary point under the iteration of a given function. His primary tool, at least in the one variable case is the Poincaré functional equation $f \circ \theta = \theta(s \cdot I)$. A solution to the Poincaré equation leads to the following expression:

$$f^n(\theta(u)) = \theta(s^n u), \tag{7.5}$$

which reduces the iteration of f to multiplication by s and yields an expression for the orbit of z under f, $O^+(z) = \{f^n(\theta(u)) = \theta(s^n u)\}$, where $\theta(u) = z$.

Lattès found a specific property of solutions θ to the Poincaré equation particularly useful. If θ is defined on a domain U where $\theta(U)$ is a neighborhood N of a repelling fixed point of f with multiplier s, then the set $\{\theta(s^n u) : u \in U\}$ covers the sphere, with the exception of at most two points. That means for any $w \in \overline{\mathbb{C}}$ that is not an exceptional value, there is a point $u_w \in U$ such that $s^n u_w$ with $f^n(\theta(u_w)) = \theta(s^n u_w) = w$. Lattès interpreted this phenomena as follows:

> *Thus one can choose, among the possible successive antecedents* [preimages] *of any point z whatsoever, a sequence of successive antecedents z_{-n} tending, as n goes to infinity, towards any point α in the set of invariant points whose multiplier is greater than one in modulus.* [**1918a**, p. 27]

That is, given w, the iteration can be traced backwards to a point $w_n = z \in N$ such that z is an nth preimage of w. Julia and Fatou observed this same fact, but arrived at it in a different manner. As we mentioned during our discussion of Fatou, he observed that Lattès was able to achieve many of the same results he obtained, including, presumably, this one, using the Poincaré equation [**1919-1920**, pp. 281–282].

Lattès called the technique he used to iterate a point $z = \theta(u)$ in (7.5) *parametric iteration*, and he believed it held the key to solving the fundamental problem of iteration:

> *So problems related to the iteration of a given substitution turn into the related problem of the growth of $\theta(u)$. In order to find classes of rational functions whose iteration is easy to study, it is convenient to choose functions which admit a Poincaré function that is among the most common meromorphic (or entire) functions. Thus, if s is an entire number, and if $\theta(u)$ is one among the functions* $\tan u, \cos u, \wp u$, *or it is a homographic transformation of these functions, one finds rational functions where the method of parametric iteration allows one to find a complete solution of the fundamental problem of iteration.* [**1918a**, p. 27]

This is similar to the approach favored by Ernst Schröder almost 50 years earlier, referred to in Chapter 2 as his "opposite approach" (see page 10), where the study of iteration is accomplished via solutions to functional equations found by exploiting well-known identities among periodic functions. Schröder, in fact, found a function very similar to Lattès' function l (see equation (6.2)), using identities among the Jacobi functions. However, he did not explore the dynamics of this particular function because, using the language of Lattès, his motivation in this instance was not to use parametric iteration to solve the fundamental problem of iteration, but rather to demonstrate how one might discover functional equations that easily gave rise to parametric iteration.[45]

It is in the context of this approach that the result that Lattès is best known for emerges.

7.5.1. The Lattès functions. While the Académie seemed most enamored with his study of iterations of functions with more than one variable, Lattès is probably best known for the function

$$l(z) = \frac{(z^2+1)^2}{4z(z^2-1)}, \tag{7.6}$$

which we discussed briefly in Section 6.6. To many researchers this function, known as a Lattès map, is best known as the first example of a function whose Julia set is the entire sphere.[46]

Both Fatou and Julia commented upon this function, as we have noted, and although it was not mentioned in either the internal or published reports on the competition, Humbert's internal report strongly suggests it was included. Not only does he allude to examples of iteration of single variable functions that were included, he also indicates that one involved the Weierstrass $\wp$ function, which as we will soon see, is intimately involved with this example.

While this is perhaps the most famous example of a function whose Julia set equals the Riemann sphere, to Lattès its importance was that it epitomized his approach to iteration. As we noted in the previous chapter, he viewed the fundamental problem of iteration as characterizing the set $O^+(z)$, or as he stated it,

> *Determine the derived set E' of the set of consequents z_n* [i.e., $R^n(z)$] *of a point z arbitrarily given.* [**1918a**, p. 27]

Lattès was able to solve this problem for the function l.

Lattès obtained l from the duplication formula for the Weierstrass $\wp$-function with $g_2 = 4$ and $g_3 = 0$,[47]

$$\wp(2u) = \frac{(\wp^2(u)+1)^2}{4\wp(u)(\wp^2(u)-1)}. \tag{7.7}$$

It follows immediately that $\wp$ satisfies the Poincaré equation

$$l(\wp(u)) = \wp(2u),$$

as well as the parametric iteration formula

$$l^n(z) = l^n(\wp(u_z)) = \wp(2^n u_z),$$

---

[45] See (7.13) in Section 7.5 and the surrounding material for more details on Schröder's approach. It should be noted that Schröder used the method of parametric iteration to demonstrate the convergence properties of Newton's method for the quadratic, as described in Alexander [**1994a**, pp. 16–19].

[46] As we will see, it was not the first such example. For more about Lattès maps and functions whose Julia set is the sphere, see Milnor [**2004**].

[47] The general duplication formula for $\wp$ is $\wp(2u) = \frac{1}{4}\left(\frac{\wp''(u)}{\wp'(u)}\right)^2 - 2\wp(u)$. Setting $g_2 = 4$ and $g_3 = 0$, a formula for $\wp''(u)$ can be deduced from the differential equation $(\wp'(u))^2 = 4\wp^3(u) - g_2\wp(u) - g_3$. Inserting both of these expressions into the duplication formula yields equation (7.7).

since $\wp(u_z) = z$ always has a solution.

To characterize the iterates of l, Lattès suggested expressing u in the form $u = 2\omega a + i2\omega b$, where a and b are expanded in base two. From this he concluded that the closure of $O^+(z)$ was in general an oval of Descartes, hence he had solved what he termed the fundamental problem of iteration for l [**1918a**, p. 28]. He did not provide details for his assertion, but his expansion of u gives us a means to see that the repelling periodic points for l are dense in a fundamental domain of $\wp$. Since a fundamental domain is mapped onto $\mathbb{C}$ by $\wp$, it follows quite easily that $J = \overline{\mathbb{C}}$.

To find repelling points of order n for l, one need only find points u that satisfy

$$\wp(u) = \wp(2^n u), \tag{7.8}$$

which can be accomplished by expressing u in binary form. If the binary expansion for u is periodic with period n, then since $\wp(2^n u) = \wp(u)$, u satisfies equation (7.8), and hence is a period n repelling point whose multiplier is 2^n. Such points are dense in the fundamental domain of $\wp$. See Color Plate 9 for a visualization of the iteration of l.

Caption for Color Plate 9: (See p. 66.) The dynamics of Lattès' example

$$l(z) = \frac{(z^2+1)^2}{4z(z^2-1)}$$

is depicted. The iterative index is set to five on the left and eight on the right. Colors correspond to the value of $f(z)$ beneath that particular color, and thus the repetition of color and pattern indicate the pervasiveness of periodic points. These graphics are rendered using the technique of *domain coloring* which is explained in Section P.3 of Appendix P.

Lattès did not describe the denseness of periodic points. Three weeks later, Julia did precisely that, remarking that in any fundamental domain of $\wp$, there is a dense set of repelling points of l [**1918b**]. Julia expressed skepticism about the import of Lattès' example, presumably because the properties of functions like l derived from the rather unique nature of the functions that gave birth to them. Repeating a quotation from our previous chapter, Julia remarked,

> *The oddities that Lattès noted seem less surprising; it is a very special case.* [**1918c**, p. 153]

How, then, did Lattès understand the iteration of l? We are certain he understood it perfectly, but within the confines of his investigation. His conceptual framework for iteration was not based on normal families, hence it did not discuss such concepts as the domain of normality and nonnormality. The import of this example for him was that it exemplified his approach to solving what he considered the fundamental problem of iteration, namely, characterizing the iterates of an arbitrary point on the sphere. Although he did not describe the periodic points of l topologically, given that he clearly understood the role of the binary expansion of a number in determining its fate under iteration, it is hard to imagine that he did not understand how to identify periodic points or that they formed a dense set on $\overline{\mathbb{C}}$. So had someone explained to him what a Julia set is, and told him that in this example the Julia set is the entire sphere, he would have most likely nodded in complete agreement. But he perhaps would have added, that while this is all very interesting, the point of his example is that it allowed him to characterize orbits of individual points, or as he put it,

> *One can thus completely resolve the various problems relative to the iteration of* (7.6). [**1918a**, p. 28]

Lattès hoped to repeat this success with other functions, and he saw in the Poincaré equation $f \circ \theta = \theta \circ (s \cdot I)$ the means to accomplish this.

He cited a similar example, noting simply how for $g(z) = 2z/(1 - z^2)$, the Poincaré function is $\tan(u)$ [**1918a**, p. 27]. Although he did not go into further detail, this follows from setting $z = \tan(u)$ in $g(z)$, which yields the the well-known identity

$$g(\tan(u)) = \frac{2\tan(u)}{1 - \tan^2(u)} = \tan(2u).$$

Schröder exploited this same identity in his paper [**1871**] in his examination of Newton's method (see Section 1.4.3), where, for an appropriate branch of the inverse tangent, he substituted $u = \arctan(iz)$ into the above expression for $\tan(2u)$, and, multiplying the result by $-i$, he obtained the function $f(z) = 2z/(1 + z^2)$, which satisfies the Schröder equation $S(f(z)) = 2S(z)$ where $S(z) = \arctan(iz)$. Indeed, Schröder used the technique that Lattès referred to as "parametric iteration" to iterate f via $f^n(z) = S^{-1}(2^n S(z))$.

Humbert's internal report states that Lattès also considered the function $f(z) = 2z^2 - 1$ [**1918**, p. 21]. Setting $z = \cos(u) = \theta(u)$ leads to the Poincaré equation, $f(\cos(u)) = \cos(2u)$.

Ritt's approach to iteration in his *Comptes Rendus* notice of March 4 [**1918a**] shared Lattès' emphasis on the Poincaré equation. We discuss Ritt's study of iteration in Section 13.2.

7.5.2. Lattès and the iteration of functions of two variables. In the introduction to his monograph of rational functions, Fatou cited several studies involving the study of iteration of functions of two variables and noted such an investigation involves additional difficulties. That was why he shied away from it in his monograph, although he would come to study it in the 1920s, in particular in [**1924**].

One of the works that Fatou quoted was Lattès' Ph.D. dissertation [**1906**] (see Sections 3.5 to 3.8), which was, at 137 pages, the longest of the works Fatou referenced. Lattès followed his thesis with several shorter works prior to his prize entry.

Due to our reliance on second-hand sources and Lattès' three short notices, we only know fragmentary details about his entry—for example, we know neither its length nor the proportion of it devoted to the iteration of two variables. Yet, Lattès' entry is clearly evidence of a continuing interest in the iteration of multivariable functions, and, as the quotation at the end of this section suggests, it was not intended as his last word on the subject.

Based on Humbert's report and Lattès' *Comptes Rendus* notice [**1918b**], one of the topics he investigated was functions of the form

$$(x, y) \mapsto (R_1(x, y), R_2(x, y)), \tag{7.9}$$

where the R_k are rational functions from $\mathbb{C}^2$ into $\mathbb{C}$.

A fixed point $A = (x_0, y_0)$ of this mapping satisfies

$$(R_1(x_0, y_0), R_2(x_0, y_0)) = (x_0, y_0),$$

whose multipliers s_1 and s_2 are given by the eigenvalues of the Jacobian for the transformations evaluated at the fixed point.[48] Lattès focused only on functions with at least one fixed point both of whose multipliers s_1 and s_2 are greater than one in modulus.

Relying on work he had done in [**1911b, 1911c**], as well as results Picard established in [**1904, 1913**], he found meromorphic solutions ψ and χ to the functional equation

$$\begin{aligned} \psi(u_1, v_1) &= R_1(\psi(u, v), \chi(u, v)), \\ \chi(u_1, v_1) &= R_2(\psi(u, v), \chi(u, v)) \end{aligned} \tag{7.10}$$

in two cases:[49]

(1) If neither s_1 or s_2 is a integral power of the other, then $u_1 = s_1 u$ and $v_1 = s_2 v$ in a neighborhood of the fixed point.
(2) If one is a power of the other, for example, $s_1 = s_2^n$, then $u_1 = s_2^n u - kv^n$, where k is zero or one, and $v_1 = s_2 v$. Lattès indicated that $k = 1$ was the general case.

Although there is no explicit reference to it in the notice [**1918b**], it is quite evident that, in carrying out this analysis for the case (1), at least, he built upon an investigation originally carried out in his Ph.D. thesis [**1906**].

Early in *Chapter I* of his thesis, Lattès focused on the iteration of the map (7.9), but as a transformation of $\mathbb{R}^2$:

> *Suppose now that the variables are real and, given a point x, y inside a domain containing an isolated double* [i.e., fixed] *point, we propose to study the arrangement of the consequents and antecedents of this point.* [**1906**, p. 20]

After citing Leau's study of functional equations of more than one variable from [**1897**], Lattès extended Kœnigs' solution of the Schröder equation in one complex variable to the case of two unspecified variables in *Chapter III*, and observed that, under the assumption that the multipliers are all less than (or respectively, greater than) one in modulus,

> *...one can obtain the solution to this equation* [i.e., the Schröder equation] *via the same method Kœnigs devised in the case of a single variable by exploiting the canonical form of the substitution given in §7 and 19. We will thus extend Kœnigs' method to the case of several variables.* [**1906**, p. 57]

At this point of his thesis, however, Lattès' interest was in solving the Schröder equation, and he did not apply his results to iteration. Thus, while it seems he may have mined results from his thesis for his *Grand Prix* entry, his interest had progressed to one centered on the study of iteration, and he covered territory that was not in his dissertation. This impression only bolsters the view of his thesis as a

[48] Compare to (3.32), where the fixed point is assumed to be the origin, $x_0 = y_0 = 0$.

[49] Notice that the complex-valued functions ψ and χ induce the change of variables

$$\left\{ \begin{array}{l} x = \psi(u, v), \\ y = \chi(u, v) \end{array} \right. \quad \text{and} \quad \left\{ \begin{array}{l} x_1 = \psi(u_1, v_1), \\ y_1 = \chi(u_1, v_1). \end{array} \right.$$

bridge between the study of ODEs and the iteration of complex functions, which we touched upon in Chapter 3. If his entry is ever found, it would be very interesting to compare its contents to those of [**1906**].

Although we have no indication that Lattès made a point of it in his entry, it is also interesting to note that, in the context of his 1918 notice, the Schröder equation associated with (7.9) assumes the general form $\varphi_k(R_1(x,y), R_2(x,y)) = s_k\varphi_k(x,y)$, for $k = 1, 2$ where the s_k are the multipliers at the fixed point, and the change of variables gives rise to the φ_k

$$\begin{cases} u = \varphi_1(x,y), \\ v = \varphi_2(x,y). \end{cases}$$

It follows, then, that (φ_1, φ_2) is the inverse of the vector function whose components are ψ and χ. Therefore, under case (1) above, equation (7.10) is the (vector) Poincaré functional equation corresponding to the two Schröder functional equations above.

Not only did Lattès derive the canonical form $(u_1, v_1) = (s_1 u, s_2 v)$ stated above at (1) in *Chapter III* of his thesis [**1906**, §26, 27], but he also provided the general expression of an invariant curve (analytical or not) in the variables (u, v), analogous to what Fatou did in his analysis of the Böttcher equation (see page 154),

$$v = u^{\frac{\log s_2}{\log s_1}} \cdot \omega_{\log s_1}(\log u) \quad \Leftrightarrow \quad v^{\frac{1}{\log s_2}} \cdot \omega_{\log s_2}(\log v) = \quad u^{\frac{1}{\log s_1}} \cdot \omega_{\log s_1}(\log u),$$

where $\omega_{\log s_1}$ is an arbitrary periodic function with period $\log s_1$. In the original variables (x, y) this becomes

$$\varphi_1(x,y)^{\frac{1}{\log s_1}} \cdot \omega_{\log s_1}(\log \varphi_1(x,y)) = \varphi_2(x,y)^{\frac{1}{\log s_2}} \cdot \omega_{\log s_2}(\log \varphi_2(x,y)).$$

Lattès alluded in [**1918b**] to a two-variable version of the technique of parametric iteration he used to analyze the iteration of a single variable that we discussed above at equation (7.5), but neither Humbert's report nor Lattès' notice included any explicit examples of it. Humbert, however, referred to an example [**1918**, p. 17] involving analytic iteration in Lattès that evidently is based on case (2) above, $u_n = m^n u - E(m^2 a)$, where he noted that E was of the form $E(v)$, continuous in v.

According to Humbert's internal report, Lattès gave four examples of iteration of functions of two variables. Two were left unstated, and one of those was evidently an example of a Cremona substitution of two variables. Humbert also reported that Lattès' entry discussed Cremona substitutions C of two variables, and proved that if C has one fixed point A' whose multipliers are less than one in modulus and another A whose multipliers are both greater than one, then, in general, the forward iterates $C^n(x,y)$ of an arbitrary point (x,y) converge to A' while the backward iterates tend towards A. Lattès noted there were exceptions to this dual convergence, such as points on invariant curves. These results are also outlined in Lattès' notice [**1918b**].

It is interesting that neither Lattès in his *Comptes Rendus* notices nor Humbert in his prize report tied Lattès' investigation of Cremona substitutions to a question posed by Poincaré in his paper [**1890a**, p. 561] regarding the possible existence of regions on which iterates of a Cremona substitution converge to a common limit, which we discussed on page 37. This further reinforces the notion that while Poincaré's paper was noteworthy as the source of the Poincaré functional equation, albeit in a multivariate setting, it was not otherwise influential in the development of the French study of iteration of complex functions.

The third and fourth examples from Lattès' prize entry involved linear fractional transformation, $L(z) = (az+b)/(cz+d)$, one where z was viewed as a function of x and y, and the other

$$L(x,y) = \left(\frac{ax+b}{cx+d}, \frac{\alpha(x)y+\beta(x)}{\gamma(x)y+\delta(x)} \right).$$

Lattès concluded his notice with the remark,

> *Certain of the preceding results can be extended immediately to the case of any number of variables whatsoever.* [**1918b**, p. 153]

Neither of the *Grand Prix* reports indicate that Lattès discussed these generalizations, so we assume that Lattès intended to address them in a future work.

7.5.3. Schröder and Böttcher functions? As has been noted in other works,[50] Lattès' l was not the first function whose Julia set is the sphere. Both Schröder in [**1871**] and Böttcher in [**1898**], and again in [**1899, 1904b**] discussed functions very similar to Lattès' l that they generated in an analogous manner. It seems that only Böttcher was aware of their dynamical properties.

Schröder's approach to solving functional equations is very similar to Lattès'.[51] However, rather than use the Poincaré equation at (7.6), he relied on the Schröder functional equation, $S \circ f = c \cdot S$, where c is a complex constant. A Schröder function, that is, a solution S to the Schröder functional equation, is an inverse of the solution θ to the Poincaré equation $f \circ \theta = \theta \circ (s \cdot I)$.[52] This brings out an interesting contrast between the Schröder and Poincaré functions. Quoting Lattès:

> *But, while the Poincaré function prolonged analytically from* $u = 0$ [the preimage under θ of the fixed point λ] *gives rise to a uniform transcendental function, either meromorphic or entire, its inverse, the Schröder function, prolonged from* λ *is itself a multiform analytic function. Thus is the advantage of using the Poincaré function over the Schröder function to study the iteration of any initial point whatsoever in the plane.* [**1918a**, p. 26]

Like Lattès, Schröder sought to solve the Schröder equation not from first principles (he thought this too difficult in general), but rather by exploiting rational relationships that arise from known identities. He gave several examples. In one, exploiting the duplication formulae for the Jacobi elliptic functions, he ended up with a function very similar to Lattès' l.

The construction is as follows [**1871**, pp. 306–307]: The functions sn, cn, and dn are Jacobi elliptic functions, and cn and dn are related to sn by $\mathrm{cn}^2 = 1 - \mathrm{sn}^2$ and $\mathrm{dn}^2 = 1 - \chi^2\mathrm{sn}^2$, where χ^2 is a parameter that is neither 1 nor zero called the elliptic modulus. Jacobi's sn satisfies the relation

$$\mathrm{sn}(2u) = \frac{2\mathrm{sn}(u)\mathrm{cn}(u)\mathrm{dn}(u)}{1 - \chi^2\mathrm{sn}^4(u)},$$

---

[50] See Milnor [**2004**, §6] and Steinmetz [**1993**, p. 31, fn.]

[51] See Alexander [**1994a**, Chap. 1] for an account of Schröder's study of iteration.

[52] See page 37.

where u is defined via an elliptic integral. Squaring both sides of the above equation yields

$$\operatorname{sn}^2(2u) = \frac{4\operatorname{sn}^2(u)\operatorname{cn}^2(u)\operatorname{dn}^2(u)}{(1-\chi^2\operatorname{sn}^4(u))^2}. \tag{7.11}$$

Setting $z = \operatorname{sn}^2(u)$ and exploiting the identities for dn^2 and cn^2 given above, Schröder induced a function f from the right-hand side of the equation (7.11),

$$f(z) = \frac{4z(1-z)(1-\chi^2 z)}{(1-\chi^2 z^2)^2}. \tag{7.12}$$

By definition, $f(\operatorname{sn}^2(u))$ equals the right-hand side of (7.11) which also equals $\operatorname{sn}^2(2u)$, so

$$f(z) = f(\operatorname{sn}^2(u)) = \operatorname{sn}^2(2u). \tag{7.13}$$

This enabled Schröder to express the nth iterate of f in the form

$$f^n(z) = \operatorname{sn}^2(2^n u), \tag{7.14}$$

where u is defined by integration. At this point Schröder concluded his example because his goal with this example was to illustrate how his methods could induce formulae for the iterates of functions such as f from known identities among periodic functions.

We can also express equation (7.13) in the form of a Poincaré functional equation where $\theta = \operatorname{sn}^2$: $f \circ \theta = \theta \circ (2 \cdot I)$. As was somewhat typical of Schröder, he did not state the relation given at equation (7.13) explicitly. However, given the details with which he treated functional equations elsewhere in both [**1870, 1871**], he almost certainly understood he had defined a functional equation that was the inverse of the Schröder equation. That he left the functional equation implicit only underscores the fact that Schröder's program in [**1871**] (whose title translates as "On iterated functions") was not to find a general solution to functional equations but rather to iterate functions by any means necessary. That the so-called Schröder equation is generally the best vehicle for his purposes does not prevent him from using its inverse in a pinch.

The reader can find a modern treatment of Schröder's function f in Milnor's paper [**2004**, pp. 19–20]; Milnor also shows that like Lattès' l, the Julia set of this function is also the entire sphere. But it must be pointed out that Schröder gave no inkling of anything remotely close to such a possibility.

That said, Schröder had what can be termed an inchoate concept of a Julia set, so it seems that had he suspected that dynamics of f were peculiar, he would have pointed this out.

Schröder's paper was really the second half of a paper he published a year earlier [**1870**], whose purpose was to study iterative equation solving algorithms based on Newton's method, and in both papers he referred to sets that contemporary researchers would recognize as Julia sets.

At the end of that first paper, he referred to the boundary of the domains of convergence for Newton's method for $z^2 - 1$:[53]

> *Thus the points of the convergence region form an infinite family of points, all of which sooner or later jump to the root and cluster*

[53]The translation is G. W. Stewart's from [**1993**].

> *about it with infinite density. It is easy to give a formula by which these operations can be extended backward or forward.*
>
> *In a subsequent treatise* on iterated functions [i.e., [**1871**]], *I will give a proof that for the above algorithms and equations the entire complex plane decomposes into two convergence regions separated by the imaginary axis, which is the only exceptional line. Moreover, I entertain the conjecture that for any algebraic equation the region of exceptional values of these algorithms is only* one *dimensional and reduces to the boundary lines of convergence domains.* [**1870**, p. 361]

Schröder thus was aware that iteration of functions such as the Newton's method function could induce a partition of the plane into convergence regions divided by a one dimensional curve.

Moreover, Schröder understood that in some cases at least, iteration on this line between convergence regions follows strange rules. At the end of his second paper, after showing that the imaginary axis divides the convergence of the Newton's method function for $z^2 - 1$,

$$N(z) = \frac{z^2+1}{2z},$$

he characterized its periodic points and indicated they were infinite. Although he did not explore the behavior of points on the imaginary axis under iteration by N in great detail, he made an observation about the nonperiodic points:

> *...for purely imaginary z, $z^{(r)}$* [i.e., $N^r(z)$] *goes through infinitely many values as r increases without end.* [**1871**, p. 320]

One might exempt the function f defined at equation (7.12) from his conjecture that the curves dividing convergence regions are one dimensional since f was evidently not intended as an example of an algorithm used to solve an algebraic equation. In particular, it is not a Newton's method function for an algebraic equation, but rather an example of how to find functions to iterate. Yet, given his statements about "exceptional lines" and his brief foray into the strange dynamics of iteration on the set of nonconvergence for the Newton's method function N, it seems reasonable to surmise that he thought the convergence regions of this function were also separated by a one dimensional curve. If not, he certainly gave no indication otherwise.

The other person to offer an example of a Lattès function prior to Lattès was Łucyan Böttcher, a young Polish mathematician born in Warsaw in whose dissertation [**1898**] from the University of Leipzig appears the function,

$$f(z) = \frac{(Az^4+B)(2z^2-1)}{B+A(2z^2-z^4)}, \tag{7.15}$$

where $0 < A < 1$, and $B = 1 - k^2$. Böttcher denoted this function "Cn 2 $\mathrm{Cn}_{-1}(z)$" [**1898**, 63], and although he provided no background on either the function or its genesis, he obviously was referring to the Jacobi cosine function, and, analogously to what Schröder did, f can be easily generated from the duplication identity

$$\mathrm{cn}(2u) = \frac{\mathrm{cn}^2(u) - \mathrm{sn}^2(u)\mathrm{dn}^2(u)}{1-k^2\mathrm{sn}^4(u)}.$$

Unlike Schröder, Böttcher by this point was coming to grips with the dynamic oddities of the function defined at (7.15). To better appreciate his conception of this function, it is helpful to backtrack and map out the dimensions of Böttcher's study of iteration.

Böttcher drew from several different strains within the study of iteration of both real and complex functions, in the process citing almost all the papers that had been written on the subject by 1898. For example, his discussion of rationally neutral fixed points referred to Leau [**1897**], Lémeray [**1897a, 1897b, 1898c**], as well as the German mathematicians Eugen Netto, whose book [**1896**] contained a brief discussion of rationally neutral fixed points, and Caspar Isenkrahe, who wrote a long, very visually detailed analysis of iteration [**1897**], which contained, among other things, several examples of graphical iteration (see his Section 8.4, and in particular Figure 8.2, for a brief discussion of graphical iteration).

Although a significant portion of his thesis focused on the iteration of a single complex variable, Böttcher, like Leau, also investigated the iteration of functions of arbitrarily many, unspecified variables. Indeed, it is not until almost half way through his thesis that he turned his attention to iteration of complex functions on the Riemann sphere.

Böttcher also was interested in iteration as an infinitesimal transformation and connected his study to Lie's theory of continuous groups (see [**1898**, §17]), as well as to papers concerning iteration by Korkine [**1882**], Farkas [**1884**],[54] and Kazimierz Żorawski [**1895**].

His dissertation reads at times like summary of the field circa 1898, especially in regards to the study of the iteration of a single variable, although, perhaps because he drew equally on traditions outside the French mainstream, his study of the iteration has a markedly different character that the French study, both in terms of emphasis and terminology. For example, the iteration of a function is "periodic" not if there is a point λ satisfying $f^n(\lambda) = \lambda$, but if for all z there is a number n with $f^n(z) = z$ that is, it satisfies the Babbage functional equation, $f^n = f$, given by Charles Babbage [**1815**, pp. 410ff.] What are commonly called periodic points of order n he referred to as points of frequency ("*Rythmus*") n.

While much of what he said rehashes what was known at the time, he seems to have made a conceptual leap that would not be seen in print in the French study until Fatou's 1906 *Comtpes Rendus* notice [**1906b**]: given a particular function f, he viewed the sphere as partitioned into convergence regions by a boundary curve.

It should not be assumed, however, that Kœnigs and Leau failed to see such a division as a possibility. If one is thinking about how iteration might partition the plane, and their quotations from the end of our Chapter 1 regarding the division of the plane problem clearly indicate Leau and Kœnigs both were, it would be natural to posit the existence of a boundary curve of the sort Böttcher described. We saw above that Schröder considered this possibility, so it seems reasonable that Kœnigs and Leau did as well. Why did they not say this? Perhaps because they did not want to advance the notion on speculation alone, and were unable to find compelling evidence that it was a general property.

In any event, Böttcher did not use normal families, which was at best an inchoate concept at this point. Although the existence of such boundary curves is assumed in his dissertation rather than rigorously established, he did nonetheless

[54] See Alexander [**1994a**, Chap. 2] for an account of Korkine and Farkas' studies.

articulate some very interesting properties of these curves, including the fact that they house what would come to be called the repelling points of f, which he viewed as attracting points of an inverse of f.

It is out of this conception that he came to his version of a Lattès function. After describing how one first finds the convergence regions for attracting points, and then for periodic orbits, until one exhausts the convergence regions, he posed a very interesting question:

> *The question now is whether there are regions in which there is no convergence. In the case of*

$$f(z) = \frac{Az^4 + B(2z^2 - 1)}{B + A(2z^2 - z^4)} = \mathrm{Cn}\ 2\ \mathrm{Cn}_{-1}(z), \tag{7.16}$$

> *there exists such an area, and it extends through the entire plane. In our case the existence of such regions is by no means impossible. ... In our case they appear as regions which are completely alien to convergence.* [**1898**, p. 63]

Unfortunately, he gave no clear indication where he got the notion there was an area throughout the plane that is "alien to convergence" or whether this area was the entire sphere or some infinite subset. He did not linger much upon functional equations in this paper, and so it is doubtful this example arose out of a search for solutions of functional equations, as was the case with Lattès and Schröder.

We do know he was aware that repelling points were contained in the regions of nonconvergence, and as we have seen, repelling points are dense in the sphere for functions f that satisfy the relation $f(\theta(u)) = \theta(2u)$ where θ is a doubly periodic function, as in the Böttcher's example. Perhaps this was how he reasoned.

In the paper he wrote in Polish a year later [**1899**], he presented a second version of this function,

$$f(z) = \frac{4z(1-z)(1-\chi^2 z)}{(1-\chi^2 z^2)^2}.$$

If this looks familiar it is because it is exactly the same example Schröder gave in equation (7.12) above.

Böttcher again cited it as example of a function whose region of nonconvergence included an "area", in this case he was clear that he meant the entire sphere. Its genesis appears to be the method Schröder used that we outlined above.

This function appears again in his 1904 paper [**1904b**, p. 13], written in Russian. We were not able to translate the Russian paper, but evidently Böttcher took his analysis one step further and characterized the iteration of f on what we would call the Julia set. Quoting Milnor:

> *He was perhaps the first to think of this example from a dynamical viewpoint, and to use the term "chaotic" to describe the behavior of the sequence of iterates of f.* [**2004**, p. 20]

Böttcher's example passed virtually unnoticed. In his synopsis of Böttcher's [**1904b**], which appeared in *Jahrbuch über der Forschritte der Mathematik*, Sintzow [**1904**], made no mention of this function. Julia, Fatou, and Lattès did not discuss Böttcher's example either, or Schröder's for that matter, instead focusing on the example Lattès gave in 1918, equation (7.6) above.

The reviewer did mention Böttcher's solution to the functional equation

$$B \circ f = (B)^k,$$

in the neighborhood of a fixed point λ satisfying $f'(\lambda) = \lambda + a_k(z - \lambda)^k + \cdots$, commonly referred to as the Böttcher functional equation. Böttcher was evidently the first to solve this functional equation, but while the fact was known in French circles in the 1910s, it seems its details were not.

Fatou remarked in [**1919-1920**, p,189, II] that Böttcher "seemed to have been" the first to solve the Böttcher equation, but he gave no explicit citation, and his remarks suggests that he had not read it. In 1920 Ritt described the details of Böttcher's paper, which suggested he had, and Julia referred to Böttcher's proof in his papers on functional equations in the 1920s.

Böttcher did publish a paper in the *Annales Scientifiques de l'École Normale Supérieure* [**1909**], concerning the solution of linear systems of functional equations in which he mentioned his solution of the Böttcher equation, but it is not cited by any of the major actors in the French study of iteration, so it is unclear whether Fatou, Julia, or Lattès were even aware of it. The probable reason for Böttcher's obscurity is the language barrier. Aside from the paper in the *Annales*, Böttcher's most interesting works were in Polish or Russian.

As we have suggested, the one paper he published in German, his dissertation, might appear at first glance not to have much relevance to Fatou and Julia, had they read it. Its subject covers ground they either knew well—the basic elements of iteration—or was beyond the scope of their respective studies, for example, the application of iteration to Lie groups or the interest in multivariable iteration. Nor did it treat the Böttcher equation. Its one evident innovation, a description of the curve bounding the domain of normality, was somewhat naive when compared to their own treatment of the domain of nonnormality. What is probably its most interesting feature, function (7.16), is buried and easily missed.

7.6. Concluding remarks

The work of a number of Parisian mathematicians, Borel, Fréchet, Jordan, Lebesgue, Montel, Picard, and Poincaré, to name just a few, revolutionized mathematics around the turn of the 20th century. Tools that were not there a few years before, at least not in a recognizable, portable form—theories of measure and sets, *analysis situs*, uniformization, normal families, a modern theory of functions—became the everyday tools of the research mathematician, especially those who came of age in the early 1900s. We do not count Fatou or Julia among these revolutionaries, but they are certainly among its very first inheritors. Using these revolutionary new ideas in ways their creators did not foresee, they reshaped a moribund study into a cutting edge theory and satisfied the promise of the new mathematics.

While there were differences in their investigations, both in the way they wrote and structured their studies and in some of the results they achieved, their approaches are marked by a remarkable (but by no means untoward) convergence: in particular, each used Montel's emerging theory of normal families to develop a manner of looking at the study of iteration of complex functions that has remained intact up to the present day, and the tools and objects they discovered remain essential to its study.

The scope of their accomplishments is magnified through comparison with the work of others who addressed the subject around this time. Böttcher, Ritt, and Lattès each made important contributions to the field; as we will see, Pincherle's entry evinced a spirit not unlike that of Fatou and Julia. Yet these works, even considered collectively, fail to achieve anywhere near the breadth and depth of Fatou and Julia's studies.

To quote again from the Académie's report on the 1918 *Grand Prix* competition, "This is not the first time in the history of science that two scientists of merit simultaneously attained the same discoveries by identical methods."

CHAPTER 8

Iteration in Italy

8.1. The Italian tradition

While not nearly as robust as the French study, the Italian interest in the iteration of functions was at least as long standing. About eight years before Kœnigs published his first paper on iteration, Carlo Formenti, who eventually settled into a career as a professor of mechanics at the University of Pavia in northern Italy, wrote a brief paper on the Abel functional equation, $A(f(x)) = A(x)+c$, [**1875**]. He related the Abel equation to the formal iteration of functions, but made no mention of dynamical properties such as fixed points, their multipliers or the convergence of iterates. Rather, he established elementary algebraic properties for iteration, for example, that iteration is commutative, and used iteration to solve a particular Abel equation. Along with papers by Korkine [**1882**] and Farkas [**1884**], Formenti's was among the first wave of papers to follow Schröder. But, as was also the case with those papers, Formenti's paper was rather narrowly focused, and he evidently published no other works on iteration.

Even though the late 19th Italian study lacked a guiding figure such as Kœnigs, interest in iteration and particularly in iterative algorithms grew during the end of the 19th and beginning of the 20th centuries, as evidenced by the work of Ettore Bortolotti [**1898**],[1] Onorato Nicoletti [**1906, 1907, 1917**], and Leonida Tonelli [**1910a, 1910b**]. These works tended towards the numerical aspects of iteration, for example, the efficiency and convergence criteria of iterative methods, rather than on dynamical properties, such as the geometry of iteration or the configuration of sets of convergence and divergence. One exception to this was Francesco Podetti, who published a very brief note related to Kœnigs' work on periodic orbits [**1897**]. Like Formenti's paper, Podetti's was a bit of a one-off which engendered no significant response.

8.2. Andreoli and Tricomi

Two years into World War I, the young Francesco Tricomi published a short notice [**1916a**] while a third-year student at the University of Naples proving the following theorem:

THEOREM 8.1. *Let $f(x)$ be a continuous, real-valued in the interval $I \equiv (\lambda - a, \lambda + ma)$, where λ is a real root of the equation $f(x) - x = 0$, and a and m are two positive numbers. Suppose also that the following conditions are satisfied:*

$$x < f(x) < \lambda + m(\lambda - x) \qquad (\lambda - a < x < \lambda),$$

[1] Bortolotti was also a historian of mathematics with a particular interest in the Italian mathematicians Paolo Ruffini and Rafael Bombelli.

$$\lambda + \frac{1}{m}(\lambda - x) < f(x) < x \qquad\qquad (\lambda < x < \lambda + ma)$$

Then, for every x_0 *in* I, *the sequence* $\{x_0, f(x_0), f^2(x_0)\dots\}$ *converges to* λ.

We find his introduction to his theorem interesting because even in its pessimism, it reflects the desire that was in the air on the eve of the 1918 *Grand Prix* to extend results from the local to the global:

> *All these* [works] *seek necessary or sufficient conditions, or even necessary and sufficient conditions, which the iterated function must satisfy, be it continuous, differentiable, etc., in order to find the initial values of the independent variable which give rise to convergent sequences of iterates. And even when someone finds conditions on the initial values, they enjoy only a very* restricted *generality.* [**1916a**]

That same year Giulio Andreoli published [**1916**], which includes a slightly generalized version of Tricomi's theorem.

Later, Tricomi extended his theorems to complex functions in [**1916b**] proving that given a simple closed curve L_1 in a sufficiently small neighborhood of an attracting fixed point λ of an analytic function f, the nested sequence $\{L_k = f(L_{k-1})\}$ shrinks to λ.

SALVATORE PINCHERLE

FRANCESCO TRICOMI

FIGURE 8.1. Photo of Salvatore Pincherle by courtesy of Archivio Storico Alma Mater Studiorum Università di Bologna. Photo of F.G. Tricomi by courtesy of the Department of Mathematics G. Peano, University of Torino, Italy.

Tricomi does not quote Fatou [**1906b**] in [**1916b**]; moreover there is no characterization of the attracting fixed point α by its derivative.

8.3. Pincherle's encyclopedia entries

The central figure in the Italian study is Pincherle. One of his colleagues, Ugo Amaldi, observed in the late 1930s that Pincherle often followed tangential directions in his researches, meaning, we presume, that Pincherle was not hesitant to explore offshoots from the main course of his research (Amaldi [**1954**, p. 14]). This would seem to be the case with his interest in the iteration of complex functions, whose genesis seems to be his assignment to write a section on functional equations and operations for the *Enzyklopädie der Mathematischen Wissenschaften* [**1906**], which he revised and expanded for the French version, the *Encyclopédie des Sciences Mathématiques Pures et Appliquées* [**1912**].

Each version is divided into two parts, one on functional operations, the other on functional equations, and it is the latter that includes his survey of the iteration of complex functions. At the beginning of the section on functional equations, he observed that the study of certain functional equations—and by a functional equation he meant any equation where a function is the unknown—have become fields of their own, such as differential, integral equations and the calculus of variations. He stated the purpose of his article as grouping together those functional equations that have not yet achieved the status of becoming their own field:[2]

> *Due to the development of these various other fields* [which already have their own entries] *the Encyclopédie decided to include this article. This is the reason why we treat particular functional equations which do not already belong to the categories already listed above; among those equations, we will only focus on those ones which have some historical interest, possess a special theoretical importance, or are distinguished by the methods used to solve them.* [**1912**, pp. 73–74]

Pincherle's survey of iteration and the related functional equations in the *Enzyklopädie der Mathematischen Wissenschaften* consisted of half a dozen sections, §21–26, over eight pages [**1906**, pp. 790–797]. He gave a standard account of the field beginning with Schröder and focusing on Kœnigs and his students Grévy and Leau. He also included brief discussions concerning Netto in Germany, Bourlet and Lémeray in France, as well as Formenti and Bortolotti in Italy.

He more than doubled the length of his survey of iteration in the *Encyclopédie des Sciences Mathématiques Pures et Appliquées*, devoting ten sections, §31–40, and over 20 pages to it [**1912**, pp. 51–72]. Most of the additional material involves the iteration of functions of several variables and seems prompted by Lattès' thesis [**1906**], which Pincherle linked to works by Poincaré and Levi-Civita in a new section, §40, entitled "*Interprétation géométrique: courbes et surfaces invariants; stabilité*",[3] whose focus is similar to our discussion in the latter portions of Chapter 3.

The treatment of the iteration of complex functions is much the same in the two versions, although he reordered it a bit for the French version and included new material on Newton's method. He also added a few more mathematicians to his discussion, including a purely bibliographic reference to Böttcher's paper [**1904b**], written in Russian, that mentioned neither the Böttcher functional equation nor

---

[2] We quote from the French entry, but it applies equally to the German version.

[3] "Geometric interpretation: invariant curves and surfaces; stability"

the fact that it can be used to treat fixed points whose multiplier is zero. Nor did he discuss Böttcher's functions (7.15) or (7.16), whose Julia sets equal the sphere.

What is perhaps even more surprising is the omission of Fatou's 1906 *Comptes Rendus* notice. It could be that Pincherle found Fatou's interest in the set of points which do not converge under iteration as running contrary to the direction of his own survey, since Fatou's study did not focus on finding solutions to functional equations, although it explicitly referred to the Schröder equation, as well as other functional relations.

Although Pincherle discussed the Poincaré functional equation, he does not refer to it as such, which suggests that Poincaré's name had not yet become attached to it. Valiron's application of the Poincaré equation to functions of a single complex variable, as we discussed in Section 3.1 and 7.3.8, would not come for another year.

Pincherle had already enjoyed a long mathematical career prior to his interest in iteration. For the last two decades of the 19th century he investigated Taylor series, analytic functions, functional operators and their calculus, both differential and difference equations, as well as continued fractions.

His emphasis on the functional equations related to iteration, rather than the dynamics of iteration, may well stem from his strong background in functional analysis; this notion is reinforced by the fact that the sections prior to his survey of iteration (we reference the French version in what follows) deal with such topics as functional operators and their calculus, linear operations on functional spaces, and applications to both integral and differential equations, including the Laplace transform and Bessel functions.

The tendency in his encyclopedia entries to use iteration as a means to study functional equations rather than vice versa, as with the case with Schröder and Kœnigs, is reflected in the opening passage of the first section on iteration from the French edition entitled "*Le calcul d'itération*",[4] §33:

> *The equations we considered in Section 32* [i.e., the Schröder and Abel equations] *belong to a chapter of analysis where one finds a large number of other functional equations each generated by the same process, whose applications are as numerous as they are varied: the iterative calculus.* [**1912**, p. 55]

In the next section we will discuss how his research program in iteration, perhaps motivated by the 1918 *Grand Prix*, evolved from a position subordinate to his interest in functional equations to one of primacy.

8.4. Early steps: Pincherle and graphical iteration

Before we investigate the body of Pincherle's work on iteration, we will begin with a bit of a mathematical curiosity: Pincherle's depiction of a technique called graphical iteration, a means of locating an attracting fixed point of a real function f that is twice differentiable, in his paper [**1914**]. Graphical iteration provides a simple visual model for the convergence of the iterates of a monotonic function to that fixed point. While interesting and used in many dynamical systems and numerical methods classrooms, its value is illustrative, and Pincherle used it to introduce a subject that his audience might not be familiar with:

[4] "The iterative calculus"

> *In the very interesting, but largely unexplored study of the iteration of analytic functions, the simplest questions are of such difficulty that one finds solutions only in special cases. Among the notable questions, one which has captured the attention of mathematicians for some time, we recall the function whose nth-fold iterate is a given function $[f^n(z) = f(z)]$. A special case of this problem is Babbage's equation ... Solutions only apply to special cases and a systematic treatment has not been attempted ....* [**1914**, p. 75]

One of the equations he looked at in conjunction with this exploration is the Pfeiffer functional equation, discussed in Section 5.7 during our discussion of Pfeiffer's paper [**1918**],

$$f[f(x)] = \phi(x). \tag{8.1}$$

To illustrate the concept of an attracting fixed point of a real function, Pincherle used the diagrams reproduced in Figure 8.2. To demonstrate convergence to the attracting fixed point of f, one first draws the graphs of $y = f(x)$ and $y = x$. Obviously, the fixed point of f coincides with the intersection the two graphs.

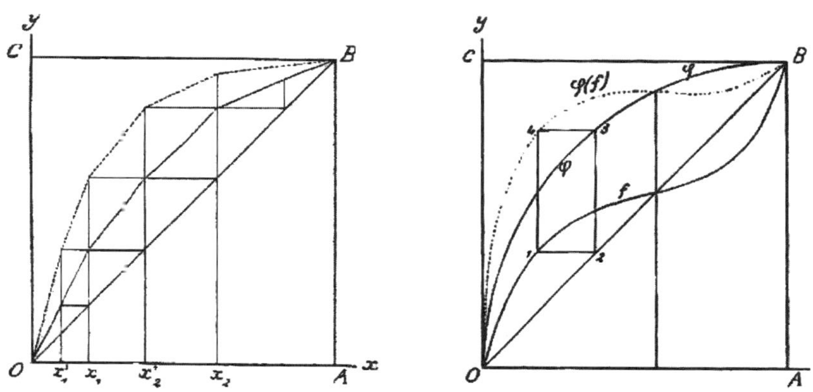

FIGURE 8.2. Graphical iteration. Two original drawings of graphical iteration in Pincherle's [**1914**, p. 75–76].

Using the left graph from Figure 8.2, follow x_1 from the real axis up to one of the two curves to find $y_1 = f(x_1)$. Next, go right to the line $y = x$ to locate $x_2 = f(x_1)$, and then upward to the graph to locate the point $(x_2, f(x_2))$. Continuing this process, one eventually arrives arbitrarily close to the attracting fixed point B.

One of the earliest depictions of this method in the context of complex dynamics occurs in 1897 in Lémeray's work [**1897a**, pp. 315–318].[5] The roots of this method, however, go back much farther and seem to have their origins in the work of Adrien-Marie Legendre and Évariste Galois. Writing in the *Bulletin des Sciences Mathématiques* in 1830, Galois remarked,

> *Legendre was the first to notice that, when an algebraic equation was written in the form*
>
> $$\phi x = x,$$

[5]See Alexander [**1994a**, pp. 64–65] for a reproduction of Lémeray's diagram.

> *where ϕx is a function in x which increases along with x, it is easy to find the root of this equation if for a nearby a, smaller than the root, $\phi a > a$, or for nearby a larger than the root, $\phi a < a$.*
>
> *To show this one draws the curve $y = \phi x$ and the line $y = x$. Given an abscissa $= a$, suppose, to fix ideas, that $\phi a > a$. I say that it will be easy to obtain the nearby root which is larger than a. In fact the roots of the equation $\phi x = x$ are nothing but the values of the intersection points of the line and of the curve, and it is clear that one approaches the intersection point by substituting ϕa for a. One will find a closer and closer value as one assumes $\phi\phi a$ $[\phi^2(a)]$, $\phi\phi\phi a, \ldots$ and so on.* [**1830**, p. 413]

The work to which Galois refers is Legendre [**1816**, p. 32ff.], which employs a method similar to graphical iteration to solve $\phi(x) = x^n$ when ϕ is increasing for $x \geq 0$. Joseph Fourier used a method very close to what Pincherle depicts in Figure 8.2 as a tool for the solution of equations [**1822**, pp. 343–347].[6]

8.5. Pincherle: From functional equations to... Pincherle-Julia sets?

In the paper [**1916**] Pincherle linked his study of the functional equation

$$\varphi(z) - k\varphi(\lambda(z)) = f(z), \tag{8.2}$$

which arises from the Fredholm equation, to the Schröder equation by assuming that $f(z) \equiv 0$, in which case equation (8.2) becomes the Schröder equation

$$\varphi(\lambda(z)) = \frac{1}{k}\varphi(z).$$

Pincherle used his results regarding (8.2) to find a solution of the Schröder equation, which he used to iterate λ via the relation

$$\varphi(\lambda^n(z)) = \frac{1}{k^n}\varphi(z).$$

The emphasis in his paper was on functional equations, rather than iteration, but he noted in [**1925**, p. 354] that [**1916**] represents the transition to his study of iteration.

An interest in iteration for its own sake seems to have taken hold over the course of Pincherle's papers [**1917a, 1918b, 1918d, 1919, 1920a, 1920b**], which treats the iteration of $\alpha_a(z) = z^2 - a$ where a is a positive real parameter.[7] What caused this change? Did the paper [**1916**] trigger a research interest in iteration, as the remark above suggests? From his encyclopedia articles it is clear he was knowledgeable about the field, so perhaps he was swept up by one of the tangential pursuits that Amaldi described above. Or was it the *Grand Prix*, that was announced around the time of the 1916 paper?

That he submitted an entry for the 1918 *Grand Prix* suggests the competition may have been a substantial motivation. Although it was never published, the entry [**1917b**] is the most comprehensive and unified treatment of his iteration research program, and we will use it as the basis of our discussion.

---

[6]See Galuzzi [**2001**] for an interesting discussion of Legendre [**1816**] and Galois [**1830**]; for more about the history of graphical iteration see Rosa and Alexander [**prep**], forthcoming.

[7]Pincherle began his study with the case where z is real, but quickly expanded it to the complex case.

One of the foci of Pincherle's attention was the set that has come to be called the Julia set. He did not view this set as domain of nonnormality, or alternatively as the closure of the repelling periodic points, as Julia and Fatou did, but rather as the boundary Γ of the set of points Ω which converge to infinity under iteration. In contemporary parlance, this set is known as the *basin of infinity*, and for polynomials, the boundary of the basin of infinity equals the Julia set. Pincherle attributed to Γ many of the properties we associate with Julia sets and, as will be discussed in the next section, it seems he was unaware that Fatou and Julia were embarking upon their studies. If so, Pincherle came to the discovery of Julia sets independently.

8.6. The Pincherle–Mandelbrot set? Pincherle's *Grand Prix* entry

Pincherle's handwritten submission to the 1918 *Grand Prix des Sciences Mathématiques*, entitled *Sur l'itération de la substitution* $x^2 - a$, was returned to him without being considered for the prize. Nor was he mentioned by name in the formal report published in *Comptes Rendus* upon the awarding of the prize in December 1918 [**CR-1918**, pp. 811–814].[8] The reference to his submission was fleeting: The Académie merely remarked that three entries had been received and only those of Julia and Lattès were retained.

It was not unusual for the Académie to refuse to consider a submission. Entries judged not to be in serious contention for an award were generally returned. As we mentioned in Section 6.3, the only other work submitted for the 1917 *Prix Bordin* besides Julia's prize-winning entry was likewise returned to its author. Perhaps the Académie felt that Pincherle's focus on just a single family of functions, $z \mapsto z^2 - a$ where $a > 0$, was far too narrow when compared with the extraordinary breadth of Julia's monograph and, therefore, it was not a suitable candidate for its *Grand Prix*.

To think that Pincherle's work was insubstantial, however, would be to underestimate it. While there is no doubt Julia's work is of far greater range and scope, Pincherle's treatment of $\alpha_a(z) = z^2 - a$ is impressively close to the kind of thinking that led to the creation of the Mandelbrot set. He also described several essential elements that are present in the studies of Fatou and Julia: the partitioning of the sphere, a classification of Julia sets, algorithms to construct them, a study of domains of convergence, a demonstration of the existence of attracting orbits, and the technique of symbolic dynamics—but only for the quadratic. Like Lattès, Pincherle did not use normal families, relying instead on direct calculation. One can make substantial progress with such an approach when the family is as elementary as $z \mapsto z^2 - a$, and Pincherle did. However, as was the case with Lattès, the very nature of his approach limited it.

Pincherle confined his treatment to $z \mapsto z^2 - a$ in reaction to what he perceived as obstacles inherent to the study of iteration:

> *This problem and the many questions which surround it present considerable difficulties, and we lack methods which permit us to approach the subject in a general fashion and obtain concrete results. In the presence of these difficulties, it is reasonable to think that the examination of a particular case, even a very elementary*

[8]See Appendix A for a translation of this report.

> *one, is not without interest; it can highlight the nature of the obstacles we need to overcome, and perhaps serve us better than a general investigation. It is with this remark that I justify the study of the iteration of the extremely simple substitution that is the subject of this short work. As simple as the example I treat is, it leads to interesting results, and is not without difficulty; indeed, several issues arise that extend to much more general cases.* [**1917b**, p. 2].

These sentiments were echoed in another paper from around this time, where he expressed the hope that his exploration of the family $\alpha_a(z) = z^2 - a$ would "indicate possible directions for more complicated cases" [**1918d**, p. 337].

He gave one example in his entry of the kind of extension to more general cases he had in mind, when, immediately following the above quotation, he noted that the results he had obtained regarding α_a could be applied to general second degree rational functions via the conjugacy $T = L^{-1} \circ \alpha_a \circ L$, where L is a linear fractional transformation. He did not specify the results he had in mind—perhaps they involved the topological properties of Ω and Γ that would be preserved under continuous mappings—and he discussed this particular example no further. Nor did he reveal how his results might apply to rational functions of higher degree.

That Pincherle thought the nature of the topic too forbidding for a general treatment is not only a classic example of bad timing, since Fatou and Julia would soon develop exactly the kind of general methods he thought were beyond reach, but it also suggests that he was unaware of the approach they were taking, assuming he was even aware they were also pursuing the study of iteration.

Given the December 31, 1917, deadline for submission, it is not surprising to find no mention of the December *Comptes Rendus* notices of Fatou and Julia, [**1917b**] and [**1917h**], respectively. What is surprising is that there is no mention of Fatou's earlier notices, either his entry from May [**1917a**] or [**1906b**], which was already over a decade old. Pincherle pointed out connections between his work and that of Kœnigs, Grévy, and Leau, and the absence of a reference to Fatou's notices is striking.

We do not know the date of completion or submission for Pincherle's entry, although it reached the Académie in time to meet its deadline so it is entirely possible he completed it prior to May 1917. But it is interesting that he did not mention the notice from 1906 since it contains a means to construct Cantor sets (see Color Plate 2) very similar to one Pincherle used, to be discussed below. Indeed, if Pincherle commenced his study prior to the appearance of Fatou's May 1917 notice, then Fatou's earlier notice would have been the only extant work to discuss such sets in the context of iteration. Surely, Pincherle would have referred to this fact had he known it, and, the fact that he did not, suggests he was unaware of it. As we noted above, Pincherle's 1912 encyclopedia article did not reference Fatou's 1906 notice either. Nonetheless, perhaps this omission struck the Académie negatively, and factored into their decision to refuse to consider his manuscript.

The principal accomplishment of Pincherle's entry was the creation of an inchoate version of the Mandelbrot set that classified the nature of the Julia set according to the magnitude of the parameter a of $z \mapsto z^2 - a$, with $a > 0$. Given the elementary nature of the segment $a > 0$, he obviously did not explore the structural richness of the Mandelbrot set that has intrigued modern researchers since its inception around 1980.

His investigation of the effect of the parameter a on the structure of Γ came in stages, and the case where $a > 2$ occupied close to half of his entry, in part because a good deal of the methodology and terminology he developed could easily be applied to the other cases. We also will focus on the $a > 2$ case, and then summarize his results for $a \leq 2$.

It is interesting that functional equations, which at one point was the motivation of Pincherle's interest in iteration, played a very subordinate role in his entry. While he did not avoid them, he confined his treatment of them to the final section, which makes it a coda to his study rather than an integral part of it.

8.6.1. Laying the groundwork: fixed points, *foyers*, *sommets*, and *chaînes*. The fixed points of α_a are given by

$$l = \frac{1+\sqrt{1+4a}}{2} \quad \text{and} \quad l' = \frac{1-\sqrt{1+4a}}{2}. \tag{8.3}$$

A quick inspection shows that l is always repelling and positive; likewise, l' is always negative and attracting when $-1/4 < a < 3/4$. Although Pincherle limited his investigation to the case where $a > 0$, he observed that he could easily extend his analysis to the case where $a > -1/4$, after which l' ceases to be an attracting fixed point. While the fixed point l is essential to his study, the role of l' is comparatively minor, and he did not discuss its nature until the second half of his entry.

Pincherle's principal concern was the structure of Γ, and he relied on two primary methodologies to describe it. One, geometric in nature, involved taking the infinite intersection of the preimages of the disc D centered at the origin and passing through the repelling point l, [**1917b**, pp. 23–27], that we will discuss below in Section 8.6.4. Indeed, this is the construction referred to above that is similar to the one Fatou used in his 1906 notice. The other involved looking at nested square roots which, depending on their construction, he called *sommets*, *foyers*, and *chaînes*, and involved an application of the technique of symbolic dynamics similar to Fatou's.[9] While perhaps not as elegant as Fatou's version, it helped him to characterize the elements of Γ (see Section 8.6.2).

A *foyer* [**1917b**, p. 8] is a preimage of zero expressed as a nested radical,

$$\epsilon_0\sqrt{a+\epsilon_1\sqrt{a+\epsilon_2\sqrt{a+\cdots+\epsilon_n\sqrt{a}}}}, \tag{8.4}$$

where ϵ_k is one or minus one. The first preimages are found by solving $z^2 - a = 0$. To find the preimages of rank two, solve $(z^2-a)^2 - a = 0$.

Following the same line of reasoning, a preimage of a point y is of the form

$$\epsilon_0\sqrt{a+\epsilon_1\sqrt{a+\epsilon_2\sqrt{a+\cdots+\epsilon_n\sqrt{a+y}}}}, \tag{8.5}$$

which he called a *chaîne* [**1917b**, p. 16]. Thus, a *foyer* can be viewed as a special case of a *chaîne*. He expressed the limit of a *chaîne* or a *foyer* in the form of an infinitely nested radical,

$$\epsilon_0\sqrt{a+\epsilon_1\sqrt{a+\epsilon_2\sqrt{a+\epsilon_3\sqrt{a+\cdots}}}}. \tag{8.6}$$

[9]Fatou represented the points of particular Julia sets as base-d sequences, where d is the degree of the rational function under consideration. See page 132.

Finally, if the point y in the expression (8.5) is replaced by negative one times the positive repelling fixed point l (see equation (8.3)), he formed what he termed a *sommet* [**1917b**, pp. 8–9]:

$$\epsilon_0\sqrt{a+\epsilon_1\sqrt{a+\epsilon_2\sqrt{a+\cdots+\epsilon_3\sqrt{a-l}}}}. \tag{8.7}$$

The study of these nested radicals is crucial to his understanding of Γ. For example, near the beginning of [**1917b**], Pincherle showed that Γ is invariant under iteration by α_a. Because $\alpha_a(-l)=l$, the *sommets* $\pm\sqrt{a-l}$ are also preimages of l, as is any *sommet*, (8.7). Since $l\in\Gamma$, the invariance of Γ implies that all *sommets* are in Γ.

The nested radicals also provided him with the means to express the repelling fixed point l as the infinite *foyer* [**1917b**, pp. 9–10],

$$\sqrt{a+\sqrt{a+\cdots}}. \tag{8.8}$$

He began with the $a>2$ case, where he showed that $\sqrt{a+\sqrt{a-\cdots}}$ is bounded from above, hence has a limit p: first, he observed that $a>\sqrt{a}$ implies that

$$a>\sqrt{2a}>\sqrt{a+\sqrt{a}}, \tag{8.9}$$

and, in a brief and informal inductive procedure, extended this to a *foyer* of n nested positive radicals, concluding that $\sqrt{a+\sqrt{a+\cdots}}$ is bounded from above by a. He pointed out that $\sqrt{a+\sqrt{a+\cdots}}$ is likewise bounded from above when $0<a\le 2$ (he did not say why but it follows from the fact that the sequence $\left\{\sqrt{a+\sqrt{a+\cdots}}\right\}$ is monotonic). Thus, for $a>0$, each infinite *foyer* $\sqrt{a+\sqrt{a+\cdots}}$ has a limit p depending on a.

To determine p, observe that

$$\sqrt{a+p}=\sqrt{a+\left(\sqrt{a+\sqrt{a+\cdots}}\right)}=p.$$

Routine calculations show that $\sqrt{a+p}=p$ implies that p is a fixed point of α_a; since $p>0$, it must equal the positive fixed point l.

He remarked that the representation of l as infinite *foyer* yields the well known identity

$$2=\sqrt{2+\sqrt{2+\cdots}},$$

when $a=2$.

8.6.2. Pincherle's method of symbolic dynamics.

Pincherle devised a method to symbolically represent a *foyer* as a sequence of the form $(r_0, r_1, r_2, \ldots, r_k)$, where the number of commas represents the number of negative signs and the r_k is the number of a's between the negative signs [**1917b**, p. 16]. Thus the *foyer*

$$\sqrt{a-\sqrt{a+\sqrt{a-\sqrt{a+\sqrt{a}}}}}$$

is represented by $(1, 2, 2)$, and

$$\sqrt{a - \sqrt{a - \sqrt{a - \sqrt{a - \sqrt{a}}}}}$$

by $(1, 1, 1, 1, 1)$. This notation was also applied to *chaînes*:

$$\sqrt{a - \sqrt{a + \sqrt{a + \sqrt{a + y}}}} = (1, 3(a + y)).$$

An infinite *foyer* is represented by an infinite sequence of r_k, and if it terminates in an infinite succession of positive radicals, the infinity symbol is used. For example,

$$\sqrt{a - \sqrt{a - \sqrt{a + \sqrt{+ \cdots}}}} = (1, 1, \infty).$$

Since $l = \sqrt{a + \sqrt{a + \cdots}} = (\infty)$, Pincherle's symbolic expressions ending with infinity are well defined since they can be rewritten as a finite expression in terms of l, as the following suggests:

$$(1, 1, \infty) = \sqrt{a - \sqrt{a - l}} = (1, 1(a - l)). \tag{8.10}$$

This also suggests that any symbolic term ending with an ∞ is a *sommet*.

His system contains one particular idiosyncrasy: he insisted that the first coefficient r_0 of any sequence be a natural number which limits his sequential representation to positive radicals [**1917b**, p. 15]. The reason for this was that his applications of his system were to show generic properties that could, without loss of generality, be restricted to positive radicals; it also simplified certain calculations, as we will see. However, this restriction cost him flexibility, and in particular, it makes it difficult to use his system to represent periodic points since not all periodic sequences of the r_k correspond to periodic points; rather, they correspond to points that are eventually periodic. For example, while routine calculations show that $(n - 1, n, n, \dots)$ corresponds to a periodic point of order n, the point represented by the sequence $r = (n, n, n, \dots)$ is eventually periodic with period n but not periodic, since $\alpha_a^k(r) = \alpha_a(r)^{k-n}$ for $k \geq n + 1$, but $\alpha_a^n(r) = -(n, n, \dots) \neq r$. When he needed to represent a negative *foyer*—for example to discuss periodic points—he switched back to the explicit radical notation given above in (8.4).

One of the virtues of Fatou's system of symbolic dynamics, to which Pincherle's is a close relative, is that a periodic base d sequence corresponds in a one-to-one manner to a periodic point.[10] Despite the self-imposed limitation of considering only positive sequences, Pincherle manipulated *foyers*, *sommets*, and *chaînes* in sequential form to model iteration, and in the process he explained properties of iteration, so his notational system represents a method of symbolic dynamics, although one not as robust as Fatou's.

Pincherle used his notation to show that any *sommet* is the limit of a *foyer*: From expression (8.8) it follows that l can be viewed as the limit of δ_n, where

[10] Modulo the base-d sequences that represent the same point, for example $(d, 0, 0 \dots)$ and $(0, d, d, \dots)$.

$\delta_n = (n) = \sqrt{a + \cdots + \sqrt{a}}$ and there are n a's. Since an arbitrary *sommet* z is of the form, $z = (r_0, \ldots, r_k(a-l))$, it follows that

$$z = \lim_{n\to\infty} \epsilon_0 \sqrt{a + \epsilon_1 \sqrt{a + \cdots + \epsilon_k \sqrt{a - \delta_n}}} \qquad \text{where } \epsilon_j \in \{0, 1\}.$$

Alternatively, z can be viewed as the limit of a sequence of *foyers* of the form

$$\left\{ \epsilon_0 \sqrt{a + \epsilon_1 \sqrt{a + \cdots + \epsilon_k \sqrt{a - \delta_n}}} \right\}. \tag{8.11}$$

8.6.3. The set Γ is the closure of the set of *sommets*. One of the most interesting applications of Pincherle's *sommets* and *foyers* is his demonstration that Γ is a Cantor set when $a > 2$ [**1917b**, pp. 12–15]. The essence of his proof is to show that the closure of the set of *sommets*, which he called $\mathcal{Z}$, is perfect and nowhere dense.

He asserted that each *sommet* is a limit point of *sommets*; if true, then $\mathcal{Z}$ would be perfect since all of its points would be limit points of $\mathcal{Z}$. To prove his assertion, Pincherle used the set of the preimages of the real open interval $q_0 = (-\sqrt{a-l}, \sqrt{a-l})$ under α_a (we know this is an interval since $a > l$ when $a > 2$; see (8.9) and the material that follows it).

First, he established that if $z \in q_0$, then $z \in \Omega$, the basin of infinity: Near the beginning of his entry [**1917b**, p. 4], Pincherle showed that for any $a > 0$, $|z| > l$ implies that $z \in \Omega$, hence converges under iteration to the point at infinity. From this, it follows that if z is interior to the circle C' of radius $\sqrt{a-l}$ centered at the origin, $z \in \Omega$. To see this, apply the triangle inequality to $z_1 = z^2 - a$, which yields $|z_1| \geq a - |z^2|$. Since z is inside the circle of radius $\sqrt{a-l}$, it follows that $|z^2| < a - l$, and we have

$$|z_1| \geq a - |z^2| > a - (a - l) = l,$$

so $z_1 \in \Omega$. By hypotheses, z_1 is the image of an arbitrary point interior to C', so all points interior to C' are also in Ω. Since $q_0 \subset C'$, if $z \in q_0$, or any of its preimages q_k, then $z \in \Omega$ as well.

Next, it is necessary to establish certain properties of the union of the preimages of q_0. Let q_k denote the kth preimage of q_0 under α_a. Since all the critical points of the inverse of α_a are exterior to q_0, each q_k consists of 2^k disjoint subintervals of uniform length contained within the interval $[-l, l]$. Since the number of intervals goes to infinity with k, their diameters must go to zero.

It also follows that because the endpoints $\pm\sqrt{a-l}$ of q_0 are *sommets*, and any *sommet* is in Γ, which, as we noted above, is invariant under iteration by α_a, they (and their preimages) are in Γ. Note that neither q_0 nor its preimages can contain a *sommet* since all *sommets* are in Γ and the points in q_0 are in Ω.

Each subinterval of q_k contains exactly one *foyer*. To see this, observe that $0 \in q_0$; since $\sqrt{a} > \sqrt{a-l}$ it follows that there are no preimages of 0 in q_0, which is another way of saying there are no *foyers* in q_0. Thus, the preimages of zero, $\pm\sqrt{a}$, are the only two *foyers* in q_1, one in each of the two intervals which compose q_1; likewise, the only *foyers* in the q_k are

$$\epsilon_0 \sqrt{a + \epsilon_1 \sqrt{a + \cdots + \epsilon_k \sqrt{a}}}.$$

By construction, the endpoints of the q_k are the *sommets*

$$\epsilon_0\sqrt{a+\epsilon_1\sqrt{a+\cdots+\epsilon_k\sqrt{a-l}}},$$

which he denoted generically as z_k, which is of the form $\pm(r_0,\ldots,r_i,\infty)$ because the preceding nested radical terminates with l. Since at each q_k, all possible *sommets* with $k+1$ terms occur, all possible finite *sommets* are an endpoint of some subinterval from the q_k. Likewise, all possible finite *foyers* are in the union of the q_k as k goes from 0 to ∞. In other words, preimages of the repelling point l occur as endpoints of the subintervals composing the q_k, and one preimage of zero occurs interior to each subinterval.

We are now ready to show that any *sommet* is a limit point of *sommets*, which will establish that $\mathcal{Z}$ is perfect. Since Pincherle showed that any *sommet* can be expressed as the limit of a sequence of *foyers* (see limit (8.11) above), near any *sommet* z_j there are infinitely many *foyers*. Since each *foyer* can be associated with a unique subinterval from the union of the q_k, there are infinitely many subintervals from the q_k near z_j. As we have seen, the endpoints of these subintervals are each a unique *sommet*. Thus, the *sommet* z_j is also limit point of *sommets*, hence in $\mathcal{Z}$ by the set's construction. Since the choice of z_j was arbitrary, it follows that each *sommet* is also a limit point of *sommets*. Because $\mathcal{Z}$ consists of the union of the *sommets* and their limit points, all points of $\mathcal{Z}$ are limit points of $\mathcal{Z}$, and it is therefore a perfect set.

To show that the closure of the set of *sommets* $\mathcal{Z}$ is nowhere dense, Pincherle proceeded by contradiction and assumed $\mathcal{Z}$ contains an interval I. Since $\mathcal{Z}$ is perfect, we can assume that the *sommets* are dense in I. Denote one of these *sommets* s; we know from above that s is an endpoint of a subinterval I' from the q_k. At least a portion of I' is in I so $I'\cap I$ is a nonempty interval with s as an endpoint. Since the *sommets* are dense in I, it follows that there is another *sommet* $s'\neq s$ in $I'\cap I$, hence s' is interior to I'. However, since I' is one of the subintervals from the q_k, this would mean that s' is interior to one of these subintervals, which, as we noted above, contain no *sommets*.

Pincherle's construction suggests that an infinite *foyer* (i.e., the limit of a *foyer* as the number of nested radicals goes to ∞) is in $\mathcal{Z}$. Using the technique of symbolic dynamics, he carefully proved that an arbitrary infinite *foyer* $\tilde{z}=(r_0,\ldots,r_k,\ldots)$ is well defined and contained in $\mathcal{Z}$ [**1917b**, pp. 20–21]. We summarize his argument:

Recall that Pincherle had already established that symbolic expressions which terminate with ∞ (i.e., *sommets* such as $(1,1,\infty)$) are well defined (see equation (8.10)). With this in mind, he defined a sequence $\{z_k\}$ of *sommets* by $z_k=(r_0,\ldots,r_k+1,\infty)$. Since $\mathcal{Z}$ is by definition the closure of the set of *sommets*, each z_k is in $\mathcal{Z}$. The subsequence $\{z_{2k}\}$ is monotonic down while its counterpart $\{z_{2k+1}\}$ is monotonic up. To bound each sequence, he defined the *foyer* $f_k=(r_0,\ldots,r_k)$, a truncated version of $\tilde{z}$. The sequence $\{f_{2k}\}$, also monotonic down, satisfies $f_{2k}<z_{2k}$ and bounds $\{z_{2k}\}$ from below, whence he concluded that $\{z_{2k}\}\searrow\mu$; the f_{2k+1} bounds the z_{2k-1} from above, hence $\{z_{2k+1}\}\nearrow\mu'$. Arguing by contradiction, he concluded that $\mu=\mu'=(r_0,\ldots,r_k,\ldots)=\tilde{z}$. Finally, since $\tilde{z}$ is the limit of a sequence of points from the perfect set $\mathcal{Z}$, $\tilde{z}$ is in $\mathcal{Z}$ as well.

Pincherle thus characterized the set $\mathcal{Z}$ in terms of *foyers* and *sommets*: infinite *foyers* as well as all *sommets* are in $\mathcal{Z}$. This also allowed him to show that periodic points were generally in $\mathcal{Z}$. Given an infinite expression of the form,

$$z = \epsilon_0 \sqrt{a + \epsilon_1 \sqrt{a + \epsilon_2 \sqrt{a + \cdots + \epsilon_n \sqrt{a + \cdots}}}},$$

Pincherle observed that if the distribution of signs—that is, the sequence $\{\epsilon_k\}$—is periodic, then z is a periodic point of α_a, as can be easily verified [**1917b**, p. 23]. He did not examine the nature of the set of periodic points further; in particular, he did not investigate their distribution in the Julia set Γ defined above, which Pincherle conceived of as the boundary of the basin of infinity.

Pincherle's construction of $\mathcal{Z}$ suggests that it is the complement of the set $\mathcal{Q}$ in $[-l, l]$, where

$$\mathcal{Q} = \lim_{k\to\infty} q_k.$$

Indeed, it is clear that $\mathcal{Z} \subseteq \Gamma$. But are the two sets equal? It not does necessarily follow from above that it is, so he had another important task to accomplish before he could assert this with certainty.

8.6.4. Constructing Γ. Pincherle began his construction of Γ with the definition of two circles, $C = \{z : |z| = l\}$ and $C' = \{z : |z| = \sqrt{a-l}\}$, which was defined above. The diameter of C' is the closure of q_0, $[-\sqrt{a-l}, \sqrt{a-l}]$, and we noted earlier that all points whose modulus is less that $\sqrt{a-l}$ are in Ω, so the interior of C'—and its preimages C'_k, whose diameters are the intervals composing q_k with their endpoints—are in Ω. Since all points exterior to C satisfy $|z| > l$, they are also in Ω, hence the set Γ must lie in C, in particular, in the closed annulus A bounded by C and C'.

Let D be the closed disc bounded by C, and D_k its preimages. In an argument that spans several pages (which we omit) [**1917b**, pp. 24–27], Pincherle showed the following:

(1) Γ is the infinite intersection of the D_k which in turn is the complement of $\mathcal{Q}$ in $[-l, l]$.
(2) The set $\mathcal{Q}$ is of full measure on $[-l, l]$ (rather than use the term "full measure". Pincherle said that they "tend towards" $[-l, l]$ [**1917b**, p. 26]).

The first confirms that $\mathcal{Z}$ is indeed in Γ since it is in the complement of the union of the q_k. Although Pincherle's justification could have been a little clearer [**1917b**, p. 27], the second point implies that $\mathcal{Z}$ equals the complement of the union of the q_k. One way to show this—and perhaps this is the argument that Pincherle had in mind but thought was sufficiently obvious—is to observe that since the union of the q_k is a union of open sets which are full measure in $[-l, l]$, the points in the complement of this union are the endpoints and their limit points, that is, the set of *sommets* and their limit points, which is $\mathcal{Z}$.

8.6.5. Pincherle's classification of Γ. Much of the theory he developed in the $a > 2$ case extends to the other cases. Since the formation of the set $\mathcal{Z}$ as the closure of the set of *sommets* does not depend on the magnitude of the parameter a, it remained central to his characterization of Γ, as did his investigation of *chaînes* and *foyers*. The sets C and D_k were used throughout his entry. For the $a = 2$ case, he formed $[-2, 2]$ by taking a nested intersection of the D_k. When $a < 2$, he viewed Γ correctly as the limit of the curves ∂D_k.

Near the end of [**1917b**], Pincherle summarized how the structure of Γ varies with a. We will adopt Pincherle's notation for the complement of the closure of Ω and refer to it as Ω'.[11]

(1) If $a < 3/4$, the fixed point l' defined above at equation (8.3) is attracting and points in Ω' converge "regularly" under iteration to l'. "Regular" convergence is a term that Kœnigs used to describe iteration, as Pincherle reminded the reader [**1917b**, p. 41], and, in the present case indicates that for any given $\epsilon > 0$, there is an N_ϵ such that $|\alpha_a^n(z) - l'| < \epsilon$ whenever $n \geq N_\epsilon$. The use of the phrase regular convergence implies a contrast between an attracting fixed point and periodic points of order greater than one, where the orbit of a point can be within ϵ of each of the periodic points infinitely often, or with a rationally neutral fixed point λ, where it is possible for a point z to be within ϵ of λ, while an iterate $f^k(z)$ satisfies $\left|f^k(z) - \lambda\right| > \epsilon$. At the conclusion of his entry, Pincherle noted that regular convergence of α_a to the attracting fixed point l' in Ω' extends to $-1/4 < a < 3/4$, [**1917b**, p. 53]. See the top image in Color Plate 7 for a depiction of Ω' and Γ when $a = 0.7$.
(2) If $a = 3/4$, $\alpha_a'(l') = 1$, and Pincherle reminded the reader that this was the case Leau treated [**1917b**, p. 42].
(3) When $3/4 < a < 2$, $\alpha_a'(l') > 1$, hence l' is no longer attracting. Instead, Ω' contains what nowadays would be called an attracting cycle of period p, that is points, $\{z_0, z_1, \ldots, z_{p-1}\}$ with $z_p = z_0$. Defining $\phi = \alpha_a^p$, Pincherle noted that $|\phi'(z_k)| < 1$. If one of the $z_k = 0$, he observed that $\phi'(z_k) = 0$,[12] noting that Grévy had treated the case where the derivative of the fixed points is zero. See the second image in Color Plate 7 for a portrayal of the dynamics when $a = 1.4$.
(4) When $a = 2$, Γ is the interval $[-2, 2]$. See the third frame in Color Plate 7.
(5) As we noted above, when $a > 2$, he showed Γ is a Cantor set. See Color Plate 7 for the case $a = 2.1$.

Caption for Color Plate 7: (See p. 65.) The family $z \mapsto z^2 - a$ for $a = 0.7, 1.4, 2, 2.1$. In all images the shades of blue indicate the rate of convergence of points in the basin of infinity. D is the circle with diameter $[-l, l]$, where l is the positive fixed point defined at (8.3).

In the top image, $a = 0.7$, the preimages of C limit upon Γ, the boundary of the region Ω'. The interval $[-l, l]$ spans Ω'. The set Ω' still has interior in the second image, corresponding to $a = 1.4$, but is shrinking down towards the interval $[-2, 2]$. We see the sets D_k (preimages of D) for the first few values of k in the third image; Γ is the segment $[-2, 2]$, formed by the infinite intersection of the D_k. In the bottommost image, the intersection of the D_k create a Cantor set contained in $[-l, -\sqrt{2.1 - l}] \cup \sqrt{2.1 - l}, l]$, similar to constructions depicted earlier in Color Plates 2 and 4.

While he did not describe the way that Γ partitioned the sphere in anywhere near the precision or detail that Fatou and Julia provided (for example, he never once indicated explicitly that there could be infinitely many components to Ω'),

[11]This set is known in the present day literature as the filled-in Julia set.
[12]This follows from the form of the derivative of ϕ at the periodic points of α_a.

like them, he invoked von Koch to describe the possible complexity of Γ, both in his entry [**1917b**, p. 40] and in another paper written around the same time, where he observed,

> *Finally, one notices that, due to the nesting of the curves C_{r}, their limit shows a singular analogy with the famous von Koch curve.* [**1918d**, p. 343]

Pincherle published several additional short papers on iteration, concluding his investigation with [**1920b, 1920a, 1920c**], but his entry to the *Grand Prix* remains his definitive treatment of iteration.

There were a small number of Italian works that followed up on Pincherle's study. Andreoli referenced him in [**1919**], Barba related holomorphic dynamics to the theory of automorphic maps in [**1929**] and [**1936**], and Andreoli revisited the iteration of real variables in [**1937**].

8.7. Concluding remarks

Clearly, Pincherle's research was at the forefront of an Italian interest in iteration that, while diverse and long lasting, did not result in the range or quality of work present in the French school.

At the beginning of the chapter we quoted Amaldi's comment to the effect that Pincherle liked to follow offshoots from his primary research. This may well account for his interest in the iteration of complex functions, which seems to have been piqued by his assignment to write about functional equations in the *Encyklopädie der Mathematischen Wissenschaften* and *Encyclopédie des Sciences Mathématiques Pures et Appliquées* in the early 1900s. His examination of iteration was only a minor facet of his overall research program, yet it culminated in a very good paper. His conceptual framework—which included a classification of Julia sets—is surprisingly close in spirit to what we see in Fatou and Julia's work, more so, in many respects, than is the case with Lattès.

Yet, as we saw with Lattès in the last chapter, Pincherle's treatment of iteration suffered from a limited perspective, at least in comparison to the works of Fatou and Julia, since it was entirely focused on a single family of functions. Nor did he exploit the theory of normal families, which perhaps would have provided the spark that moved him beyond degree two rational functions.

It should be kept in mind, however, that while, in hindsight, the study of normal families is perfectly suited for the study of iteration, its utility was by no means obvious. When he was searching for ways to demonstrate the power of his theory of normal families, it evidently did not occur to Montel to apply it to iteration, and it was not until some time after they embarked upon their respective studies of iteration that Julia and Fatou realized its extraordinary usefulness.

Perhaps the absence of normal families in Pincherle's study was generational. He was 25 years older than Fatou and almost 40 years Julia's senior, and it could be that he was not as attuned to the mathematics of the new century as they were. We saw evidence of this in his omission of Fatou's 1906 notice from both his 1912 encyclopedia survey and his entry in *Grand Prix*. Certainly, he was geographically removed from the seat of creation of normal families, and one should probably not ignore the value of quotidian contact with the creator of a new theory in coming to realize its value.

CHAPTER 9

The Giants Fall

9.1. Introduction: Centers... and frictions

Despite the many successes of Fatou and Julia's long monographs on rational functions, they left important issues unresolved. The most salient of these concerns the nature of iteration in neighborhoods of an irrationally neutral point, and it turned on the existence of centers:[1] both Fatou and Julia failed in their attempt to determine whether or not centers exist for rational functions of degree two or more.

The term "center" has its roots in the work of Poincaré, where he used it to describe a particular kind of equilibrium point, as shown in Figure 3.1. Julia applied this terminology to iteration of complex functions, noting that if a solution to (9.1) exists, the fixed point would be "analogous to the centers studied by Poincaré" [**1918c**, p. 156].[2]

Centers exist for linear transformations as well as for nonrational functions, such as algebraic functions, as we will see below in Section 9.2, so it was natural to ask if they also exist for nonlinear rational functions. It is just as natural to seek their existence through the canonical Schröder equation,

$$S \circ f = f'(\lambda) \cdot S. \tag{9.1}$$

However, equation (9.1) does not yield easily to a direct approach when λ is an irrationally neutral fixed point.

Kœnigs and Leau made no progress in their attempts to explain iteration near irrationally neutral fixed points in the late 1800s. As we saw in Chapter 5 during our discussion of Pfeiffer's 1917 investigation of (9.1), it is relatively easy to use the method of undetermined coefficients to find a formal solution for S in terms of the coefficients of f. It is quite another thing, however, to determine the actual convergence or divergence of the series. This is because the coefficients of S are of the form

$$c_{n+1} = \frac{c_1[a_1^{n-1}(1-a_1)\cdots(1-a_1^{n-1})a_{n+1} + P_{n+1}(a_1,\ldots,a_n)]}{a_1^n(1-a_1)\cdots(1-a_1^n)}, \tag{9.2}$$

[1]Recall that if iteration near an irrationally neutral point λ is conformally equivalent to rotation of a disc, λ is said to be a center. Conformal equivalence is established via the existence of an analytic solution S of the canonical Schröder equation $S \circ f = f'(\lambda) \cdot S$ on a neighborhood D of λ. If such an S exists, then f is conjugate to the linear mapping $z \mapsto f'(\lambda)z$ on D, and iterates of a point on $z \neq \lambda$ in D lie on an invariant curve that is conformally equivalent to a circle centered at the origin. The neighborhood D is referred to as a rotation domain or a Siegel disc. We have discussed centers previously in Sections 1.6.2 and 4.4, and in Chapter 5. See Figure 1.2 for a depiction of a center.

[2]We discuss the connection between the centers that occur in complex dynamics and those in the work of Poincaré on pages 83 to 84.

where $a_1 = f'(\lambda) = e^{i\theta}$, and the denominators can be quite close to zero making the c_n potentially very large. This is a classic small divisors problem of the sort discussed in detail in Chapter 4, and it is notoriously difficult to solve. Indeed, it was by making the c_n arbitrarily large that Pfeiffer showed that formal solutions to the Schröder equation (9.1) diverge (see Section 5.4.5), and thus he became the first mathematician to demonstrate that there are irrationally neutral fixed points that are not centers.

Fatou and Julia investigated the center problem in the monographs they prepared in anticipation of the 1918 *Grand Prix* competition, unaware of Pfeiffer's accomplishment. As we will see, Julia unsuccessfully tried to solve the center problem via the Schröder equation, and no doubt Fatou was also well aware of the difficulties involved in small divisors problems. For these reasons we speculate that both Fatou and Julia thought that their best chance for success lay in the direction of normal families, which had been so helpful elsewhere in their studies of rational functions.

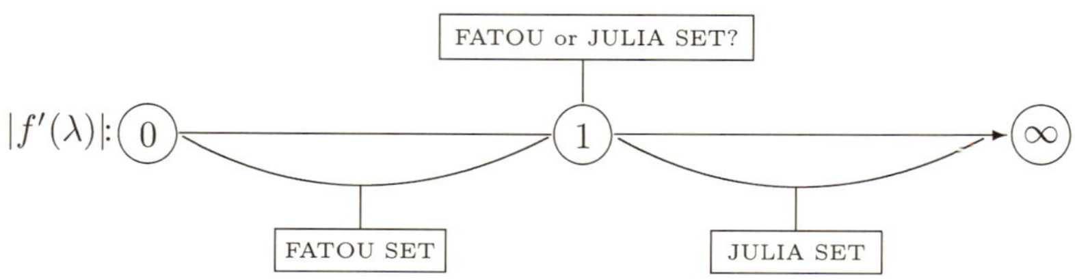

FIGURE 9.1. A global view of fixed points circa 1920. This updates Figure 1.4 and reflects what was learned about the relation between periodic points and the partitioning of the sphere from the monographs of Fatou and Julia. The question they left open is, Do irrationally neutral fixed points belong to the Fatou set or the Julia set?

In this chapter we focus on Fatou and Julia's respective explorations of the center problem in the late 1910s. In subsequent chapters we explore the answers to the questions they posed that emerged over the subsequent two decades. Those answers were perhaps surprising since they did not lie in the direction of normal families, as Fatou and Julia seemed to have hoped. Instead, it was shown that whether or not λ is a center depends upon delicate number theoretic conditions involving the multiplier $f'(\lambda)$. To the observant eye, Pfeiffer's work hinted at this (see Section 5.5), and Cremer and Siegel made Pfeiffer's hint corporeal.

9.2. Julia's (non?) centered approach

Julia's study of irrationally indifferent fixed points in [**1918a**] is part of a larger discussion of neutral fixed points that comprises the final chapter of his monograph. A portion of his investigation of rationally neutral fixed points is devoted to a review of established theory, including a reworking of Leau's Flower Theorem and a demonstration that the canonical Schröder equation (9.1) does not have even a formal solution on neighborhoods of rationally neutral points. But he also recast the study of neutral fixed points in light of the theory of normal families, showing that rationally neutral fixed points are in the domain of nonnormality, E'.[3] We discussed his argument that such points are in E' at the end of Section 7.2.

As we have noted in earlier chapters, Julia generally avoided the use of functional equations in his monograph since he did not feel that they were essential to the study of iteration. The one exception is with neutral fixed points where the existence of a solution to the Schröder equation would imply that iteration is either periodic in the case of a rationally neutral fixed points, or quasi-periodic in the case of an irrationally neutral fixed point. That no solution to the Schröder solution exists for rationally neutral fixed points shows that iteration cannot be conformally equivalent to a periodic rotation. At the time of his monograph, the possibility of quasi-periodic rotation remained open.

While the goal of Julia's exploration of irrationally neutral fixed points was similar to that of his study of rationally neutral points—to determine if such points are in E' or not, and to investigate solutions to the Schröder equation—an air of uncertainty pervades it, as the following quotation suggests:

> *Finally, I will outline a means of approaching the case where $\lambda = f(\lambda)$, $f'(\lambda) = e^{i\theta}$, θ being any real whatsoever, that reduces the possible hypotheses to two, and the realization of one in a given situation excludes the possibility of the other. I have not had success up to now when θ is incommensurate with 2π in determining if one can construct examples of fractions $f(z)$ for each of the two possibilities. For θ commensurate with 2π, and $f(z)$ rational,* it is always one of the two hypothesis, *and always the same, that is verified and never the other.* [**1918a**, pp. 222–223]

The two mutually exclusive possibilities to which he refers are that λ is in the domain of normality and the Schröder equation $S \circ f = f'(\lambda) \cdot S$ has a convergent solution S, or that λ is in the Julia set E' and no such solution exists. Since Julia had already showed that all rationally indifferent points are in the Julia set, there might be a certain appeal to him if this were also true for irrationally neutral fixed points.

The uncertainty Julia expressed towards the irrationally indifferent case is given additional context by a footnote which appears near the end of the above quotation:

> *I have succeeded, subsequent to the deposit of this memoir, in excluding one of the two possibilities; this will be the object of a subsequent memoir.* [**1918a**, p. 237]

Perhaps to add drama to his treatment, he did not tell the reader which of the two hypotheses he has dismissed, but as we will see, he declared in the *Comptes Rendus* notice [**1919a**], mistakenly as it turns out, that all indifferent fixed points are in the Julia set; in other words, centers do not exist at all for any rational function.

Since Julia could prove neither the existence nor nonexistence in his monograph [**1918a**], he attempted to delineate conditions that would imply that λ is a center, hence in the domain of normality. If λ is a center for a function f, it would be surrounded by a rotation domain D on which iteration is conformally equivalent to

[3] We will use Julia's notation for the Julia set for the remainder of our discussion about Julia; when we discuss Fatou we will adopt his notation.

a rotation of a disc. Thus f can be viewed as a self-map of a disc. This allowed him use Schwarz's Lemma (he also invoked work of Koebe) to prove the following:[4]

THEOREM 9.1. *Let D surround a fixed point of f, which we can assume to be zero. Let ψ be a branch of the inverse of f fixing 0. If $\psi^n(D)$ contains no critical points of ψ (i.e., critical values of f), then the origin "is the center of a region D furrowed by analytic curves conserved by $z_1 = f(z)$ as well as $z_1 = \psi(z)$". One can also find a solution S to the Schröder equation $S \circ f = f'(0) \cdot S$ on D.* [**1918a**, p. 234]

Julia observed that if S is analytic on D, then D is conformally equivalent to a disc, and the invariant curves on which iterates of a given $z \in D$ lie are equivalent to curves $|S(z)| = c$. Thus, the fixed point zero cannot be a limit under iteration for any point in the plane, much less the critical points of the inverse ψ. This last statement is important since he showed that when the fixed point is rationally indifferent, it is a limit point of the iterates of the critical value.

Thus, iteration near a center would be quite different than what occurs with both attracting fixed points and rationally neutral fixed points, which not only attract points, but critical values of f. Another difference is that rationally neutral fixed points are in the Julia set. So if a center exists, its behavior would be *sui generis*.

His deducing an analytical solution to the Schröder equation is implicit and is reminiscent of the geometric argument that the American mathematician Bennett gave (recounted in Chapter 5 on page 109). Julia used the existence of conformal equivalence of the neighborhood D of λ to a disc surrounding the origin to induce a solution S to the Schröder equation which maps the invariant curves on which iterates of $z \in D$ lie to circles surrounding the origin.

To give a sense of what a center might look like, Julia gave an example of an algebraic function. His construction is similar to Schröder's "opposite approach" to solving $S \circ f = h \cdot S$, where, rather than starting with a function f and attempting to find S, Schröder instead fixed an invertible function S and solved for functions f satisfying $S^{-1} \circ (h \cdot S)$, where h is a constant (see equation (1.7)).

EXAMPLE 9.2 (An algebraic function with a center). Julia assumed the existence of a Schröder equation $S \circ f = e^{i\theta} \cdot S$, where θ is arbitrary, with a solution $S(z) = 2z - z^2$ and solved for f. He chose this particular S because the curve $|2z - z^2| = 1$ produces a lemniscate, one loop of which contains the origin, which will be a center for the map f he constructs. To find f, he set $S(f(z)) = e^{i\theta} S(z)$, which results in the equation $2f - f^2 = e^{i\theta}(2z - z^2)$. Viewing this as a quadratic in f, he found a solution $f(z) = 1 - \sqrt{1 - (2z - z^2)e^{i\theta}}$ with a rotation domain contained within one loop of the lemniscate. Routine calculations verify that $f'(0) = e^{i\theta}$, hence iteration is equivalent to a rotation through θ [**1918a**, p. 238].

Julia summarized his findings up to this point:

> *Given a rational function f, the problem of whether $\lambda = 0$ is a center reduces to determining, (1) if the equation*
>
> $$S(f(z)) = e^{i\theta} S(z)$$

[4]It is known that if λ is a center, then the connected component of the Fatou set which includes it—known as a Siegel disc—is bounded by the closure of the post-critical orbit of f (see Definition 1.12); if an irrationally neutral fixed point is not a center, then it is contained in the closure of the post-critical orbit.

has a solution $S(z)$ holomorphic around the origin, or else, (2) if it does not, in which case $\lambda = 0$ will be a limit point of the consequents of a critical point of the branch $\psi(z)$ that is zero at the origin. ... We know that if f is rational, the second is the only possibility if θ is commensurate with 2π, and we have seen in the preceding paragraphs that the impossibility of the first hypothesis when θ is commensurate with 2π results in $\lambda = 0$ being in the set *E'. ... One can also say, that, in any case where the first hypothesis is verified, the area Γ interior to the curves $|S(z)| = const.$ (the constant being sufficiently small) will be transformed into itself bi-univalently by $z_1 = f(z)$ and $z_{-1} = \psi(z)$, thus the sequence f_k will be normal inside Γ. 0 will not be in E'.* [**1918a**, p. 239]

Julia next proved that if a fixed point 0 of f satisfies $f'(0) = e^{i\theta}$ and is in the domain of normality, then a solution to Schröder equation exists. He used the uniform boundedness of normal families, and then showed that normality implies the existence of a center. To construct the rotation domain, he formed a small region surrounding the origin bounded by a close simple curve γ. Since $|f'(0)| = 1$, he deduced that the images of the regions under f^k are simply connected regions bounded by simple curves γ_k that remain bounded away from the origin.

One can deduce the existence of an area D interior to all the γ_k, simply connected, surrounding 0. D will be transformed bi-univalently into itself by $z_1 = f(z)$ and by the branch $z_{-1} = \psi(z)$, null at 0. The conformal representation of D on a circle with center 0, along with the conservation of the origin, prove that again the origin is a center. [**1918a**, p. 241]

Central to his examination of the normality of the iterative family on a rotation domain D is the nonconstancy of its limit functions, which he deduced from the fact that if $|f'(\lambda)| = 1$ and λ is in the domain of normality, then one can bound the derivatives of the iterative family $\mathcal{I}$ away from zero on compact subsets of D. That the limit functions could be nonconstant seems to be a concern to him. Centers would be the only place where this occurs, and a year later he listed the nonconstancy of all limit functions as one of the important consequences of his assertion from [**1919a**] that centers do not exist.

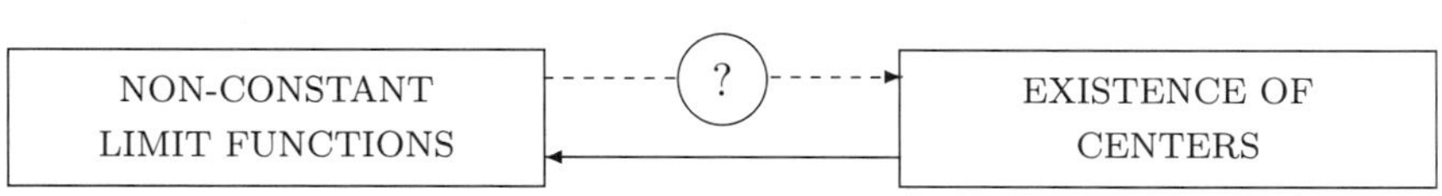

FIGURE 9.2. Julia's exploration of centers. The solid arrow indicates that Julia knew if a center exists around an irrational neutral fixed point then so do nonconstant limit functions. The broken arrow suggests that proving the existence of nonconstant limits—which he was unable to do—would show centers exist.

Given the hindsight provided by Julia's declaration soon after his monograph appeared that centers could not exist for rational functions, one might wonder if Julia had already made up his mind on the matter in [**1918a**]. His examination of centers displayed no indication that this was the case. What one finds instead is uncertainty:

> *I have not yet been able to decide between the two hypothesis, which, for a specific case, are mutually exclusive. To pursue the question more thoroughly, I need more time and knowledge than I now have. I will return to the subject later, if there is a reason, and it could be that the utilization of the Schröder equation will settle the question.* [**1918a**, pp. 242–243]

This utilization of the Schröder equation could be in the negative: Immediately after the above quotation came an observation that suggests he found a direct approach to showing the existence of series development satisfying the Schröder equation prohibitively difficult. Noting that one can find formal solutions to $S \circ f = f'(0) \cdot S$, he stated,

> *I tried to show the convergence of the development by the method of majorants, but all the majorants I found were divergent! ... It remains to be seen if, for a rational function $f(z)$, they* [centers] *can occur; it would be interesting to decide either in the affirmative, by constructing an example, or in the negative.* [**1918a**, p. 243]

Julia's difficulties in finding series solutions to the Schröder equation were most likely due the fact that determining the coefficients c_n is a small divisors problem as was pointed out during our discussion following equation (9.2). In fact, one can view Pfeiffer's paper [**1917**], as a way to use the divergence of majorants to prove the divergence of certain formal solutions to the Schröder equation. That Julia was not able to find a way to create divergent formal solutions indicates that he was almost certainly unaware of Pfeiffer's work.

One almost gets the sense, too, that Julia is closing his otherwise remarkably successful monograph with a sign of resignation regarding a problem that he could not solve.[5]

9.3. Julia's (not so) final verdict on centers

Early in 1919 Julia resolved the center problem—incorrectly, as it turned out:

> *In my note of January 28th 1918, there was a pending question, namely if a rational substitution $z_1 = f(z)$ with a fixed point $\lambda = f(\lambda)$, satisfying $f'(\lambda) = e^{i\theta}$, θ incommensurate with 2π, can have a center. At present, I am in position to settle this question in the negative:* for any rational substitution there are no centers. Every fixed point $\lambda = f(\lambda)$, $|f'(\lambda)| = 1$, is a point in the perfect set which I call E'. [**1919a**, p. 147]

Julia gave the following argument. If a center exists, then $\lambda \notin E'$, and the Schröder equation $S \circ f = f'(\lambda) \cdot S$ has an invertible solution. From this he deduced that the inverse ϕ of S satisfying $\phi(f'(\lambda) \cdot z) = f(\phi(z))$ is analytic on the Riemann sphere which in turn implies that for all z (aside from the fixed point) the closure of the set $\{f^n\}$ is closed curve. In other words, the quasi-periodic rotational behavior of the center would extend to the entire plane, precluding the existence of periodic points aside from λ.

[5] While this quotation was indeed the last in Julia's memoir proper, he included an addendum he wrote after he had submitted the paper correcting an error he had made earlier in the paper (see page 140 of our book for details).

But, as Julia noted,

> *. . . this is contradicted by the existence of a denumerable infinity of roots of $z = f^n(z)$ which have only a finite number of consequents.* [**1919a**, pp. 147–148]

Given the nature of *Comptes Rendus*, his note did not include a full proof. Nor did he ever publish one. However, sometime prior to 1934 he evidently realized that his argument contained an error. When he referenced this result again in his *Notice de candidature académique* prepared in anticipation of his 1934 election to the French Académie des Sciences, he remarked that his belief that centers did not exist followed from a misassumption he made regarding the accessibility of boundary points of rotational domains:[6]

> *I affirmed, in the note* [**1919a**], the impossibility of centers for rational substitutions $z_1 = f(z)$, *from which I deduced several important consequences, notably* the absence of nonconstant limits *of the family of iterates f_n of any rational function* whatsoever. *A recent examination of the method that I had envisioned for the proof of this claim shows that I had implicitly supposed the existence* [in the Julia set] *of* accessible *repelling points or cycles, which are in no way a priori certain, despite properties of E' I had already established, and thus my demonstration needs to be completed. Having devoted my efforts to other areas of research, I have not yet returned to the subject.* [**1934**, p. 22][7]

Note that this statement falls short of saying his results is incorrect; rather it indicates that his argument needs to be corrected.

In addition to the fact that all limit functions for the iterative family would be constant, the other important consequence that Julia deduced from the impossibility of centers was that the only possibility for the derived set of the backward orbit $O^-(z)$ (assuming z is not an exceptional value of f) would be that it equals E'. If a center exists, this would not be true since for points z in the rotation domain D, the derived set of $O^-(z)$ would also include one of the invariant curves surrounding the center λ. Julia found this noteworthy:

> Therefore, the derived set of the set of the antecedents of any point in the plane *(except the exceptional value, if any exist)*, does not just include points in E': It is identical to E'.
>
> *This statement can only be firmly established if one achieves what was stated in* [paragraph] *II, namely, that every limit function of the sequence of f^n is a constant; without that, the possibility of centers would imply the existence of nonconstant limit functions and would contradict the previous statement. On realizes from II*

---

[6]The issue of accessibility is evidently challenging, and during the preparation of this manuscript, questions about the accessibility of points in the Julia set were still being explored. See, for example, the introduction to Petracovici [**2006**].

[7]Julia's description is suggestive of what occurs with certain irrationally indifferent fixed points that are noncenters, points that have the so-called small cycles property, meaning that any neighborhood of such a point has infinitely many *complete* repelling periodic orbits, not just infinitely many repelling points. Perhaps Julia discovered such points but assumed the property extended to all irrationally indifferent points. If so, Julia would have been close to proving the existence of irrationally indifferent points that are not centers.

> *and III the importance of the impossibility of centers in the general theory of iteration of rational fractions.* [**1919a**, p. 148]

For Julia, then, not only would the nonexistence of centers mean the dispatching of an important question he had left open in his entry for the *Grand Prix*, but would also allow for certain categorical statements, for example, that the limit functions of the iterative family are *always* constant and the limit set of the preimages of any point in $\overline{\mathbb{C}}$ is, apart from the exceptional points, *always* the Julia set E'.

We will see in subsequent chapters that even prior to 1934, Julia's statement that centers did not exist was by no means universally accepted.

9.4. Fatou's singular approach

Fatou discussed the possibilities for irrationally neutral fixed points in detail at several points in his monograph [**1919-1920**], including §§14, 28–31, and 54. Like Julia, he established no concrete results about the behavior of rational functions with irrationally neutral fixed points.

Nonetheless, there is quite a different flavor to Fatou's investigation. Julia focused on the possible existence of centers, and his investigation was an explicit exploration of the possibilities that their existence might imply, for example, the solution to the associated Schröder equation or nonconstant limit functions of the iterative family. Fatou, on the other hand, posited the possible existence of domains that he called "singular domains", where the limit functions are nonconstant, and then asked questions such as, What would iteration on such domains be like? What are necessary conditions of their existence? How could we rule out their possibility? Is there a limit on their number?

While singular domains encompass centers, they hold open the possibility of other kinds of behaviors, for example annular rotations now called *Herman rings* that would not be discovered for many years.[8] Fatou was careful, however, not to be too explicit about what a singular domain might look like, but it is clear that he considered the possibility of rotation domains, as well as flower-like regions similar to those that occur with rationally neutral fixed points.

Fatou introduced his investigation of irrationally neutral fixed points in §14, following a lengthy discussion of the rational case:

> *It remains to study fixed points whose multiplier is of the form $e^{i\theta}$, θ being incommensurate with 2π. We know very little about these fixed points, whose study, from our point of view, appears to be very difficult.* [**1919-1920**, p. 220, II]

His comment "from our point of view" is a bit of an understatement given that his French predecessors—Leau and Kœnigs in particular — made no progress whatsoever.

[8]A Herman ring is a domain on which iteration is conformally equivalent to an annular rotation. See Color Plate 1 for a depiction. Michel Herman showed that such domains exist for rational functions around 1980. In 1932 Cremer showed that it is possible for such regions to exist, which we will discuss in Chapter 11, but he did not show that they actually do.

To underscore this difficulty, he introduced a linear rational function with an irrationally neutral fixed point whose behavior under iteration was well known:

> *Consider first a substitution of the first degree,* $z_1 = e^{i\theta} z$*; iteration yields* $z_n = e^{i\theta n} z$*; the points* z_n *are dense throughout the circumference of a circle centered at* O *and passing through* z*, and the sequence of functions* $f^n(z)$ *admits all the functions* $e^{i\beta} z$*, where* β *is any real number whatsoever, as limit functions.* [**1919-1920**, p. 220, II]

Observing next that "it is not easy to recognize" whether a given function of the form $z_1 = e^{i\theta} z + \cdots$ is conjugate to $z \mapsto e^{i\theta} z$, he considered a possibility that Julia did not:

> *...do there then exist domains where iterates tend toward such a fixed point? At present we neither provide an example where it does, nor prove that is impossible.* [**1919-1920**, p. 221, II]

He seems to be asking if iteration near an irrationally indifferent point can be similar to that of a rationally indifferent fixed point, where points in a "petal" of the flower converge to the rationally indifferent point. We will come back to this point.

As brief as Fatou's remarks in §14 are, they serve as a fitting prelude to the investigations of irrationally neutral fixed points that take place in later sections of his monograph, where he expands upon themes he raised here, and, in particular, explores the implications of the existence of such points as members of the domain of normality.

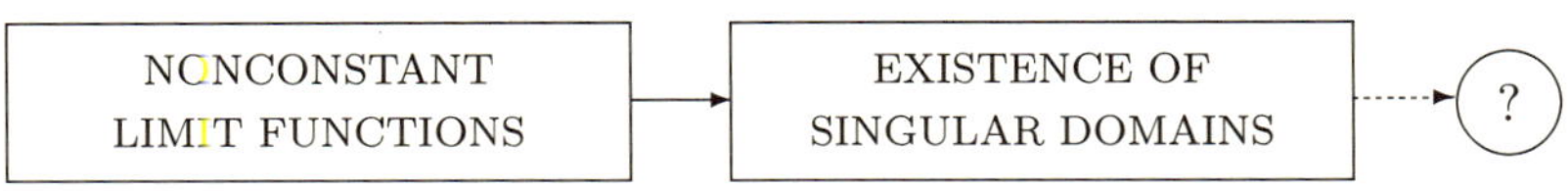

FIGURE 9.3. Fatou's exploration of singular domains. Fatou posited the existence of *singular domains* where the limit functions of the iterative family are nonconstant and explored the implications of their existence.

The heart of Fatou's examination of singular domains occurs with Chapter IV §§28–31 as part of his exploration of the domain of normality. He showed that if an irrationally neutral fixed point of a function f is in the domain of normality, then the limit functions of its iterative family are nonconstant. His argument is similar to Julia's: he bounded the derivatives of the functions in the iterative family away from zero which implies that the derivative of any limit function is not uniformly zero. Since limit functions are constant in the rationally neutral, attracting or superattracting periodic cases, nonconstant limit functions are inextricably linked to the existence of irrationally neutral fixed points.

At this point he began to focus on the nature of a singular domain, and one of the important concepts he articulated is that of a forwardly invariant singular domain D, that is, a singular domain D in which $f(D) \subseteq D$. Another concept he employed is that of an "antecedent domain", a domain D' with $f^n(D') \subseteq D$, where $n \in \mathbb{N}$. An antecedent domain is disjoint from D unless otherwise stated.

He established several basic properties which for convenience we will group together as a theorem, something he did not do.

THEOREM 9.3 (Properties of Singular Domains). *Let D be an invariant singular domain.*

(1) *If ϕ is a limit function from the iterative family in an antecedent domain D', and if $z \in D'$, then $\phi(z) \in D$.* [**1919-1920**, p. 57]
(2) *The identity function is always a limit function of the iterative family on an invariant singular domain; it can never be a limit function on an antecedent domain.* [**1919-1920**, pp. 55–56, IV]
(3) *An invariant singular domain is not backward invariant.* [**1919-1920**, pp. 56–57, IV]

He used this last item to state sufficient conditions for there not to be a singular domain. He earlier proved two fundamental properties: The first is that the Fatou set has zero, one, two, or infinitely many connected components. The second is that if a domain D, singular or otherwise, has one antecedent domain, it has infinitely many. Thus, if there is a singular domain, it has an antecedent domain, hence it has infinitely many:

> *It is thus established that if the number of regions contiguous to $\mathcal{F}$*[9] *is finite, that is to say, equal to* 1 *or* 2, *there can only be constant limit functions.* [**1919-1920**, p. 57, IV]

Points (1) and (2) from Theorem 9.3 show that the behavior of limit sets of the iterates of points in antecedent domains is different than in invariant domains. For example the limit set of a point in an antecedent domain D' includes no points in D'; on the other hand, in an invariant domain D, the limit set of any point $z \in D$, other than the fixed point λ, includes points in D, in particular z itself.

Fatou collected these differences in a classification of the limit set of the orbit $O^+(z)$, where z is in the domain of normality.

(1) The limit set of $O^+(z)$ is characterized by the values of the limit functions of the iterative family, and if ϕ is a limit function, then $\phi(z) = a$ is a member of the limit set of $O^+(z)$.
(2) If $\phi \equiv a$, then the limit set of $O^+(z)$ is an attracting limit cycle containing a either in the domain of normality if the cycle has a multiplier strictly less than one, or in $\mathcal{F}$ if the multiplier is a root of unity.
(3) If $\phi(z)$, is nonconstant, then the limit set of $O^+(z)$ consists of infinitely many points.

While he proved the first two cases, he noted the third was "*hypothetical*". Perhaps the most intriguing comment from his classification is the following, which fleshes out the hypothetical case:

> *If z belongs to a region contiguous to $\mathcal{F}$ where the limit functions are nonconstant, the point a which depends on the position of z* [i.e., $\phi(z) = a$] *belongs to a contiguous region* [forward] *invariant by any power of the given substitution, or to the boundary of this region* [i.e., $\mathcal{F}$]. ... *This is a hypothetical case that, according to the*

[9] $\mathcal{F}$ is Fatou's notation for the Julia set, a notation we will adopt for the remainder of the section. Fatou referred to components of the domain of normality as being contiguous, or bounded by $\mathcal{F}$.

> *recent research by G. Julia, cannot occur if $f(z)$ is rational.*[10] *In any event, we will soon give a sufficient condition—and one which has been present in all the examples we have studied—for the non-existence of such domains (which we will call singular domains).* [**1919-1920**, pp, 57–58, IV]

The "hypothetical" case, then, is the case where limit functions are nonconstant, the limit set of an orbit could be in the Fatou set, or the Julia set.

It is hard to resist thinking that Fatou's footnote evinces a certain anticipatory glee at the possibility of catching Julia in an error. On the other hand, his announcement of the soon-to-follow sufficient condition shows a prudent hedging of his bets: After all, despite his extended speculations about the existence of singular domains, in none of the many examples he investigated has he shown that they actually exist. It is also interesting that he waited until this point to give the name "singular domain" to a concept he had been talking about for several pages already. It is as if by naming it, he asserted the possibility of an object whose existence Julia denied.

In any event, the sufficient condition to which he referred would come as a consequence of an exploration of the preimages of points in the domain of normality whose greater goal was not just the exploration of singular domains but rather to establish a limit on the number of attractive and neutral orbits, as well as the number of invariant domains. A secondary goal was to explore the normality of the family of inverses of f, that is, for a fixed inverse ψ, the set $\{\psi^n\}$ where ψ^n is the n-fold composition the branch ψ. Nevertheless, out would fall several interesting properties of the hypothetical singular domains.

The first property concerns the limit sets of preimages of points in an invariant singular domain D; again, the preimages of points in a singular domain, where limit functions were nonconstant, had different properties than those from domains where the limit functions were constant:

> *In summation, the derived set of preimages of a point coincides with $\mathcal{F}$ except where it is a periodic attracting point or a point belonging to a singular domain which is invariant by any power* [i.e., a forward iterate] *of the given substitution.* [**1919-1920**, p. 58, IV]

To show that the number of attractive cycles (including cycles around rationally neutral fixed points) is finite, Fatou focused on the post-critical set E_c', the limit set of the critical values of f, showing that all attracting fixed points must attract a critical value, hence the number of such orbits is limited by the number of critical values, which is at most $2d-2$ counted with multiplicity, where d is the degree of f.

He next showed that this limit on attracting orbits also applies if the rationally indifferent periodic points are also included. Adding the possibility of singular domains to the mix, he showed that the number of invariant singular domains is no greater than $4d-4$, but expressed disappointment he was not able to find a sharper estimate:

[10]Here, Fatou inserted the following footnote: "*Julia announced this without demonstration*". Julia's *Comptes Rendus* notice evidently appeared in print prior to the completion of Fatou's monograph.

> *It is probable that this same limitation* [i.e., $2d-2$] *persists when one considers periodic points whose multiplier is at most one, with the case $s = e^{i\theta}$ (θ real and incommensurate) not excluded.*[11] *...But until now I have only succeeded in showing, and this is the essential point, that the number of cycles of this last category is itself finite.* [**1919-1920**, p. 66, IV]

Singular domains also act differently regarding normal families of the inverse of f. If ψ is a fixed inverse, the inverse iterative family $\mathcal{P} = \{\psi^n\}$ is normal on compact subsets of $\mathbb{C}$ which do not contain points from E_c. Moreover, their limit functions are also constant—except on invariant singular domains, and it is from this last fact that Fatou draws the sufficient condition for the nonexistence of singular domains alluded to above:

> *It is easy to deduce from this a sufficient condition for there not to be singular domains: that the derived set E_c' of the consequents of critical points does not divide the plane.* [**1919-1920**, p. 70, IV][12]

He showed this as follows: Suppose $E_c{}'$ does not divide the plane, that is, it is not a closed curve, yet there is an invariant singular domain D. Let z be a point in D. Let D' be an antecedent domain of D (as noted, Fatou showed there must be infinitely many) or another noninvariant component of the domain of normality, and let $z' \in D'$. Since z' is not in an invariant singular domain, the limit functions of the inverse iterative family $\mathcal{P}$ for any fixed branch ψ of the inverse are constant.

One can join the points z and z' with a polygonal line l which intersects neither $E_c{}'$ nor E_c. Because l contains no critical values of f, a fixed branch of the inverse ψ can be defined in a neighborhood N of l, and its inverse iterative family $\mathcal{P}$ is normal on N. Since z is in the invariant singular domain D, there are nonconstant limit functions for $\mathcal{P}$ on N. The limit functions of the inverse iterative family near z' must then be nonconstant as well. This, however, contradicts the fact that there cannot be nonconstant limit functions for $\mathcal{P}$ in a domain that is not invariant, so singular domains do not exist if E_c does not divide the plane.

While Fatou indeed articulated a sufficient condition for there not to be singular domains for a given rational function, he noted it is not one that is necessarily easy to apply:

> *One must however note that the problem of finding the limit points of the consequents of critical points is not susceptible to a general solution.* [**1919-1920**, p. 71, IV]

9.5. The possibilities for singular domains

Fatou was very careful not to outline explicit possibilities for singular domains. Perhaps he wanted to avoid statements that might be later turn out to be mistaken, or perhaps he recognized that forces were at work he did not fully comprehend. Nonetheless, he considered centers as a possibility, as well as points in the domain of normality whose iterates might have as a limit an irrationally neutral fixed (or

[11] Fatou's speculation was indeed correct, as was finally shown by Shishikura in [**1987**].

[12] This is consistent with the fact that the boundaries of Siegel discs are contained within E_c'. Since these boundaries are closed, a Siegel disc, and with it a center, cannot exist if E_c' is not a closed curve.

periodic orbit) residing in the Julia set $\mathcal{F}$. We included a quotation on page 205 that suggests this behavior. There is a similar statement in §28:

> *The case where a is at the same time a point in $\mathcal{F}$ and a limit point of consequents of certain domains of the plane presents itself, as we have seen, when a is a periodic point whose multiplier is a root of unity. I do not have an example where this exact circumstance occurs, without a being a periodic point of this nature, but it is infinitely possible that one could exist.* [**1919-1920**, p. 54, IV]

However, he later excluded the possibility that all points from an invariant domain D would converge to an irrationally neutral fixed point in D under iteration in a lengthy discussion replete with technical minutiae [**1919-1920**, pp. 242–247, VI]. Fatou also seemed to find this discussion quite long. Following a heuristic at the conclusion of the proof wherein he compared iteration near an irrationally neutral fixed point in $\mathcal{F}$ to that of a rationally neutral fixed point $e^{(2\pi i)/n}$ as $n \to \infty$, he observed:

> *But this is only a heuristic method, and the preceding demonstration, which moreover takes longer to explain than to understand, cannot be avoided.* [**1919-1920**, p. 247, VI]

Did he also consider the possibility of what is called a *wandering domain*, a connected component D of the Fatou set with infinitely many forward images $f^k(D)$, all of which are pairwise disjoint?[13]

The nonexistence of wandering domains is sometimes referred to as "Fatou's conjecture", and the section of Sullivan's paper [**1985**] where he showed wandering domains do not exist for rational functions is entitled "Solution of the Fatou-Julia problem on wandering domains". However, neither Sullivan's papers or the others we consulted contain a direct citation to a passage where he made such a conjecture.

Nonetheless, the title Sullivan gave the problem is apt, since it is certainly a problem implied by the work of Fatou and Julia, but we could find no evidence that it is one they explicitly posed. If they indeed considered it, they did not emphasize or pursue it in their respective monographs, as they did with the center problem, nor did those who immediately followed them remark upon it. None of this is meant to minimize the importance of the problem of wandering domains, or the skill it required to solve it, and it would be interesting to know when this problem achieved the status of a significant, open problem.

9.6. Concluding remarks

Fatou and Julia's respective investigations of the center relied predominantly on normal families, perhaps because, as we have noted, finding a direct solution S to the Schröder equation $S \circ f = f'(\lambda) \cdot S$ results in a difficult small divisors problem. It would not be until Siegel's paper [**1942**] that a direct approach to the center problem was successful, and, as we will see in Chapter 12, Siegel's solution of the Schröder equation relied on rather intricate estimates.

[13] Irvine Noel Baker showed in 1976 that this can happen for certain transcendental functions, [**1976**]; Dennis Sullivan showed that wandering domains do not occur with rational functions [**1985**]. It was shown that wandering domains can only exist if there are infinitely many "singular values", which for rational functions amount to orbits of critical values. Since there are only finitely many critical values for a rational function, wandering domains cannot exist. See Section L.1 of Appendix L for additional details and references.

Despite their reliance on normal families, the approaches of Fatou and Julia are quite different from one another. Julia's is more concrete, since he followed the implications of the explicit assumption that a rotation domain exists. Fatou, in turn, made no explicit speculations about how iteration might behave near an irrationally neutral fixed point; instead he showed that if such a fixed point is in the domain of normality, then there exist what he called singular domains, where the limit functions of the iterative family are nonconstant.

So, rather than assume a singular domain is a rotation domain—an assumption about the behavior of iteration—he focused instead on the behavior of the limit functions. This is a much less restrictive assumption that Julia's, and perhaps it is because of this flexibility that Fatou saw deeper into the situation, showing, for example, a limit on the number of singular domains and giving two sufficient conditions for them not to exist.

One of the ironies of their studies was the man who eschewed functional equations for most of his monograph studied the canonical Schröder equation for irrationally neutral fixed points, while the man who used the phrase "functional equations" in the title of his monograph did not.

In the end, though, neither mathematician made headway with the fundamental problem of determining whether rotation domains—or singular domains in the case of Fatou—exist or not. Progress would await a new approach, one which was already underway in the United States with Pfeiffer—of whom Fatou and Julia were evidently unaware—and would achieve fruition in the hands of two Germans, Cremer and Siegel.

Why were Fatou and Julia not aware of Pfeiffer? One reason we suggested in Section 5.4 is that the context of Pfeiffer's work was not iteration, although he certainly did reference several key works in the 19th century study, and it is clear his result can be applied to the study of iteration of complex functions.

Another issue is involves placement: in [**1915a**, p. 407] Pfeiffer announced, without proof, the existence of a class of functions with irrationally neutral fixed points for which no convergent solution to the Schröder equation exists, but it is in a footnote in the midst of a 36-page paper in the *American Journal of Mathematics*.

Pfeiffer also presented this result at the October 1915 meeting of the American Mathematical Society, and an announcement appeared in the *Bulletin of the American Mathematical Society* in 1916 [**1916b**], under the title "Existence of divergent solutions of the functional equations $\Phi[g(x)] = a\Phi(x), f[f(x)] = g(x)$, where $g(x)$ is a given analytic map, in the irrational case".

Although articles from the *Bulletin* were indexed in *L'Enseignment Mathématique*, as were those from the *Annals of Mathematics* and the *American Journal of Mathematics*, notices regarding presentations at meetings generally were not. So, if Fatou and Julia were in the habit of browsing through *L'Enseignment Mathématique* they would not have seen an announcement of Pfeiffer's result. Nor was it mentioned in the one American journal article that was indexed in *L'Enseignment*—Bennett's paper from the *Annals* entitled "The iteration of functions of one variable" [**1915**], that we discussed in Section 5.8. Ritt did not reference Pfeiffer's result in his March 1918 *Comptes Rendus* article, either, most likely because it had no relevance to the subject matter he was presenting.

A third possibility is that even though the *Transactions of the American Mathematical Society* was available in Europe, perhaps neither Fatou nor Julia paid close attention to American journals.

Part 3

After-Maths (1920–1942)

CHAPTER 10

Branching Out: Fatou and Julia in the 1920s

10.1. Introduction

Our treatment of the subject matter of this chapter differs somewhat from our approach in other chapters. We are not attempting a detailed treatment of each subject that Fatou and Julia covered in the 1920s. Rather, we hope to sketch developments in hopes of encouraging further discussion.

Our reasons for this are both practical—space and time—and historical: Unlike the case with Fatou and Julia's researches in rational functions near the end of World War I, which prompted responses in the 1920s and 1930s from many mathematicians—among them Joseph Fels Ritt in the U.S., Valiron, Arnaud Denjoy and Paul Lévy in France, Cremer in Germany, and Julius Wolff in Holland—the work of Fatou and Julia in the 1920s by and large did not bear mature fruit during the time period we cover. We therefore asked several contemporary researchers to link what happened in the 1920s to current research, and to convey a sense of how Fatou and Julia's ideas ripened over time. These responses are collected in our appendices. Throughout this chapter we will refer readers to their contributions.

After the completion of their long and exhaustive monographs on the iteration of rational maps, Julia and Fatou answered the question "What's next?" by developing coherent and robust dynamical programs where they extended their explorations into areas they thought ripe for investigation.

We summarize Fatou and Julia's programs below in Tables 1 and 3, respectively; Fatou also submitted several communications to the Société de Mathématiques de France in the 1920s, listed in Table 2 which give us further insight into his interests. Evidently, there is official record of these presentations other than their titles.

Our groupings in the tables are based primarily on the contemporaneous categorizations used in their research notices—Fatou [**1929a**] and Julia [**1934**]—and naturally some of the categories overlap. For example, their respective studies of permutable functions focused on a functional equation involving two rational functions f and g where $f \circ g = g \circ f$.

While their respective programs took them into common areas, most notably the study of permutable functions and the iteration of functions of two variables, there were significant differences. Fatou studied the dynamics of various classes of functions. Julia, on the other hand, was drawn to functional equations—which he intentionally neglected in his *Grand Prix* entry—as well as to analysis. In analysis he applied tools developed in his study of the iteration of rational functions, such as normal families, to function theory, in particular Picard theory, which led to the so-called "curves of Julia", which are the subject of Appendix E. His last research effort involving complex dynamics concerned series of iterates of a rational functions (that is, series of the form $\sum_{k=0}^{\infty} a_k f^k$, where f is a rational function) and was concluded by 1932.

TABLE 1. Fatou's dynamical program of the 1920s

Year	Topic	Papers	Output
1921–22	rational maps	[**1921a, 1921b, 1921c**] [**1921d, 1922b, 1923a, 1923b**]	50 pages
1921 & 1923	applications to point-set topology	[**1921a, 1923b**]	10 pages
1922–24	iteration in two variables	[**1922c, 1922d, 1924**]	84 pages
1923–24	permutable maps	[**1923-24**]	91 pages
1922	algebraic maps	[**1922a**]	11 pages
1926	transcendental maps	[**1921a, 1926**]	34 pages

TABLE 2. Fatou's Société de Mathématiques de France communications between 1921–1926.

Season	Communication title
November 9th 1921	*Remarques sur la méthode d'approximation de Newton* (BSMF [**1921**, p. 50])
April 26th 1922	*Sur le mouvement de planètes en milieu résistant* (BSMF [**1922**, p. 21])
February 24th 1926	*Sur les valeurs asymptotiques des fonctions uniformes* (BSMF [**1926**, p. 29])
March 24th 1926	*Sur l'itération des fonctions uniformes dans un domain de connexion quelconque* (BSMF [**1926**, p. 30])
June 23th 1926	*Sur des problèmes de géométrie se rattachant à l'itération des fonctions rationnelles* (BSMF [**1926**, p. 40])
October 27th 1926	*Sur l'applicabilité de deux aires planes par reprèsentation conforme* (BSMF [**1926**, p. 42])
November 10th 1926	*Sur les transformations birationnelles des courbes algébriques en elles-mêmes* (BSMF [**1926**, p. 43])
November 24th 1926	*Sur les transformations birationnelles des courbe à courbe* (BSMF [**1926**, p. 43])

There is thus a wide variety of mathematics in the dynamical programs of Fatou and Julia. But it found limited response.The Japanese mathematician Tatsujiro Shimizu (whom we discuss in Section 13.5) wrote two papers on the iteration of algebraic functions in response to Fatou's paper, but it was not until after the rebirth of the complex dynamics in 1980s that the study was significantly advanced.

Responses to Fatou's paper on the iteration of transcendental functions came slowly: first was the German mathematician Hans Töpfer, a student of Cremer (see Section 13.7), in [**1939, 1949**]. Paul Charles Rosenbloom published a short notice in *Comptes Rendus* on the iteration of entire functions [**1948**] and a second, relatively short paper [**1952**] four years later. The situation would change in the mid-1950s with Irving Noel Baker's inaugural effort in [**1955**] in what became a life-long investigation into the dynamics of entire transcendental functions, where

TABLE 3. Julia's dynamical program of the 1920s and early 1930s.

Year	Topic	Papers	Output
1919	rational maps	[**1919a**]	3 pages
1922–23	iteration in two variables	[**1922e, 1923a**]	7 pages
1921–22	permutable maps	[**1921a, 1922d**]	89 pages
	algebraic maps	none	
	entire transcendental maps	none	
1919-21	function theory	[**1919b, 1919-21, 1921c**]	111 pages
1920	extension of Schwarz's Lemma	[**1920**]	6 pages
1921–24	functional equations	[**1921b, 1921c, 1921d, 1922b**] [**1922c, 1922a, 1923b, 1924**]	101 pages
1925	quasi-analytic functions and series of iterates of rational functions (Part I)	[**1925a, 1925b, 1925d, 1925e**]	13 pages
1930–32	quasi-analytic functions and series of iterates of rational functions (Part II)	[**1930a, 1930b, 1931a, 1931b**] [**1931c, 1931d, 1932**]	138 pages

he examined the functional equation $g^2 = f$,[1] where f is a given entire function. Appendix L discusses the dynamics of entire maps.

Fatou's work with the iteration of functions of two variables likewise met with a desultory response. Bieberbach modified an example from Fatou's work in [**1933**], but there was little follow-up until the 1980s and 90s. In current expositions Fatou's paper seems to be quoted more frequently than Julia's. We do not know the reason for this, but it may simply be that Fatou's investigation was carried out in a full-length research paper, whereas Julia only published two short *Comptes Rendus* notes [**1922e, 1923a**].

We will begin our exploration of Fatou and Julia's respective dynamical programs with a discussion of permutable functions since that is where their dynamical programs of the 1920s have the greatest overlap. While there was some interest at the time in something called permutable functions in the study of integro-differential equations, in particular in the works of Volterra [**1888, 1910, 1911a, 1911b**], they are a different kind of mathematical beast, and there is no evidence of cross-fertilization between them. Rather, the interest in complex dynamics stems from the fact that permutable functions share the same Julia set. However, the converse is not true. As Steinmetz points out, Blaschke products often have the unit circle as a Julia set but do not commute [**1993**, p. 177].

Permutable functions obviously also share the same Fatou set; however, the periodicity of their behavior under iteration may differ. This is easily seen by considering a function f with an attractive period two cycle. Defining $g = f^2$, it is

[1] This equation, referred to earlier in our book as the Pfeiffer functional equation, was also studied by Lattès in [**1906**] (see Section 3.5), Pfeiffer in [**1916b**] (see Section 5.7), and Pincherle in [**1914**] (see equation (8.1)).

clear that f and g permute but the period two points of f become attracting fixed points for g.

10.2. The trilogy of permutable rational maps

Ritt, Julia, and Fatou each published works of 50 or more pages on permutable maps between 1922 and 1924. Julia's came first in 1922, and was cited by Ritt when his own article appeared in 1923. Fatou's paper appeared in two installments, one in 1923 and the other in 1924. In each, he remarked in footnotes evidently added after the main text had been typeset that Ritt's article appeared after he had written his. These three articles are discussed in Appendix J.

As that appendix shows, the individual investigations of Ritt Julia, and Fatou had their own character, although Julia and Fatou proceeded along somewhat similar lines. One significant difference in Ritt's approach was that he did not focus on the dynamics of permutable functions, but rather on their classification and their relations among one another. Appendix K discusses several post-1980 developments.

There is a fourth paper that the reader may wish to consult, an unpublished paper by Kiyoshi Oka [**1930-W**], that is based on both Julia and Ritt's approaches and is briefly discussed in Section 13.5. Oka studied in Paris between 1929 and 1932, and it is thought he may have met with Julia.

This trilogy of papers was not the first time permutable functions were addressed in the context of the theory of iteration. Seeds of 20th century developments are often found in the works of Kœnigs, so it should be no surprise that the origins of the study of permutable functions lie in his work. Indeed, in the opening chapter of his paper on permutable functions, Fatou reviewed Kœnigs' approach and remarked that he used it as the basis of his own work [**1923-24**, pp. 344–345]. Keep in mind, however, that the only condition that Kœnigs imposed on functions f is that they are analytic on some neighborhood of a fixed point, so, unlike the case with Fatou, Julia, and Ritt, we can not conclude they are rational functions. In fact, we will see an explicit example below from Kœnigs' work where f is not rational.

Kœnigs' investigation of the permutable functional equation was part of an extended application of the canonical Schröder functional equation to several other functional equations that he showed were related to it, in particular,

$$g \circ f = f \circ g, \qquad \Phi^n = f, \qquad P \circ f = g \circ P. \tag{10.1}$$

This examination was rooted in two questions:

(1) Given an analytic function f with an attracting, but not superattracting, fixed point λ, what functions Z besides f satisfy $S_f \circ Z = k \cdot S_f$, where $|k| < 1$ and S_f is the solution to the canonical Schröder equation $S_f \circ f = f'(\lambda) \cdot S_f$?

(2) Given a function S satisfying $S(v) = 1$ and $S'(v) = 1$, where S is analytic on a region surrounding an arbitrary point v in $\mathbb{C}$, what functions Z satisfy $S \circ Z = k \cdot S$, where k is also arbitrary?

We review Kœnigs' examination of these questions in order to provide context for his solutions of the permutable equation, as well as the equation

$$P \circ f = g \circ P, \tag{10.2}$$

from (10.1) above, where f and g are both given, which he viewed as a generalization of the permutable equation.

In order to streamline things, our review will not follow the exact sequence of Kœnigs' investigation, which spanned several pages [**1885b**, pp. 386–404]. Observe also, that while Kœnigs did not relate question (2) to Ernst Schröder, it is a formalization of Schröder's "opposite approach" that we have discussed at several points in our work.[2]

Kœnigs introduced question (2) in order to treat the first, and before addressing either, he reviewed what was known about S_f, recalling that it is analytic on a neighborhood D of λ and satisfies $S_f(\lambda) = 0$ and $S_f'(0) = 1$. He also implicitly used the fact that S_f is a bijection of D and a neighborhood N of the origin.

Denoting Z as $Z(z, k)$, he gradually established its general properties, three of which we list as a theorem:

THEOREM 10.1 (Properties of $\mathbf{Z(z, k)}$). *Suppose that $S \circ Z(z, k) = k \cdot S(z)$. If $S = S_f$, the solution to the canonical Schröder equation for f, then*

$$f(z) = Z(z, f'(\lambda)). \tag{10.3}$$

Under the hypotheses of either question (1) *or* (2),

$$Z(Z(z, k), k')) = Z(z, kk') = Z(Z(z, k'), k) \quad \textit{and} \quad Z'(\lambda, k) = k. \tag{10.4}$$

In addressing the first question, he implicitly defined $Z(z, k)$ as

$$Z(z, k) = S_f^{-1} \circ (k \cdot S_f), \tag{10.5}$$

and assumed an analogous definition in question (2). The properties in equations (10.3) and (10.4) follow from routine calculations. The definition $Z(z, k) = S_f^{-1} \circ (k \cdot S_f)$ implies that

$$f^n(z) = S_f^{-1} \circ (f'(\lambda)^n \cdot S_f) = Z(z, (f'(\lambda)^n),$$

for $n \in \mathbb{N}$, so among the Z, are the forward iterates of f. Moreover, the commutation property in (10.4) becomes a generalization of the trivial identity $f^n \circ f^m = f^m \circ f^n$.

At the end of examination of the first question, he was not able to say what explicit functions Z besides f (and its forward iterates) satisfy $S_f \circ Z = k \cdot S_f$. This issue, among others, evidently prompted the second question. Nonetheless, using the properties listed in Theorem 10.1, he gave what amounts to a formal solution for $g \circ f = f \circ g$, where g is unknown.

To solve the permutable equation, observe that $f = Z(z, f'(\lambda))$, and let $k = f'(\lambda)$. Setting $g = Z(z, k')$ and then applying equation (10.4), we get

$$g \circ f = Z(Z(z, k), k') = Z(Z(z, k'), k) = f \circ g. \tag{10.6}$$

The solution is quite elementary, and at this point it is not clear what functions g besides the forward iterates of f satisfy equation (10.6). However, his examination of question (2) will yield additional solutions.

Kœnigs answered the second question over the next several pages, showing that given S under the hypotheses of question (2), the function $Z(z, k)$ satisfying $S \circ Z(z, k) = k \cdot S$ is well defined for all z in a region D with v in its interior

[2]See page 10 (in particular, (1.7)) for a discussion of Schröder's "opposite approach". For applications, see page 168 and Example 9.2.

for complex values $|k| < 1$. We will omit his discussion, but the interested reader should consult [**1885b**, pp. 390–393].

Rather than using his investigation of question (2) to show that solutions other than the iterates of f exist to the permutable equation $g \circ f = f \circ g$, Kœnigs instead applied it to the generalized permutable equation (10.2), where f and g are two given analytic functions whose domains of definition will be specified momentarily.

To simplify his exposition, he restricted the conditions on f and g so that f is analytic on a disc D surrounding an attracting fixed point λ, g is analytic on a disc D' surrounding an attracting fixed point μ, and the multipliers for λ and μ are equal, that is, $f'(\lambda) = a = g'(\mu)$.[3]

His method of solution to (10.2) was to seek a general solution $V(z, k)$, analogous to the solution $Z(z, k)$ of the permutable equation $g \circ f = f \circ g$, that is defined in terms of the solutions to the canonical Schröder equations for f and g, [**1885b**, pp. 395–398]. Denoting these as S_f and S_g, he defined the function $V(z, k)$ implicitly by

$$S_g(V(z, k)) = kS_f(z). \tag{10.7}$$

Since the Schröder function S_g is a bijection, an equivalent definition is given by $V = S_g^{-1} \circ (k \cdot S_f)$, which is well defined in D, and moreover, is a generalization of his definition of $Z(z, k)$ at equation (10.5).

Letting z become $f(z)$ and fixing k, it follows that

$$S_g(V(f(z), k)) = kS_f(f(z)) = akS_f(z),$$

where a is the common multiplier, and also that

$$S_g(V(z, ak)) = akS_f(z). \tag{10.8}$$

Since the right-hand sides of the preceding two equations are equal and the Schröder functions are bijections, comparing their left-hand sides yields the identity

$$V(f(z), k) = V(z, ak).$$

Using the same k and defining

$$P(z) = V(z, k) \tag{10.9}$$

yields

$$P(f(z)) = V(f(z), k) = V(z, ak). \tag{10.10}$$

Likewise, it follows from the definition of S_g and equations (10.7) and (10.8) that

$$S_g(g(V(z, k))) = aS_g(V(z, k)) = akS_f(z) = S_g(V(z, ak)),$$

yielding another identity,

$$g(V(z, k)) = V(z, ak). \tag{10.11}$$

Applying g to the definition of P at (10.9) and using (10.11),

$$g(P(z)) = g(V(z, k)) = V(z, ak). \tag{10.12}$$

Finally, since the right-hand sides of (10.10) and (10.12) are equal, we obtain

$$P \circ f = g \circ P.$$

---

[3]He remarked, in effect, that these restrictions are more a matter of convenience than a necessity. His reasoning involved a further generalization that is beyond the scope of our examination. See [**1885b**, pp. 399–401] for details.

Now the results from the generalized permutable equation can be applied to $g \circ f = f \circ g$: simply let $f = g$, and $V(z,k)$ becomes $Z(z,k)$ (which is also a consequence of their definitions). Kœnigs left this implicit.

Kœnigs' overall approach to the solution of the functional equations at (10.1) is somewhat formal. He made no assumptions about the nature of f, aside from it being analytic on a disc, usually containing a fixed point, and was careful to specify that the above calculations take place on suitable domains of f and g. Nor did he draw any conclusions about how permutable functions behaved under iteration.

Kœnigs did, however, provide several examples, including a solution to $P \circ f = g \circ P$ [**1885b**, pp. 403–404]. Let $f(z) = \sqrt[m]{az^m + c}$ and $g(z) = \sqrt[m']{az^{m'} + c'}$, where c, m, c', and m' are unspecified constants, and the modulus of a is strictly less than one. The fixed points of f and g are, respectively, given by x and x', where

$$x^m = \frac{c}{1-a} \quad \text{and} \quad x'^{m'} = \frac{c'}{1-a}.$$

Routine calculations verify that $f'(x) = a = g'(x')$, so the fixed points are indeed attracting, and that the respective Schröder equations are given by $S_f(z) = z^m - x^m$ and $S_g(z) = z^{m'} - x'^{m'}$. The fixed points of f and g share a common multiplier, so using the implicit definition of P from (10.7), $V(z,k)$ satisfies

$$S_g(V(x,k)) = V^{m'} - x'^{m'} = k(z^m - x^m) = kS_f(z).$$

From the definition of P above at equation (10.9), we have $P = V$ and $P \circ f = g \circ P$. To deduce the solution to $g \circ f = f \circ g$, equate f and g (Kœnigs treated this case in an earlier example on [**1885b**, pp. 402–403]). It follows that

$$V^m - x^m = k(z^m - x^m),$$

hence $V(z,k)$ coincides with $Z(z,k)$.

It is interesting that Fatou and Julia's researches crossed paths again with permutable functions. (And not only here—Fatou and Julia each published *Comptes Rendus* notices on the iteration of functions of two variables in 1922.) This time, however, there was no controversy, and they generously cited each other, although Julia did point out in his *Notice de Candidature Académique* that his brief study of multivariable iteration treated something that "escaped" Fatou's attention, because, according to Julia, Fatou's work was "oriented to other goals" [**1934**, p. 22].

10.3. Seeds of dynamics for algebraic functions

In 1922 Fatou undertook a brief study on the iterations of algebraic functions. His goals were modest:

> *We will examine here examples of iteration for algebraic or rational maps, relatively simple examples which can be treated completely. But certain properties will emerge which are quite different from those we have already seen in our study of iteration.*
>
> *When a given algebraic function is an inverse of a rational function, the study of the configurations of the limits of* consequents *of a point in the complex plane is evidently identical to the problem concerning the* antecedents *of a point of rational substitution, and this problem should be considered virtually solved in the memoir already cited* [his rational functions monograph]. *This is*

> *not the case, on the other hand, with the iteration of a symmetric biquadratic relation among two variables, ... To find new examples which are sufficiently simple, we recall that the simplest rational functions to study are those which possess a fundamental circle.*[4] [**1922a**, pp. 188–189].

His intent evidently was to develop a more comprehensive study of algebraic functions, but the difficulties involved got in the way. He closed the paper by observing

> *... a general theory of this problem, even if limited to substitutions with a fundamental circle, does not seem easy. We hope to return to it later.* [**1922a**, p. 198]

He never did.

Fatou used the following example throughout his paper. Let $0 < k' < k < 2$, and let z and t be two complex variables. Define

$$\begin{aligned} z &= R(t), \\ z_1 &= kR(t) + k't, \end{aligned}$$

where $R(t) = t - \frac{1}{t}$. He viewed t in terms of z, and since each z corresponds to two t, $t(z)$ is multifunction, thus z_1 has two branches, which Fatou denoted z_1 and z_1', with infinity a fixed point of each.

To confront the issue of multivalued functions, he developed the concept of a chain, defined as follows, to track orbits of a point under a multifunction:

DEFINITION 10.2 (Chains of iterates). Let $S = \{n_1, n_2, \dots\}$ be a sequence of nonnegative integers, whose terms are not necessarily distinct. A chain $\mathcal{C}_S$ is the infinite composition

$$\mathcal{C}_S = \cdots z'_{n_4} \circ z_{n_3} \circ z'_{n_2} \circ z_{n_1},$$

where z_{n_k} and z'_{n_k} denote the n_k-th iterate of z and z', respectively.

This same concept under a different guise cropped up in Fatou's monograph where he found a correspondence between base-d sequences $\{d_1, d_2, \dots\}$ and points in the Julia set for a rational function f of degree d (see page 158). These base-d sequences are ways of coding the backwards orbit $O^{-1}(z)$ of a point under a particular composition of inverses of f, which of course are algebraic functions. That is exactly the concept behind Fatou's chains of iterates. Pincherle used essentially the same idea in his entry to the *Grand Prix* (see Section 8.6.2) to code his nested radicals, which can be viewed as chains of an algebraic map with two branches.

Fatou considered the action of the set of chains on points z. The mapping $R(t)$ leaves the real line, and both the upper and lower half of the plan invariant, thus z_1 and z_1' both have the real axis as a fundamental circle. From here it follows after routine calculations that if z is not on the real line, it converges under iteration by either branch (and therefore under all possible chains) to ∞. He extended this convergence to all points on the real axis except for an interval $(-\xi, \xi)$ where ξ depends on the choice of k and k'.

The interval $(-\xi, \xi)$ contains sets $E(C_S)$ which Fatou pointed out are analogous to the Julia set for a rational function. To define $E(C_S)$, let $z \in (-\xi, \xi)$, and let $\mathcal{C}_S$ be a chain of iterates. The set $O^{+}_{\mathcal{C}_S}(z)$ denotes the orbit of z under the partial

[4] Recall that the concept of a fundamental circle is an integral part of his investigation of the iteration of rational functions. See, for example, page 155.

compositions of $\mathcal{C}_S$, namely $z_{n_1}, z'_{n_2} \circ z_{n_1}$, and so forth. Finally, define $E(C_S)$ to be the the set of all the points in $(-\xi, \xi)$ whose entire orbits $O^+_{\mathcal{C}_S}(z)$ remain in $(-\xi, \xi)$.

As is the case with the Julia set of a rational function, such as $z \mapsto z^2/(z^2+2)$ where the only component of the Fatou set is the domain of attraction of a fixed point (see Color Plate 2), $E(\mathcal{C}_S)$ is the set of points which does not converge to the attracting fixed point.

Fatou noted both parallels and significant differences between $E(\mathcal{C}_S)$ and a Julia set J. Neither J nor $E(\mathcal{C}_S)$ is empty, but where J is perfect, the most one can say about $E(\mathcal{C}_S)$ is that it must be closed. Moreover, it can consist of just a single point. He observed, however, that the union of the $E(\mathcal{C}_S)$ over all possible chains $\mathcal{C}_S$, which represents the set of all possible orbits of points in $(-\xi, \xi)$ is, like the Julia set, a perfect set, in this case the interval $(-\xi, \xi)$.

As with Julia sets, preimages of open sets in $\mathbb{C} - (-\xi, \xi)$ limit upon $(-\xi, \xi)$, and all points in $(-\xi, \xi)$ are the limit of preimages of arbitrary points in the domain of convergence of ∞. Finally, the limit set of the periodic points of the multifunction Φ is the interval $(-\xi, \xi)$.

10.4. Fatou's solo in transcendental dynamics

Fatou embarked upon what turned out to be the final stage of his dynamical program in [**1926**], a 40-page work devoted to the iteration of entire transcendental functions.

This article is organized into two parts. The first consists of a mostly theoretical exposition of the degree to which properties from rational functions extend to transcendental entire functions. In the second part, he treated two examples in great detail, devoting over five pages to each. Fatou may have hoped to include additional examples, but noted that generating transcendental examples is quite laborious [**1926**, p. 358]. While the scope of this particular paper has some of the exploratory character of Fatou's paper on algebraic functions—especially in its development of exposition via examples—he was more successful in developing a theoretical framework.

This was not Fatou's first encounter with the dynamics of entire transcendental mappings. In the *Comptes Rendus* notice [**1921a**] he presented an example involving the exponential mapping

$$E(z) = z + 1 + e^{-z}. \tag{10.13}$$

Fatou observed that points on the right half-plane converge to infinity under iteration by E, and that within the domain of convergence D, the Abel equation $A \circ E = A + 1$ has an uniform solution A.

Fatou's goals with [**1921a**] seem to be two-fold. One is obviously the iteration of E, an interest he would pursue again in his paper [**1926**] where he delved more deeply into its dynamics in hopes of gaining insight into how the iteration of rational functions compares to that of transcendental functions.

The other goal was to show how a recent development in point set topology, the concept of an inaccessible boundary point, might be applied to complex dynamics. A point p on the boundary ∂D of an open set D is inaccessible from D, if, given any point z in D, there is no curve of finite length connecting p to z that, aside from its endpoint p, is contained entirely in D. Fatou remarked,

> *Mathematicians who study sets of points have, in recent years, introduced the notion of an inaccessible boundary point of a domain, a notion that has, until now, remained unapplied.* [**1921a**, p. 344]

By "unapplied" he seemed to mean that examples of their occurrence with actual functions were rare. He speculated further that the iteration of rational functions might be a rich source of similar examples, but evidently he had not made much progress towards finding any. He would take up inaccessible boundary points again in the short paper [**1923b**], in which he also followed up on remarks from his monograph on rational functions concerning the accessibility of points in the domain of nonnormality.

His notice [**1921a**] is also noteworthy since it contains perhaps the first description of a so-called Baker domain, that is, a region D in $\overline{\mathbb{C}}$, forward invariant under f with a point $a \in \partial D$ that is an essential singularity of f, with the property that while no orbits $O^+(z)$ originating in D have an accumulation point in D, for all $z \in D$, $f^{pk}(z) \to a$ as $k \to \infty$ for some p. See Figures 10.1 and Color Plate 10 for depictions of Baker domains.

FIGURE 10.1. The function $E(z) = z + 1 + e^{-z}$ (10.13) iterated 11 times over the sphere. The Baker domain is the region in white.

Caption for Color Plate 10: (See p. 66.) The Baker domain for $E(z) = e^z + z$ viewed from three different vantage points. Using the left and middle images from Color Plate 11 as reference images, the lightish green areas represent points with large moduli in the upper half-plane while those in blue correspond to large values in the right-half plane.

For polynomial functions, analogous behavior is the norm (for example, the set outside the unit circle for the $f(z) = z^d$ has no finite accumulation point), but the dynamics are different since for a polynomial of degree d, the point at infinity acts like a superattracting point rather than an essential singularity, and iteration is not a translation but is instead conjugate to the mapping $z \mapsto z^d$. However, for a transcendental map with a Baker domain, iteration near infinity is analogous in some respects to that of a function with a rationally neutral fixed point. Fatou highlighted this resemblance when he revisited the function E in [**1926**]:

> *By applying the method that I have already developed to study substitutions having a double point at infinity with multiplier +1, one*

> *easily obtains the domains where the iterates of $E_n(z)$ converge to ∞.* [**1926**, p. 358]

The method he referred to involves the Abel function equation, which he used to model iteration near a rationally neutral fixed point in his monograph on rational functions [**1919-1920**, §9–12], and he showed, as he did in [**1921a**], that $A \circ E = A + 1$ has a solution A on the right half-plane.

The opening remarks of his paper [**1926**] highlight both its triumphs and its shortcomings:

> *In a series of articles published in the* Bulletin de Société de Mathèmatiques *we established the general principles regarding the iteration of rational functions and studied the properties of the functional equations to which they are related. In this article we propose to extend these results to entire transcendental functions. This extension presents considerable difficulties which we did not completely overcome; it is easily understood that, due to the presence of an essential singularity at infinity, the phenomena which is already very complicated in the case of the iteration of polynomials, acquires a much greater complexity here. This is why we will confine ourselves to establishing a few general theorems and to studying relatively simple particular examples. We will intentionally leave aside the case of meromorphic functions which give rise to transcendental functions with infinitely many essential singularities.* [**1926**, p. 337]

In §9 of [**1926**] Fatou explored the nature of the domain of nonnormality for a transcendental map f. He began his discussion by defining the set J, which is the Julia set f, as the domain of nonnormality. For Fatou, this was the right definition for J since he was unable to show not only that J is in general the closure of repelling points for a transcendental function f, but that even a single repelling cycle exists. This led him to observe:

> *The elegant method Julia (Memoire sur l'itération des fonctions rationnelles ...) used to study the set F* [Fatou's notation for J], *regarding it a priori as the derived set of the points belonging to the repelling cycles, cannot be applied here, until the existence of at least one such cycle has been established.* [**1926**, p. 358]

So, armed with his definition of J as the domain of nonnormality, Fatou listed several properties that extend to the iteration of transcendental mappings [**1926**, p. 348]:[5]

(1) J is closed.
(2) J is invariant under f^n.
(3) Repelling and rationally neutral periodic points are in J (if they exist).
(4) If an irrationally neutral fixed point is not in J, its limit functions are nonconstant.
(5) If the multiplier of a periodic point is strictly less than one in modulus, it is attractive.

[5] Appendix L discusses the dynamics of entire maps, including the surprising nature of many of the Julia sets.

On the other hand, he noted that he was unable to show that the following carryover from the iteration of rational functions [**1926**, pp. 369–370]:

(1) The existence of a function f whose domain of nonnormality is $\mathbb{C}$.[6]
(2) A function all of whose points converge to ∞ aside from a perfect, discontinuous set.
(3) In addition, he showed that, if $f(z) = h\sin(z)$, where $0 \leq h$, all points in $\mathbb{C}$ converge to 0 under iteration by f except for a set J which contains a family of unbounded analytic curves whose points diverge to infinity under iteration by f. Fatou wondered if "this property must not also extend" to a very general family of functions [**1926**, p. 369].

These last three issues, as well as the question of whether or not repelling periodic points were dense in J, remained unanswered for over 40 years. The preface to Rippon and Stallard [**2005**, pp. x–xiii] details the resolution of these four questions by Baker and others, beginning with Baker's paper [**1968**], where he showed that the repelling points are dense in J. Michal Misiurewicz in [**1981**] showed that the Julia set for $f(z) = e^z$ is $\mathbb{C}$ (it was known not too long before then that for some functions $J = \mathbb{C}$). Baker showed in [**1975**] that J cannot be a totally discontinuous set and must contain a continuum. Later, he showed that J does not always contain the type of unbounded analytic curves described in (3) above (see [**1984**]).

Perhaps the questions Fatou left unanswered provide a reason for the gap between his initial exploration of transcendental maps in [**1921a**] and the publications of [**1926**]: It could be that he hoped to resolve these open questions, but was unable to, and decided to publish what research he had. It could also be that he was moving on to other things. This was the last paper in what we have called his dynamical program. Subsequent publications involved perturbation theory and celestial dynamics and his redaction of the second volume of Appell and Goursat's *Théorie des fonctions algébriques d'une variable et des transcendantes qui s'y rattachent, Fonctions automorphes*, which contained passages relating iteration to Kleinian groups (for example, [**1930**, pp. 95–96]).

Julia did not write a paper on transcendental dynamics, but he did allude to the iteration of transcendental functions at several points in his monograph on rational functions, for example, at [**1918a**, p. 82] where he observed that iteration cannot be well defined at an essential singularity, at [**1918a**, p. 89], where he compares a rational function with a single exceptional value[7] to the behavior of a transcendental function under iteration, and, finally, at [**1918a**, p. 137], where he notes the possibility of infinitely many superattracting points. These obviously do not constitute a theory, but they do signal an awareness of the differences between transcendental and rational dynamics.

---

[6]He suspected, correctly as it turns out, that this was the case for $f(z) = e^z$. All he could demonstrate, however, was that unlike the mapping $E(z) = z + 1 + e^{-z}$, there is no neighborhood N of infinity (i.e., a set $N = \{z : |z| > M\}$ for some constant M) on which iterates "converge regularly to ∞", as he put it, (i.e., for all $z \in N$, $|f(z)| > |z|$) and that there are no fixed points satisfying $|f'(z)| < 1$ [**1926**, p. 370].

[7]A point is a unique exceptional value w on $\mathbb{C}$ if in neighborhoods of points in J the iterative family takes on all values in $\mathbb{C}$ except for w.

10.5. Julia's dynamical program

Julia's program in the 1920s has three main threads. One was functional equations. As we have noted (see page 133 in Chapter 7), Julia decided to focus on iteration in his 1918 monograph and explicitly noted that he was deferring a study of functional equations to some unspecified future date. He inaugurated that study in 1921 with a study of permutable functions, [**1921a, 1922d**], and extended it into what he later called "quasi-permutable" equations [**1921b, 1921c, 1921d, 1923b**], which are of the form $P \circ f = g \circ P$, similar to those Kœnigs discussed at equation (10.2). Julia also studied the relations between functional equations and the conformal representations of their domains of solution on the Riemann sphere [**1922a, 1922b, 1924**].

Julia's study of functional equations is to some degree analogous to Chapter VII of Fatou's monograph on rational functions, where he studied the domains of solutions to the various functional equations used in iteration and then explicitly related his findings to results he obtained in the first six chapters.

However, while Julia's investigation of functional equations is rooted in iteration, it is not always its central focus. For example, while at several points he acknowledged that the genesis of his interest in functional equations was his study of iteration, his references to his own monograph on rational functions were of a general character, and he generally did not—at least in comparison to Fatou in Chapter VII of his monograph—apply his results to specific passages from his monograph [**1919-1920**].

With his second thread—which actually was the first major study Julia undertook after the completion of his monograph—Julia applied techniques and ideas he used to study iteration to function theory. Beginning in 1919 he published an extensive and ground-breaking three-part exposition applying the theory of normal families to the Picard theory [**1919b, 1919-21, 1921c**]. The focus of these papers is on the values of the family $\{f_t(z) = f(z\sigma(t))\}$, where $\sigma(t)$ is a parameterized path, the so-called lines or curves of Julia. In particular, he was interested in determining where such families were not normal. While these families were not iterative, there are parallels between the sets of nonnormality and Julia sets. Appendix E discusses these papers in much more detail.

Related to the second thread is Julia's paper "*Extension nouvelle d'un lemme de Schwarz*" [**1920**], which grew out of an extension of Schwarz's Lemma he discussed in §5 of [**1918a**]. We examined Julia's extension as presented in his monograph in Section 7.2 at Theorem 7.4.

This paper is part of an ongoing and, at the time of this book, still active body of work, which is discussed in several appendices: Appendix F gives a contemporary version of Julia's extension and traces responses from 1922 through the beginning of the 21st century. Appendix G analyzes one these responses, the Denjoy-Wolff theorem. Finally, in Appendix H, several mathematicians discuss Julia's extension as well as the Denjoy-Wolff theorem from various points of view, including its history and its role in present-day mathematics.

The last part of Julia's dynamical program involves what he referred to as a series of iterates of a rational function,

$$g(z) = \sum_{k=0}^{\infty} a_k f^k(z), \tag{10.14}$$

where f is a rational function satisfying $f(\lambda) = \lambda$ with $|f'(\lambda)| < 1$ or a root of unity.

He took up the subject in 1925 with several *Comptes Rendus* notices [**1925a, 1925b, 1925c, 1925d, 1925e**], and for reasons we do not know set it aside for several years before picking it up again with the notice [**1930a**]. His only full-length paper on the subject is [**1931b**].

In his *Notice de Candidature Académique* he remarked that this study was motivated in part by Borel's study of quasi-analytic functions,

$$\sum_{k=0}^{\infty} \frac{C_k}{z - c_k}$$

defined in a domain C. In one case, Borel determined a way to find a determination for this function beyond C which was bounded by a curve in which the c_n were dense and did not allow a Weierstrassian continuation. Julia wondered if he could do something similar in a "natural problem":

> *I proposed this problem in 1924–1925, and, to be precise, sought to link the classical analysis of Cauchy, Riemann,* [and] *Weierstrass to Borel's new chapter,* by an application of a *REGULAR ALGORITHM GIVEN IN ADVANCE,* by forming, as a Taylor series is formed in terms of successive powers of the variable and arbitrary coefficients a_n, general classes of series whose sum is sometimes an analytic function and sometimes a quasi-analytic function.
>
> *My research on iteration has provided me with an algorithm that allows the* condensation of singularities *which I envision: the poles of the iterates of* f^n *of a rational fraction* f *are condensed in the* perfect set, $\mathcal{E}_f$. [**1934**, p. 8]

The series he had in mind is (10.14), given above, and his hope was to determine if it is convergent in components of the Fatou set beyond the immediate domain of convergence D of an attracting fixed point of f. By performing a root test on the series of coefficients

$$\sum_{k=0}^{\infty} a_k,$$

he determined that the answer was sometimes yes, and sometimes no. The conditions under which the series converges beyond D are both detailed and lengthy, and they are omitted since our aim here is to give a flavor of his investigation, rather than an exhaustive account. The interested reader, however, can find a detailed description in Julia's paper [**1931b**, pp. 192–195].

10.6. Self-maps of the unit disc

Ever since Kœnigs conformally mapped Schröder domains to discs centered at the origin via the Schröder functional equation in the mid-1880s, self-maps of a disc[8] D centered at the origin have been essential to the study of complex dynamics. One of the earliest recognitions of this was Fatou's *Comptes Rendus* notice [**1917a**], discussing maps with a fundamental circle, where he discussed a mapping f which fixes a circle Γ and leaves its interior invariant (see Section 7.3).

[8]A self-map on a simply connected domain D is a function f, where $f(D) \subseteq D$.

Julia, too, paid close attention to self-maps. Like Fatou, he made liberal use of the fact that an open set could be conformally mapped to a disc to study iteration on attractive domains. Such maps were so important to Julia that he devoted a substantial portion of his *Preliminaries* to Schwarz's Lemma, and proved (although, as Audin points out, he did not state it as concisely as he might) what has come to be known as Julia's extension of Schwarz's Lemma, which we discuss above, as well as in Section 7.2. Appendix F discusses Julia's extension as well as related developments.

In the context of iteration, in 1926 Wolff and Denjoy independently proved versions of what has come to be known as the Denjoy–Wolff theorem, as did Valiron in 1931. After the 1920s, self-maps became an object of study in their own right. The results of Denjoy, Wolff, and Valiron, as well the mathematics which they stimulated, are discussed in Appendices G and H.

Wilhelm Blaschke developed the concept of a Blaschke product in [**1915**], but not in the context of iteration. Although he did not refer to Blaschke, Fatou [**1923a**] stated conditions under which holomorphic maps of the disc might be stated in terms of a Blaschke product. After the discovery of the existence of annular rotation domains called Herman rings around 1980, it was discovered that Blaschke products of odd degree three or greater with Diophantine rotation numbers could be used to construct functions with Herman rings.

10.7. Concluding remarks

Fatou and Julia each followed up their monographs on the iteration of rational functions with a series of papers in the 1920s, and, in Julia's case, the early 1930s, that we refer to as their respective dynamical programs. While none of the these papers achieved the heights of their work with rational functions, they each nonetheless made significant contributions to the study of iteration and to the theory of functions.

One aspect of their dynamical programs could be viewed as the typing up of loose ends, including functional equations, in particular the permutable functional equation $f \circ g = g \circ f$, which they both investigated in lengthy papers in the early 1920s. Permutable functions have the same Julia set, although their dynamics are not always the same on the Fatou set. Ritt also published a study of permutable functions from a somewhat different perspective.

Fatou and Julia both broke new ground in the 1920s, but Fatou is to be considered more successful in developing and executing a wide-ranging dynamical program. For one thing, he seemed to proceed with the conscious agenda of extending the theory developed in the pursuit of rational functionals to other kinds of functions such as algebraic functions, transcendental functions and functions of two variables. Indeed, Fatou's studies of the dynamics of various classes of functions are foundational.

Julia, on the other hand, successfully applied methods and tools developed in his study of the iteration of rational functions to various to complex function theory and developed fundamental ideas, including the so-called "curves of Julia" and an extension of the Schwarz Lemma, which remains of interest to researchers. These topics are the subject of several appendices which treat self-maps of the disc, essential lines of singularities and the so-called curves of Julia.

CHAPTER 11

The German Wave

11.1. Introduction

As the French study of iteration was winding down in the late 1920s, Hubert Cremer wrote a series of papers beginning in 1927 and ending in 1938 whose principal focus was the center problem. Although he was clearly influenced by the French study, he took a different route from Fatou and Julia, and, taking his cue from Pfeiffer, proved the existence of a wide class of rational functions with irrationally neutral fixed points that were not centers.

In 1917 Pfeiffer constructed functions f with irrationally neutral fixed points λ such that all formal solutions to the canonical Schröder equation $S \circ f = f'(\lambda) \cdot S$ diverge, hence λ is not a center. His construction was not practical, however, because it provided no way to determine if an irrationally neutral fixed point was a center or not unless you could construct it using his method, which was rather convoluted.

In 1927 Cremer found an explicit number-theoretic condition on $\gamma \in (0,1)$ which allowed him to find rational functions f with an irrationally neutral fixed point λ satisfying $f'(\lambda) = e^{i2\pi\gamma}$ that is not a center. With this result, he rendered Pfeiffer's work into a more useful and elegant form. Over the next decade he extended his investigations from rational functions to other kinds of functions, including entire functions and gap series, and in so doing modified the conditions under which an irrationally neutral fixed point is not a center.

Cremer also examined the frequency of noncenters for rational functions, asserting that the set of multipliers for $f'(\lambda)$ for which centers do not exist is an uncountable, dense subset of the unit circle. Since he could not show the existence of centers, it was possible, he noted, that the set of these multipliers could be a set of full measure on the unit circle.

Despite admitting the complete nonexistence of centers for rational functions of degree two or more as a possibility, Cremer, did not seem to share Julia's conviction that centers do not exist in the general case. He generally referred to Julia's assertion as if it were a conjecture, noting in [**1927**, p. 154] that Julia had given the result "without demonstration", and that he could see "no reason" why it should be true". At another point, he remarked that he thought the sketch of the proof Julia had offered was not accurate [**1932a**, p. 317].

He also studied the Fatou set, and, very much in the spirit of Fatou's investigation of singular domains described in Section 9.4, Cremer explored the implications that would follow should centers exist to a degree that would be unusual had he disbelieved their existence (see his papers [**1927**, pp. 159–163] and in particular [**1932a**, §5-6]).

Cremer, however, evidenced no progress towards demonstrating the existence of centers. That was left to Siegel, who in 1942 not only found a sufficient number-theoretic condition for an irrationally neutral fixed point to be a center, but also showed that centers are ubiquitous in the sense that multipliers for which a center exists form a subset of full measure on the unit circle.

We will discuss Cremer in this chapter and Siegel in the next.

11.2. Iteration in Germany: Historical overview

The contributions of Cremer and Siegel in some senses represent a return of the study of the iteration of complex functions to its German roots. Yet, despite the fact that German mathematicians were responsible for major progress in the center problem, one should be cautious about thinking of a "German school" of iteration theorists because unlike the situation in France with Kœnigs' and those who followed him, Kœnigs' students Grévy, Leau, and Lattès, as well as Fatou and Julia, the story in Germany was not one of a progressive, linked growth. Nor did Cremer and Siegel work in concert.

Schröder, to whom one can date the first series works on the iteration of complex functions in [**1870, 1871**]), certainly loomed large in the German study. There were several papers by German mathematicians involving iteration that followed his, for example, works by Max Koppe [**1909**], Eugen Netto [**1887a, 1894**], Otto Rausenberger [**1881**], Ernst Reichenbächer [**1908**], Hermann Schapira [**1887a, 1887b**], and Otto Spiess [**1902, 1906**]. Unlike the French study, these works do not form a deep, unified corpus. They instead tend to be ad hoc and episodic, treating specific problems or examples that do not naturally coalesce into a general study.

The only German paper on iteration after Schröder that Cremer cited is by Rausenberger [**1881**] on the functional equation $f^n(z) = z$, a topic covered by several mathematicians dating back to Charles Babbage in [**1820**]. It is clear that Cremer saw his study as an extension of the American (Pfeiffer and Kasner) and French (Fatou and Julia) developments in complex dynamics; his analysis, however, was rooted in German function theory as evidenced by references to Hermann Amandus Schwarz and applications of recent developments by Paul Koebe and Hermann Weyl.

Cremer and Siegel's respective engagement in the subject are also quite different. Siegel's short paper [**1942**] was his only foray into complex dynamics while Cremer wrote several related papers over a more than ten-year span. They also worked in different mathematical circles. Siegel's doctorate was from Göttingen in 1920 under the direction of Edmund Landau. He taught at Frankfurt soon thereafter and was at the Institute for Advanced Study at Princeton University during World War II. Cremer was working towards his Ph.D. under Ludwig Bieberbach in Berlin, where he first encountered the subject in the middle of the 1920s. Later he taught in quick succession at Münster and Leipzig, where, he served as Koebe's assistant in the late 1920s, before settling in Cologne for a number of years. In 1940 he became a full professor at Breslau University of Technology[1] and remained in Germany throughout the war.

There was one German who followed up on Cremer's work: his student Hans Töpfer wrote two papers on iteration, his thesis [**1939**] and a postwar follow-up

[1]Breslau was the name of Wrocław prior to World War II when it was part of Germany.

[**1949**]. He cited Cremer's study of centers in both papers, and explicitly acknowledged his adoption of Cremer's terminology. We briefly discuss Töpfer in section 13.7.

11.3. Transalpine transitions: Placing Cremer in post-war Germany

With the defeat in World War I and the concomitant break-up of the Austro-Hungarian Empire into several independent nations, Germany was in turmoil, beset by economic, social, and political crises that would culminate in the rise of the Nazis. Germany lost over two million soldiers and almost a half-million civilians in World War I. As was the case with France, many mathematicians died in the conflict. "*Im Kriege gefallen*" (fallen in war) often followed the death notices of the Deutscher Mathematiker Vereinigung (German Mathematical Society).

Political tensions bled into mathematics. The International Research Council (IRC) was formed in 1919. Scientists from those countries the Allies had fought—including Germany—were forbidden membership. Under the auspices of the IRC, the International Union of Mathematicians (IMU) was established in 1920. Among its first actions was to assume responsibility for the International Congress of Mathematicians, held every four years during peacetime. The IRC upheld the embargo against Germany, Austria, Bulgaria, and Hungry, hence mathematicians from these countries were excluded from the 1920 and 1924 Congresses.

Neither the embargo nor the societal problems dampened mathematical inquiry in Germany. Indeed, according to Paul Forman, the following remarks by Max Planck from a mid-November 1918 plenary session at the *Preußische Akademie der Wissenschaften* in Berlin[2] (as it appeared in translation in Forman's paper [**1973**]) suggest the international embargo may have sparked it:

> *If the enemy has taken from our fatherland all defense and power, if severe domestic crises have broken in upon us and perhaps still more severe crises stand before us, there is one thing which no foreign or domestic enemy has yet taken from us: that is the position which German science occupies in the world. Moreover, it is the mission of our academy above all, as the most distinguished scientific agency of the state, to maintain this position and, if the need should arise, to defend it with every available means.* [**1973**, p. 163]

Mathematical activity is certainly part of what Forman suggests was the German scientific community's attempt to wield science as a "surrogate for political and economic power" [**1973**, p. 165].

The IRC offered membership to the scientists of the excluded countries in 1926, which conferred membership into the associated scientific organizations including the IMU, and all but Germany accepted. This resulted in a curious situation at the 1928 International Congress of Mathematicians involving two mathematicians we have discussed previously. Pincherle, who had been elected president of the IMU in 1924, felt that the Congress should be open to all mathematicians, even those from Germany, and he invited them to attend. Kœnigs, who had been the IMU's Secretary General since its inception, argued that the Germans should not be invited since they refused membership in the IRC.

[2]Prussian Academy of Science.

In the midst of this, the French study of iteration continued through the 1920s (as recounted in Chapter 10), but by the mid-1920s new papers came a slower rate, and no one had come along in France to push the study Fatou and Julia created into its next phase.

It is against this background that Cremer became involved in the study of iteration of complex functions as a graduate student, a study heretofore dominated by the French. Was he hoping to assert German interests in a field in which the French were preeminent? Did Julia's declaration that centers did not exist hold such sway over the French study that it took a German to reopen the center problem? Or did Cremer work in a spirit of scientific indifference, where what mattered was not nationality but the new knowledge he had to offer concerning the nonexistence of centers?

We do not know, but suspect the latter. According to Sanford Segal, although Cremer was a student of Bieberbach and remained in German territory throughout the war, he seemed to have avoided political involvement with Nazis. Referring to Litten [**1996**], Segal characterized Cremer as a "tacit anti-Nazi" [**2003**, p. 491].

11.4. The bequest of Fatou and Julia

The mathematical backdrop to Cremer's study is of course the work of Fatou and Julia. Julia showed that the existence of a center was equivalent to a solution of the Schröder equation, hence iteration in a neighborhood of the the fixed point is conformally equivalent to a rotation of a disc. Moreover, he observed that if a center exists, the iterative family was locally normal, and its limit functions are nonconstant.

In the *Comptes Rendus* notice [**1919a**], Julia mistakenly thought he had proved centers do not exist for rational functions except in the trivial case where the degree was one. As was customary in *Comptes Rendus*, the results were given without an accompanying proof. In 1934, several years after the publication of Cremer's 1927 paper, Julia acknowledged in his *Notice de Candidature Acadëmique* [**1934**], that he had based his argument on a mistaken assumption, without making any further claim on the matter, except to note he had turned his attention to other matters. It is unclear whether or not Cremer knew of Julia's recantation, but as late as [**1938**, p. 575, fn.] he referred to Julia's assertion not as a theorem but as speculation.

Fatou, on the other hand, remained agnostic towards the existence of centers and explored the implications of the existence of nonconstant limit functions on a domain, calling such domains "singular". This is the approach Cremer took as well, and like Fatou, seemed not to advance a claim on their existence. The approaches of Fatou and Julia, including the latter's mistaken claim, are discussed in Chapter 9.

11.5. Cremer: Who is he? What did he do?

Hubert Cremer was born to an academic family in Munich. His father Max was a physicist who many consider to be the inventor of the glass electrode. His brother Lothar was a professor at the Technical University of Berlin who studied acoustics, and his sister Erika was a chemist who made significant contributions to the study of gas chromatography.[3]

[3]See Bobleter [**1990**] for more information on Erika Cremer.

HUBERT CREMER

FIGURE 11.1. Cremer (at right) with the German mathematician Erich Kamke. Photo courtesy of the Archives of the Mathematisches Forschungsinstitut Oberwolfach.

Cremer completed his dissertation, entitled *Zum Zentrumproblem* (The Center Problem) in 1927 under the supervision of Bieberbach at the University of Berlin. During his time as a graduate student Cremer published a survey of known results on the iteration of rational functions entitled "*Referat über neuere Arbeiten von Julia*" [**1925**],[4] which grew out of a presentation he gave on May 27, 1924, at the weekly Mathematics Colloquium at the University of Berlin.[5] In his introduction to Cremer's survey, which he wrote in the form of a nine-line footnote, Bieberbach said

> *I believe this work to be an accessible explication of the beautiful results and methods of Julia and Fatou which should be of interest to a wider audience. (Cremer* [**1925**, p. 185]*)*

As this quotation suggests, this seems to be the first detailed description of the work of Fatou and Julia in the German language.

Cremer followed-up this initial interest with a sustained investigation of the center problem which consists of a series of papers from 1927 to 1938, the highlights of which are "*Zum Zentrumproblem*" [**1927**] and "*Über die Schrödersche Funktionalgleichung und das Schwarzsche Eckenabbildungsproblem*" [**1932a**], which at over 30 pages, is his longest and most extensive work on the subject. After the war,

[4] "*Referat*" often refers to a seminar paper, so the title can be translated as "A seminar paper on the recent work of Julia".

[5] It is sometimes suggested that the entire colloquium was devoted to the study of iteration, but in reviewing the list of presentations during the summer of 1924, only Cremer's title has an obvious link. It seems unlikely that the other presentations were about the iteration of complex functions; for example, Gábor Szegő's "*Über einen Satz von A. Markoff*", Rolf Nevanlinna's "*Über meromorphe Funktionen*", or Bieberbach's "*Uniformisierungsproblem*", Jahresbericht [**1925**, p. 119].

Cremer's interests began to shift towards computer science and applied mathematics. His study of centers produced no additional significant results, although he did publish two short papers on the center problem, [**1948, 1951**].

Cremer's overall approach to the study of iteration was in some senses that of a classicist whose results improve upon existing theory, and are often based on applications of contemporary developments rather than radical theoretical reimaginings. He was quite conscious of those who came before him, and a sense of history permeates his papers, which often begin with a review of what had been done before him, and contain numerous footnotes referencing past results.

This is not meant to denigrate Cremer's accomplishments in any fashion. If he is indeed a classicist, he used the classics as a base from which to explore frontiers. Contemporary expositors, such as Milnor, Beardon, Carleson, and Gamelin, and others, not only credit fundamental results to Cremer, but use statements that are quite similar to those he originally gave. In the case of Pfeiffer's study, Cremer improved upon what he saw and developed not only a more refined approach, but one which also strikes closer to heart of the matter.

As we noted earlier in this chapter, Cremer's investigation can be seen as a resumption of German tradition instigated by Schröder. If we adopt this view, then Cremer's work was closer to Schröder's than to the other Germans, in the sense that his intent was to investigate a fundamental issue rather than approach a particular problem. He gave the first clear condition for centers not to exist, and, much like Schröder, opened the door for those who followed him.[6]

Two principal themes emerge from Cremer's papers on the center problem. The first explores conditions under which a center does not exist. One might call this the Pfeiffer approach since it uses Pfeiffer's 1917 work as a point of departure, and, among its many accomplishments, it is a significant improvement of Pfeiffer's result on the existence of noncenters.

Pfeiffer approached the center problem as a small divisors problem. When $f'(\lambda) = \beta$ is one in modulus but is not a root of unity, he showed that formal series solutions to the Schröder equation $S \circ f = \beta \cdot S$ always exist and are of the form

$$\sum_{k=1}^{\infty} c_k z^k \quad \text{where} \quad c_{k+1} = \frac{c_1 Q_k(\beta, a_2, \dots, a_{k+1})}{\beta^k (1-\beta)\cdots(1-\beta^k)}, \tag{11.1}$$

where Q_k is a polynomial in the coefficients a_j of the series expansion of f around the fixed point. Since $|\beta| = 1$, $1 - \beta^k$ is arbitrarily close to zero infinitely often, and it is quite difficult to construct β so that the series converges—indeed, it was Siegel's singular accomplishment to construct such a β. It is easier, but by no means trivial, to find a β which forces the formal series to diverge.

As we saw in Chapter 5, Pfeiffer constructed β via a sequence $\{t_n\}$ which forces $1-\beta^{k_n-1}$ to be arbitrarily close to zero yet at the same time bounds the Q_{k_n} away from zero. The c_{k_n} then become arbitrarily large, which causes the Schröder function to diverge on any neighborhood of the fixed point. Pfeiffer's approach is constructive, but he never gave a precise algorithm for producing such a β, so in some senses his result is impractical.

---

[6]While one can debate whether Schröder's work had any effect on the French school, there is no doubt of his continuing interest, not only in iteration, but in numerical algorithms.

While he built upon Pfeiffer's approach—for example his multipliers bear some similarities to Pfeiffer's—Cremer took a different route to finding irrationally neutral fixed points that are not centers. Rather than show all formal solutions to the Schröder equation diverge, he directly showed that a fixed point satisfying his condition is not a center because it is surrounded by infinitely many repelling points.

The second strand of Cremer's œuvre—the Fatou approach—was to posit the existence of a singular domain, along with the concomitant nonconstant limit functions, and explore the nature of these functions. The highlight of this approach was showing that if a singular domain exists, the degree of connection is at most two.

We will examine each of these strands in detail below.

11.6. Cremer's noncenters

The key to Cremer's investigation of noncenters is the limit

$$\liminf_{n\to\infty} \sqrt[\sigma(n)]{|f'(\lambda)^n - 1|}, \tag{11.2}$$

where $n \in \mathbb{N}$ and $\sigma(n)$ is a positive function that measures the growth of f and varies according to the nature of $f(z)$, for example, if it is rational or entire. Irrationally neutral fixed points of f whose multiplier β satisfies (11.2) will be said to satisfy a *Cremer condition*; such a β will be called a *Cremer multiplier*; the real numbers γ satisfying $\beta = e^{2\pi\gamma i}$ will be called *Cremer numbers.*

In both [**1927, 1932a**] he showed that if the above limit is zero, then the fixed point is not a center. In his paper [**1932a**] he relaxed this condition for polynomials, demonstrating a center could not exist if the above limit is strictly less than one.

Cremer was also interested in finding out how prevalent noncenters are: he showed in [**1927**] that the Cremer multipliers are uncountable yet of measure zero, and in [**1932a**] he asserted without proof that they are dense.

His 1927 paper focused on rational functions and he proved the following:

THEOREM 11.1. *Let* $\beta = e^{2\pi i\gamma}$.

(1) *If*

$$\liminf_{n\to\infty} \sqrt[d^n]{|\beta^n - 1|} = 0, \tag{11.3}$$

then any rational function f *of degree* $d > 1$ *with* $f(\lambda) = \lambda$ *and* $f'(\lambda) = \beta$ *does not have a center at* λ.

(2) *The set* $\mathcal{M}$ *of Cremer multipliers* β *is uncountable.*[7]

(3) *The set* $\mathcal{M}$ *is of Lebesgue measure zero.*

If $\beta = e^{2\pi i\gamma}$ *satisfies condition* (11.3), *we will sometimes refer to* β *as a Cremer multiplier of degree* d *and* γ *as a Cremer number of degree* d.

During the course of his ten-year investigation of centers, his approach to the above theorem evolved both in terms of the kinds of proofs he offered and the types of function he applied it to. But the essence of his pursuit remained the same: to find conditions under which an irrationally neutral fixed point λ of a given type of function was not a center.

The linchpin of his 1927 proof of the above theorem was the following lemma.

[7] Cremer spoke of the set $\mathcal{M}$ as having the "power of the continuum". In [**1927**] he proved the equivalent result that the set of γ which produces noncenters was uncountable, but in subsequent papers he stated this result in terms of the set $\mathcal{M}$ of Cremer multipliers.

LEMMA 11.2. *Let $g(n) > 0$ for all natural numbers n. There are "continuously many" $\beta = e^{i\phi}$ satisfying*

$$|\beta^n - 1| < g(n) \tag{11.4}$$

for infinitely many n.

Out of Cremer's proof of the lemma comes a method for constructing such β which satisfy the first condition of Theorem 11.1, allowing the construction of functions with noncenters.

11.6.1. Outline of Cremer's construction of β. The gist of his method is this (more precise details will follow): Let a function $g(n)$ be positive but very small for all $n \in \mathbb{N}$. Construct a number $\gamma \in (0,1)$ in binary form with sufficiently large gaps between the ones so that shifting γ to the left k_r places makes $2^{k_r}\gamma - \lfloor 2^{k_r}\gamma \rfloor < g(2^{k_r})/2\pi$. This creates sequences $m_r = \{\lfloor 2^{k_r}\gamma \rfloor\}$ and $n_r = \{2^{k_r}\}$ such that $|2\pi n_r\gamma - 2\pi m_r| < g(2^{k_r})$.

Viewing the difference $|2\pi n_r\gamma - 2\pi m_r|$ as a small angle of the unit circle, it follows that the arc it sweeps out has the same magnitude, that is

$$||e^{n_r 2\pi\gamma i} - e^{m_r 2\pi i}|| = |2\pi n_r\gamma - 2\pi m_r| < g(2^{k_r}), \tag{11.5}$$

where we understand "||" to indicate the distance between the respective points along the circle.

Since the chordal distance $\left|e^{2\pi n_r\gamma i} - e^{2\pi m_r i}\right|$ is less than the corresponding distance along the arc $||e^{n_r 2\pi\gamma i} - e^{m_r 2\pi i}||$, it follows that

$$|e^{n_r 2\pi\gamma i} - e^{m_r 2\pi i}| < |2\pi n_r\gamma - 2\pi m_r| < g(2^{k_r}). \tag{11.6}$$

Setting $\beta = e^{2\pi\gamma i}$, the above inequality becomes

$$|e^{n_r 2\pi\gamma i} - e^{m_r 2\pi i}| = |\beta^{n_r} - 1| < |2\pi n_r i\gamma - 2\pi m_r| < g(2^{k_r}). \tag{11.7}$$

Assuming we can actually construct the sequences $\{n_r\}$ and $\{m_r\}$, the lemma follows (modulo the claims about the uncountability of the set $\mathcal{M}$).

11.6.2. Constructing β. What follows is a slightly expanded and formalized version of what Cremer did that is nonetheless very faithful to his construction. Let $g(n) > 0$ be given. To prove Theorem 11.1, $g(n)$ will need to be quite small.

We will want to express $\gamma \in (0,1)$ as a binary number in the form

$$\gamma = .\underbrace{000\cdots000}_{l_0 \text{ zeros}}1\underbrace{000\cdots000}_{l_1 \text{ zeros}}1000\cdots0001\underbrace{000\cdots000}_{l_r \text{ zeros}}100\cdots = \sum_{r=1}^{\infty}\frac{1}{2^{k_r}}, \tag{11.8}$$

where l_r is the number of zeros following the rth one, which is the k_r-th place; this induces the recursive relation, $k_{r+1} = k_r + l_r + 1$, where $k_1 = l_0 + 1$ with l_0 arbitrary.

To construct γ, define l_r in terms of $g(2^r)$ by requiring that $l_r > -\log_2(g(2^{k_r})) + 3$, for $r = 1, \dots,$ and let k_{r+1} be defined as in the previous paragraph. Since $g(n)$ is close to zero, the condition we imposed on l will make the gaps between the nonzero terms in γ quite large. Note, also, that l_r and k_{r+1} are well defined since l_r depends on $g(2^{k_r})$, and k_r is recursively defined in terms of l_{r-1} and k_{r-1}. Our definition of l_r yields

$$2^{-l_r} < \frac{g(2^{k_r})}{2^3}, \tag{11.9}$$

which we will use below.

Let $n_r = 2^{k_r}$ and $m_r = \lfloor 2^{k_r}\gamma \rfloor$. Multiplying γ by n_r shifts its binary expansion leftwards towards the decimal k_r places, hence

$$\begin{aligned} n_r\gamma &= m_r.\underbrace{000\cdots 0001}_{l_r \text{ zeros}}\cdots \\ &= m_r + 2^{-l_r-1} + \cdots \\ &< m_r + 2^{-l_r}. \end{aligned} \tag{11.10}$$

Subtracting the first and last terms of (11.10) and combining the result with (11.9) yields

$$|n_r\gamma - m_r| < 2^{-l_r} < \frac{g(n_r)}{2\pi},$$

from whence it follows that

$$|2\pi n_r\gamma - 2\pi m_r| < g(n_r). \tag{11.11}$$

Let $\beta = e^{i2\pi\gamma}$. Recall from equation (11.6) that the chordal distance is less than the distance along the arc, that is,

$$|e^{2\pi n_r\gamma i} - e^{2\pi m_r i}| < \left|\left|e^{2\pi n_r\gamma i} - e^{2\pi m_r i}\right|\right| = |2\pi n_r\gamma - 2\pi m_r|.$$

Substituting β into the leftmost term of the inequality directly above and recalling (11.11), we get

$$|\beta^{n_r} - 1| < g(n_r), \tag{11.12}$$

and there is at least one β for which equation (11.4) from Lemma 11.2 is satisfied.

The details of the construction of γ are important since, coupled with a suitable function g, they not only provide a direct means to construct a noncenter (something Pfeiffer did not accomplish) but also because, as we will see later, Cremer linked his construction of γ to Liouville's construction of transcendental numbers.

It is also instructive to compare Cremer's construction to Pfeiffer's. If Cremer's γ terminates with a one in the rth place, it is an rth root of unity, so we can look at γ as being constructed as limit of roots of unity whose degrees go to infinity very quickly. This is the essence of Pfeiffer's construction as well (see Section 5.4.2).

11.6.3. Showing the lim inf is satisfied. To obtain the lim inf at (11.3) from Theorem 11.1, let $d \in \mathbb{N}$ and let

$$g(n) = \frac{1}{n^{d^n}}. \tag{11.13}$$

Since $g(n)$ satisfies the hypotheses of Lemma 11.2, there is a β and a sequence $\{n_r\}$ satisfying

$$|\beta^{n_r} - 1| < g(n_r) = \frac{1}{n_r^{d^{n_r}}}.$$

Taking the appropriate roots of each side yields

$$\sqrt[d^{n_r}]{|\beta^{n_r} - 1|} < \frac{1}{n_r},$$

which implies that the following Cremer condition is satisfied:

$$\liminf_{n\to\infty} \sqrt[d^n]{|\beta^n - 1|} = 0.$$

Cremer left the definition of g at equation (11.13) and the subsequent calculations implicit.

11.6.4. The set $\mathcal{M}$ is uncountable. We quote in its entirety Cremer's argument that given any $g(n)$ the set of β satisfying equation (11.4)—the phrase "resulting numbers"—refers to the set of γ that can be constructed via the process:

> *Thus, the resulting numbers have the power of continuum because the inequality* (11.5) *remains valid if any digit* 1 *within* γ *is replaced by a* 0, *and the set of numbers derived from an irrational quantity in this manner has the power of continuum.* [**1927**, p. 155]

Cremer seems to have invoked a version of Cantor's diagonal argument which can be made rigorous by mapping the set of all such Cremer numbers γ to the set of binary sequences.

Since the γ are uncountable, it follows that so are the corresponding angles $\phi = 2\pi\gamma$ and the Cremer multipliers, $\beta = e^{i2\pi\gamma}$, for which a Cremer condition exists. At different points during his various papers on the center problem, Cremer stated this result in terms of the γ, the angles ϕ, and the multipliers.

11.6.5. The numbers in $\mathcal{M}$ are not algebraic. Before completing the proof of Theorem 11.1, Cremer made an observation regarding the construction of a Cremer number γ:

> *It is based on Liouville's famous proof of the existence of transcendental numbers in which he constructs numbers which are well approximated by rational fractions.* [**1927**, p. 155]

The large number of zeroes between the ones in the binary expansion of γ is reminiscent of the Liouville constant, and it can be shown that γ is indeed transcendental.

We know by its construction that a Cremer number is irrational, so to show that it is transcendental, we need to prove that it is not algebraic. Recall that an irrational number x is algebraic of degree k if the degree of the minimal polynomial with integer coefficients having x as a root is also k, in which case there exists a positive constant $c(x)$ such that for all rational p/q, normalized so $q > 0$,

$$\left| x - \frac{p}{q} \right| > \frac{c(x)}{q^k}. \tag{11.14}$$

Let γ be a Cremer number of degree d (see Theorem 11.1). It follows from Sections 11.6.2 and 11.6.3 that we can find infinite sequences of natural numbers $\{n_r\}$ and $\{m_r\}$ such that

$$|n_r\gamma - m_r| < \frac{g(n_r)}{2\pi} = \frac{1}{2\pi n_r^{d^{n_r}}},$$

which implies that

$$\left| \gamma - \frac{m_r}{n_r} \right| < \frac{1}{2\pi n_r^{d^{n_r+1}}}. \tag{11.15}$$

Suppose that γ is algebraic of degree k. Then there is a $c(\gamma)$ and k such that γ satisfies equation (11.14). But since the n_r go to infinity, $c(\gamma)$, d, and k are constant, we can find an N such that for any $n_r \geq N$

$$\left| \gamma - \frac{m_r}{n_r} \right| < \frac{1}{2\pi n_r^{d^{n_r+1}}} \leq \frac{c(\gamma)}{n_r^k}. \tag{11.16}$$

Since k is fixed, the rightmost inequality holds because we have the freedom to pick n_r as large as needed to ensure that $2\pi n_r^{d_r^n+1} \geq n_r^k/c(\gamma)$. Hence,

$$\left|\gamma - \frac{p}{q}\right| > \frac{c(\gamma)}{q^d}$$

fails for the rational number m_r/n_r, and γ is not algebraic. Readers who are familiar with the center problem will recognize in this fact that Liouville numbers[8] can be used to construct rational functions with irrationally neutral fixed points that are not centers.

We get a better indication of why Cremer found this connection to Liouville important in his next sentence:

> *Whereas Liouville only required of his numbers that given arbitrary A and n the inequalities*
>
> $$\left|\gamma - \frac{p}{q}\right| > \frac{1}{Aq^n} \text{ or } |p - qx| > \frac{1}{Aq^{n-1}}$$
>
> *do not hold for all rational numbers $\frac{p}{q}$, we shall make much stricter demands* [schärfere Forderung] *of our β.* [**1927**, p. 156]

Although he did not say what these demands were, if one compares the rightmost inequality from the conditions stated in his quotation to inequality (11.15), one sees that the former need only satisfy $|p - q\gamma| < \epsilon/q^k$, while the latter satisfies $|p - q\gamma| < \epsilon'/q^{c^q}$, a far more stringent condition. In other words, while a Cremer number of degree d satisfies the Liouville condition, which makes Cremer numbers a subset of Liouville numbers, the converse does not necessarily follow. Perhaps this is what Cremer had in mind.

Why did he emphasize the fact that his numbers satisfy stricter conditions than Liouville numbers? One reason might be that they imply that Cremer numbers are a subset of Liouville numbers, which comes into play with his proof that the set of Cremer numbers has zero Lebesgue measure, as we will see in Section 11.6.6. Moreover, the demands he places upon Cremer numbers allow him to prove that irrationally neutral fixed points of rational functions exist which are accumulation points of repelling periodic points, a crucial observation in his proof that noncenters exist (see Section 11.6.7). Another possibility might be that because it takes "stricter demands" to produce noncenters, he is signaling that he is not assuming that all irrational numbers produce noncenters, which, again, would be tantamount to drawing a distinction between his investigation of centers and Julia's blanket statement that centers do not exist.

Although it lurks underneath the surface in Pfeiffer's work, Cremer's statement that his numbers are well approximated by rationals was the first explicit mention in the literature of the fundamental connection between the center problem and Diophantine approximation. This connection, as we pointed out in Section 4.4, was

---

[8] Irrational numbers which are not Diophantine (see Definition 5.2) are said to be *Liouville numbers*. One way to define a Liouville number γ is that for any positive constant K and natural number n there exists a rational number p/q satisfying $|\gamma - p/q| < K/q^n$. Although the definition requires that only one rational number satisfies the preceding inequality for each n, there are in fact infinitely many that do. As is evident from Cremer's construction of an irrational multiplier for which centers do not exist (and to a lesser extent Pfeiffer's), such numbers can be constructed in a manner that recalls Liouville's construction.

later shown to extend beyond complex dynamics into KAM theory. We will touch upon this matter again in Chapter 12.

Cremer, however, only found success in exploiting one side of this connection, namely the connection between Liouville numbers and noncenters. As he remarked late in his study, "a particularly good approximation of 1" produces noncenters [**1938**, p. 574]. If Liouville numbers are noncenters, perhaps algebraic or even Diophantine numbers yield centers. The possibility of a connection between Diophantine numbers and centers must have occurred to Cremer. That it did is suggested by his listing the question of whether algebraic numbers are centers as one of the questions left open by his investigation [**1927**, p. 158, fn.], but there is no overt recognition of it in his published papers. One is left to speculate that while he suspected where centers might lie, he found nothing substantial enough to publish, or perhaps, he simply chose to avoid such speculation to prevent saying something that he might later regret.

11.6.6. The measure of $\mathcal{M}$ is zero. Cremer's argument that the Lebesgue measure of $\mathcal{M}$ is zero is more a sketch than a complete proof, made no less so by being given in a footnote which he introduces by observing that it can be shown that the set of angles with "center enmity" or "*Zentrumfeindschaft*" (by which he means angles of the form $2\pi\gamma$, where γ is Cremer number) is of Lebesgue measure zero.

Cremer at times displays a tendency towards brevity at the expense of detail. His proof that the measure of the Cremer numbers is zero is no exception. Central to his argument is a set, which we will denote C_d, equivalent to the following: let $d \in \mathbb{N}$, and let $\gamma \in (0,1)$ be such that there exists a positive constant K and sequence of rational numbers $\{p_n/q_n\}$ satisfying

$$\left|\gamma - \frac{p_n}{q_n}\right| < \frac{K}{n_n^{d+1}}. \tag{11.17}$$

Cremer shows that the set C_2 is of measure zero, and from there he asserts that this implies that what we are calling the Cremer numbers is as well.

He does not specify his reasoning, but he seems to be employing one of two arguments. The first is that since all Liouville numbers satisfy inequality (11.16) for all natural numbers $d \geq 2$ (see footnote 8 on p. 239), if C_2 is of measure zero, then so is the set of all Liouville numbers. We have also observed that Cremer numbers are a subset of Liouville numbers, so the Cremer numbers would also be a set of zero Lebesgue measure. That this might be the case is suggested by the introductory language to his proof observing that Cremer numbers are also Liouville numbers.

On the other hand, the introductory language regarding angles with "center enmity" that we referred to at the beginning of this subsection suggests that he may be arguing directly that the Cremer numbers are of measure zero since they also satisfy (11.16). To see this, modify the function $g(n)$ so that the exponent of the n_r in the denominator of the rightmost term of (11.15) is $d+1$.

Cremer's actual proof that the set C_2 is of measure zero, which we omit, is quite similar to the standard argument that Liouville numbers are of zero Lebesgue measure.

11.6.7. Finishing the proof of Theorem 11.1. With the lemma proved, Cremer's strategy was to show that if the conditions of Theorem 11.1 hold, then f will have infinitely many periodic points in any neighborhood of the fixed point, implying that the fixed point is not in the domain of normality, which, as Julia had noted, implied that a center could not exist in this instance.

He began with the assumption that the rational function $f = \frac{p}{q}$ has an irrationally neutral fixed point at 0 with $f'(0) = \beta$ and $\deg q > \deg p$. Neither of these cause any loss of generality. Obviously, if the fixed point λ is not zero, mapping $z \mapsto z - \lambda$ shifts the fixed point to zero without altering its derivative.

If $\deg q \leq \deg p$, then Cremer pointed out one could use a linear transformation which maps a preimage of the fixed point to infinity. Although Cremer, as always eschewing statements of the obvious, did not explicitly give such a map, one can easily find M satisfying

$$\begin{array}{ccc} \overline{\mathbb{C}} & \xrightarrow{f} & \overline{\mathbb{C}} \\ {\scriptstyle M}\downarrow & & \downarrow{\scriptstyle M} \\ \overline{\mathbb{C}} & \xrightarrow[f^*=\frac{p^*}{q^*}]{} & \overline{\mathbb{C}} \end{array} \tag{11.18}$$

with $f'^*(0) = \beta$ and $f^*(\infty) = 0$. The latter condition implies that $\deg q^* > \deg p^*$, and since $f'^*(0) = \beta$, he pointed out that we can assume that $p^*(z) = \beta z + \cdots + a_r z^r$ and $q^*(z) = 1 + \cdots + b_d z^d$ (since the lower degree coefficients of p and q are, respectively, $c\beta$ and c, the c can be factored out). Since f is always of degree at least 2, and the derivative is nonzero, he noted we are assured that there will be a preimage of the fixed point not equal to the fixed point.

Describing the argument as "*leichte*" (slight or easy), Cremer simply asserted that iteration not only preserves the property that the denominator is of higher degree than the numerator, but that the numerator is of the form $\beta^n z + \cdots + A_n z^{r_n}$ (where r_n is not in general equal to r^n) and the denominator equals $1 + \cdots + b_d^{d^n} z^{d^n}$. The argument is not difficult, but it is a little closer to tedious than slight as it necessitates care with a few nagging details.

Once these preliminaries are out of the way, Cremer hunted for potential periodic points of order n near zero by setting $f^n(z) = z$ and then dividing by z terms to exclude the irrationally neutral fixed point at zero. This yields

$$\beta^n z + \cdots + A_n z^{r_n} = z + \cdots + b_d^{d^n+1} z^{d^n} \Rightarrow (\beta^n - 1) + \cdots - b_d^{d^n} z^{d^n} = 0. \tag{11.19}$$

It follows that the product of the moduli of the roots of $f^n(z) = z$ equals $\left|\frac{\beta^n - 1}{b_d^{d^n}}\right|$. Letting $\tilde{z}_n$ denote the smallest period n fixed point,

$$|\tilde{z}_n|^{d^n} \leq \left|\frac{\beta^n - 1}{b_d^{d^n}}\right| \Rightarrow |\tilde{z}_n| \leq \frac{\sqrt[d^n]{|\beta^n - 1|}}{|b_d|}. \tag{11.20}$$

As we increase n, the number of $\tilde{z}_n$ satisfying (11.20) increases, but are there infinitely many on an arbitrary neighborhood of the origin? If $\liminf_{n\to\infty} \sqrt[d^n]{|\beta^n - 1|} = 0$, then there are, and by setting $g(n) = n^{-d^n}$ and applying Lemma 11.2, we can then find a β with $\liminf_{n\to\infty} \sqrt[d^n]{|\beta^n - 1|} = 0$. But this is not what Cremer did. Rather, with his customary brevity, he upped the ante and found a way to make the nonexistence of a center independent of the degree of the rational function by setting $g(n) = 2^{-n!}$ and simply observing that any fixed point whose multiplier is β with $|\beta^n - 1| < 1/2^{n!}$ cannot be a center of any rational function regardless of its degree.

The reasoning behind his statement is probably something like this: Let $d \in \mathbb{N}$ with $d > 1$. Then

$$|\beta^n - 1| < \frac{1}{2^{n!}} \tag{11.21}$$

implies that

$$\sqrt[d^n]{|\beta^n - 1|} < \frac{1}{2^{\frac{n!}{d^n}}}.$$

Since for fixed d, $\frac{n!}{d^n}$ goes to infinity with n, it follows that for any d,

$$\liminf_{n\to\infty} \sqrt[d^n]{|\beta^n - 1|} = 0.$$

The arc of Cremer's proof that there are infinitely many periodic points near a fixed point with a Cremer multiplier may yield insight into how he thought to use condition (11.3) from Theorem 11.1 to prove that a rational function of degree d with a Cremer multiplier does not have a center. Clearly, one way to establish that a rational function with an irrationally neutral fixed point does not have a center is to show that arbitrarily near the fixed point are infinitely many periodic points. Following that implication, one is led to inequality (11.20) and from there to condition (11.3).

The construction of noncenters is the principal accomplishment of Cremer's examination of the center problem. In his subsequent works, he linked his efforts to function theory and investigated the center problem for entire maps as well as special cases of functions, such as power series with large gaps between nonzero terms.

He also extended his efforts to doubly connected domains. As will be discussed below, Cremer showed that if singular domains exist, their order of connection is at most two. With this in mind, he extended Theorem 11.1 to doubly connected regions in his paper [**1934**], where he showed if a degree d rational function f possesses an irrationally neutral fixed point whose multiplier satisfies limit (11.3) and in addition maps an annulus univalently to itself, then $d = 1$; that is, f is a Möbius mapping. He also developed a new method to map doubly connected regions onto annuli.

We will postpone our discussion of Cremer's study of annuli until after we discuss the existence of centers for nonrational functions.

11.7. Cremer's 1932 paper I: The existence of noncenters

Although Cremer is most famous for his construction of noncenters, the range of his study of the center problem was considerable. The paper [**1932a**] is perhaps his *opus magnum*. It is certainly his longest, and it has the broadest scope. In it he expanded his study to include entire functions, as well as other types of functions; he also followed implications of the existence of centers, analyzing the possible structure of the Fatou set should a singular domain exist. He tied his investigation to function theory, applying the theory of Riemann surfaces to the Schröder functional equation, and he described a connection between the center problem and an unpublished problem he claimed Schwarz often posed to his students.

The paper [**1932a**], consisting of six parts, is entitled, "*Über die Schrödersche Funktionalgleichung und das Schwarzsche Eckenabbildungsproblem*",[9] and the

[9] "On the Schröder functional equation and the Schwarz angle-mapping problem".

Schröder equation is the element that unifies the paper. The first four sections concern noncenters where the appropriate Schröder equation diverges. The fifth concerns the structure of the Fatou domain in the event a solution to the Schröder equation exists. The sixth discusses connections between convergent Schröder equations and the mapping of a simply connected domain to the unit disc provided by the Riemann Mapping Theorem.

The reference to Schwarz in the title refers to a proto-version of the center problem that Schwarz used to challenge his students, reminiscent of the questions Kasner posed in his lectures (see page 99 in Section 5.3), and which Koebe related to Cremer. During a review of the historical background of the study of the center problem, Cremer remarked that the center problem is a version of the problem of mapping a polygonal region onto a disc, a problem he referred to as the *Eckenabbildungsproblem.* In the context of the *Eckenabbildungsproblem*, Cremer stated the center problem as follows: Let L_1 and L_2 be curves which form an angle intersecting in a fixed point of f. When can the angle formed by $f(L_1)$ and $f(L_2)$ be considered a rotation of the former angle?

> *In this form, the question caught the interest of H.A Schwarz* [immediately after which occurred this footnote:] *H. A. Schwarz published nothing concerning this, but—according to a friendly conversation I had with Professor Koebe—he frequently posed this problem to his students.* [**1932a**, p. 294]

Cremer then noted that if a center exists, an appropriate iterate of a solution to the Schröder equation provides a means to map the angle through the fixed point onto the unit disc; moreover, he claimed that Schwarz tied the possibility of mapping function acting like a rotation to the linear term of its power series—which in the case of the center is the Schröder function S—and noted that if the angle between L_1 and L_2 is an irrational multiple of π, the formal series expansion of the mapping function would be possible. The question then would be whether the formal series converges. However, Cremer observed, the formal solutions of the mapping functions considered in his present paper would diverge (presumably because they could not possess a center—unlike Pfeiffer, his strategy was not to show the divergence of formal series). With that comment, he remarked, he would close his discussion of the connection to Schwarz and mention it no more [**1932a**, p. 295].

Recall that the form of the limit from Cremer's construction of a noncenter in Theorem 11.1,

$$\liminf_{n\to\infty} \sqrt[d^n]{|\beta^n - 1|} = 0,$$

depends on the degree d of the rational function, although he did produce multipliers that would preclude the existence of a center for a rational function of arbitrary degree (see (11.20)). In [**1932a**] Cremer dealt with entire functions, and so he replaced the dependence on degree with a dependence on the analogous concept of growth. This led to him to consider general forms of the above limit, such as the following from Section I [**1932a**, p. 299]:

THEOREM 11.3. *Let $F(r)$ be an arbitrary positive function, and let $f(z) = e^{i\gamma 2\pi}z + \cdots$ be a nonlinear entire function which grows slower than the function $F(r)$ in the sense that*

$$\sup_{|z|\le r} |f(z)| < F(r)$$

for sufficiently large r. In addition let $\beta = e^{i\gamma 2\pi}$ *satisfy*

$$\liminf_{n\to\infty} \sqrt[\log F^n(r)]{|\beta^n - 1|} = 0 \tag{11.22}$$

for sufficiently large r, where F^n *is the nth iterate of* F*. The irrationally neutral fixed point at the origin is not a center.*

Cremer altered F depending on the nature of f. For example, in applying Theorem 11.3 to the case when f is an entire function of finite order, he set $F(r) = E^2(r)$ where E_k is the kth iterate of e^z and proved the following corollary:

COROLLARY 11.4. *If* f *is an entire function of finite order with a center at the origin satisfying*

$$\liminf_{n\to\infty} \sqrt[\log E^{2n}(r)]{|\beta^n - 1|} = 0,$$

where $f'(0) = \beta$ *and*

$$\sup_{|z|\leq r} |f(z)| < E^2(r),$$

then f *is a linear function.*

His proof of Theorem 11.3 relies on a careful estimate developed in concert with condition (11.22) which allowed him to construct a sequence n_k for which

$$\sqrt[\log F^{n_k}(r)]{|\beta^{n_k} - 1|}$$

and $|f^{n_j}(z) - \beta^{n_j} z|$ go to zero, the latter uniformly on $D_r(0)$ of the origin. From this he concluded that f^{n_j} converges uniformly to the identity. This allows him to show that f is one-to-one and hence a Möbius mapping: Let $z_1, z_2 \in D_r(0)$ with $f(z_2) = f(z_1)$. Since f^{n_j} converges to the identity function,

$$\begin{aligned} z_1 &= \lim_{n_j\to\infty} f^{n_j}(z_1) \\ &= \lim_{n_j\to\infty} f^{n_j-1}(f(z_1)) \\ &= \lim_{n_j\to\infty} f^{n_j-1}(f(z_2)) \\ &= \lim_{n_j\to\infty} f^{n_j}(z_2) \\ &= z_2, \end{aligned} \tag{11.23}$$

and f is one-to-one.

The method of proof is particular to entire functions, but his strategy, to show that if $f(z)$ satisfies the conditions of the theorem and has a center, then it is a Möbius function, is one he used again, both in [**1932a**] and in subsequent works.

He concluded his proof with the assertion that the set of multipliers satisfying condition (11.22) are uncountable (to which he referred the reader to the argument from [**1927**] discussed above in Section 11.6.4) and dense, for which he offered no proof [**1932a**, p. 305].

An alternate proof of Theorem 11.1 on the nonexistence of centers from [**1927**] applied to polynomials follows from Theorem 11.3. Suppose p is a polynomial of degree d satisfying the conditions of Theorem 11.1, in particular,

$$\liminf_{n\to\infty} \sqrt[d^n]{|\beta^n - 1|} = 0.$$

Set $F(r) = r^d$, which implies that $F^n(r) = r^{d^n}$ and observe that we can assume that $\sup_{|z|\leq r} |p(z)| < F(r)$ for sufficiently large r; if not, choose a suitable constant

c and use $c^{-1}f(cz)$. It follows then that $\log F^n(r) = d^n \log(r)$. We may apply Theorem 11.3 after observing that

$$\liminf_{n\to\infty} \sqrt[\log F^n(r)]{|\beta^n - 1|} = 0$$
$$\Leftrightarrow \liminf_{n\to\infty} \sqrt[d^n \log(r)]{|\beta^n - 1|} = 0$$
$$\Leftrightarrow \liminf_{n\to\infty} \sqrt[d^n]{|\beta^n - 1|} = 0,$$

since the term $\log(r)$ will not affect convergence of the lim inf.

In [**1932a**], Cremer offered an alternate proof of Theorem 11.3 in §II. In §III he improved the result for polynomials of degree two or greater by showing that the condition

$$\liminf_{n\to\infty} \sqrt[d^n]{|\beta^n - 1|} = 0$$

can be replaced by

$$\liminf_{n\to\infty} \sqrt[d^n]{|\beta^n - 1|} < 1.$$

Finally, in §IV he found analogs to the condition

$$\liminf_{n\to\infty} \sqrt[\sigma(n)]{|\beta^n - 1|} = 0$$

for functions f whose series expansion near the fixed point is a gap series, that is, a series whose nonzero terms are very infrequent. Although he linked his work to existing studies of gap series by Jacques Hadamard, Alexander Ostrowski, George Pólya, and others, the functions that Pfeiffer constructed, whose irrationally neutral fixed points are noncenters, are also gap series. Cremer did not seem to have that connection in mind, and his focus was instead to establish a connection between Cremer multipliers and existing theory involving gap series.

The final two sections of his paper examine implications of the existence of centers. We will cover one of these in Section 11.8, which concerns the degree of connection of a rotation domain, when we discuss Cremer's study of the Fatou set.

Cremer's last paper of import on noncenters [**1938**] reversed the approaches we have discussed above where, for a particular kind of function, he sought the conditions on the multiplier which implied centers did not exist. Instead, he started with particular conditions on a multiplier a_1 of an irrationally neutral fixed point (that may or may not be a center) and constructed a function with a noncenter by perturbing coefficients of the existing function.

He noted that his paper is "a counterpart" (*Gegenstück*) [**1938**, p. 574] to a theorem of George Pòlya and Adolf Hurwitz from [**1916**], in which, proving a conjecture Fatou made in [**1906a**] (but not in the context of the iteration of complex functions), they showed that one could construct a noncontinuable series from a continuable series by multiplying certain coefficients by -1. It is not quite clear what Cremer meant by saying his paper is a counterpart, but he used an analogous technique to perturb his functions, and perhaps he wanted to highlight the similarity.

The varying approaches to showing the nonexistence of centers that Cremer used throughout his œuvre reflect the complexity of irrationally neutral fixed points. This was in clear contrast to the one-size-fits-all state of affairs with fixed points whose multiplier was not one in modulus. From the birth of the study of the dynamics of rational functions (whether you date it to Schröder's papers or to Kœnigs), it was abundantly clear that a multiplier less than one in modulus implied attraction and a modulus greater than one indicated repelling behavior. If the modulus of the multiplier of a fixed point of a rational function is a root of unity the Flower Theorem describes the dynamics explicitly.

One of the perplexing qualities of the study of irrationally neutral fixed points at the time of Cremer's study was the uncertainty of the degree to which irrationally neutral multipliers determined the dynamics of a function. Did the existence of noncenters depend just on the irrationality of the multiplier, as Julia thought, or were there also deeper links, say between the degree of the irrationality of the multiplier and the growth of the function?

11.8. The 1932 paper II: Doubly connected singular domains

Less, perhaps, is known of Cremer's study of the Fatou set. A principal motivation in [**1932a**], one that carries over to other papers, seems to be the question, Does iteration ever act like a rotation of an annulus? This is a natural extension of the center problem to an annulus, and anyone studying the center problem would likely ask it. It seems that Cremer was the first to explore this question explicitly in journal pages.

His only definitive answer was to extend Theorem 11.1 to doubly connected regions and state conditions under which iteration is not conformally equivalent to a rotation of an annulus; this appeared in [**1934**].

However, he also answered some interesting speculative questions about the iteration on doubly connected domains. Despite Fatou's study of singular domains,[10] discussed in Section 9.4, no one prior to Cremer investigated the question of whether or not there was anything a priori that would preclude the iteration of a rational function from being equivalent to an annular rotation. As we will see, Cremer's short answer (contained in §V of [**1932a**]) was "No".

Although it was not, properly speaking, a paper on iteration since it does not reference either iteration or related studies, such as functional equations, Cremer also explored the nature of annular mappings in [**1930a**], where he developed a new proof of Koebe's result from [**1915**, p. 202], stating that a doubly connected open domain with at least two boundary points could be mapped onto an annulus. We will discuss these papers in thematic order: first the extension of to Theorem 11.1 to doubly connected regions, then his proof regarding conformal equivalence of doubly connected regions and annuli, and finally his answer to the question of whether iteration could act like an annular rotation.

11.8.1. Cremer's extension of Theorem 11.1 to the annulus. Cremer explicitly observed at the onset of his paper [**1934**, pp. 276–277], that if an invariant doubly connected singular domain D exists, iteration would act like an irrational rotation through an angle θ. Moreover, there would be a function S mapping D conformally to an annulus. This function S would also satisfy $S \circ f = \beta \cdot S$ where $\beta = e^{i\theta}$. However, in this case, β is not the multiplier of a fixed point, but what is called a *rotation number*.

He summarized his investigation with the following theorem [**1934**, pp. 276–277]:[11]

[10]Recall that a singular domain D is one where the limit functions of the iterative family $\mathcal{I}$ are nonconstant.

[11]Our statement of it is a more direct version of the one he gave. The proof we offer is his, although we have added some additional explanation.

THEOREM 11.5. *Let f be a rational map of at most degree d which univalently maps a doubly connected open set D into itself. Let $|\beta| = 1$ but not be a root of unity, and let $\text{Arg}(\beta) = \theta$ be the rotation angle of f on D. In addition, suppose*

$$\liminf_{n=1,2,\ldots} \sqrt[d^n]{\beta^n - 1} = 0. \tag{11.24}$$

The only possibility for f is that it is a rational map of degree one.

The close relation between this theorem and Theorem 11.1 from his paper **[1927]** is underscored by a footnote in which he observed that he had also developed a new proof of Theorem 11.1 based on his proof of Theorem 11.5.

The overall strategy will be to show that if f indeed univalently maps D onto itself, then the Cremer condition at equation (11.24) does not hold unless f is a Möbius mapping.

His proof begins with several assumptions similar to those he imposed on f in his 1927 theorem: $f = p/q$ is at least of degree two with $\deg p = r < \deg q = d$. In this case q is normalized so the coefficient of the highest term is one, whereas in the 1927 proof it was normalized with the constant term one. He also assumed that f maps D onto itself and is equivalent to a rotation by β, and that there is a function S mapping D onto an annulus A. Finally, the nth iterate of f is denoted $f^n = p_n/q_n$.

He assumed without loss of generality that D is bounded. Since $f(D) \subseteq D$, f and S are bounded, as are the polynomials p and q. Thus there is a C such that

$$|q_n(z)| < C^{d^n}. \tag{11.25}$$

If S conformally maps D to an annulus A, its inverse $S^{-1} : A \to D$ will also be bounded; hence, for any $w_1, w_2 \in A$, there is a positive constant M such that $|S^{-1}(w_1) - S^{-1}(w_2)| < M\,|w_1 - w_2|$.

That f is conjugate to a rotation through an angle ϕ on A permitted him to conclude that $S \circ f = \beta \cdot S$, where $\beta = e^{i\theta}$, so a solution S to the canonical Schröder equation exists. Since for all $z \in D$ there is a $w \in A$ with $S(z) = w$, it follows that $f^n(z) = S^{-1}(\beta^n w)$. Thus, since $|S^{-1}(w_1) - S^{-1}(w_2)| < M\,|w_1 - w_2|$ for all $w_1, w_2 \in A$, it follows that

$$\begin{aligned} |f^n(z) - z| &= |S^{-1}(\beta^n w) - S^{-1}(w)| \\ &\leq M|\beta^n w - w| \\ &= M|w| \cdot |\beta^n - 1| \\ &< M_1|\beta^n - 1|, \end{aligned} \tag{11.26}$$

where $M_1 > M \sup |w|$ as w ranges over A.

Next, he viewed $|f^n(z) - z|$ in much the same manner he did in his 1927 paper:

$$\begin{aligned} |f^n(z) - z| &= \left|\frac{p^n(z)}{q^n(z)} - z\right| \\ &= \left|\frac{p^n(z) - zq^n(z)}{q^n(z)}\right| \\ &> \left|\frac{p^n(z) - zq^n(z)}{C^{d^n}}\right|, \end{aligned} \tag{11.27}$$

the last inequality following from equation (11.25).

If we compare the chain of equalities ending at (11.26) with those from (11.27) it follows that

$$(11.28)\qquad M_1|\beta^n-1|>\left|\frac{p_n(z)-zq_n(z)}{C^{d^n}}\right|\Rightarrow C^{d^n}M_1|\beta^n-1|>|p_n(z)-zq_n(z)|\,.$$

Defining $g(z)=p_n(z)-zq_n(z)$ and $M^*=C^{d^n}M_1|\beta^n-1|$, we can restate (11.28) as

$$M^*>|g(z)|\,,$$

that is, M^* is an upper bound for $g(z)$ on D.

From above, we know that the highest degree coefficient of $q(z)$ equals one, so the highest degree coefficient of $g(z)$ is also one. Since the highest order term in g is of degree d^n, Taylor's theorem implies that for arbitrary z_0

$$1=\frac{g^{(d^n)}(z_0)}{d^n!}.$$

Cauchy's inequality (see (3.2)) applied to g in turn implies that for an arbitrary $z_0\in D$ with $|z-z_0|\leq R$ contained in D

$$1=\frac{g^{(d^n)}(z_0)}{d^n!}\leq\frac{M^*}{R^{d^n}},$$

where M^* is the upper bound for g given below (11.28). We have the following:

$$1\leq\frac{M^*}{R^{d^n}}=\frac{M_1C^{d^n}|\beta^n-1|}{R^{d^n}}.$$

Cross multiplication of the first and last terms implies that

$$\frac{R^{d^n}}{M_1C^{d^n}}\leq|\beta^n-1|.$$

Taking d^nth roots of both sides yields

$$\frac{R}{M_1^{\frac{1}{d^n}}C}\leq\sqrt[d^n]{|\beta^n-1|}.$$

Since the d^nth root of M_1 goes to one as $n\to\infty$, we can find a positive Δ independent of n for which $\Delta\leq\sqrt[d^n]{|\beta^n-1|}$ on D. For example we can set

$$\Delta=\frac{cR}{C},$$

where

$$c=\min\left\{1,\left(\frac{R}{M_1}\right)^{\frac{1}{d^n}}\right\},$$

for n greater than an appropriately chosen N.

From the above it follows that $\sqrt[d^n]{|\beta^n-1|}$ is bounded away from zero independently of n so the Cremer condition at (11.24) is not satisfied for a rational function of degree strictly greater than one. It follows then that f must be linear. What is remarkable is that Cremer's theorem shows that numbers for which it is impossible for iteration to be conformally equivalent to an annular rotation are the same numbers he found in Theorem 11.1. Cremer does not delve into this connection but this connection is real, and as it turns out, the rotation numbers for which there are centers are exactly the same as those which allow for a doubly connected domain to be conformally equivalent to an annular rotation (see, for example, Milnor [**2006a**, p. 165]).

11.8.2. A new proof: Koebe and complex dynamics. Koebe evidently influenced Cremer's decision to study mappings of doubly connected regions into an annulus, a subject Cremer first explored in [**1930a**]. Cremer served as Koebe's assistant at the University of Leipzig in the 1920s. As was noted above, it seems they had become close, and Cremer's papers contain references to informal mathematical conversations he had with Koebe. We described in Section 11.7 how the reference to Schwarz in the title of his paper [**1932a**] is tied to an anecdote that Koebe told him about Schwarz.

Likewise, Cremer motivated his interest in annular mappings by remarking that Koebe had often expressed an interest in a method for mapping doubly connected open sets onto an annulus using an approach similar to the "Fejér-Riesz method" for finding a conformal map from a simply connected domain D with at least two boundary points into a disc centered at the origin [**1930a**, p. 190].

Cremer was referring to the common proof of the Riemann Mapping Theorem utilizing the family $\mathcal{F}$ of one-to-one and bounded maps f from D into the unit disc $\mathbb{D}$, which have the property that they map some specified point $z_0 \in D$ onto the origin with $f'(z_0) = 1$. For maps f from the family $\mathcal{F}$, let $\mu(f)$ denote the supremum of $|f(z)|$ for $z \in D$, and let ρ represent the greatest lower bound of the set of these supremum. The family $\mathcal{F}$ is both nonempty and normal. One finds a sequence of maps f_n from $\mathcal{F}$ with the property that $\mu(f_n)$ converges to ρ and from this sequence extracts a uniformly convergent sequence which converges to a function F.[12] The mapping

$$f = \frac{F}{\rho}$$

is the desired map from D to $\mathbb{D}$.

Evidently in response to Koebe, Cremer extended this method to doubly connected domains. He observed that the conformal mapping of a doubly connected region with two boundary points to an annulus had already been achieved by Koebe in [**1915**, p. 202], but that his was the first proof analogous to the Fejér-Riesz method for simply connected domains. Cremer also pointed out that another student of Koebe's, Ernst Gräser, who received his doctorate from theUniversity of Leipzig in 1930 and taught at Göttingen from 1937 to 1955, offered yet another proof in his dissertation [**1930**] also based on Koebe's approach. Gräser's proof was evidently not published in a mathematical journal.

Cremer's proof is based on maps from what he referred to as a *normalized open set* D which contains no points of $\mathbb{D}$ and has an unspecified boundary γ that includes the unit circle, but no points of D on its interior. An example of such a set is $\text{Int}(C) - \overline{\mathbb{D}}$, where C is a simple closed curve whose interior contains $\overline{\mathbb{D}}$. Cremer, again quoting Koebe, noted that any bounded doubly connected open domain E with at least two boundary points could be mapped to a bounded normalized open set, so he set his task as finding a simple mapping from such sets to an annulus.

Cremer's proof is concise. There is nothing incorrect about it, yet important details seem to be left unstated. Nonetheless, his proof is referenced in several modern works and Hans-Peter Hoidn in [**1982**, p. 640] credits both Gräser and Cremer with originating this method of mapping doubly connected regions into an annulus, which he refers to as the *method of osculation*, since it produces a family of domains whose boundary curves converge to a circle. Bieberbach's proof

[12] See Bieberbach [**1953**, §18] for the mathematical details of this method.

of the existence of a mapping from doubly connected domains to an annulus in [**1953**, pp. 209–211] is quite similar to the proof Cremer gave, although he does not reference Cremer.

Even in 1930 Cremer's interest in annular mappings probably went a little deeper than just trying to extend the osculating method to normalized open sets. Given his the long-standing interest in the center problem, it would be natural for him to consider the analogous question for doubly connected singular domains: Given a doubly connected singular domain D, does iteration ever act like a rotation of an annulus through β? That is, does a Schröder mapping S of D to an annulus satisfying $S \circ f = \beta \cdot S$ exist? If so, his paper on the existence of maps from a normalized domain to an annulus would provide insight into such Schröder maps.

This result also set the stage for Cremer's most notable achievement in his study of singular domains.

11.8.3. The connection of invariant singular domains. In §V of his paper [**1932a**], Cremer showed that if an invariant singular domains exists for a rational function, its degree of connection is at most two. It was not until around 1980 that rational functions with doubly connected invariant singular domains were shown to actually exist. Such domains are called Herman rings, and iteration is conformally equivalent to an irrational rotation of an annulus. Michel Herman is credited with their discovery.[13]

It is important to note that Cremer never explicitly speculated that anything resembling a Herman ring might exist. That he had a strong sense what such domains would be like should they exist, is made clear by his description of iteration on a doubly connected domain in his proof of Theorem 11.24 (discussed in Subsection 11.8.1). He posited the existence of an analytic solution of the Schröder equation $S \circ f = \beta \cdot S$ and used phrases like "rotation angle". Thus, it is reasonable to assume that he thought such domains might exist. However, with his characteristic understatement regarding things he could not prove, if he entertained such speculation, he kept them to himself.

Cremer's result can be summarized as the following theorem:

THEOREM 11.6. *Let f be an entire or a nonlinear rational function. Let D be an invariant singular domain. If f is entire, D is simply connected. If f is rational, then D is at most doubly connected.*

Cremer began his proof by showing that among the limit functions of the iterative family of f in an invariant singular domain D is the identity, a result he would use midway through his proof of Theorem 11.6. He attributed this result to Fatou, although Fatou's proof was for rational functions (see Theorem 9.3). Since the identity is a limit function on D, every point of D is an accumulation point of some sequence of iterates of points in D.

Cremer first considered the case where f is entire. If f has an invariant singular domain D whose degree of connectivity is more than one, then somewhere within

[13]See Herman's paper [**1979**]. An example of a function with a Herman ring is

$$f(z) = \frac{e^{i2\pi\theta} z^2(z-4)}{1-4z},$$

where θ is chosen so that iteration by f is conjugate to the rotation $z \mapsto ze^{i2\pi\phi}$. See Color Plate 1. For more details, see either Herman's paper or one of the standard introductory monographs on the iteration of rational functions, for example, Milnor [**2006a**, Chapter 15].

the interior of the "inner boundary" of D is a point in the Julia set. Cremer showed that this is impossible. He evidently conceived of D as being annular, in the sense it has both an outer and an inner boundary, and that both boundaries contain points in the Julia set.

Since f is normal interior to D, the sequence $\{k_n\} = \{f^n(z) - z\}$ is also normal; hence it is uniformly bounded on compact subsets of D, in particular on all closed simple curves Γ inside D. Let Γ be such a curve whose interior contains a point in the Julia set, for example a curve interior to D whose interior contains the inner boundary of D.

In neighborhoods of points in the Julia set, for example those interior to Γ, the sequence $\{k_n\}$ is not uniformly bounded despite the fact that it is uniformly bounded on Γ. However, since f is entire, so are the k_n, and the maximum modulus theorem implies that the maximum of $|k_n(z)|$ on $\Gamma \cup \mathrm{Int}(\Gamma)$ occurs on Γ which in turn implies that its iterates are uniformly bounded on the interior of Γ so it can contain no points in the Julia set. Thus D is simply connected.

We found no earlier proof that Herman rings do not exist for entire maps, although Cremer did not express the result in this fashion. Rather, he simply said the degree of connection could not be two or higher in this case. Cremer's result implies as well that a polynomial cannot have a Herman ring, although, again, Cremer did not make this observation, even implicitly.

He then argued that D must be simply connected and that f must be conformally equivalent to a rotation on D. Suppose f is not. Then, since D is an invariant singular domain, f is a univalent self-map of D. Quoting an unnamed "well-known theorem" [**1932a**, p. 318] (perhaps the Denjoy-Wolff Theorem?[14]), $f^k(z)$ must converge to a boundary point on D for all z in D. This implies that no point interior to D is an accumulation point for iterates of points in D. However, as Cremer pointed out, the identity function is one of the limit functions of the iterative family on D, hence, as we noted above, every point in D is accumulation point for some of sequence of iterates, so we have a contradiction.

For rational functions, the above holds when D is simply connected, as Cremer observed, but the argument that D must always be simply connected fails, since the maximum modulus theorem does not apply if there is a pole inside Γ. Another approach is needed.

Cremer dipped into function theory and, applying the "*Uniformisierungstheorie*" of Herman Weyl [**1932a**, p. 439, fn.], offered only the following quotation from Weyl's monograph [**1913**, p. 163] as proof:

> *The group of conformal maps of the Riemann sphere is always discontinuous, aside from the following seven exceptions: an entire sphere, a singly or doubly punctured sphere, a spherical cap with a point removed, a spherical calotte (the portion of a sphere between parallels), and a Riemann surface of genus* 1.

This was followed by:

> *From this theorem it follows immediately that D is at most doubly connected.* [**1932a**, p. 319]

[14]See see Appendix G) for a discussion of the Denjoy-Wolff Theorem.

We will let the reader decide if the result follows immediately. For a more complete version of the proof that singular domains for rational functions are at most doubly connected, see Milnor [**2006a**, pp. 160–161].

Cremer then noted that if D were doubly connected, then for either f or an appropriate iterate f^k, iteration was tantamount to a rotation of an annulus by $w \mapsto az$, where $|a| = 1$ but was not a root of unity.

In §VI of his paper [**1932a**], Cremer used the Schröder equation to solve a version of the Riemann mapping problem on simple open domains D.

11.9. Concluding remarks

Cremer investigated several aspects of the center problem in a series of papers between 1927 and 1938. His most renowned accomplishment is a proof of the existence of rational functions with irrationally neutral points that are noncenters. His condition for their existence—the inequality given in Theorem 11.1—is much more practical than Pfeiffer's, and it is often given in the present-day literature as a sufficient condition for an irrationally neutral fixed point to be a noncenter. Less well known are his other accomplishments, such as his anticipation of Herman rings and the implicit hint in his study of annular mappings that for rational functions, the conditions for nonexistence of centers are identical to those for regions on which iteration is conformally equivalent to a rotation of an annulus. However, he did not prove the existence of centers nor did he speculate on the existence of Herman rings.

Cremer's study is characterized by a rich mixture of techniques. He was conversant with German function theory, and his anticipation of Herman rings relied on Weyl's uniformization theory. His study of annular mappings was influenced, and possibly suggested, by Koebe, with whom he was evidently close, and again, prompted by Koebe, he connected the center problem to a mapping problem that Schwarz would sometimes give his students.

The importance of Cremer using number theory to analyze the center problem should not be overlooked. Fatou and Julia's approaches were both characterized by their reliance on normal families. While the observant eye could see an implicit hint in Pfeiffer's work that number theory might be useful—and since Cremer knew Pfeiffer's work well, perhaps he acted upon it—it was Cremer who made the connection explicit. However, his application of the connection was one-way: while he showed that the multipliers that produced irrationally neutral fixed points that were noncenters were Liouville numbers, he was unable—if indeed he saw the connection—to exploit the converse linkage of Diophantine numbers to the possible existence of centers. This would be left to Siegel, who entered the door that Cremer left open and showed that the existence of centers is indeed linked to Diophantine numbers.

CHAPTER 12

Siegel, the Center Problem, and KAM Theory

12.1. Full circle!

The outward appearance of Carl Ludwig Siegel's groundbreaking paper [**1942**] in the *Annals of Mathematics* is unassuming: It is only six pages long, and its prosaic title "Iteration of Analytic Functions" gives no hint of what it contains. Yet there are few papers as short that have as significant an impact on not one, but two fields of mathematics.

Once inside the paper, one notices it provides almost no context for itself. It does not begin with an announcement heralding a solution to a long-standing problem in the iteration of functions that also happens to be a *bona fide* small divisors problem of the sort that has defied the best attempts of mathematicians as august as Poincaré. Rather, it begins with a prefatory definition,

> *Let*
>
> $$f(z) = \sum_{k=1}^{\infty} a_k z^k \tag{12.1}$$
>
> *be a power series without constant term and denote by $R > 0$ its radius of convergence. The fixed point $z = 0$ of the mapping $z \mapsto f(z)$ is called stable if there exist two positive finite numbers $r_0 \leq R$ and $r \leq R$, such that for all points z of the circle $|z| < r_0$ the set of images points $z_1 = f(z)$, $z_{n+1} = f(z_n)$ $(n = 1, 2, \dots)$ lies in the circle $z < r$.* [**1942**, p. 607]

It continues with introductory material for another page or so before stating its intent to solve the Schröder equation for irrationally neutral fixed points.

12.2. Stable once more

Siegel's concept of stability concerns future stability in the sense it involves forward iterates of the system, but he seems to have a slightly weaker concept than the classical definitions of Tullio Levi-Civita and Aleksander M. Lyapunov (presented in Chapters 3 and 4), which, at least in regards to future stability, can be summarized as follows:

DEFINITION 12.1. The fixed point $z = 0$ of the mapping $z \mapsto f(z)$ is called stable if for any positive number $r \leq R$ there exists a positive number $r_0 \leq R$ such that for all points z in the circle $|z| < r_0$ the set of image points $z_1 = f(z)$, $z_{n+1} = f(z_n)$ $(n = 1, 2, \dots)$ lies in the circle $z < r$.

In particular, if all orbits near one equilibrium point, which we will assume to be the origin, were to converge to another attractor contained in the disc $|z| < R$, then the origin would be considered stable under Siegel's definition, but not in the study of ODEs that we discussed.

Examples of behaviors which are stable under Siegel's conditions but are not under the classical definitions do occur within ODEs, for example, with a van der Pol equation, where all orbits beginning near the origin (which is an unstable equilibrium point) converge to an attracting periodic orbit relatively close but bounded away from the origin. Since such behavior cannot occur in the iteration of complex rational functions of a single variable,[1] Siegel's definition poses no problems. Nonetheless, it is curious. Siegel conducted research in celestial mechanics (see, for example, [**1941**]), so he was no doubt familiar with the standard ODE definition.

One does not expect Siegel to do anything without due consideration, so there probably is a good reason he choose the definition he did, which is crafted to allow attracting fixed points and centers to be considered stable, but not repelling or rationally neutral fixed points. Perhaps this was all Siegel was concerned about. In any event, his proof of the existence of centers indirectly shows that in the context of complex dynamics, his definition of stability is equivalent to the classical one.

12.3. Solved at last

Noting that it is "easy" to prove that 0 is stable if $|a_1| < 1$, he initially confined himself to the case where $|a_1| \geq 1$. He used Schwarz's Lemma to reduce the possibility of stability to the case where $|a_1| = 1$, and noted that if a_1 is a root of unity, stability can only occur if f satisfies the so-called Babbage functional equation[2] (a term he did not use), $f^n(z) = z$.

This further reduces his study to the case where $|a_1| = 1$ but is not a root of unity, which we have termed the irrationally indifferent (or irrationally neutral) case, and he asserted that the stability of 0 is equivalent to the solution of the Poincaré equation,[3]

$$f(\varphi(z)) = \varphi(a_1 z). \tag{12.2}$$

Siegel assumed a solution, if it exists, would be of the form

$$\varphi(z) = z + \sum_{k=2}^{\infty} c_k z^k. \tag{12.3}$$

That $\varphi(0) = 0$ follows from setting $z = 0$ in (12.2); differentiating both sides of (12.2) shows we have free choice for $\varphi'(0)$. Siegel took the standard route of normalizing his solution φ so $\varphi'(0) = 1$.

He then observed that Pfeiffer constructed the first function for which convergent solutions to the Schröder equation did not exist, and that Cremer constructed

[1] As Theorem 7.6 implies, in any neighborhood of a repelling point, which is in the set E referred to in the theorem, there are points whose orbits are unbounded.

[2] For rational functions the Babbage equation has no solutions if the degree of f is greater than one.

[3] Siegel referred to this equation as the "Schröder equation" but the version he gave is actually the inverse of the usual Schröder equation. Here, solving the Poincaré equation is equivalent to solving the Schröder equation. See page 37 for a discussion of the relation between the Schröder equation and the Poincaré equation.

a_1 which produced only divergent solutions. He ends his brief review of the literature with the remark,

> *Until now, however, it was not known if there exists a number a_1 of absolute value 1, such that every convergent power series $f(z) = a_1 z + \cdots$ has a convergent Schröder series. Julia tried to prove the erroneous hypothesis that the Schröder series is always divergent.* [**1942**, p. 608]

It is at this point he announces with no fanfare (in Theorem 12.2 below), that his is the first solution in the $|a_1| = 1$ case. Siegel was thus the first to be able to say with surety that the Schröder equation (or in his case, the equivalent Poincaré equation) is not always divergent, and in the next five pages he proves that centers exist—a result that eluded Fatou, Julia, Kœnigs, Cremer, and Pfeiffer (although, it is not certain that Pfeiffer sought a solution). We give the details of his proof below.[4] Later, we will discuss the importance of this paper to KAM theory.

Siegel proved the following theorem stated with elegant simplicity:

THEOREM 12.2 (Siegel's Theorem). *"Let*

$$\log |a_1^n - 1| = O(\log n) \qquad (n \to \infty), \tag{12.4}$$

then the Schröder series is convergent." [**1942**, p. 608]

Relation (12.4) reveals the extent to which the sequence a_1^n is allowed to approach one as n grows to infinity. Noting that the left-hand side is bounded from above and negative for a_1 near 1, the hypothesis (12.4) my be recast as

$$\log |a_1^n - 1| \geq -\nu \log n + \log \gamma,$$

for some positive constants ν and γ, or equivalently[5]

$$|a_1^n - 1| \geq \gamma n^{-\nu}. \tag{12.5}$$

Under this guise, it is easy to see that (12.4) states that the rotation angle θ associated with a_1 must be Diophantine. In fact, writing $a_1 = e^{2\pi i \theta}$ yields $|a_1^n - 1| = |e^{2\pi i(n\theta - m)} - 1|$ for any integer m. The addition of the integer m allows us to interpret the condition that a_1 is close to 1 as $n\theta - m \simeq 0$, for some m, and we can replace the exponential by its linear part, $e^{2\pi i(n\theta - m)} \simeq 1 + 2\pi i(n\theta - m)$. Substituting into (12.5) and setting $\mu = \nu + 1$ yields

$$\left|\theta - \frac{m}{n}\right| > \gamma q^{-\mu}, \tag{12.6}$$

for all nonnegative integers p, q, with $q \neq 0$, where γ and μ depend on θ, which is equivalent to the definition of Diophantine numbers we have used elsewhere (for example, in our discussion of Pfeiffer, Definition (5.2)). We have scaled γ to make inequality (12.6) strict.

Siegel did not provide the above details, rather, following his statement of Theorem 12.2, he immediately stated equation (12.6) and observed,

---

[4]For another analysis of Siegel's paper, see Pérez [**2011**].

[5]Note that the equation preceding (12.5) gives us some freedom in choosing ν. For example, if a relatively small ν produces a very sharp bound, we can replace it by a larger one. In this sense the choice of ν is a matter of convenience, much as one can choose a convenient value for δ in a $\delta - \epsilon$ proof. This point is important since in specific applications we present below, different choices of ν are made, either explicitly, or implicitly, through the value of μ. However, the value of ν also determines the radius of convergence for φ: the larger ν is, the smaller the radius of convergence. Hence, the determination of a specific value of ν is also a balancing act.

> *It is easily seen that* (12.4) *holds for all points on the unit circle* $|a_1| = 1$ *with the exception of a set of measure* 0. [**1942**, p. 608]

With characteristic understatement, he did not provide a proof of this astonishing result, which modulo the proof of Theorem 12.2, implies not only that centers do indeed exist, but that they are far more prevalent than noncenters.[6]

Shortly thereafter he stated another equivalency to equation (12.4), one that he will use in his proof:[7]

$$\varepsilon_n \equiv \frac{1}{|a_1^n - 1|} < (2n)^\nu. \tag{12.7}$$

If we examine the formal solutions to Siegel's Poincaré equation (12.2), we can develop an understanding of why φ converges. The function f is defined at equation (12.1) and has the usual Taylor expansion in the a_k. Other than the stipulation that $|a_1| = 1$ is not a root of unity, the conditions on f imposed by Siegel were minimal—only that it be analytic on a disc surrounding the fixed point. His conclusions apply to rational functions, but also to any locally defined analytic function.

The Poincaré equation thus becomes

$$\sum_{k=2}^{\infty} c_k(a_1^k - a_1)z^k = \sum_{l=2}^{\infty} a_l \left(z + \sum_{r=2}^{\infty} c_r z^r\right)^l, \tag{12.8}$$

---

[6]We offer a proof from Broer, Huitema, and Sevryuk from [**1996**] to show that the set of a_1 satisfying (12.6) is of full measure on the unit circle. It is a variant of the standard proof used to show that Liouville numbers are of measure zero, and perhaps Siegel had a similar argument in mind. If so, this could explain why he did not include it. Cremer, in fact, used a similar argument to show that his numbers—numbers in the unit interval corresponding to values of a_1 for which centers do not exist—are of measure zero.

For a given $\mu > 2$ and $\gamma > 0$ denote by $[0,1]_{\mu,\gamma}$ the set of all θ satisfying (12.6). Following the argument presented in [**1996**], we show that the union $\bigcup_{\gamma>0}[0,1]_{\mu,\gamma}$ is of full measure, as Siegel asserted. It suffices to show that the Lebesgue measure of $[0,1] \setminus [0,1]_{\mu,\gamma}$ is $O(\gamma)$ as $\gamma \to 0$. This property follows from observing that $[0,1]_{\mu,\gamma}$ may be obtained from $[0,1]$ by removing infinitely many intervals, according to the following procedure. Fix a positive integer $q \geq 2$. Then, for all integers $p = 1, \dots, q-1$, delete the interval centered at p/q and with radius γ/q^μ which, clearly contains points that do not satisfy (12.6). The intervals $[0,\gamma]$ and $[1-\gamma, 1]$ must also be deleted for the same reason. It follows that

$$\text{measure}([0,1] \setminus [0,1]_{\mu,\gamma}) \leq 2\gamma + \sum_{q=2}^{\infty}(q-1)\frac{\gamma}{q^\mu} < 2\gamma \sum_{q=1}^{\infty} \frac{1}{q^{\mu-1}} = O(\gamma),$$

since the generalized harmonic series converges because $\mu > 2$. For this reason, the assumption that $\mu > 2$ is often incorporated into the definition of Diophantine condition (12.6), and the union

$$\bigcup_{\mu>2,\gamma>0} [0,1]_{\mu,\gamma}$$

is often referred to as the set of Diophantine (or Siegel) numbers (see, for example, Milnor [**2006a**]). Broer, Huitema, and Sevryuk also show that $[0,1]_{\mu,\gamma}$ is a Cantor set.

[7]This relation is readily deduced from (12.5) by posing $\gamma = 2^{-\nu}$ after suitably tuning the constant ν.

from which, equating the coefficients of the terms with the same power at both sides, we get the recursion relation ($c_1 \equiv 1$):[8]

$$c_2 = \frac{1}{a_1^2 - a_1} c_1^2 a_2,$$
$$c_3 = \frac{1}{a_1^3 - a_1}(c_1 c_2 + c_2 c_1 + c_1^3),$$

and, in general,

$$c_k = \frac{1}{a_1^k - a_1} \sum_{r=2}^{k} a_r \left(\sum_{l_1 + \cdots + l_r = k} c_{l_1} \cdots c_{l_r} \right), \tag{12.9}$$

with $l_j \geq 1$ for $1 \leq j \leq k$. Siegel's proof exploited the method of majorants to demonstrate convergence of (12.3).

Note the $a_1^k - a_1 = a_1(a_1^{k-1} - 1)$ term in the denominator of the right-hand side of equation (12.9), which defines the unknown coefficients c_k of the Poincaré function φ. Siegel used the Diophantine condition in equation (12.7) to bound the c_k; in essence, this is another small divisors problem, similar to the one Pfeiffer considered (examined in Section 5.4), except that unlike Pfeiffer, who used the fact that his numbers were well approximated by rationals to make his small divisors arbitrarily large to force divergence, Siegel kept the c_k under control. This was a groundbreaking achievement, and not just in complex dynamics. In Chapter 4 we discussed the persistence of small divisors in ODEs and perturbation problems related to them. Siegel's proof is considered by many to be the first successful solution of a significant small divisors problem; see, for example, Marmi [**1999**, p. 28].

12.3.1. Siegel vs. Cremer numbers. It will be helpful to give a descriptive account of rotation by Diophantine multipliers, and compare it to what happens when, for a given function, only divergent solutions exist to the Schröder functional equation. As has been pointed out above and in earlier chapters that the points $a_1^n = e^{2\pi i\theta n}$ represent successive rotations by θ along the unit circle. Both (12.4) and (12.7) may be interpreted as a constraint on the way the a_1^n are distributed near 1 on the unit circle: equivalency (12.7) tells us that the quantities ε_n (the reciprocals of the distance between a_1^n and 1) grow at most at a polynomial rate, that is, as a power of n. It is this constraint which controls the size of the c_k.

This is in contrast to the Cremer numbers β, which produce divergent solutions to the Schröder equation (described in equation (11.3)), that satisfy

$$\liminf_{n\to\infty} \sqrt[d^n]{|\beta^n - 1|} = 0.$$

In Chapter 11 we obtained the above result via Cremer's Lemma 11.2 by showing that if β is a Cremer number,

$$|\beta^{n_r} - 1| < \frac{1}{n_r^{d^{n_r}}} \tag{12.10}$$

[8]These are analogous to expressions that Pfeiffer derived for the coefficients of the formal solution to the Schröder equation at equation (5.8).

for an infinite sequence $\{n_r\}$ (see equation (11.13) and subsequent equations). If we put equation (12.10) in the same form as equation (12.7), we have

$$\frac{1}{|\beta^{n_r} - 1|} > n_r^{d^{n_r}}, \tag{12.11}$$

that is, the reciprocals of the distances grow at least at a factorial rate for an infinite sequence. This is what causes the c_k to grow at a rate which forces divergence of the Schröder equation. The Cremer numbers act like rational numbers, for which no linearization exists; in contrast, the Diophantine multipliers are sometimes said to be highly irrational. The differences between Cremer and Siegel's results are highlighted by their approaches to bounding $|a^n - 1|$. Cremer sought conditions which allowed him to bound it from above while Siegel took the opposite approach. Compare, for example, equations (11.7) and (12.5).

12.3.2. Taming the coefficients. Lemma 3 from Siegel's paper[9] provides the means to bound the c_k:

LEMMA 12.3 ([**1942**, p. 610]). *"Consider now the sequence of positive numbers $\delta_1 = 1, \; \delta_2, \ldots,$ recurrently defined in the following way: For every $k > 1$, let μ_k denote the maximum of all products $\delta_{l_1}\delta_{l_2}\cdots\delta_{l_r}$, with $l_1 + l_2 + \cdots + l_r = k > l_1 \geq l_2 \geq \cdots \geq l_r \geq 1,\; 2 \leq r \leq k$, and put*

$$\delta_k = \varepsilon_{k-1}\mu_k. \tag{12.12}$$

Then

$$\delta_k \leq k^{-2\nu} N_2^{k-1} \tag{12.13}$$

[with $N_2 = \left(2^{5\nu+1}\right)$]" [**1942**, p. 610].

Recall from equation (12.7) that $\varepsilon_{k-1} = |a_1^{k-1} - 1|^{-1}$. The δ_k form an upper bound for the individual terms of the series of the c_{l_i} at equation (12.9). Finally, as we will see below, Siegel normalized the a_k so that they are at most one in modulus.

Siegel arrived at the crucial insight—if the numbers δ_k are dominated by an exponential growth rate (the contribution of the polynomial infinitesimal factor $k^{-2\nu}$ being negligible), then the coefficients of the formal solution (12.9) are such that the corresponding series (12.3) has a nonzero radius of convergence, and thus linearization holds. In other words, Siegel could bound the c_k by a geometric series.

The goal is then to find a suitable majorant function. In general, one does this via Cauchy's inequality, which implies that any power series converges if and only if it is dominated by a geometric series, that is,[10]

$$|a_k| = \frac{|f^k(0)|}{k!} \leq a^{k-1}, \tag{12.14}$$

for a suitable positive constant a. One can then substitute (12.14) in (12.9) to obtain a bound of c_k independent of the coefficients a_r.

However, Siegel, following an alternate but equivalent approach, without loss of generality assumed $|a_k| \leq 1$ and observed,

[9]Lemmas 1 and 2 assert technical results in support of Lemma 3 and will be not discussed.

[10]The reader can find details about the origin of the method of majorants and its applications in Section 3.2. The Cauchy inequality is given at (3.2).

Since the power series (12.1) *has a positive radius of convergence, there exists a positive number* a, *such that* $|a_k| \leq a^{k-1} = \quad (k = 2, 3, . \; .)$. *The functional equation* (12.2) *remains true under the transformation* $f(z) \to af(z/a)$, $\varphi(z) \to a\varphi(z/a)$; *hence we may assume* $|a_k| \leq 1$ $(k = 2, 3, \dots)$. [**1942**, p. 611]

Siegel then defined the positive sequence σ_k where $\sigma_1 = 1$ and for $k > 1$,

$$\sigma_k = \varepsilon_{k-1} \sum_{r=2}^{k} \left(\sum_{l_1 + \cdots + l_r = k} \sigma_{l_1} \cdots \sigma_{l_r} \right), \qquad k = 2, \dots,$$

which is the formal solution of

$$\sum_{k=2}^{\infty} \sigma_k \varepsilon_{k-1} z^k = \sum_{l=2}^{\infty} \left(z + \sum_{r=2}^{\infty} \sigma_r z^r \right)^l . \tag{12.15}$$

The two equations above are to be compared to (12.9) and (12.8), respectively. Applying complete induction yields $|c_k| \leq \sigma_k$, which moves the problem of convergence from (12.8) to (12.15). The final stage was to neglect ε_{k-1} in (12.15) and consider the simplified functional equation

$$\sum_{k=2}^{\infty} \tau_k z^k = \sum_{l=2}^{\infty} \left(z + \sum_{r=2}^{\infty} \tau_r z^r \right)^l \tag{12.16}$$

in the unknown $\psi(z) = \sum_{k=1}^{\infty} \tau_k z^k$. Proceeding analogously, its solution is $\tau_1 = 1$,

$$\tau_k = \sum_{r=2}^{k} \left(\sum_{l_1 + \cdots + l_r = k} \tau_{l_1} \cdots \tau_{l_r} \right), \qquad k = 2, \dots .$$

Again, a complete induction argument yields $\sigma_k \leq \delta_k \tau_k$:

$$\begin{aligned} \sigma_k &= \varepsilon_{k-1} \sum_{r=2}^{k} \left(\sum_{l_1 + \cdots + l_r = k} \sigma_{l_1} \cdots \sigma_{l_r} \right) \leq \varepsilon_{k-1} \sum_{r=2}^{k} \left(\sum_{l_1 + \cdots + l_r = k} \delta_{l_1} \tau_{l_1} \cdots \delta_{l_r} \tau_{l_r} \right) \\ &\leq \varepsilon_{k-1} \mu_k \sum_{r=2}^{k} \left(\sum_{l_1 + \cdots + l_r = k} \tau_{l_1} \cdots \tau_{l_r} \right) = \delta_k \tau_k . \end{aligned}$$

The right-hand side of (12.16) is a geometric series in ψ, so we can rewrite (12.16) as the functional equation

$$\psi(z) - z = \frac{\psi^2(z)}{1 - \psi(z)}.$$

If we view this as a quadratic equation in ψ it admits a solution which fixes the origin[11]

$$\psi(z) = \frac{1}{4} \left(1 + z - \sqrt{1 - 6z + z^2} \right).$$

Finally, since $\psi(z)$ converges in the circle $|z| < 3 - 2\sqrt{2}$, it follows from Siegel's Lemma 3 (i.e., our Lemma 12.3) that the series for $\varphi(z)$ at (12.7) converges in the circle $|z| < (3 - 2\sqrt{2}) 2^{-5\nu - 1}$ and linearization holds.

[11] The second solution $(1 + z + \sqrt{1 - 6z + z^2})/4$ is discarded since it does not admit the origin as a fixed point.

CARL LUDWIG SIEGEL

FIGURE 12.1. This picture was taken in Göttingen in 1975; courtesy of the Archives of the Mathematisches Forschungsinstitut Oberwolfach.

12.4. Siegel and the center problem

Why did Siegel care about this problem? He wrote no other works in complex dynamics, although he was certainly involved in its sister field, celestial mechanics. One possibility is suggested by his paper [**1941**], "On the integration of canonical systems", in which he states, in the context of solving a differential equation,

> The question of convergence has never been settled. *If we define sum, difference, product, quotient, and derivative of power series in a formal algebraic manner, we can perform these operations with divergent power series, and then we can construct by straightforward calculation a transformation ... which reduces H to a power series.... But no proof for the convergence of this contact transformation has been given,.... On account of the small divisors appearing in the coefficients of the transformation, it seemed to be probable that the series will diverge in general, but no single example had hitherto been found.* [**1941**, pp. 807–808]

The center problem is nothing if not a small divisors problem. Perhaps Siegel became interested in it precisely because of this. Moreover, Diophantine analysis was a field he excelled in, having proved fundamental results earlier in his career. We wonder, then, if he found out about the center problem, read the literature and became intrigued when he noticed the clear connection to a field of number theory in which he was expert. Perhaps he pondered the obvious question: if among the Liouville numbers are noncenters, then maybe Diophantine numbers produce centers. He had superb computational skills, experience in solving problems involving series, a strong awareness of the longstanding problem of small divisors in celestial mechanics, and expertise in number theory. He was, in many respects, the perfect person to successfully solve the center problem.

We want to make it clear that when we refer to solving the center problem in complex dynamics, we mean only in terms of how it was viewed at the time. To solve this problem in 1942 was to determine if centers exist and if so, to somehow give a means to find out if a given irrationally indifferent multiplier is a center. At the time that we wrote this book, the problems related to fixed points with irrational multipliers had mushroomed to proportions well beyond what our protagonists might have imagined to include objects such as hedgehogs and small cycles, which present considerable challenges. From a contemporary perspective, it might be said that Siegel did not so much as solve the center problem, as open it up enough for mathematicians to see what the real problems are.

We also wonder how Siegel found out about the center problem. Both of his papers that we have mentioned were published in the *Annals of Mathematics* in 1941 and 1942. Several past members of the editorial board of the *Annals* were involved with or aware of developments of complex dynamics. Pfeiffer served in the late 1920s, long before Siegel's papers. But as recently as 1939, G. D. Birkhoff (who helped Pfeiffer with his small divisors problem) was on the board, along with Ritt, who was quite active in complex dynamics in the 1920s (see Section 13.2). We are suggesting that there are people associated with the *Annals* who were also quite knowledgeable about the iteration of complex functions who could have read Siegel's 1941 paper, noticed the quotation about small divisors and perhaps conversed with Siegel in some fashion about it. We do not know, of course.

What we do know is that Siegel's solution of the center problem played a role in the development of a new perspective on small divisors problems in celestial mechanics. This is the subject of out next section.

12.5. KAM theory: A new approach to small divisors

Siegel's solution to the center problem in complex dynamics played an inspiration role in the development of Jürgen Moser's approach to what is nowadays called KAM theory (named for Andrei Kolmogorov, Vladimir Arnold, and Moser) since it served as a proof of a concept for solvable small divisors problems.

Moser, who studied celestial mechanics during the 1951–1952 winter semester at Göttingen under Siegel's direction (Russman [**1983**, p. 190]), was very familiar with Siegel's paper. His interest in perturbation grew with the assignment to review Kolmogorov's paper [**1957**], the print version of an address he gave at the 1954 International Congress of Mathematicians, in which Kolmogorov reiterates his assertion from [**1954**] that invariant tori in a completely integrable analytic Hamiltonian system may persist under small analytic perturbations. However, Moser was not fully persuaded by Kolmogorov's argument, and in his review remarked, parenthetically,[12]

> *The proof of this theorem was published in Dokl. Akad. Nauk SSSR* ***98*** *(1954), 527–530, but the convergence discussion does not seem convincing to the reviewer.* [**1959**]

[12]According to Hasselblatt and Katok [**2002**], Moser also sent a letter to Kolmogorov asking for the details of his proof.

JÜRGEN MOSER

FIGURE 12.2. This picture is from Tokyo, 1969; courtesy of the Archives of the Mathematisches Forschungsinstitut Oberwolfach.

Moser's first published paper [**1962**] related to KAM theory treated twist maps and four years later he published a two-part paper, the second part of which [**1966**] contains a technique to overcome small divisors problems.

Moser acknowledged Siegel's role near the beginning of his paper, noting,

> *Such problems occur in celestial mechanics and are closely related to the so-called small divisor difficulty. We shall discuss the simplest model problem in which this difficulty occurs, the center problem. This problem was treated successfully by C. L. Siegel in* [**1942**]. ...
>
> *We shall describe this problem in the following section but approach it with a different method which allows other applications. In fact, this is exactly the method which was suggested by Kolmogorov* [**1954**] *for a small divisor problem and was used by Arnold* [**1963**] *in the proof of Kolmogorov's theorem. The main point will be to construct a rapidly convergent sequence of approximations* .... [**1966**, pp. 499–500]

Shortly thereafter, he became more specific about his approach:

> *Siegel's original proof depends on delicate estimates which take into account that the number* $|a^q - 1|^{-1}$ *is usually much smaller than* $c_0 q^2$.[13] *The proof that will be presented here is much cruder and*

[13]This is analogous to the quantity $(2n)^\nu$ that Siegel used in his proof (see equation (12.7)). From the text surrounding equation (12.6), it is clear that the only condition that Siegel imposed on ν is that it is greater than one.

therefore is applicable to the more difficult problems of celestial mechanics. [**1966**, p. 501]

The proof that Moser gave is of the efficacy of his method as a means to overcome small divisors. Following his proof, he showed how it could be used to solve the complex dynamics center problem which he referred to as Siegel's theorem. We will follow in his footsteps, first by reviewing his method, then discussing how it can be used to prove Siegel's Theorem.[14]

Moser emphasized that he wanted to describe his method in a general setting in order to bring out its important features. For example, he applied it to the functional equation

$$f \circ \varphi = \varphi \circ \Lambda,$$

where f and Λ are given,[15] or equivalently

$$\mathcal{F}(f, \varphi) = \Lambda, \tag{12.17}$$

where the functional $\mathcal{F}$ is defined by

$$\mathcal{F}(f, \varphi) = \varphi^{-1} \circ f \circ \varphi.$$

The functional $\mathcal{F}$ satisfies the group properties (I denotes the identity map)

$$\mathcal{F}(f, I) = f, \qquad \mathcal{F}(f, \varphi \circ \psi) = \mathcal{F}(\mathcal{F}(f, \varphi), \psi), \tag{12.18}$$

which, as we will see, turn out to be crucial in devising the procedure. The partial differential operators of $\mathcal{F}(f, \varphi)$ are defined as

$$\mathcal{F}_1(f, \varphi)\, g = \lim_{\epsilon \to 0} \frac{1}{\epsilon} \left[\mathcal{F}(f + \epsilon g, \varphi) - \mathcal{F}(f, \varphi)\right]$$

and

$$\mathcal{F}_2(f, \varphi)\, g = \lim_{\epsilon \to 0} \frac{1}{\epsilon} \left[\mathcal{F}(f, \varphi + \epsilon g) - \mathcal{F}(f, \varphi)\right],$$

respectively. The Newton's method–like iteration that summarizes Kolmogorov's idea, as described by Moser, generates two sequences of functions $\varphi_0, \varphi_1, \ldots, \varphi_n, \ldots$ and $f_0, f_1, \ldots, f_n, \ldots$ such that, $\varphi_0 = I$, $f_0 = \mathcal{F}(f, \varphi_0) = f$. Under suitable assumptions

$$\lim_{n \to \infty} \varphi_n = \varphi \qquad \text{and} \qquad \lim_{n \to \infty} f_n = \Lambda,$$

which are reminiscent of methods using infinite compositions (described in Sections 4.3 and 4.4).

As with Newton's method, the nonlinear equation (12.17) will be solved via a sequence of linear equations obtained by replacing the functional $\mathcal{F}$ with its linear part: In a neighborhood of (Λ, I), we have

$$\mathcal{F}(f, \varphi) \simeq \mathcal{F}(\Lambda, I) + \mathcal{F}_1(\Lambda, I)\,(f - \Lambda) + \mathcal{F}_2(\Lambda, I)\,(\varphi - I). \tag{12.19}$$

Note that from (12.18), $\mathcal{F}(\Lambda, I) = \Lambda$, and

$$\mathcal{F}_1(\Lambda, I)\, g = \lim_{\epsilon \to 0} \frac{1}{\epsilon} \left[\mathcal{F}(\Lambda + \epsilon g, I) - \mathcal{F}(\Lambda, I)\right] = g,$$

therefore (12.19) becomes

$$\mathcal{F}(f, \varphi) \simeq f + \mathcal{F}_2(\Lambda, I)\,(\varphi - I). \tag{12.20}$$

---

[14]We introduced Moser's method in Section 4.4 at the end of our discussion of small divisors problems.

[15]If $\Lambda(z) = f'(0)z$ we obtain the Poincaré functional equation Siegel considered in his proof.

One difference with the classical Newton method is that the partial differentials of $\mathcal{F}$ are always evaluated at (Λ, I) rather than on the last computed approximation (f_n, φ_n). Concerning this point, we quote Moser:

> *This is crucial for the success in small divisor problems in which small changes of the linear operator $\mathcal{F}_2$ may destroy invertibility, since the spectrum of $\mathcal{F}_2(\Lambda, I)$ comes arbitrarily close to* 0. [**1966**, p. 505]

As is explained below at (12.22), Moser was referring to the solvability of the linear systems associated with his iterative procedure: The assumption that the point around which $\mathcal{F}(f, \varphi)$ is linearized remains the same at each step guarantees that the sequence generated by the procedure is indeed well defined.

Suppose $\varphi_0, \varphi_1, \ldots, \varphi_n, f_0, f_1, \ldots, f_{n-1}$ have been already defined. The definition of the next iterate, which allows us to construct both f_n and φ_{n+1}, is

$$f_n = \mathcal{F}(f, \varphi_n) = \varphi_n^{-1} \circ f \circ \varphi_n \tag{12.21}$$

and solves the linear part of the system $\mathcal{F}(f_n, \psi) = \Lambda$ in the unknown ψ. As indicated by (12.20), this yields

$$\mathcal{F}_2(\Lambda, I)(\psi - I) = \Lambda - f_n. \tag{12.22}$$

Under the assumption that $\mathcal{F}_2(\Lambda, I)$ is invertible, we set

$$\varphi_{n+1} = \varphi_n \circ \psi_n, \qquad \text{with } \psi_n = I + (\mathcal{F}_2(\Lambda, I))^{-1}(\Lambda - f_n). \tag{12.23}$$

Notice that (12.18) implies

$$\mathcal{F}(f_n, \psi) = \mathcal{F}(\mathcal{F}(f, \varphi_n), \psi) = \mathcal{F}(f, \varphi_n \circ \psi).$$

The ψ_n acts as a correction on φ_n which will tend quadratically to the identity I. To see this, it is enough to relate the error $\Lambda - f_{n+1}$ after the step n, to the error $\Lambda - f_n$ at the preceding step, and to show that the reduction of the error is of quadratic type. Hereafter $||\Lambda - f_n||$ denotes the distance, in some norm of the functions Λ and f_n; notice that from (12.23) it follows that $\psi_n - I = O(||\Lambda - f_n||)$, meaning that $\psi_n - I$ is of the same order as $\Lambda - f_n$. We have

$$\begin{aligned}
\Lambda - f_{n+1} &= \Lambda - \mathcal{F}(f, \varphi_{n+1}) = \Lambda - \mathcal{F}(f, \varphi_n \circ \psi_n) \\
&= \Lambda - \mathcal{F}(\mathcal{F}(f, \varphi_n), \psi_n) = \Lambda - \mathcal{F}(f_n, \psi_n) \\
&= \Lambda - \Big[\mathcal{F}(\Lambda, I) + \mathcal{F}_1(\Lambda, I)(f_n - \Lambda) \\
&\qquad + \mathcal{F}_2(\Lambda, I)(\psi_n - I) + O(||\Lambda - f_n||^2)\Big] \\
&= \Lambda - f_n - \mathcal{F}_2(\Lambda, I)(\psi_n - I) + O(||\Lambda - f_n||^2) \\
&= O(||\Lambda - f_n||^2).
\end{aligned}$$

Moser successfully used the above version of Kolmogorov's method to provide an alternative proof of the existence of centers for holomorphic germs. On the other hand, the first successful proof of the convergence of the series arising in the study of stability of nearly integrable Hamiltonian systems was evidently Eliasson's paper [**1988-96**].[16]

If f satisfies $f(0) = 0$ and $f'(0) = e^{i2\pi\theta}$, where θ is Diophantine, choosing $\Lambda(z) = a_1 z$ and $\varphi(z) = z + O(z^2)$ turns equation (12.17) into the Poincaré functional equation given at (12.1), which provides a solution to the center problem in complex dynamics.

[16] See Chierchia and Falcolini [**1994**] for a discussion of the background to Eliasson's paper.

To see the very first step of the Newton's method–like iteration for this choice of Λ, observe that

$$\begin{aligned}\mathcal{F}(\Lambda, I+\epsilon(\psi-I)) &= [I+\epsilon(\psi-I)]^{-1}\circ\Lambda\circ[I+\epsilon(\psi-I)]\\ &= \left[I-\epsilon(\psi-I)+O(\epsilon^2)\right]\circ\left[a_1I+\epsilon a_1(\psi-I)\right]\\ &= a_1I+\epsilon a_1\psi-\epsilon\psi(a_1I)+O(\epsilon^2),\end{aligned}$$

and we obtain

$$\mathcal{F}_2(\Lambda, I)\,(\psi-I) = a_1\psi-\psi(a_1I).$$

Hence, because $f_0 = f$, equation (12.22) becomes

$$\psi(a_1z)-a_1\psi(z) = f(z)-a_1z = \sum_{k=2}^{\infty}a_kz^k, \tag{12.24}$$

providing the first step.

Moser noted that equations such as (12.24) in the unknown $\psi(z)$ can be conveniently solved by power series expansion, yielding a solution that we denote ψ_0:

$$\psi_0(z) = z+\sum_{k=2}^{\infty}\frac{a_k}{a_1^k-a_1}z^k.$$

Assuming that R is the radius of convergence of $f(z)$, then $|a_k| < cR^{-k}$, and equation (12.7) yields

$$\left|\frac{a_k}{a_1^k-a_1}\right| < \frac{c}{|a_1|}R^{-k}(2k)^{\nu},$$

which implies also that $\psi_0(z)$ converges for $|z| < R$. One then sets $\varphi_1 = \varphi_0\circ\psi_0$, $f_1 = \varphi_1^{-1}\circ f\circ\varphi_1$, and continues with the second step.

Moser carried out the proof of the existence of centers in the $\mu > 2$ case by showing that the sequence $f_n = \varphi_n^{-1}\circ f\circ\varphi_n$ converges on a nonvoid open set containing the fixed point zero. The proof relies upon showing that if $f_n = a_1z+\hat{f}_n$ with

$$|\hat{f}_n'(z)| \le \epsilon_n, \quad \text{in } |z| < r_n,$$

then f_{n+1} satisfies

$$|\hat{f}_{n+1}'(z)| \le \epsilon_{n+1} = O(\epsilon_n^2), \quad \text{in } |z| < r_{n+1},$$

where the radii r_n decrease but are bounded away from zero. From the above estimates, he easily concluded that the sequence ϵ_n will tend to zero with quadratic rate.

12.6. Concluding remarks

Siegel was perhaps uniquely positioned to solve the center problem in complex dynamics. Its solution requires very strong computational skills: One needs the ability to make delicate estimations with difficult series, and one needs to be facile with Diophantine approximations. Most of all, one needs a sense of how to solve the problem: The center problem is a small divisors problem, a kind of problem—unsolvable prior to this point—with which Siegel was already quite familiar. It perhaps also helped that Siegel was not from the field of complex dynamics, because he brought a fresh approach to the problem and was not blinded by its standard techniques. It seemed that Julia and Fatou felt their best hope in solving the center problem was with normal families, and it is telling that the concept of normal families was never applied in Siegel's paper.

We do not know if Pfeiffer or Cremer tried to show that centers exist, so obviously it is a bit absurd to talk of their approaches to doing so. All we can say is that if they did try—and we think Cremer might have—they did not make enough headway to even speculate about possible approaches. Nonetheless, perhaps Siegel took a hint from their approaches to noncenters. It would have been reasonable for him to suspect that Diophantine forces might be at work since it was evident from their work (less so in Pfeiffer's perhaps) that Liouville numbers beget noncenters. So, might not Diophantine numbers give rise to their existence?

Siegel's solution to the center problem serves as a "proof of concept" that small divisors could be tamed. No doubt the rarified nature of complex functions of a single variable, where the assumption of analyticity comes with a rich set of hypotheses, helped make it possible to overcome the small divisors issues that arise in the center problem. Certainly, it can be easier to work with complex functions than with real functions of more than one variable.

Our quotations from Moser [**1966**] suggest he thought that this was so. Yet, it is also clear that once Moser realized that Kolmorogov was correct (and he initially had his doubts), Siegel's solution provided him with inspiration. Moser developed what he called a "cruder" approach to small divisors than the one Siegel applied to the center problem, one he felt more suitable to celestial mechanics, that grafted Siegel's assumption on the Diophantine character of the rotation number to the Newton's method–like procedure that Kolmogorov developed to handle small divisors in ODEs. At the same time, Moser showed that this perhaps less refined approach nonetheless afforded him an alternative means to prove what he referred to as Siegel's Theorem (our Theorem 12.2). So, for Moser, at least, Siegel's result was fundamental.

That Moser's approach to the convergence of the infinite compositions that arise in perturbation theory relies on an iterative technique so reminiscent of Newton's method brings us full circle back to the algorithm that played a pivotal role in the genesis of complex dynamics.

CHAPTER 13

Iteratin' Around the Globe

13.1. Introduction

The story we have told so far takes place largely in France and Germany, although important events took place in the United States and Italy. But an interest in the iteration of complex dynamics took root across the world following the works of Fatou and Julia.Mathematicians in Japan, Greece, Romania, and Scandinavia took up the study of iteration, and mathematicians in both the United States and Germany responded, not just to the monographs on rational functions, but also to the subsequent dynamical programs that Fatou and Julia developed in the 1920s.

We make no pretense that we are discussing everyone around the Mercator sphere who iterated during the 1920s and 30s. We omit, for example, papers in the 1920s by the French mathematician Paul Lévy [**1927a, 1928b**] and in 1930s by the Americans Morgan Ward [**1936**] and F. B. Fuller [**1934**] (coauthored with Ward), because these studies are tangential to the works we discuss in this book.

Our intent here is to briefly sketch the works of these mathematicians, so mathematical details are generally omitted. Capsule biographies of the mathematicians we discuss in this chapter, with the exception of Töpfer, about whom we could find very little, are found in Appendix O.

13.2. Ritt: Iterate on, Columbia, iterate

Joseph Fels Ritt was born in 1893, six months after Julia. His doctoral thesis on differential operators was written under the supervision of Edward Kasner at Columbia University. He was awarded his Ph.D. in 1917 and, following World War I, taught at Columbia, where he remained until his death in 1951. Although he is probably best known for his work on algebraic differential equations, he wrote a number of papers on the iteration of complex rational functions in the 1910s and 1920s, including a *Comptes Rendus* notice that appeared in early 1918 during the run-up to the *Grand Prix.* More information about Ritt and his career can be found in Lorch's necrology, [**1951**].

Ritt's publications involving the iteration of complex functions are indicative of the emerging American interest in the field in the 1910s (discussed in Chapter 5) that predated the Académie's December 1915 announcement of its 1918 *Grand Prix.* Ritt's paper on the Babbage functional equation $f^n(z) = z$ [**1915**] was no exception to this, and it was read before the American Mathematical Society about nine months prior to the Académie's announcement. Thus, it is unlikely that Ritt's interest in iteration was entirely inspired by the *Grand Prix.*

Ritt was nonetheless a victim of bad timing occasioned by the events surrounding the *Grand Prix.* His 1918 *Comptes Rendus* notice, *Sur l' itération des fonctions rationnelles*, was overshadowed not only by the ambitious and remarkably broad works Fatou and Julia prepared in anticipation of the *Grand Prix*, but also by Lattès, whose approach to iteration in three notices [**1918a, 1918b, 1918c**] was quite similar to Ritt's (see Section 7.5).

Because of this, Ritt's first full-length work on the iteration of complex functions [**1920**], evidently conceived as a longer paper than what ultimately appeared, was abridged for publication:

> *Notes on the iteration of rational functions have been published in the Comptes Rendus during the last few years by Julia, Fatou, Lattès, and myself. Many of the results announced were obtained independently by more than one writer, the priority in such cases seeming to belong to Julia. I wish to present here that small part of my work has not been covered by other writers, postponing until after the appearance of their memoirs the publication of such of my other results as may be of interest.*[1] [**1920**, p. 348]

It is not entirely clear what papers among those he later published are referred to, but very strong candidates would be Ritt's generalization of Schwarz's Lemma to multiply connected regions [**1921**]; his paper [**1924**], *Equivalent Rational Substitutions*, which treats three rational functions, ϕ, α and β, satisfying the functional equation, $\phi \circ \alpha = \phi \circ \beta$; and especially his investigation of permutable functions [**1923c**], which took a different route than the approaches of Fatou and Julia, and was perhaps more comprehensive. Ritt was not, however, focused on the dynamics of permutable maps (for example, he did not discuss the relation between the Julia sets of such maps), but rather on the structure and genesis of permutable functions. Similar comments apply to Ritt's paper [**1924**], which concerned the functional equation $\phi \circ \alpha = \phi \circ \beta$. For example, he showed that if $\deg \phi \leq 2 \cdot \deg \alpha$, then $\beta = \alpha \circ L$ where L is a linear function satisfying the Babbage equation $L^2 = z$. Appendix J discusses his study of permutable maps in much greater detail.

The version of Ritt's study of the iteration of complex functions that finally appeared in [**1920**] is in many respects a focused examination of superattracting fixed points. He was one of the first researchers to explicitly cite the details of Böttcher's Russian language paper [**1904b**], which contained a solution to the Böttcher functional equation, $B \circ f = (B)^n$, where 0 is a superattracting fixed point of $f(z) = a_n z^n + \cdots$.

Ritt developed the concept of associate points in [**1920**], that is, points which are mapped to same point by an iterate of f. For example, if $f(z) = z^2$, then any z in the punctured open disc, $\mathbb{D} - \{0\}$ would have 2^p associates under $f^p(z)$. Associate points evidently held the most interest for Ritt near superattracting fixed points and provided him with the following means to distinguish the attracting fixed points from those that are superattracting.

The Böttcher equation implies that iteration near a superattracting fixed point is an n-to-one mapping. Analogous to what we noted for the case $z \mapsto z^2$, if w is near an attracting fixed point λ, the number of points associated to w via f^p is 2^p,

[1]In a footnote attached at the end of the above quotation, Ritt remarked, "While this paper was in the hands of the editors of the *Transactions* [of the AMS], papers appeared by Julia [**1918a**] and by Fatou [**1919-1920**]. Fatou's paper contains some of the details of §III of the present paper, but as far as I can see, these results were first stated in my note in the *Comptes Rendus*." The details to which he referred are results pertaining to what he called "associates," that is points $\{\alpha_k\}$ which, for an arbitrary p, satisfy $f^p(\alpha_1) = f^p(\alpha_2)$, a terminology he coined. Fatou used the term "*points associés*" in [**1919-1920**, p. 189, II], without attribution.

so as p approaches infinity, the number of the associated points does as well; for w sufficiently close to λ, the closure of the set of associated points contains a simple closed curve surrounding the fixed point and passing through w on which points converge to λ

For a nonsuperattracting fixed point, iteration is one-to-one near the fixed point, and it is not until a point nears the boundary of the basin of attraction that the set of points associated to given point begins to accumulate, and so, unlike the superattracting case, the limit points of the set of associated points are not attracted to the fixed point. Ritt assembled these facts, and others, into a theorem about limit sets of associated points in [**1920**, p. 353], which is stated as Theorem J.14 in Appendix J.

The final section of his paper utilized the Böttcher equation on a neighborhood of infinity to develop conditions for two polynomials p and P to have a common iterate on a neighborhood of infinity.

13.3. Remoundos: Iterated algebraic functions in Greece

One of the important events in modern Greek mathematics was the founding of the Hellenic Mathematical Society in 1918; see Rassias [**2004**]. Like other mathematical societies, its purpose was to encourage research, education, and the use of mathematics. Nikolaos Hatzidakis' energy was instrumental in its founding, as was that of George Remoundos and Panayotis Zervos. This society helped consolidate the already strong mathematical activity at the University of Athens. One year after its founding, the Society began to publish the *Bulletin of the Greek Mathematical Society*.

The work of the Greek mathematician Constantin Carathéodory has been of lasting importance in the study of iteration. Fatou and Julia cited his work on the theory of functions and uniformization, and his work has continued to be instrumental to the post-Siegel study of complex dynamics. However, he was not a direct participant in the study of the iteration of complex functions. Our interest, then, is in Remoundos who, in addition to his role as a founder of the Society, was a member of the editorial board of the *Bulletin* and served as its second president from 1925–27.

Remoundos had strong ties to the French mathematical community and published widely. One of his chief research interests was the extension of Picard theory to algebroid maps. In the 1910s and 1920s, he was a frequent contributor to French journals, publishing over 25 *Comptes Rendus* notices, as well as seven articles in the *Bulletin de la Société Mathématique de France*. Our bibliography lists a representative sample of his work.

His only work to directly concern iteration, [**1923**], appeared in *Comptes Rendus* and focused on the iteration of algebraic multivalued maps. He applied the theory of normal families and discussed three properties these maps share with rational maps:

(1) The existence of attracting fixed points or limit cycles to which orbits of nearby points converge.
(2) If an attracting fixed point is also a critical point, then the map is locally conjugate to $w = z^k, k > 1$, where k depends on the number of branches near the limit cycle.
(3) Repelling fixed points can be characterized as essential singularities.

Iteration on neighborhoods of a repelling fixed point omits at most two values. He gave no indication of his proofs, and since this is his only work to explicitly treat iteration that we are aware of, we do not know their nature, or whether he carried his study any further.

It is interesting that Remoundos quoted Julia and Kœnigs' results exclusively, omitting any mention of Fatou. Neither Fatou nor Shimizu quoted his notice, and since he did not refer to Fatou's 1922 paper on the iteration of algebraic functions, it is unclear if he even knew about it, so it is doubtful that Fatou had any influence on his study.

13.4. Tambs-Lyche: Centers in Norway

Echoes of French and German studies of iteration were also heard in Scandinavia. Paul Johansen studied the iteration of both single-variable complex functions and real functions of two variables in [**1939**]. However, the most important Norwegian for our purposes is the American-born Ralph Tambs-Lyche, who focused on the center problem and was cited by Cremer in his paper [**1932a**].

During the late 1920s, Tambs-Lyche examined iteration in several papers, including [**1927, 1929a, 1929b, 1930**] as well as those we discuss below.

Our discussion focuses on his *Comptes Rendus* notice [**1928b**], since it involves an application of a computational technique he developed to attack the center problem. He published related works in *Comptes Rendus* [**1928a**] and in the *Bulletin de la Société Mathématique de France* [**1927**].

His study of the center problem focuses on the values assumed by the coefficients of a given analytic map with an irrationally neutral fixed point as it is iterated,

$$f(z) = sz + a_2 z^2 + \cdots + a_p z^p + \cdots ,$$

where $|s| = 1$ but not a root of unity.

Expressing the nth iterate as

$$z_n = f^n(z) = s^n z + a_2^{(n)} z^2 + \cdots + a_p^{(n)} z^p + \cdots , \tag{13.1}$$

he introduced an intriguing symbol analogous to the binomial coefficients,

$$\begin{bmatrix} n \\ r \end{bmatrix}, \quad \text{where} \quad \begin{bmatrix} n \\ 0 \end{bmatrix} = 1,$$

and

$$\begin{bmatrix} n \\ r \end{bmatrix} = \frac{(s^n - 1)(s^{n-1} - 1)\cdots(s^{n-r+1} - 1)}{(s-1)(s^2-1)\cdots(s^r - 1)},$$

that he used to derive the formula for a_p^n,

$$a_p^{(n)} = \sum_{r=1}^{p-1} (-1)^{p-r-1} s^{n-r+\frac{(p-r)(p-r-1)}{2}} \begin{bmatrix} n \\ r \end{bmatrix} \begin{bmatrix} n-r-1 \\ p-r-1 \end{bmatrix} a_p^{(r)}. \tag{13.2}$$

Tambs-Lyche's goal was to show that centers exist. Ironically, he cited Julia's monograph [**1918a**] for inspiration, perhaps unaware of Julia's claim a year later in [**1919a**] that centers do not exist. In any event, he assumed conditions based on (13.2) which he observed, if true, would imply not only that the iterative family for f is normal on compact subsets of a neighborhood of the fixed point, but would show the existence of invariant curves surrounding it [**1928a**, pp. 103–104]: in other words, it would solve the center problem.

However, novel as this approach is, he was not able to establish the veracity of his assumptions, and so it did not bear fruit, as the following quotation from Hubert Cremer indicates:

> *Here, among the other things, he used a generalization of the binomial coefficients to develop a remarkable recursion formula between the coefficients of the iterates, as well as theorems on the properties of the determinants that arise. On the other hand, a proof for either centers or noncenters along the road Lyche indicated has not yet been achieved.* [**1932a**, p. 298]

13.5. Complex dynamics in Japan: Shimizu and Oka[2]

Tatsujiro Shimizu was a prominent Japanese mathematician both before and after World War II. He was evidently the first—and perhaps only—mathematician of his time to respond to Fatou's short paper [**1922a**], introducing the iteration of algebraic functions, publishing two papers in English in the Japanese journal, *The Proceedings of the Physico-Mathematical Society of Japan* [**1931a, 1931b**].

Near the beginning of the first paper, he remarked,

> *The theory of the iteration of rational functions is deeply developed by Fatou, Julia and others.*
>
> *On the other hand, we know little literatures on that of algebraic functions.*
>
> *We shall give here and in subsequent papers some properties of such an iteration.* [**1931a**, p. 255]

He observed that Fatou's paper developed results via interesting examples, but took a different approach, first stating general results and then concluding his first paper with a few short examples.

Shimizu considered polynomials $P(z, w) = 0$ and developed the branches of the function $w = \Phi(z)$ in a power series around a fixed point satisfying $P(z, z) = 0$ or $\Phi(z) = z$. Mapping the fixed point to the origin, it becomes a branch point, and each branch of Φ is of the form $\phi_k(z) = \alpha_{k,0} + \alpha_{k,1}(z - \alpha_{k,0})^{r_1} + \alpha_{k,2}(z - \alpha_{k,0})^{r_2} + \cdots$, where the r_k are positive rational numbers and $\alpha_{k,j}$ is the jth coefficient of the expansion of the kth branch.

To give a flavor of Shimizu's work, we state the following theorem, which he stated in both papers. We take the more concise version from his second paper:

THEOREM 13.1 ([**1931b**, pp. 292–293]). *Let $\phi(z) = w = a_1 z^{r_1} + \cdots$ be a branch of an algebraic function $w = \Phi(z)$ satisfying $\Phi(0) = 0$, where 0 is the only branch point of Φ in $N_\epsilon(0)$. Assume as well that $0 < |a_1| < 1$ and $r_1 \geq 1$. We can then find a neighborhood $N_{\epsilon'}(0)$ of the origin, where $|\epsilon'| \leq |\epsilon|$, such that the iterates ϕ^k also have no branch points on $N_{\epsilon'}(0)$.*

He then proved the following corollary,

COROLLARY 13.2. *If $|a_1| > 1$ and $0 < r_1 \leq 1$, then we have an analogous result for the inverse of $w = a_1 r_1 + \cdots$.*

---

[2]While the Shimizu papers we quote from are written in English, and while unpublished works that Oka wrote in French are available, none of the authors know Japanese, and so our study is limited by the language barrier. Our hope is that someone with the appropriate language skills will take up the study of iteration in Japan during this time period, since we suspect we are only scratching the surface.

Shimizu's results thus develop an analogy between fixed points of an algebraic function and those of an analytic function.

The case where $1 = r_1$ and $a_1 = e^{i2\pi\theta}$ is more complicated since ϕ_k and its inverse can be multivalued if a_1 is not a root of unity. Much like Fatou's example from [**1922a**], in which he discussed the concept of a chain of iterates of a multifunction (see Definition 10.2 from Section 10.3), Shimizu reduced the iteration of a multivalued function to n-fold compositions of single-valued functions, that we denote $\{y_n : n \geq 0\}$ [**1931a**, p. 260].

Using what he referred to as results from Julia's studies of families of iterates of functions, Shimizu induced a solution to the Schröder functional equation $S \circ u = e^{-i2\pi\theta} \cdot S$, where u_n is a particular transformation of the y_n, which are defined in the previous paragraph. That centers can exist with algebraic functions should not be surprising, since, as we showed near the end of Section 9.2 in Chapter 9, Julia gave an example of an algebraic function with a center in Chapter IV of his monograph [**1918a**, pp. 242–243].

Shimizu also included conditions under which the iteration of the algebraic function $y = \Phi(x)$ forms a finite discontinuous group (see [**1931b**, pp. 294–295]).

Kiyoshi Oka[3] is another prominent Japanese mathematician who became interested in complex dynamics in the 1920s. His doctorate is from Kyoto Imperial University in 1925. Evidently, he was initially interested in physics, but decided upon mathematics in 1923. After receiving his degree, he stayed on for a while as a lecturer at the university.

Oka spent the bulk of his research career studying complex functions of several variables and made fundamental contributions to the field. However, prior to that, Oka was interested in the works of Julia, Ritt, and others. There is an unpublished notebook in his hand dating from 1927 entitled "Julia", and it contains notes concerning such topics as Picard theory, Schwarz's Lemma, and the Weierstrass product theorem on the representation of functions with infinitely many zeros (see the website [**1919-29-W**]).[4] He travelled to France in 1929 and returned to Japan in 1932. During his stay, he wrote a long unpublished paper which grafted Julia's study of permutable functions to Ritt's, entitled, *Fonctions algébriques permutables avec une fonction rationnelle non-linéaire*. There are several versions of this paper, including two drafts from 1930 in notebooks and a more complete, handwritten version [**1930-W**] in French, available online.

It is thought that Oka connected with and perhaps studied under Julia, and presented him with a version of his paper on permutable functions before he returned to Japan in 1932. It is unclear why Oka never published his paper on permutable functions. Based on draft versions from 1930, it is reasonable to suspect that the handwritten version [**1930-W**] dates from 1930 or 1931. We were unable to determine whether he considered this a final version of his paper or a draft.

---

[3]The authors wish to express our profound gratitude to Professor Mitsuhiro Shishikura who alerted us to Oka's interest in iteration and directed us to his unpublished paper on permutable functions [**1930-W**]. He also related what he knew of Oka's time in France, which has informed our discussion of that period of Oka's life.

[4]The Nara Women's University Library maintains both Japanese and English language websites devoted to Oka. There is a wealth of unpublished material including Oka's journals and notebooks, as well as drafts and longer papers that were never published. Our Bibliography contains web addresses where the material to which we refer can be found.

It was around this time that Oka became interested in what would be his life's work—the study of complex functions of several variables. (There is a 1931 draft of a paper on meromorphic functions at the website [**1930-34-W**], and it could be that as his interest in functions of more than one variable subsumed him, he never returned to the task of finishing [**1930-W**].)

These two men may not have been the only two mathematicians in Japan interested in the study of iteration. Shimizu notes that a "Mr. Shinji Nakashima" assisted him with the statements of theorems applying finite discontinuous groups in the study of iteration [**1931b**, p. 294], but we were unable to find more information about him.

It would be interesting to know if Shimizu and Oka were in contact in the 1930s while they were each studying iteration. We have no evidence that they were. The most likely time period for this to have happened would be before Oka went to Paris, since he was in Kyoto while Shimizu was in Tokyo. After his return, Oka was farther to the south in Hiroshima, while Shimizu remained in Tokyo. It would also be interesting to know if they influenced younger mathematicians to study the iteration of complex functions later in their careers.

13.6. Radu Bădescu: Fatou, Julia, and the Abel equations

In 1897 two recently formed Romanian scientific societies, the Societaţea Amicii Ştiinţelor Matematice and the Societatea de Ştiinţe Fizice din Bucuresci[5] merged to form the Societaţea de Ştiinţe din Bucuresci, which soon changed its name to the Societaţea Române de Ştiinţe.[6] After its formation in 1897, the *Societaţea* began to publish the *Buletinul Societaţii Române de Ştiinţe.*[7] After World War I, the society was restructured with several sections, and it put out three scientific journals, including the *Buletinul de Ştiinţe Matematice Pure si Aplicate*,[8] which is still published; see Stefanescu [**2005**].

Spiru Haret (for whom the Haret crater of the moon is named) was instrumental in the development of Romanian mathematics around this time. He was educated in France and is said to be the first Romanian to obtain a Ph.D. from the Sorbonne. His doctoral dissertation, earned in 1878, involved celestial mechanics, and he made important contributions to the n-body problem.

Haret's work is linked to Poincaré's study of celestial mechanics, and the article [**2005**] by Mioc and Stavinschi discusses relations among Poincaré, Haret, and Nicolae Coculescu, another Romanian who received his doctorate in France,[9] asserting that Haret's work was the "beginning of the end ... of exclusively quantitative approaches" to the n-body problem. The authors of [**2005**] also describe a "special" relationship that developed between France and Romania, both mathematical and otherwise, which included post–World War I cultural exchanges.

[5]Respectively, The Friends of the Mathematical Sciences, and The Bucharest Society of Physical Sciences.

[6]Respectively, The Bucharest Society of Science, and the Romanian Society of Science.

[7]The Bulletin of the Romanian Society of Sciences.

[8]The Bulletin of Pure and Applied Mathematics.

[9]Poincaré was evidently listed as a reader of this thesis, which involved the study of the sort of perturbations we discussed in Chapter 4.

Haret returned from France soon after he earned his doctorate and evidently gave up on mathematical research, devoting himself instead to politics and to improving education in Romania. He was the Minister of Public Education three separate times between 1897 and 1910.

Out of this milieu came Radu Bădescu, who published a series of six notices in *Comptes Rendus* between 1929 and 1932, [**1929a, 1929b, 1929c, 1930a, 1930b, 1931**], involving the integro-functional equation, which he treated as a Fredholm of type two, equation

$$\Phi(z,\lambda) - \lambda\left\{F(z)\Phi(\alpha z,\lambda) + \int_a^b K(z,s)\Phi(zs,\lambda)ds\right\} = \Psi(z), \tag{13.3}$$

where F, K, and Ψ are given analytic functions.

He devoted the first three notices to finding a convergent function Φ satisfying (13.3), and the work of Fatou and Julia does not enter in to his solution. However, in the next three notices Bădescu turned his attention to the functional equation

$$\Phi(z) = \lambda F(z)\Phi[\theta(z)] + \Psi(z), \tag{13.4}$$

where Φ is unknown and λ and the other functions are given.

He noted that (13.4) becomes the canonical Abel equation, $\Phi \circ \theta = \Phi + 1$ when $\Psi(z) \equiv -1$ and $\lambda F(z) \equiv 1$. Bădescu observed that where Abel originally solved it via difference equations, Fatou and Julia's work leads to a solution Φ holomorphic on a deleted neighborhood of the fixed point of θ [**1930a**, p. 430]. Equation (1.11) shows how the solution of the Abel equation is derived from the Schröder equation via a logarithm, which causes the singularity at the fixed point.

In [**1931**] Bădescu defined $T_k(z) = \lambda F(z)T^{k-1}(\theta(z))$, where $T_0(z)$ is a given function denoted $\Psi(z)$, which gives rise to the function

$$\Phi(z) = \sum_{k=0}^{\infty} T_k(z) = \sum_{k=0}^{\infty} \lambda^k F(z)F[\theta^1(z)]\cdots F[\theta^{k-1}(z)]\Psi[\theta^k(z)]. \tag{13.5}$$

The reader can verify that Φ is a formal solution to

$$\Phi(z) = \lambda F(z)\Phi[\theta(z)] + \Psi(z).$$

Local convergence follows if the domain of Φ is within a neighborhood of an attracting fixed point of θ, and, under certain conditions, Φ can be extended to a meromorphic function with a countable infinity of simple poles.

In his final paper [**1931**], Bădescu linked his work to that of Fatou and Julia in three additional ways. He first presented the equation $T_k(z) = \lambda F(z)T^{k-1}(\theta(z))$, used to solve (13.4), as a generalization of the concept of iteration. Although he did not say why, setting $\lambda F(z) \equiv 1$ again and letting $T = \theta$ yields conventional iteration, so perhaps this is what he had in mind.

The second link was to extend solutions to the functional equation (13.4) to the case where the attracting cycle of θ is periodic. The final connection was the somewhat cryptic observation that if one replaces λ^k by a coefficient a_k, the series (13.5) is "generalized as envisioned by Julia" [**1931**, pp. 601–602]. We assume that he referred to the "series of iterates" formed in the neighborhood of an attracting fixed point of f,

$$g(z) = \sum_{k=0}^{\infty} a_k f^k(z),$$

that Julia investigated first in a series of notices around 1925, and then again in 1930–31, as described in Section 10.5.

One wonders if Bădescu went a bit out of his way to tie his work to that of Fatou and Julia. While the use of iteration to construct the solution (13.5) to the generalized Abel equation (13.4) is interesting, his mention of Julia and Fatou's solution of the canonical Abel equation is probably unnecessary since it has no bearing on his solution to (13.4). Nonetheless, his invocation of Fatou and Julia suggests that even ten years after its appearance, their work on the iteration of functions was still held in high regard and had not faded from the public view.

13.7. Back to Germany: Hans Töpfer

In 1939 Hans Töpfer wrote what is generally thought to be the first significant mathematical response to Fatou's 1926 paper, on the iteration of transcendental functions. Töpfer's paper, *Über die Iteration der ganzen transzendenten Funktionen insbesondere von* sin z *und* cos z [**1939**],[10] is his doctoral dissertation. He acknowledges the "support and encouragement of Prof. Dr. H. Cremer", who served as his thesis advisor. As we will see, while very insightful, the paper was not error free.

Töpfer wrote a second paper [**1949**] on the iteration of both rational and transcendental functions. These two papers seem to represent not only his entire study of iteration, but the lion's share of his research output. Evidently, he taught at the University of Köln for a portion of his career: according to the Mathematics Genealogy Project, Cremer supervised only one successful Ph.D. candidate, Bruno Klingen, whose dissertation [**1962**] concerned the inverse Schröder function. Klingen also published a paper a few years later on the iteration of entire functions [**1968**].

While our focus in this section is on events prior to 1942, we will give a brief synopsis of his 1949 paper, which is essentially a survey of iteration. In the first half he reviews what is known about iteration, including, as Walter Bergweiler noted,[11] an early version of the Classification Theorem for the Fatou set which catalogs the possibilities for components of the domain of normality.[12] Töpfer's version had the benefit of being written after Siegel's solution of the center problem, so he included the existence of centers as a possibility, but obviously not Herman rings or wandering domains, since their existence would not be discovered for quite some time.

---

[10]"On the iteration of entire transcendental functions, in particular, sin(z) and cos(z)."

[11]We refer the reader to Bergweiler's excellent paper, *Iteration of meromorphic functions* [**1993**, p. 12], which, despite its title, is also an excellent survey of the literature, the mathematics, and the history of the study of the iteration of entire transcendental functions.

[12]In [**1985**] Sullivan showed that every nonempty component of the Fatou set for a rational function is preperiodic; that is, the forward orbit $O^+(z)$ of an arbitrary point in the Fatou set eventually lands in a periodic component of the Fatou set (a component D with $f^p(D) \subseteq D$, with $p \geq 1$). The Sullivan Classification Theorem lists the five possibilities for a periodic component D for a rational function, stated here for the $p = 1$ case (analogous results hold for $p > 1$): (1) all points in D converge to a superattracting fixed point; (2) all points in D converge to an attracting (but not superattracting) fixed point; (3) all points in D converge to a rationally neutral fixed point on its boundary; (4) D contains an irrationally neutral fixed point (i.e., a center) called a Siegel disc, and iteration is conformally equivalent to a rotation of a disc; (5) it is a Herman ring, and iteration is conformally equivalent to an annular rotation. For transcendental functions there is the possibility of a wandering domain (see Section 9.5) and a Baker domain (page 222), but a Herman ring cannot occur (Section 11.8.3). See Blanchard [**1984**, p. 116] for additional details.

The second portion of the paper consists of an investigation of functional equations culminating in a discussion of an iteration operator it that models the behavior of the solution A to the canonical Abel equation $A \circ f = A(z) + 1$ [**1949**, pp. 217–222]. For example, $it(f(z)) = it(z) + 1$.

Töpfer's 1939 paper is similar in structure to Fatou's study of transcendental functions, with the first half being devoted to formal exposition and the second to two examples, in Töpfer's case, the complex sine and cosine functions. He began his formal exposition with a brief review of Fatou's results, although he credited Fatou with a bit more than he actually showed, stating that Fatou proved that the Julia set for transcendental functions is "the derived set of the fixed points" [**1939**, p. 66], as it is for rational functions, which was not proven until Irvine Noel Baker did so in [**1968**].

The first part of Topfer's paper [**1939**] focuses on the structure of the Fatou and Julia sets.[13] His goal for the Fatou set was to explore conditions under which a component of the Fatou set is invariant and/or simply connected. To give a sample of what he did, we will state two of his results. Töpfer viewed the first as an extension of Fatou's theorem that there can be at most two completely invariant domains for a rational function [**1939**, p. 69]:

THEOREM 13.3 (Invariant domains of the Fatou set). *There are at most two completely invariant components of the Fatou set.*

While his theorem is true, it is not optimal: Baker showed in [**1970**] that there can only be one such domain (also see Bergweiler [**1993**, p. 23]).

Since the result is known for rational functions, and therefore polynomials, Töpfer's assumed f was transcendental. Töpfer's argument was very brief, with few details, but it relied upon an application of the criteria for an inverse function f to be hyperbolic given by Lars Ahlfors:[14]

$$\sum_{k=1}^{q} \frac{1}{m_k} < q - 2, \tag{13.6}$$

where the $\{D_k : 1 \leq k \leq q\}$ are simply connected domains, and m_k is the number of branches of the inverse of f in D_k [**1932**, p. 203].

Since the inverse of f is parabolic, Töpfer did not use equation (13.6) directly; instead he reversed it, noting he was "tailoring" Ahlfors' summation to his own needs [**1939**, p. 69] to obtain

$$\sum_{k=1}^{q} \frac{1}{m_k} \geq q - 2, \tag{13.7}$$

which he simply asserted implied the existence of no more than two completely invariant domains.

The scant details he provided indicate that he had in mind a theorem that Ahlfors gave later in the paper [**1932**, p. 206], which states that, given a parabolic

[13] His characterization of these sets was in terms of "compactness" not "normality", claiming that Fréchet's term "compact" had become the preferred term.

[14] Ahlfors said that, informally speaking, inverse functions which have a few singularities are parabolic, while those with many are hyperbolic, and as an example, he stated that an inverse with a line of singularities is hyperbolic. He also gave Rolf Nevanlinna's definition of a parabolic inverse as one that had a finite number of algebraic singularities, and he noted such functions are inverses of entire or rational functions [**1932**, pp. 199–200].

function and three distinct simply connected domains, D_1, D_2, and D_3, there exist no more than three branches of the inverse in at least one of them. Viewing the D_k as components of the Fatou set, he evidently assumed that in one, there were infinitely many branches. His reasoning for this was not clear, but perhaps he assumed one had to include a neighborhood of infinity, on which an infinite number of inverses exist. Using the notation from the summation (13.7), one of the m_k must equal infinity, and one of the possibilities is that the other m_k both equal one, implying the corresponding D_k are completely invariant.

If this is what Töpfer had in mind, he did not apply Ahlfors' results correctly. Töpfer's explicit hypothesis is that the domains of the D_k share a boundary, and, as Baker points out in [**1970**, p. 34], the correct application of Ahlfors' theorem requires that the closures of the domains be distinct.[15] In fairness to Töpfer, Ahlfors' article was intended as a introductory survey, and perhaps for that reason, he did not present his theorem with full precision. In particular, he did not state that the domains could not share a boundary point. That said, clearer statements of Ahlfors' result were available to Töpfer at the time he wrote his paper.

Even if Töpfer was incorrect in his application, he was correct in thinking that Ahlfors' work was applicable to complex dynamics, and many years later it was successfully applied by Baker in [**1968**], albeit to prove a different result, namely that the Julia set for a transcendental function is indeed the closure of the repelling periodic points. See Bergweiler's papers [**1993, 2000**] for more details.

We close with an important result from Töpfer's paper [**1939**, p. 69],

THEOREM 13.4. *The Julia set for an entire transcendental function does not contain isolated Jordan arcs.*[16]

This implies that the Julia set for a transcendental function cannot be a circle or a segment which, as we have seen, occurs with both polynomial and rational functions (see Bergweiler [**1993**, p. 24]).

13.8. Concluding remarks

The impact of the work of Fatou and Julia reached far beyond western Europe, stimulating responses in Scandinavia, Eastern Europe, the U.S., and Japan. Not all the works we discuss in this chapter represent significant contributions to the field: Ritt's work certainly does; Oka's long, unpublished paper on permutable functions may; and Shimizu's and Töpfer's works are referenced in contemporary studies.

But what is perhaps more striking is the variety of the responses, many of which follow up on the dynamical program that Fatou undertook in the 1920s (in particular, the iteration of algebraic and transcendental functions), as well as Bieberbach's paper [**1933**] on the iteration of functions of two variables, a topic (and response) that is beyond the scope of this book. There are also references to the center problem and to Julia's study of series of iterates. One problem that does not seem to have garnered attention during this time is the problem of wandering

[15] Baker attributed this insight to his doctoral student Prodipeswar Bhattacharya without citation. Bhattacharya, an Indian mathematician, received his Ph.D. [**1969**] from the University of London. He also authored [**1971**] and in the early 1970s was associated with the University of Minnesota.

[16] A Jordan arc is isolated in J if there is an open set which contains the entire arc (except its endpoints) but no other points in J.

domains, which may not yet have come to be articulated as a problem in its own right.

Just as the early works of Farkas, Korkine, Formenti, and Rausenberger underscore the fertility of the ideas that Schröder developed, and the French papers of the 1890s reflect the insight of Kœnigs' study, the papers we discuss in this chapter form a measure of the field's vibrancy in the early and mid-20th century, and helped set the stage for subsequent developments.

These papers perhaps serve as an indicator of something else: It is curious that Tambs-Lyche referenced Julia's work on the center problem rather than Fatou's, which was more robust. We also saw that Remoundos, who studied the iteration of algebraic functions, did not cite Fatou's paper, which had been published the year before and was of obvious import to his own study. Ritt of course noted that Julia had earned priority, as the result of the Academy's ruling. This prompts a question: To what degree—if at all—does this reflect the prominence bestowed upon Julia for winning the 1918 *Grand Prix*? A propensity to quote Julia over Fatou was by no means universal, and we saw that Shimizu and Bădescu each quoted both, while Töpfer's cited Fatou almost at the exclusion of Julia.

Part 4

Tying the Future to the Past

APPENDIX A

Report on the *Grand Prix des Sciences Mathématiques* in 1918

Translated from
Comptes rendus hebdomadaires des séances de l'Académie des Sciences, 167, pp. 811–814.

Daniel S. Alexander
Mathematics Department
Drake University, Des Moines, Iowa, U.S.A.

Alessandro Rosa
Software Developer
Brindisi, Italy

SESSION OF DECEMBER 2, 1918

PRIZES AND GRANTS AWARDED IN 1918

REPORTS

MATHEMATICS

PRIZE ENDORSED BY THE STATE
GRAND PRIX OF MATHEMATICAL SCIENCES

(Members of the Commission: Jordan, Appell, Painlevé, Hadamard, Boussinesq, Lecornu; Referees: Émile Picard and Humbert)

The Académie staged a competition for the study of the *iteration* of a substitution and, recalling that only the *local* point of view had been studied, invited the contestants to focus on the *general* point of view.

Existing works about a one variable substitution S, $z_1 = \varphi(z)$, in particular the fundamental works of Kœnigs, developed the notion of an *attracting point*: if ζ is a point fixed by S or one of its iterates (*the invariant point*) and if the corresponding value, called the *multiplier*, is *less than* one in modulus, then the successive transformations (*consequents*) of a point z very close to ζ tend to ζ, or tend periodically to p points, one of which is ζ, and all the others are its first $(p-1)$ consequents.

These initial results lead quickly to several problems: Are the points of attraction themselves limited in number? What is the exact domain of attraction of such a point? What is the division of the plane associated to a given function $\varphi(z)$?

These fundamental questions, which had before been addressed only in a notice by Fatou (October 1906) where the author showed through examples that the regions of the division could be bounded by nonanalytic curves, highlight the difficulties and the complexity of the question.

Finally, from a different point of view Poincaré established that in some cases, one could associate with S a meromorphic function $\theta(u)$ defined in the entire plane, such that if $z = \theta(u)$, then $z_1 = \theta(su)$, where s is a constant whose modulus is *larger than* 1, which reduces the study of the iteration to that of $\theta(u)$; but no application of this method of *parametric iteration* had been given.

Three entries for the contest were deposited with the Secretary; the Commission had kept only those by LATTÈS, professor at the University of Toulouse and by JULIA, Lieutenant of 34th Infantry Regiment, winner of the Bordin Prize in 1917.

The work by Lattès is an application of Poincaré's ideas, and at the same time is based on a theory by E. Picard.

Lattès established the existence of the Poincaré function for a *rational* substitution of one variable for a very extended class of functions but without drawing any general conclusions about iteration; then, abandoning this general point of view, he examined the particular cases where the Poincaré function was $\cos z$, $\tan z$ or $\wp u$. The equation $z_1 = \varphi(z)$ is then an integer *multiple* of one of these functions and, in each case he determined and analyzed the *set* of the consequents of a point, as well as its derived set, by an ingenious method.

He then turned his attention to the case of a rational function of *two variables* and showed that if the roots of a quadratic *equation involving the multipliers* are distinct and larger than one in modulus at an invariant point, then parametric iteration can be achieved with the help of two *meromorphic* functions in two variables.

Applying these results to a Cremona transformation, it follows, at least in some cases, that the limit of the consequents of any point in the plane is a fixed point, and that its antecedents likewise converge to another point.

The portion of the memoir about substitutions of two variables is without a doubt its most interesting and novel part; it contains elegant results, notably about Cremona transformations, which although a little fragmentary, open up a field of research that could become very rich; the author proved to be a skillful and wise spirit, and he could have taken his discoveries further if earlier this year, death had not taken him still young and in full possession of his talent.

Julia's memoir concerned only rational functions in one variable; he systematically examined not only the attracting invariant points, but also the invariant points where the modulus of the multiplier is *larger* than one; their fundamental property is that they are *repelling points.* More precisely, if one of these points is surrounded by an arbitrarily small domain, the consequents of the interior of this domain include all the points of the plane except at most one or two.

The analogy of this statement with a theorem of Picard is nothing mysterious: Julia's proof is based here, as so often in the rest of the memoir, upon the beautiful theory of the normal families of Montel, a theory which is known to be closely linked to Picard's theorem.

The perfect set E', the derived set of the set E of the repelling points, has a fundamental role in the study of iteration: Julia studied its properties and observed that it could be discontinuous or a linear continuum, and if it one of its parts is *superficiel*[1] it must cover the entire plane.

The most interesting case is where E' is linear: The set E' then partitions the plane into regions D which enjoy the characteristic property that if the point z is

[1] A rough translation of "*superficiel*" is "contains an area".

inside one of the regions, then the limit points of the *set* of its consequents depends *analytically* on z. From this property, and a reciprocal one, the author deduced the fundamental result that the points of E' are the *essential singularities* of the limit functions of the *sequence* $\varphi(z)$, $\varphi[\varphi(z)]$, ..., etc., and that, if the *behavior* of this sequence is known when z is in the interior of an arbitrarily small domain inside D, the related behavior is known in all of D as well.

The last statement serves as well as a link between the local study of iteration and the general one: An *attracting* point z_0 belongs to a region D_0, and the consequents of any point in the region D_0 have z_0 as a limit point; D_0 is the *immediate* domain of convergence of the attracting point.

Another fundamental consequence is one that Julia established by two distinct methods: *every immediate domain includes at least one critical point of the algebraic inverse of* $\varphi(z)$, and, as a corollary, *the number of the attracting points is finite.* This is the response, possibly unexpected, to a question posed by Kœnigs.

Julia also determined the *total* domain of convergence for an attracting point, and he found a necessary and sufficient condition that it is also the immediate one; finally, he illustrated this theory through well-chosen examples. Usually, the set E' must be a simple Jordan curve; sometimes it is a closed continuous line with fixed points everywhere dense, a special case the author described by geometrically constructing, *a priori*, a continuous set which enjoys analogous properties.

The last part of the memoir investigates the case of invariant points whose multiplier is one in modulus, which establishes in full generality the proposition that the number of attracting points is *finite.*

In the course of his research Julia consigned his results in sealed envelopes deposited at the Académie; after depositing the last envelope in December 1917, a well-known geometer, Fatou, to whom the theory of iteration was already in debt due to his progress along new avenues, published many of the same results in *Comptes Rendus*, which he obtained on his own by the same methods, utilizing Montel's theory of normal families: This is not the first time in the history of science that two scientists of merit simultaneously attained the same discoveries by identical methods.

The memoir by Julia bears the mark of a mathematical spirit of the highest order, one with the vigor to understand problems in all their generality and to follow their consequences no matter where they lead; he displays as well a profound knowledge of the results and methods of modern analysis, together with a remarkable aptitude in applying them. He made decisive progress in his investigation of iteration and showed once again how the most subtle properties from both the theory of sets and the study of Jordan curves can be naturally introduced into the study of analysis, and especially rational functions.

Thus, the Commission decides unanimously to award the *Grand Prix of Mathematical Sciences* to GASTON JULIA; the Commission also confers a most honorable mention to the work of SAMUEL LATTÈS.

The Académie adopted the statement of the Commission.

APPENDIX B

A History of Normal Families

Joel L. Schiff
Department of Mathematics
University of Auckland, New Zealand

The concept of normal families has its roots in the Arzelà–Ascoli Theorem of the late 19th century whereby from any locally bounded sequence of equicontinuous functions $\{f_n\}$, a subsequence can be extracted that converges locally uniformly to a limit function f. The Frenchman, Paul Montel, realized that if the family of functions $\{f_n\}$ were analytic, then the property of local boundedness already implies equicontinuity and in his first of many papers [**1907**] on the subject, proved what has become known as Montel's Theorem:

If $\mathcal{F}$ is a locally bounded family of analytic functions on a domain Ω in the complex plane $\mathbb{C}$, then every sequence $\{f_n\} \subseteq \mathcal{F}$ contains a subsequence that converges either to a limit function f or to ∞ uniformly on compact subsets of Ω.

In this case, the function f is necessarily analytic by a theorem of Weierstrass and Montel's Theorem implies that if the family $\mathcal{F}$ is locally bounded and closed in the topology of uniform convergence on compact subsets, it is sequentially compact. At the heart of the proof lies the Bolzano–Weierstrass property coupled with the diagonalization method. The theorem was independently established by Koebe in 1908. The words "*famille normale*" appeared later in [**1911**] to describe any family of analytic functions that satisfied the conclusion of Montel's Theorem. This theorem was the beginning of a very rich vein in mathematical thinking that was mined by Montel with gusto so that 20 years later, his influential tome, *Leçons sur les familles normales des fonctions analytiques et leurs applications* [**1927**], spanned nearly 300 pages. Many well-known families turn out to be normal owing to an inherent locally bounded growth property: the *schlicht* functions $\mathcal{S}$ defined on the unit disk U by

$$\mathcal{S} = \{f : f \text{ analytic, univalent in } U,\ f(0) = 0,\ f'(0) = 1\},$$

as well as the analytic functions defined on U satisfying the

$$m(r, f) = \frac{1}{2\pi} \int_0^{2\pi} \log^+ \left|f(re^{i\theta})\right| d\theta \leq M, \quad 0 < r < 1,$$

among others.

What then does a nonnormal family look like? If one considers the family $\mathcal{F} = \{f_n(z) = nz : n = 1, 2, 3, \dots\}$, then $f_n(0) \to 0$, however $f_n(z) \to \infty$ for any $z \neq 0$. Therefore $\mathcal{F}$ cannot be normal in any domain containing the origin.

One of the great virtues of the concept of a normal family is its power in proving many deep results. For example, the Riemann Mapping theorem, which is part of the foundation of complex analysis, has a standard proof based on normal families.

PAUL MONTEL

FIGURE B.1. Photo courtesy of Académie des sciences—Institut de France.

Montel's Theorem turned out to be equivalent to the Vitali–Porter Theorem which established uniform convergence on compacta from merely pointwise convergence of a locally bounded sequence of analytic functions. The local boundedness condition in Vitali–Porter was replaced by Carathéodory and Landau [**1911**] with the condition that each function omitted two given fixed values. Herein lies the seed of the *Critère fondamental* of Montel's celebrated paper [**1912a**], which has turned out to have so many wonderful and far-reaching consequences:

Let $\mathcal{F}$ be the family of analytic functions on a domain Ω which omit two fixed values. Then $\mathcal{F}$ is normal in Ω.

There are many proofs of the theorem from differing platforms of complex function theory. The original proof of Montel employs the elliptic modular function. From this seminal result easily flow Picard's Little and Big Theorems, as well as the Schottky and Julia theorems, covering theorems of Valiron, Fekete, Montel, and so many more. Of course the notion of normal families need not be restricted to analytic functions but carries over to meromorphic functions by simply switching metrics from the usual metric in the former case to the chordal metric $\chi(\cdot,\cdot)$ in the latter, given by

$$\chi(z_1, z_2) = \frac{|z_1 - z_2|}{\sqrt{1+|z_1|^2}\sqrt{1+|z_2|^2}}, \qquad z_1,\ z_2 \in \mathbb{C},$$
$$\chi(z_1, \infty) = \frac{1}{\sqrt{1+|z_1|^2}}.$$

Converging uniformly in the χ metric is termed *spherically uniformly*, and a family $\mathcal{F}$ of meromorphic functions is *normal* if every sequence $\{f_n\} \subseteq \mathcal{F}$ contains a subsequence that converges spherically uniformly on compacta. The magic number in the *Critère fondamental* for families of meromorphic functions is now three, that is to say if the family $\mathcal{F}$ of meromorphic functions omits three distinct values, then it is normal. However, for families of meromorphic functions, the concept of local boundedness is not entirely relevant. But what we have instead is a condition on the spherical derivative

$$f^{\#}(z) = \frac{|f'(z)|}{1+|f(z)|^2}$$

(with an obvious modification when $f(z) = \infty$), so that if $f^{\#}(z)$ is locally bounded, then the family is normal (Marty's theorem from [**1931**]). This condition was refined by Royden and subsequently generalized by Chen and Gu [**1993**], and Wilhelm Schwick [**1989**] as well as others. For the last word on this topic, see the survey paper by Lawrence Zalcman [**1998**] which also contains a remarkably simple proof of the *Critère fondamental* among other gems. A guiding principle in the study of normal families has been the Bloch Principle, namely that a family of analytic (meromorphic) functions that have a common property $\mathcal{P}$ in a domain Ω will *in general* be a normal family if an analytic (meromorphic) function in the entire complex plane possessing property $\mathcal{P}$ is reduced to a constant. The property of omitting two (three) values in the *Critère fondamental* is one such example by Picard's Little Theorem. Indeed, the Bloch Principle has been very successful in leading the way to the discovery of new criteria for normality. For example, a meromorphic function $f(z)$ in $\mathbb{C}$ satisfying the condition

$$\sum_{\upsilon=1}^{q}\left(1 - \frac{1}{m_\upsilon}\right) > 2,$$

where each zero of $f(z) - a_\nu$ ($a_\nu \in \mathbb{C}$) has multiplicity m_υ (≥ 2), $\upsilon = 1, 2, \ldots, q$, reduces to a constant by Nevanlinna's Second Fundamental Theorem. The family of all such functions on a domain $\Omega \subseteq \mathbb{C}$ is normal there. Another example is the classical result that any entire function for fixed $n \geq 1$ that satisfies $f(z) \neq 0$, $f^{(n)}(z) \neq 1$, reduces to a constant, with Hayman [**1959**] establishing the same for meromorphic functions in $\mathbb{C}$. It was proved by Miranda [**1935**] that a family of analytic functions on a domain D that satisfy this condition is normal, with the meromorphic version for normal families being established by Gu in [**1979**].

Of course the Bloch Principle as stated is not always true. A number of nontrivial counterexamples were presented by Rubel [**1986**], but perhaps the simplest is the property that "f is bounded". Then any entire functions with this property is constant by Liouville's Theorem, and each function of the family $\{nz\}$ satisfies this condition on the unit disk, but as we have seen above the family is not normal there. However, there is a formalized version initiated by Abraham Robinson and completed by Zalcman that *is* true. This so-called *Robinson–Zalcman Heuristic Principle* can establish the normality of numerous families with merely routine verification. The proof of the normality in the aforementioned example is just one of many. The Robinson–Zalcman Heuristic Principle depends on a lemma by Zalcman [**1975**], to wit:

Let $\mathcal{F}$ be a family of analytic (meromorphic) functions in the open unit disk U. Then $\mathcal{F}$ is not normal in U if and only if there exist (i) a number r with $0 < r < 1$; (ii) points z_n satisfying $|z_n| < r$; (iii) functions $f_n \in \mathcal{F}$; (iv) positive numbers $\rho_n \to 0$ as $n \to \infty$; such that

$$f(z_n + \rho_n\zeta) \to g(\zeta) \quad \text{as} \quad n \to \infty,$$

uniformly (spherically uniformly) on compact subsets of $\mathbb{C}$, where g is a nonconstant entire (meromorphic) function in $\mathbb{C}$.

In spite of its technical appearance, the Zalcman Lemma has become one of the cornerstones of the modern treatment of normal families. Indeed, Bergweiler [**1998**] gives a stunning new proof of the famous Ahlfors Five Islands Theorem (as well as Nevanlinna's theorem that a meromorphic function on $\mathbb{C}$ has at most four totally ramified values) based upon the Zalcman Lemma. Moreover, the latter has been extended by Pang [**1989, 1990**] in order to handle conditions involving derivatives, making it even more versatile. Just prior to Zalcman's insight, David Drasin [**1969**] brought the formidable powers of Nevanlinna theory to bear upon the question of normality. Within this context he gave another proof of the *Critère fondamental* as well as such delicious items as the Bloch Principle: *Let $\mathcal{F}$ be a family of analytic functions in the open unit disk. If, for each $f \in \mathcal{F}$,*

$$f'(z) - af(z)^n \neq b, \quad (a \neq 0,\ n \geq 3)$$

for some $b \in \mathbb{C}$, then $\mathcal{F}$ is a normal family.

This is another example of the Bloch Principle in action as Hayman [**1959**] had shown that any entire function satisfying the above condition was constant. This same is true for meromorphic functions and $n \geq 5$ (again [**1959**]) and the normal family equivalent was established by Langley [**1984**], S. Li [**1984**], and X. Li [**1985**]. Moreover, in spite of the fact that Mues [**1979**] gave counterexamples to the Hayman result when $n = 3, 4$, Pang [**1990**] proved the normal family version for $n \geq 4$, with Chen and Fang [**1995**] as well as Zalcman [**1995**] contributing normal family versions for $n = 3$. Further developments saw Schwick [**1989**] couple the Nevanlinna theory of the Drasin approach to the Zalcman Lemma to prove further results suggested by the Bloch Principle. Some consequences of the Bloch Principle have long histories. For example, if $f(z)$ is entire and satisfies

$$f'(z)f(z)^n \neq 1,$$

for all $z \in \mathbb{C}$, then f is identically constant. This was established by Hayman [**1959**] for $n \geq 2$ and Clunie for $n = 1$. The normal family equivalent for analytic functions was established by Yang and Chang [**1965**] for $n \geq 2$, with the case $n = 1$ considered by Oshkin [**1982**], and it is also a consequence of a result of Schwick. An extension was provided by Hua and Chen as well as Pang and Zalcman. The story for meromorphic functions satisfying this condition is similar, with Yang and Chang establishing the case $n \geq 5$, Gu doing $n = 3, 4$, and Pang [**1989**] giving a proof for $n \geq 2$ via the Heuristic Principal with the case $n = 1$ being polished off simultaneously by Bergweiler and Eremenko [**1995**] as well as Chen and Fang [**1995**] and Zalcman [**1995**]. Much of the preceding work as well as that surrounding the foregoing example answered questions posed by Hayman in his short monograph on outstanding problems in function theory [**1967**]. Many normal family consequences of the Bloch Principle have been established by the "Chinese School" which has its origins in the works of Hiong and Chuang, with the principal players being Yang, Chang, Chen, Li, Hua, Pang, Gu, and Fang. One of the most significant applications of the theory of normal families is in the field of complex dynamical systems initiated by Julia [**1918a**] and Fatou [**1919-1920**]. In fact, the iterations of a rational map

$$R(z) = \frac{P(z)}{Q(z)} : \widehat{\mathbb{C}} \to \widehat{\mathbb{C}},$$

where $P(z)$ and $Q(z)$ are polynomials with complex coefficients and no common factors, form a family of meromorphic functions, and the *Julia set* (J_R) is exactly the set of points where this family is *not* normal (*normal at a point* means normal in a neighborhood of the point). The *Fatou set* (F_R) is the complement of J_R with respect to the Riemann sphere. The structure of the Julia set has provided endless fascination for mathematicians and lay persons alike. The theory of iterations of entire functions was initiated by Fatou [**1926**], with one of the highlights of the theory being the 1968 theorem of Baker that the repulsive fixed points are dense in the corresponding Julia set (as is also the case for rational maps). Of course, at the heart of the proof we find normal families. A subsequent proof by Schwick [**1997**] also employs normal families in the form of the Zalcman Lemma. Normal families turn up in the theory of discontinuous groups and have spawned the notion of *normal* (meromorphic) *functions* due to Noshiro [**1938**] that are characterized by the growth of their spherical derivative,

$$\sup_{z \in U} \left(1 - |z|^2\right) f^{\#}(z) < \infty, \qquad |z| < 1.$$

Of necessity, this is only an abbreviated version of the multitude of events that unfolded since Montel's original inspiration in 1907, and new results are still forthcoming. Most recently, Bargmann, Bonk, Hinkkanen, and Martin proved [**1999**] the interesting variant of the *Critère fondamental* that if a family of meromorphic functions avoids the values of three continuous functions (that avoid each other), then the family is normal. So the story continues.

According to Roger Penrose, there is some mathematics that is artificially created by Man, and other mathematics that is an intrinsic part of the universe that is revealed by Man. I would like to think that normal families are the latter.

APPENDIX C

Singular Lines of Analytic Functions

Aimo Hinkkanen
Department of Mathematics
University of Illinois at Urbana–Champaign,
Urbana, Illinois, U.S.A.

In 1913, Gaston Julia considered analytic functions that cannot be analytically continued across certain arcs. So Julia's paper [**1913**] appeared well before the global theory of complex dynamics based on normal families was developed by Pierre Fatou and by Julia, which began to appear in late 1917. Here we discuss Julia's paper and the developments leading to it.

We first review what was known about the issues of interest by the time Julia wrote the paper [**1913**]. Julia referred to a number of papers by Borel [**1894, 1895**], Goursat [**1887, 1893**], Picard [**1881**], Poincaré [**1892b**], and Pringsheim [**1893, 1894a**] We next explain what those papers had to offer.

A series of rational functions may define different analytic functions in different domains. In case of a power series, the situation is simple: the series converges in an open disk centered at the basepoint and diverges outside the corresponding closed disk; it may also converge at some or all points on the boundary circle. In the late 19th century, several authors were interested in situations related to more general series. They used the term "*espace lacunaire*" (*lacunary space*) to indicate the subset of the complex plane where the series defines no analytic function. So for a power series, the exterior of the disk of convergence is a lacunary space. Of course, this does not prevent the analytic function defined by the series in its disk of convergence from being analytically continuable beyond the disk. It may be represented, for example, by a Taylor series with a basepoint in the original disk of convergence close to its boundary in another disk that partially extends outside the original disk of convergence. The consideration of such Taylor series is an essential part of the theory of analytic continuation and also plays a central role in Julia's paper [**1913**] and in its predecessors.

Several authors were interested in what may happen in the case of more general series than a power series. For this purpose, they considered series of the following form (only Picard [**1881**] considered a Weierstrass-type product instead). Suppose that a_n is a sequence of nonzero complex numbers with $\sum_n |a_n| < +\infty$, and that b_n is a sequence of distinct complex numbers. Let L be the set of all cluster points of the sequence b_n. Then the series

$$f(z) = \sum_{n=1}^{\infty} \frac{a_n}{z - b_n}$$

defines a meromorphic function in each component of the complement Ω of L.

These authors considered the situation where L is a bounded Jordan arc or Jordan curve, or a collection of such arcs; for simplicity, let us assume now that L is connected. Suppose that γ is a Jordan curve such that either $L = \gamma$ or L is a proper closed subarc of γ. Let the components of the complement of γ be D_1 and D_2, with D_1 being the interior component. Let the series f define a meromorphic function f_j in D_j, for $j = 1, 2$. The authors considered the question of whether one of f_1 and f_2, say f_2, can be analytically continued across L, or a point of L, from D_2 into (the union of D_2 and a subdomain of) D_1. If so, then the continuation certainly need not be the same as f_1. The question is whether such continuation is possible at all. If not, one would call L a *singular line* for f_2. (For many considerations, the distinction between the bounded and unbounded components of the complement of γ is irrelevant. Pringsheim [**1893**] considered the possibility of continuing f_1 into D_2 when L is the unit circle.)

Goursat [**1887**, pp. 110–112] developed a basic technique for studying this question. He assumed that z_0 lies in a component D_2 of Ω and assumed that there is $b_k \in L$ with $|z_0 - b_k|$ equal to the distance of z_0 from L and, further, that the disk centered at z_0 and of radius $|z_0 - b_k|$ is contained in D_2. He expanded each term $a_n/(z - b_n)$ as a Taylor series at z_0. When considering the possibility of analytic continuation, finitely many values of n can be omitted. The terms $a_n/(z - b_n)$ are not quite in a symmetric position since $\sum_n |a_n| < +\infty$. Thus one may pick one value of n, say $n = k$, omit finitely many values of n, and get to a situation where, in a suitable sense, the sum of the Taylor series of the remaining terms $a_n/(z - b_n)$ is dominated by the Taylor series of $a_k/(z - b_k)$. One can now show that the radius of convergence of the Taylor series of f at z_0 cannot exceed the distance of z_0 from L as it would, at least for some such z_0, if there were a point $z_1 \in L$ with a disk neighborhood B such that f can be continued analytically to $B \cup D_2$.

Poincaré [**1892b**] assumed that all the points b_n lie in $D_1 \cup L$ and that, in addition, the points b_n are dense in the arc L. We consider $z_0 \in D_2$ and may obviously assume that z_0 is close to L. Poincaré expanded each term $a_n/(z - b_n)$ as a Taylor series at z_0 as Goursat [**1887**] did and then made use of the assumption that $\sum_n |a_n| < +\infty$. The remaining details differ slightly from Goursat's argument. But again, Poincaré [**1892b**, pp. 203–220] showed that if there is $b_k \in L$ with $|z_0 - b_k|$ equal to the distance of z_0 from L and if the disk centered at z_0 and of radius $|z_0 - b_k|$ is contained in D_2, then f_2 cannot be continued beyond b_k.

The above extra assumptions mean that among the points b_n, there is one, say b_k, that is closest to z_0. What if there is no such b_k for a given $z_0 \in D_2$? Poincaré [**1893**, p. 206], repeating a reasoning of Goursat [**1887**, pp. 112–113], offered an argument based on the assumption that L is a sufficiently nice arc: for each $z_1 \in L$, we must be able to draw a normal to L from z_1 into D_2 so that if $z_2 \in D_2$ is on this normal close enough to z_1, then the open disk centered at z_2 and of radius $|z_1 - z_2|$ is contained in D_2. Then if $z_1 \in L$ is a point of L closest to z_0, choose some b_k in L close to z_1, draw a normal to L at b_k and pick a point $z_2 \in D_2$ on that normal close to b_k. Now the previous argument shows that the Taylor series of f_2 at z_2 has radius of convergence $|b_k - z_2|$. Since b_k could have been arbitrarily close to z_1, it follows that the Taylor series of f_2 at z_0 has radius of convergence $|z_1 - z_0|$.

The problem remains if, for example, L contains no points b_n (or if L is not a sufficiently nice arc; but we omit that problem). Pringsheim [**1893**, pp. 167–169] claimed that even then f_2 cannot be analytically continued beyond L. However, he effectively assumed that the hypothetical continuation of f_2 beyond a point $z_1 \in L$ would be given by the same series f, which need not be the case. This gap invalidates his argument. Pringheim's principal achievement in [**1894a**, pp. 170–176] is to show that if D_1 is the unit disk, $L = \gamma$ is the unit circle, and all the $b_n \in D_2$ (so $b_n \notin L$) and if the a_n tend to zero fast enough, then the series $f = f_1$ and all its derivatives extend continuously from D_1 to L and the corresponding series converge uniformly in the closed unit disk.

Julia [**1913**] referred to, but did not discuss in detail, the criticism of Pringsheim's paper [**1893**] presented by Borel [**1895**, pp. 32–34]. Borel started [**1895**, p. 2] by noting that the integral in the Cauchy integral formula gives an example of an *analytic expression* that defines an analytic function g inside a Jordan curve and defines the constant function zero outside that curve. The function zero can be analytically continued to the plane but does not (usually) coincide with g inside the curve. Borel suggested that there exist such examples given by series of the form f. By Goursat's argument, we then must have the points b_n clustering to L but not lying on L. Borel thought that such an example could be constructed as follows. Choose positive real numbers r_1 and r_2 with $0 < r_1 < r_2 < +\infty$. Choose the numbers a_n and b_n so that $\sum_n |a_n| < +\infty$, $r_1 < |b_n| < r_2$, and, most importantly, $\sum_n a_n/b_n^p = 0$ for every positive integer p. Then $f(z) \equiv 0$ for $|z| < r_1$, and Borel showed that also $f(z) \equiv 0$ for $|z| > r_2$. Thus f restricted to appropriate components of Ω has an analytic continuation to the whole plane even if the series defining f has infinitely many singularities. But Borel did not explain exactly how one may choose the a_n and b_n so as to satisfy these conditions. Thus his discussion leaves something to be desired, to say the least. We mention in passing that Borel [**1895**] also studied series of the form

$$\sum_{n=1}^{\infty} \frac{a_n}{(z - b_n)^{m_n}},$$

where the m_n are positive integers, as well as functions that have singularities due to the presence of logarithms.

In the paper [**1913**], Julia considered the case that has remained open, where the b_n lie in D_1 but not on L. (It is irrelevant in Julia's paper whether D_1 is bounded or not.) Now every point of L is a cluster point of the poles of f_1 so that f_1 cannot be continued beyond any point of L. The question is whether f_2 can be analytically continued beyond some point of L. Julia first gave a heuristic argument showing that f_2 can be analytically continued beyond L only under circumstances that appear rather exceptional. If this happens, a slight change in the series f will cause f_2 to no longer be analytically continuable beyond L. For example, we may keep the points b_n unchanged and multiply certain numbers a_n by the same complex number different from zero and one. Then Julia constructed particular examples where he rigorously proves that such a continuation is not possible beyond any point of L. Here he assumed that L is at least a C^2-arc. Some parts of the construction would be more carefully discussed these days, but it seems clear that his idea works.

Julia also noted in the spirit of Pringsheim [**1893**] that if, in his construction, the numbers a_n are chosen so that they tend to zero sufficiently rapidly as $n \to \infty$, in a precise sense that he quantifies, then the function f_2 and each of its derivatives $f_2^{(n)}$ extend continuously to each point of L, and are represented on L by the series obtained by differentiating the series for $f = f_2$. In fact, these series then converge absolutely and uniformly in $D_2 \cup L$. Roughly speaking, $|a_n|$ must be small enough in relation to the distance of b_n from L (recall that all the $b_n \in D_1$)

The method of Julia starts with that of Goursat [**1887**, pp. 110–112] but is much more careful and involved. He expanded each term $a_n/(z - b_n)$ as a Taylor series at a point z_0 in D_2 close to L. When considering the possibility of analytic continuation, finitely many values of n can be omitted. Again one may pick one value of n, say $n = 1$ after reindexing, omit finitely many values of n, and get to a situation where, in a suitable sense, the sum of the Taylor series of the remaining terms $a_n/(z - b_n)$ is dominated by the Taylor series of $a_1/(z - b_1)$. To achieve this even when b_1 need not be the point b_n closest to z_0 (such a closest point need not exist at all), Julia chooses carefully the points b_n as well as the numbers a_n. The latter choice is a novel feature due to Julia, and the arguments of the numbers a_n are also of special importance. With this choice, Julia showed that the radius of convergence of the Taylor series of f_2 at z_0 cannot exceed the distance of z_0 from L as it would, at least for some such z_0, if there were a point $z_1 \in L$ with a disk neighborhood B such that f_2 can be continued analytically to $D_2 \cup B$.

In the study of complex dynamics of rational functions F of degree at least 2, analytic functions defined in a domain D that cannot be analytically continued beyond D make an appearance. Suppose that F has an attracting fixed point α and that D is the component of the Fatou set of F containing α. Suppose that D is simply connected. As is well known, there is an analytic function h, defined initially in a neighborhood B of α, such that with $\lambda = F'(\alpha)$, $0 < |\lambda| < 1$, we have $h(F(z)) = \lambda h(z)$ for all $z \in B$. Thus if $n \geq 1$, we have $h(F^{\circ n}(z)) = \lambda^n h(z)$ for all z close enough to α. This last equation can be used to continue h analytically to all of D since $F^{\circ n}(z) \to \alpha$ as $n \to \infty$, locally uniformly for $z \in D$. However, h cannot be continued meromorphically to any neighborhood of any point of ∂D. For if $z \in \partial D$, there is a sequence w_n of points of D tending to z such that $h(w_n) \to \infty$. For example, the w_n can be inverse image points under suitable iterates of F of the same point in $B \setminus \{\alpha\}$. The set ∂D is usually a fractal and not a nice arc.

In the paper [**1913**], Julia did not offer any direct motivation for studying functions with singular lines, and the series he (and others before him) considered do not seem to be exactly what one would end up with in complex dynamics. It seems that the possibility of constructing analytic functions with singular lines was a topic of some intellectual interest and challenge at the time, with other, natural examples showing up later as a consequence of new branches of complex analysis being created.

APPENDIX D

Kleinian Groups

Katsuhiko Matsuzaki
Department of Mathematics, School of Education
Waseda University, Japan

A study of the dynamics for a discontinuous group of linear fractional transformations on the Riemann sphere, often called Kleinian groups, was begun by Robert Fricke and Felix Klein in [**1911**]. Fricke and Klein studied the set of accumulation points (the limit set) of the orbit under the action of a discontinuous group. Its complement, where the action is discontinuous, is called the ordinary set. In a recent monograph [**2002**], *Indra's Pearls, The Vision of Felix Klein*, by Mumford, Series, and Wright, many beautiful fractal pictures of limit sets are exhibited with detailed expositions including Fricke and Klein's original ones.

The properties of the limit set and the ordinary set, for Kleinian groups, have strong similarities with those for the Julia set and the Fatou set for the iterations of a rational map. For example, chaotic behavior and density of the orbit on the limit (Julia) set are the same. Actually, by means of the concept of "normal family" introduced by Montel, these two sets can be defined in a unified way. The ordinary (Fatou) set is a set of points in a neighborhood of which the linear fractional transformations $\{g_n\}$ in a discontinuous group G, or the iterated maps $\{f^n\}$ of a rational map f, constitute a normal family. Hence several properties of these holomorphic dynamics can be explained simultaneously by the normal family arguments.

The theory of discontinuous groups developed in complex analysis as a branch of the theory of Teichmüller spaces. A related powerful tool is the quasi-conformal map which gives deformation on the dynamics. One of the monumental results in this field was the Ahlfors finiteness theorem, obtained in the middle 1960s. About two decades later, a new revolution in the theory of iteration of rational maps was brought about by Dennis Sullivan, who imported quasi-conformal maps from discontinuous group theories and proved the nonwandering domain theorem with a method similar to that used in the proof of the Ahlfors finiteness theorem.

After this success, the similarity or the analogy between discontinuous groups and the iteration of rational maps was recognized again through more sophisticated analytic concepts, and it provided a sort of principle for new researches in these fields. A list of the correspondence of concepts and theorems between the two holomorphic dynamics is often called *Sullivan's Dictionary* (Sullivan [**1985**]). The number of items in this dictionary is still increasing.

APPENDIX E

Curves of Julia

Sanford L. Segal

Department of Mathematics
University of Rochester, New York, U.S.A.

Gaston Julia's three memoirs "*On some new properties of entire and meromorphic functions*" [**1919-21**] followed immediately after his long and famous memoir [**1918a**] on the iteration of rational functions.

The first memoir begins by asking a new question in the "Picard circle of ideas": What could one say about values taken on in an angle with vertex at an isolated essential singularity? The results of course are what we today call *lines of Julia, curves of Julia*, and *Julia sequences.* Picard first proved his result on omitted values in [**1879**] using the elliptic modular function and Landau's generalization appeared for the first time in [**1904**]. However, Montel in [**1912a**] and [**1916**] developed the theory of normal families of functions and used it to prove Picard's theorems.

Julia's memoirs are essentially explorations of what one can obtain using Montel's (recent) theory.

The first memoir proves the existence of lines of Julia and curves of Julia for entire and meromorphic functions. As far as the author knows, these are still essentially the only proofs of these results. The emphasis in the first memoir is on the existence of points in every "ring-shaped region" (*couronne*—not necessarily circular) on which a certain family of functions is not normal which leads to the creation of a "privileged curve". The origin O is taken as the isolated essential singularity and the ring-shaped regions are around O; the family of functions is $f_t(z) = f(z\sigma(t))$, where $f(z)$ is an arbitrary entire function. Julia observes that this result on ring-shaped regions depends on their being around the origin and the family may be normal in various regions separated by points where it is not normal. For example $e^{zt} = f_t(z)$ is normal to the right of the imaginary axis where it goes to ∞, and to the left of the imaginary axis where it goes to zero, but is *not* normal on the imaginary axis. On the other hand, it is also possible that the family of functions may not be normal in any point of the z-plane. An example is the Weierstrass σ-function.

If we call the set of points where the family $f_t(z)$ is not normal, E, then if the entire function has an omitted value, E is a perfect set.

Iversen had shown three years before in [**1918**] (refer also to his thesis [**1914**]) that every omitted value of an entire or meromorphic function is an asymptotic value.

For meromorphic functions $\phi(z)$ the analogous family of functions $\phi_t(z)$ is formed, and it is assumed ϕ has an asymptotic value. Analogous results and examples hold for meromorphic functions with an asymptotic value as for entire functions.

However, a meromorphic function may not have any asymptotic values. In this case appealing to the Riemann surface of the inverse function, it is shown that the plane can be decomposed into infinitely many polygons in each of which the meromorphic function takes every value, for example, for an elliptic function without asymptotic values the polygons are the period-parallelograms.

The first memoir, by using elliptic functions, closes with an example of a function for which 0 and ∞ are isolated essential singularities but that has no asymptotic values as z tends towards 0 or ∞, and for which the family $f_t(z) = f(zt)$ is normal in every point of the plane except the origin. Thus the hypothesis of an asymptotic value is necessary for the existence of curves of Julia for meromorphic functions.

The second memoir continues the first by studying the set ε_σ of points where the family of functions $f_n(z) = f(z\sigma^n)$ (here σ is complex, $|\sigma| > 1$, and f is entire) is not normal. ε_σ is closed and if f has an omitted value, then in fact it is a perfect set. However, the function

$$f(z) = \prod_{n=1}^{\infty} \left(1 - \frac{z}{\sigma^n}\right) \quad (|\sigma| > 1),$$

which is entire of order 0 shows that ε_σ can be closed and countable and consist of isolated points (except O and ∞).

This cannot happen if f has an asymptotic value. However, it can happen that *some* subsequences of the family of functions may be a normal family. In fact, ε_σ depends on the value chosen for σ. For example, if $f(z) = e^z$, and so $f_n(z) = e^{z\sigma^n}$, various cases are easily demonstrated; in particular if $\sigma = ae^{i\theta}$, where $\frac{\theta}{2\pi}$ is irrational, then ε_σ is the whole plane, whereas if σ is real, ε_σ is the imaginary axis.

It is in this memoir that a connection is made between this set of three memoirs and complex dynamics by defining $f(z)$ as the entire solution of the functional equation $f(\sigma z) = P(f(z))$, where $|\sigma| > 1$, $P(z)$ is the polynomial $P(z) = \sigma z + a_2z^2 + \cdots + a_nz^n$, with $f(0) = 0$, $f''(0) \neq 0$. Once again defining the family $f_n(z)$ by $f(z\sigma^n)$, referring to his memoir written two years prior, Julia shows (by reference) that in this case ε_σ can be a nonanalytic simple Jordan curve with no tangents anywhere.

The functional equation in question goes back to Poincaré in [**1890a**].[1] Julia credits Fatou with other examples and with showing that the inverse image of ε_σ (under f) is analytic only if it is a circle. It is striking that while the connections between lines of Julia and what we call today Julia sets, and between "classical" complex analysis and complex dynamics, were prominent in Julia's mind, they ended up divorced from one another.

The second memoir also studies further geometric and analytic properties of ε_σ and also (like the first memoir) approximation to the roots of $f(z\sigma^n) = a$.

All results apply to meromorphic functions with one finite omitted value α (since then $\frac{1}{f(z)-\alpha}$ is entire). All the results established for entire functions, hold

[1]Authors' note: The functional equation $f(\sigma z) = P(f(z))$ is often referred to as the Poincaré functional equation, and its use in complex dynamics seems to have its roots in an application by Valiron of the equation found in Poincaré [**1890a**] to the case of a complex function of a single variable. See p. 37 in Section 3.1 of our work.

for meromorphic functions with at least one asymptotic value. If the meromorphic function ϕ has two distinct omitted values, then ε_σ is a perfect set. But, in general ε_σ consists of a perfect set and a countable set, and just as in the entire case, the perfect set part may be empty (an example is given). On the other hand, if f is the Weierstrass ζ-function, then ε_σ is the whole plane (note that $\zeta(z) = \frac{\sigma''(z)}{\sigma(z)}$, where σ is the Weierstrass σ-function). Also, just as in the entire case, one can have sets ε_σ which are very complicated ("Julia sets").

However, for meromorphic functions without asymptotic values, it can happen that ε_σ may fail to exist or consist only of O and ∞. While these ε_σ are without interest, Julia analyzes the conditions that a meromorphic $f(z)$ must fulfill for this to happen. In particular, the number of zeros of $f(z) - a$ in a circle of radius r grows less fast than $A \log r + B$. One can compare the fact known today that meromorphic functions without a line of Julia must satisfy

$$T(r, f) = O((\log r)^2),$$

where $T(r, f)$ is the Nevanlinna characteristic.

Both of Julia's first two memoirs ([**1919a**] and [**1920**], respectively) contain further results of the sort indicated.

The third, much briefer, memoir [**1921c**] deals with "Julia sequences"; what one may call the "general case", in which we are given a sequence $\{\sigma_k\}$ of complex numbers of increasing modulus to ∞, and form the family $f_n(z) = f(z\sigma_n)$. The set E where the family is not normal (0 and ∞ are of course eliminated from all considerations) is studied from the point of view of the regularity of growth of $f(z)$, or of the σ_n. E will exist provided the zeros of f are not "too infrequent". Thus for a set E to be reduced to 0 and ∞, what is necessary is an entire function, without an omitted value, with "very infrequent" zeros and also a sequence $\{\sigma_k\}$ growing more rapidly than $\{\sigma^k\}$. A suitable function is one of order 0 (functions of order 0, already appear in the second memoir as providing counterexamples). Nevertheless, given any sequence $\{\sigma_k\}$, one can find a more rapidly growing sequence $\{S_k\}$ and a suitable (but arbitrary) f such that a nontrivial E exists.

The study of Julia sequences was further pursued by Ostrowski in [**1925**] who characterized *Julia exceptional* meromorphic functions (those which have no line of Julia) and showed also that questions about Julia curves can be reduced to questions about Julia sequences. Ostrowski's memoir already indicates the bifurcated direction taken by follow-ups to Julia's three memoirs. For Julia, they all concerned normal families of functions.

APPENDIX F

Progress in Julia's Extension of Schwarz's Lemma

Peter R. Mercer
Department of Mathematics
S.U.N.Y. College at Buffalo, New York, U.S.A.

In contemporary terms, Julia's Lemma is as follows:

Let f be an analytic self-map of the unit disk Δ in $\mathbb{C}$. Suppose that p_n is a sequence of points converging to $x \in \Delta$ in such a way that $f(p_n)$ converges to $y \in \partial\Delta$ and

$$(*) \qquad \frac{1-|f(p_n)|}{1-|p_n|} \to \alpha < +\infty.$$

Then

$$\frac{|y-f(z)|^2}{1-|f(z)|^2} \to \alpha \leq \alpha\frac{|x-z|^2}{1-|z|^2},$$

with equality holding at any $z \in \Delta$ if and only if f is an automorphism.

Geometrically, this says that f maps any disk with radius $\frac{k}{1+k}$ tangent to ∂D at x to the disk with radius $\frac{\alpha k}{1+\alpha k}$ tangent to ∂D at y. As such, Julia's Lemma may be regarded as a boundary version of the Schwarz–Pick Lemma. (Just as Wolff's Lemma may be regarded as a boundary version of the Schwarz Lemma.) Indeed, for z, $p_n \in \Delta$ the Schwarz–Pick Lemma says that $f(z)$ lies in a disk Δ_n whose center and radius depend on z, p_n, and $f(p_n)$. One then moves p_n out to Δ as described in (*) above. This hypothesis forces the convergence of the disks Δ_n, as desired. Julia's Lemma is a cornerstone of geometric function theory. It plays an essential role (for example) in standard proofs of the Julia–Carathéodory Theorem, towards which F. and R. Nevanlinna [**1922**], Wolff [**1926c**], Landau and Valiron [**1929**], and Carathéodory [**1929a**] all made contributions. Minialoff [**1935**] generalized Julia's Lemma to the unit ball in $\mathbb{C}^2$ and Hervé [**1963**] to the unit ball in $\mathbb{C}^n$. Hervé also obtained some of the Julia–Carathéodory Theorems in this setting; Rudin [**1980**] supplied the complete set.

Abate generalized these results much further. First, P. Yang [**1978**] observed that (*) can be written as $\rho(0, p_n) - \rho(0, f(p_n)) \to \frac{1}{2}\log\alpha$, where ρ is the Poincaré distance on Δ. (Indeed, the Schwarz–Pick Lemma is best stated in terms of Poincaré distance.) Now the Poincaré distance can be extended to general domains in $\mathbb{C}^n$ via the Kobayashi (pseudo-) distance and so a hypothesis analogous to (*) can be recast in these terms.

As such, Abate [**1990, 1991**] proved generalizations of Julia's Lemma for various bounded domains in $\mathbb{C}^n$, as well as significant generalizations of Julia–Carathéodory Theorems, Wolff's Lemma, and beyond. Many of the excellent details, as well as comprehensive background material, can be found in Abate [**1989**]. We make a short list of other contributors, with no claim that it is complete:

Lewittes [**1968**] and Sarason [**1988**] for Hilbert space settings; Glicksberg [**1976**] for function algebras; Fan [**1979**], Ando and Fan [**1979**], Kovalishina [**1990**] and Z. G. Tao [**1988**] for operators; Włodarczyk [**1986, 1987**] and Mellon [**1996**] for J^* Algebras; Jafari [**1993**] and Abate [**1998**] for polydisks; Reich and Shoikhet [**1996, 1997a**] andElin and Shoikhet [**2001**] for semigroups of analytic mappings; Serrin [**1956**] and Mercer [**1999, 2000**] for related one variable results.

As a final point, we add that a number of fundamental results in the field of composition operators on spaces of analytic functions rely on versions of Julia's Lemma. See the books by Shapiro [**1993**] and Cowen and MacCluer [**1995**] for examples.

APPENDIX G

The Denjoy–Wolff Theorem

Katsuhiko Matsuzaki
Department of Mathematics, School of Education
Waseda University, Japan

The Denjoy–Wolff Theorem considers the iteration of a holomorphic function $f : \Delta \to \Delta$ on the unit disk and asserts that, excluding a trivial exception where f is the identity or an elliptic automorphism of Δ, there exists a unique point ξ in the closure $\widehat{\Delta}$ of the unit disk such that the orbit of every point $z \in \Delta$ under the n-times iteration $f^n(z) \to \xi$, as $n \to \infty$. Since f is contracting with respect to the hyperbolic metric, as the Schwarz–Pick Lemma states, if f has a fixed point in Δ, the statement is easily obtained by taking the fixed point of f as $\xi \in \Delta$. Hence, the problem lies in the treatment of the case where f has no fixed points and especially the uniqueness of the point ξ on the boundary $\partial\Delta = \widehat{\Delta} - \Delta$ is not so obvious.

The works of Fatou and Julia classify the periodic components D for the iteration of rational maps g into several types, and if D is simply connected and m is the period of g for D, then either:

(1) $f = g^m$ has an attracting fixed point in D;
(2) f has a rationally indifferent fixed point on the boundary ∂D; or
(3) f is a rotation of D with the fixed point inside.

By the Riemann Mapping Theorem, the situation can be transferred to the unit disk Δ, and these three possibilities correspond to the above-mentioned cases for the iteration of a holomorphic function $f : \Delta \to \Delta$. However, the rational map g is originally defined on the Riemann sphere, and $f = g^m$ is of course extendable to the closure $\bar{D}$. Hence, the existence of the fixed point of f is easily seen, and the behavior of the orbit as in the assertion of the Denjoy–Wolff Theorem is then clear. In other words, the Denjoy–Wolff Theorem localizes the problem on simply connected periodic components of rational maps and generalizes the arguments of Fatou and Julia. In fact, the statement of the theorem first appeared in Wolff's paper [**1926a**] under the assumption that $f : \Delta \to \Delta$ is continuously extendable to the boundary.

Soon afterwards, Denjoy [**1926**] and Wolff [**1926b**] independently succeeded in removing this assumption. Wolff utilized a general fact due to Fatou that a bounded holomorphic function on Δ has a nontangential limit at almost every point on ∂D, and he proved that the full statement of the theorem follows even from this weaker continuity condition. In contrast, Denjoy's approach is more geometric: f is approximated by $(1-\varepsilon)f$ with $\varepsilon \to 0$ in order to appeal to the stronger contraction property for $(1-\varepsilon)f$. This method later extended to an argument given by Wolff [**1926c**], proving the existence of a horoball tangent at $\partial\Delta$ that is mapped by f into itself.

ARNAUD DENJOY

FIGURE G.1. Photo courtesy of Académie des sciences—Institut de France.

A historical survey of the proofs of the Denjoy–Wolff Theorem as well as a concise exposition on the related topics can be found in an article by Burckel [**1981**]. The second proof due to Wolff [**1926c**] was refined by Beardon [**1990**] in a very simple geometric form. In a standard textbook on complex dynamics written by Carleson and Gamelin [**1993**], the Denjoy–Wolff Theorem was introduced with this proof attributed to Beardon; however, its origin is in the 1926 papers by Denjoy and Wolff. Beardon's paper also recognized the crucial points of the Denjoy–Wolff Theorem as the investigation on the contraction and the ideal boundary of metric spaces, and Beardon proposed the way to generalize the theorem to assertions for metric spaces regardless of the analyticity of the mappings. The references for other such generalizations, such as to multiply connected domains, Riemann surfaces, higher dimensional manifolds, and so on, can be found in the bibliographies of Burckel [**1981**] and Beardon [**1990**]. See also a survey given by Reich and Shoikhet [**1997b**].

The Denjoy–Wolff Theorem was a prototype of hyperbolic dynamics and contributed much to modern studies on dynamical systems even if the relationships are not explicitly mentioned. Among them, we introduce two recent results on one-dimensional holomorphic complex dynamics, which are directly related to the Denjoy–Wolff Theorem.

The first one is the so-called *Snail Lemma*, which asserts that the Denjoy–Wolff Theorem forces an indifferent fixed point $\xi \in \partial D$ where the orbit converges, to be rationally indifferent for a holomorphic map $f : \bar{D} \to \bar{D}$ on a general domain D. This can be used to prove the classification of Fatou components for rational maps as well as to treat irrationally indifferent, nonlinearizable fixed points. See, for example, Pérez-Marco's paper [**1997**].

The second one is a characterization of certain geometric properties of subdomains D in the unit disk Δ in terms of the convergence of the compositions of holomorphic functions in a family $F = \{f | f : \Delta \to D\}$. The Denjoy–Wolff Theorem states that the iteration of a single function $f : \Delta \to \Delta$ (nonelliptic) always converges to a constant function. Moreover, this happens for any sequence of compositions taken from the family F if the range D is degenerate in some sense. One can review classical and recent results on this problem in an article by Lorentzen [**1999**].

APPENDIX H

Dynamics of Self-Maps of the Unit Disc

Lawrence A. Harris, David Shoikhet,
Mark Elin, and Simeon Reich

Lawrence A. Harris
Mathematics Department
University of Kentucky, Lexington, Kentucky, U.S.A.

The theorem of Denjoy and Wolff may be thought of as complementary to a theorem of Kœnigs in which an analytic function is assumed to have a fixed point where its derivative is a contraction. Specifically, Julius Wolff showed in [**1926a**] that if an analytic mapping f of the open unit disc Δ into itself has no fixed points, then the sequence of iterates $\{f^{(n)}\}$ converges uniformly on compact subsets of Δ to a constant of modulus one. There was an additional hypothesis that f has a continuous extension to the closed unit disc. This was eliminated shortly afterward in independent publications by Wolff [**1926b**] and Arnaud Denjoy [**1926**]. Georges Valiron in [**1931**] gave a proof some years later that, like Wolff's, relied on Fatou's theorem on the existence of radial limits.

Shortly after Denjoy's publication, Wolff published a second proof in [**1926c**] that relied instead on a consequence of the Schwarz–Pick Lemma to eliminate the additional hypothesis. This key result states that if an analytic mapping f of Δ into itself has no fixed point in Δ, then there exists a unique point α on the unit circle such that f maps every open disc in Δ that is tangent to the unit circle at α into itself.

Wolff's extension of Schwarz's Lemma to boundary points obtains the existence of the angular derivative of a fixed point free holomorphic mapping of the open unit disc of the complex plane into itself. For a modern exposition of Wolff's ideas, see Chapter 1 of Shoikhet [**2001**] and especially Proposition 1.4.2, where the main result of this paper is deduced from the Julia–Carathéodory Theorem. See also Chapter 4 of Joel H. Shapiro [**1993**]. A version of the Julia–Carathéodory Theorem is given for polydiscs by Marco Abate in [**1998**] and for other finite dimensional bounded domains in Abate and Tauraso [**1999**]. In the original paper, Wolff found it convenient to transform the problem from the open unit disc to the right half-plane. This form has been extended in Fan [**1986**] to operator-valued analytic functions and extended further by Mackey and Pauline Mellon [**2002, 2003**] to JB^*-triples, where a result is obtained for any bounded symmetric domain in a Banach space. An infinitesimal version of Fan's result is given in Theorem 2.4.3 of Elin, Harris, Reich, and Shoikhet [**2002**].

Formulations for the Denjoy–Wolff results now exist for multiply connected regions in $\mathbb{C}$, Riemann surfaces, the Hilbert ball, polydiscs, strongly pseudoconvex domains, Riemannian manifolds, and for bounded symmetric domains and nonexpansive mappings in Banach spaces. References for these extensions are given in the commentaries below.

The connection with nonexpansive mappings is that any holomorphic mapping of a bounded domain into itself is nonexpansive with respect to a function-theoretic metric that is a generalization of the Poincaré metric (such as the Kobayashi distance). For the basic ideas behind this, see Goebel and Reich [**1984**] and for a survey of Denjoy and Wolff's results in Banach spaces including counterexamples, see sections 12 and 13 of Kuczumow, Reich, and Shoikhet [**2001b**].

David Shoikhet
Department of Mathematics
Ort Braude College, Karmiel, Israel

In a long history associated with the problem of iterating holomorphic functions and their fixed points, the two notes of Wolff [**1926c, 1926d**] both entitled "On a generalization of a lemma of Schwarz", are very important. Combining the results of these articles, one can formulate the following statement which is nowadays variously called "the boundary Schwarz–Wolff Lemma" or just "Wolff's Theorem":

> *If $f : \Delta \to \Delta$ is a holomorphic function on the unit disk Δ, with no fixed point in Δ, then there is a unique unimodular point $w \in \partial\Delta$, such that:*
> (1) *each disk in Δ, internally tangent to $\partial\Delta$ at w, is f-invariant;*
> (2) *the angular limits*
> $$\lim_{z\to w} \frac{f(z)-w}{z-w} \quad \text{and} \quad \lim_{z\to w} f'(z)$$
> *exist and are both equal to a positive real number $\alpha \le 1$.*

This assertion is a development of a previous result by Gaston Julia [**1920**] (the so-called "Julia's Lemma") and can be interpreted as an analogue of the Schwarz–Pick Lemma, where the role of the fixed point is taken over by a point on the boundary of D. This result was a breakthrough in understanding the behavior of iterates of holomorphic self-mappings on bounded domains as discrete dynamic systems in the complex plane. Historically, one of the first applicable models of the complex dynamic systems, related to the iteration theory on the unit disk, arose at the end of 19th century in the study of the so-called "Markov-branching processes". In [**1874**] Francis Galton and Henry William Watson, in treating the problem of the extinction probability of a branching process, formulated an analytical model in terms of the probability generating function

$$F(z) = \sum_{k=0}^{\infty} p_k z^k, |z| < 1, \qquad p_k \ge 0, \sum_{k=0}^{\infty} p_k = 1,$$

and its iterations $F^{(n)}(z) = F(F^{(n-1)}(z)), n = 1, 2, \ldots, F^{(0)}(z) = z$.

In a parallel way the problem of asymptotic behavior of the semi-group of iterates of a holomorphic function F, mapping the unit disk into itself, appeared in the works of Ernst Schröder [**1871**] and Gabriel Kœnigs [**1884b**] who studied the solvability of the functional equation

$$h(F(z)) = F'(z_0)h(z),$$

where $z_0 = F(z_0)$ is a fixed point of F. These and other problems led to the following general question: Does the sequence of iterates $\{F^{(n)}(z)\}_{n=1}^{\infty}$ converge to some holomorphic mapping on Δ?

In 1926 Denjoy [**1926**] and Wolff [**1926a, 1926b**] simultaneously solved this problem and showed that if $F : \Delta \to \Delta$ is a holomorphic self-mapping of Δ which is not an elliptic automorphism of Δ and is not the identity, then $\{F^{(n)}(z)\}$ converges uniformly on compact subsets of Δ to a point $w \in \bar{\Delta}$.The Wolff boundary extension of the Schwarz Lemma is an essential complement of this result in the further development of the geometrical function theory. Later, this theorem also gave a powerful thrust to the development of dynamical systems in higher dimensions.

It was extended to the finite-dimensional Hilbert ball by Michel Hervé [**1963**] and to the infinite-dimensional Hilbert ball by Goebel [**1982**]. In the 1980s Abate [**1988, 1989**] tackled similar problems for certain bounded domains in $\mathbb{C}^n$. In the 1990's Wolff's Theorem was extended to J^*-algebras by Włodarczyk [**1987**] and Mellon [**1996**]. A special class of holomorphic mappings defined by the Riesz–Dunford integral on the space $L(H)$ of bounded linear operators on a Hilbert space is of great interest in operator theory. A powerful tool in the development of the functional calculus are Fan's [**1979, 1982, 1983**] extensions of Julia's Lemma and Wolff's Theorem to the function theory of proper contractions on Hilbert spaces. Finally, note that a natural analogue of the Wolff Theorem (as well as the Denjoy–Wolff Theorem) for continuous time dynamic systems was given by Earl Berkson and Horacio Porta [**1978**] and has been used to study semigroups of composition operators in Hardy spaces (see also Abate [**1989**], Carl C. Cowen and Barbara MacCluer [**1995**], Reich and Shoikhet [**1997a**], and Elin and Shoikhet [**2001**]).

Mark Elin
Department of Mathematics
Ort Braude College, Karmiel, Israel

One of the main problems of complex dynamical systems is the study of iterations f_n of a holomorphic function f which is a self-mapping of the open unit disk $\Delta_0 = \{z : |z| < 1\}$ or the right half-plane $x = \Re(z) > 0$. (By the Riemann Mapping Theorem, the consideration of any simply connected domain can be transferred to the open unit disk Δ.)

The convergence problem for iterations of a function f, different from an elliptic automorphism, was studied in the works of Wolff, Julia, Denjoy, and others, preceding the paper [**1931**] of Valiron. It was shown that iterations converge uniformly either to an interior point a which is, actually, the unique fixed point of the considered function, $f(a) = a$, or to a boundary point b which can be referred to as boundary attracting fixed point $\lim_{r\to 1^-} f(rb) = b$. In this context the question of finding the rate of convergence arises naturally.

In the case of an interior fixed point, the Schwarz Lemma and the Schwarz–Pick Lemma guarantee the power convergence of the iterations in the Poincaré metric. Moreover, Kœnigs proved in [**1884b**] that if f has an interior fixed point a and $c = f'(a) \neq 0$, then the limit

$$K = \lim_{n\to\infty} \frac{f_n(z) - a}{c^n} \tag{H.1}$$

exists. The limit function K is called the "Kœnigs function for f". So, in this situation, the rate of convergence is of power type:

$$f_n(z) - a \sim c^n K(z). \tag{H.2}$$

It is worthwhile to note that Kœnigs developed the limit technique and derived (H.1) as a solution of the Schröder functional equation

$$K(f(z)) = cK(z). \tag{H.3}$$

(If the limit (H.1) exists, then (H.3) follows.) Therefore, the asymptotic behavior of the iterations can, in principle, be expressed in terms of the Kœnigs function and then by the original function f by means of (H.3). Considerations become more complicated if f has no interior fixed points. Let f be a holomorphic self-mapping of the right half-plane and have an attracting fixed point at ∞. Then, by the celebrated Julia–Wolff–Carathéodory Theorem, the convergence is at least of power type. More precisely,

$$\Re[f_n(z)] \geq c^n \Re(z), \qquad \text{where } c = \lim_{x\to\infty} \frac{f(x)}{x} > 1 \tag{H.4}$$

is an angular derivative of f at ∞. In this case the relation (H.1) has the form

$$K = \lim_{n\to\infty} \frac{f_n(z)}{c^n}. \tag{H.5}$$

It turns out that, in general, the Kœnigs function defined by (H.5) can be equal to infinity.

Valiron investigated the limit (H.5) in the hyperbolic case ($c > 1$) and found the criterion of its existence. This result was improved by Pommerenke in [**1981**] (for functions holomorphic in the unit disk). To overcome the problem, Valiron proved that the limit

$$K = \lim_{n\to\infty} \frac{f_n(z)}{f_n(z_0)}, \tag{H.6}$$

where z_0 is an arbitrary point in the right half-plane, exists finitely. This function is also called the Kœnigs function for f. Obviously it is also a solution of the Schröder equation (H.3). Moreover, Valiron showed that the sequence f_n/c^n in (H.5) is bounded if and only if the angular derivative of the Kœnigs function (H.6) is greater than 0. In the final part of his paper, Valiron studied the exceptional parabolic case ($c = 1$). Investigations of Valiron stimulated a utilization of a functional equation of Schröder type and its solutions, which are called Kœnigs functions in complex dynamics (see, for example, Milnor [**2006a**]).

The generalized Abel–Schröder equations were used by Cowen [**1981**] to construct the so-called "fractional linear model". A historical sketch as well as applications of these items to the theory of composition operators can be found in the books of Shapiro [**1993**] and Cowen and MacCluer [**1995**]. Kœnigs functions also have diverse applications to the theory of branching processes (see, for example, Athreya and Ney [**1972**] and Harris [**1963**]). Concerning some geometrical properties of these functions as well as the Kœnigs embedding problem (the embeddability of iterations into a continuous-time semi-group), the reader is referred to Shoikhet's book [**2001**]. During more recent years, the study of the Abel–Schröder equations and corresponding Kœnigs functions (in spirit of Valiron) has become a subject of interest even in infinite-dimensional Banach spaces (see, for example, Khatskevich and Senderov [**2001**] and Khatskevich, Reich, and Shoikhet [**2001**]).

Simeon Reich
Department of Mathematics
Technion – Israel Institute of Technology
Haifa, Israel

In the first of his remarkable series of four papers from 1926, Wolff [**1926a**] proved what is now known as the Denjoy–Wolff Theorem under the additional assumption that the holomorphic self-mapping of the open unit disk $\mathbb{D}$ is continuous on the closed unit disk $\bar{\mathbb{D}}$.

In his second note [**1926b**] Wolff showed that his theorem remains true without this superfluous assumption. In the meantime, Denjoy also established the Denjoy–Wolff Theorem without this redundant continuity assumption. Denjoy's paper [**1926**] is devoted to his elementary proof. This proof, in turn, inspired Wolff to develop his boundary Schwarz Lemma in the third and fourth notes of his 1926 series of articles [**1926c, 1926d**].

With regard to that theorem, if F has a fixed point $a \in \mathbb{D}$, then the convergence of its iterates to this fixed point follows from the Schwarz–Pick Lemma and Banach's fixed point theorem. Perhaps the more interesting case is when F is fixed point free. In this case the convergence of the iterates can be established by observing that $\{F^n\}$ is a normal family and using invariant horocycles. The limit of the iterates is the so-called *sink point* of F.

Since 1926, the Denjoy–Wolff Theorem has motivated higher (both finite- and infinite-) dimensional results. See the two surveys by Reich and Shoikhet [**1997b, 2001**] and the many references mentioned therein.

Consider first the open unit ball B in a finite-dimensional complex Euclidean space C^m. It is known (Hervé [**1963**], Kubota [**1983**], MacCluer [**1983**]) that if F is a fixed point free holomorphic self-mapping of B, then there is a unique point ξ_F on the boundary of B such that the iterates $\{F^n\}$ converge to ξ_F uniformly on compact subsets of B.

On the other hand, in the case of an infinite-dimensional complex Hilbert ball, Stachura [**1985**] gave an example which shows that the Denjoy–Wolff Theorem fails even for biholomorphic self-mappings. However, if a fixed point free self-mapping F of the Hilbert ball is, for example, either firmly holomorphic (Goebel and Reich [**1982**]) or an averaged mapping (Reich [**1985**]), then for each point x in the Hilbert ball, the sequence of iterates $\{F^n(x)\}$ does converge to a unique boundary point.

More recently, Cho-Ho Chu and Mellon [**1997**] have shown that the Denjoy–Wolff Theorem is still valid for a *compact* holomorphic fixed point free self-mapping of the Hilbert ball. Using a new class of horospheres, the authors Kapeluszny, Kuczumow, and Reich [**1999**] prove that such results are also valid for holomorphic self-mappings of the open unit ball of a strictly convex Banach space and for those self-mappings of the open unit ball of a uniformly convex Banach space which are nonexpansive with respect to the Kobayashi distance.

Recent continuous versions of the Denjoy–Wolff Theorem, which provide information on the asymptotic behavior of one-parameter continuous semigroups of holomorphic and nonexpansive mappings can be found, for instance, in Reich

[**1991**], Kuczumow, Reich, and Shoikhet [**2001a**], and Elin, Harris, Reich, and Shoikhet [**2002**]. In this connection, we also mention the very recently established Julia–Carathéodory Theorem and boundary Schwarz–Wolff Lemma for hyperbolically monotone mappings in the Hilbert ball by Elin, Reich, and Shoikhet [**2008**]. Implicit continuous versions of the Denjoy–Wolff Theorem, which pertain to the asymptotic behavior of nonlinear resolvents of semigroup generators, have also been established by Reich and Shoikhet [**1997a**]. They can be used, for example, to show that if a semigroup generator on the Hilbert ball has a continuous extension to its closure, then it has a null point there. This provides a positive answer to a generalization of a problem raised by Harris [**1979**].

APPENDIX I

Koebe and Uniformization

John W. Milnor
Department of Mathematics
Stony Brook University, New York, New York, U.S.A.

The theory of Topfer has important applications in holomorphic dynamics. In fact the classical work of Montel, Fatou, and Julia depends on the Uniformization Theorem, applied to the universal covering of a region in the Riemann sphere, and is combined with the Schwarz Lemma or the closely related Theorem of Pick.

Let me quote Hermann Weyl:[1]

> *The proof of the possibility of uniformization was provided simultaneously by P. Koebe and H. Poincaré in 1907, based on the idea of covering surface. From that time on, Koebe spent his entire research life in thinking through the problem of uniformization in all directions and with the most varying methods. He above all is to be thanked that the theory of uniformization stands before us today as a mathematical structure of particular harmony and grandeur, which deserves a central role in complex function theory.* [**1913**, p. 137]

In modern terminology, we would define a *Riemann surface* to be a connected complex manifold of complex dimension one, and we formulate the *Uniformization Theorem* as the statement that the universal covering of any Riemann surface is conformally isomorphic either to the complex plane, to the open unit disk, or to the Riemann sphere. In the very special case of a simply connected open subset of $\mathbb{C}$, this result is known as the "Riemann Mapping Theorem", although Riemann himself provided only an intuitive proof.

Our modern formulation was not available to Koebe. In fact the concept of an abstract manifold was not introduced until a few years later in Hermann Weyl's "*Idee der Riemannschen Fläche*". Classically, a "Riemann surface" was the object obtained by starting with an analytic function $y = y(x)$, described locally by a power series, and then considering its analytic continuations along all possible paths. The resulting object has a finite or infinite number of "sheets" lying over any point in the x-plane. A "uniformization" was a kind of global parametrization, analogous to the way that we can parametrize the circle $x^2 + y^2 = 1$ by setting $x = \cos(\theta)$ and $y = \sin(\theta)$, where the parameter θ ranges over the universal covering space $\mathbb{R}$. The problem was to construct some analogous kind of parametrization for an arbitrary Riemann surface. Already in 1883, Poincaré had taken the first step by

[1] "Die Idee der Riemannschen Fläche", Teubner 1913. (This is my unpublished translation from the 1955 edition, p. 137.)

giving a rough description of what we would now call the "universal covering" of such a Riemann surface [**1883a**, p. 113]. However, he did not give this construction a name; thus Koebe had to use awkward phrases[2] to refer to it. In fact, Weyl was probably the first to provide a clear and solid foundation for the concept of "universal covering", and to give it a distinctive name.

Koebe's "General Principle of Conformal Mapping" asserts that if a Riemann surface (in this older sense) can be topologically embedded in the complex plane $\mathbb{C}$, then it can be conformally embedded in $\mathbb{C}$. Furthermore, in the simply connected case it is always conformally isomorphic either to $\mathbb{C}$ itself or to the open unit disk. (His Riemann surfaces are never compact, so he does not mention the Riemann sphere case.) Combining this with Poincaré's topological construction, he obtains a "General Principle of Uniformization". His papers [**1907a, 1907b, 1910, 1915, 1927, 1941**] provide complete proofs and the basis for the more modern formulation described above.

[2]For example, Koebe describes "the generation of a Riemann surface relative to the Riemann surface F of the given function $y(x)$, so that such surface Φ in the sense of *Analysis situs* is equivalent to a one-sheet Riemann surface".

APPENDIX J

Permutable Maps in the 1920s

Xinhou Hua and Rémi Vaillancourt
Department of Mathematics and Statistics
University of Ottawa, Ontario, Canada

J 1. Permutability and iteration (I): Julia's point

Julia's research interest on permutable rational functions was mainly based on Poincaré's functional equation [**1890a**],

$$G_j(sz) = R_j[G_1(z), \dots, G_n(z)], \quad j = 1, \dots, n,$$

where G_j are meromorphic functions, R_j are rational functions, and s is a complex constant with modulus $|s| > 1$.

Any rational function R, with a repelling fixed point at $z = z_0$,

$$R(z_0) = z_0, \quad |R'(z_0)| > 1,$$

will always be associated to a meromorphic function G with

$$G(0) = z_0, \quad G'(0) = 1,$$

and which satisfies the functional equation

$$G(sz) = R[G(z)], \quad s = R'(z_0).$$

The function G is said to satisfy a theorem of rational multiplication and is called the fundamental meromorphic function associated to R and the point z_0.

The interesting point to Julia[1] was to seek conditions for a single-valued function G that admits two multiplication theorems related to two different rational functions:

> *An easy observation to make is that there exist interesting classes of single-valued maps with two distinct theorems of multiplication. For example, all circular functions, the elliptic functions. Given two rational maps, if one looks for the conditions of a single-valued map $G(z)$ to admit two theorems of multiplication corresponding to these two functions, one necessary condition, the permutability of two functions, comes out immediately and it is remarkable, as it will be proven further, that such condition is sufficient.* [**1922d**, p. 132]

[1]Authors' note: Julia wrote about permutable maps in the *Comptes Rendus* notice [**1921a**] and in the paper [**1922d**].

Let G be a meromorphic function satisfying the two relations

$$G(s_1 z) = R_1[G(z)], \quad G(s_2 z) = R_2[G(z)],$$

where R_1 and R_2 are nonlinear rational functions, s_1 and s_2 are two numbers such that $s_2 \neq s_1^k$ for any integer k, and $G(0) = 0$, $G'(0) = 1$. For the two relations to hold, Julia proved the following: *It is necessary and sufficient that R_1 and R_2 be permutable, that is $R_2[R_1(z)] \equiv R_1[R_2(z)]$.* Obviously, if two functions R_1 and R_2 are permutable, then every iterate of R_1 is permutable with any iterate of R_2. This result shed a new light on the iteration itself and on related problems on functional equations. For these two reasons, Julia researched permutable rational functions in the early 1920s. At first, he classified all the linear rational functions which are permutable with a nonlinear rational function. Then he studied relations between the dynamics of two permutable rational functions.

Let R_1 and R_2 be two permutable rational functions. Julia showed that their periodic points have the following relations:

- The finite set of points consisting of attracting fixed points and cycles of R_1 is identical to the analogous set of R_2.
- The finite set of points consisting of indifferent fixed points and cycles of R_1 is identical to the analogous set of R_2.
- The cycle of R_2, obtained by indefinitely iterating a repelling cycle of R_1 by means of R_2, is repelling for R_2. Furthermore, if we denote the set of points of repelling cycles of R_1 and R_2 by $E(R_1)$ and $E(R_2)$, respectively, then their derived sets $E'(R_1)$ and $E'(R_2)$ are perfect and identical, i.e., $E'(R_1) = E'(R_2)$.

The derived set for inverse maps here is homonymous with the so-called Julia set J. Thus one can state that *two permutable functions have the same J.*[2]

Julia analyzed the properties of the derived set E' of a rational function R:

- $E'(R)$ is the set of points where the family of iterates of R fails to be normal in the sense of Montel.
- E' is either the complex plane $\overline{\mathbb{C}}$ or an analytic curve which either consists of analytic arcs, often overlapping themselves, or which gives rise to cusps.

We denote by "fundamental disk" a circle or an infinite straight line. When two rational functions are permutable, their derived sets E' have a special form:

- The only permutable independent rational functions with a fundamental disk are those coming from the theorems of multiplication of $G(z) = \tan z$. In this case, the fundamental disk is the real axis.
- The only independent permutable rational functions with a fundamental circular arc which turns into the canonical form $(-1, 1)$ by a transform are those coming from the theorems of multiplication of $G(z) = \cos z$.
- The only entire functions G with two distinct theorems of rational multiplications are linear functions of $e^{\lambda z}$ and $\cos \lambda z$, where λ is a constant.

As applications, Julia studied the influence of permutability of rational functions on the solution of some corresponding functional equations. For a rational function R, we consider three kinds of equations:

[2] A reader interested in permutable maps might consult works by Atela and Hu [**1996**], Baker, Kotus, and Lu [**1991**], Hua and Vaillancourt [**2005**], Hua, Wang, and Yang [**2000**], Hua and Yang [**1998**], Langley [**1999**], K.K. Poon and Yang [**1988**], Schmidt and Steinmetz [**1995**].

- When R has an attracting fixed point z_0 and $R'(z_0) \neq 0$, then we have the Schröder equation, $F[R(z)] = sF(z)$, where $s = R'(a)$.
- When R has an attracting fixed point z_0 and $R'(z_0) = 0$, we can expand R, near $z = z_0$, in the form $R(z) = a(z - z_0)^k + \cdots$, with $a \neq 0$. In this case, we have the Böttcher equation, $F[R(z)] = aF(z)^k$.
- When R has an indifferent fixed point z_0 and near $z = z_0$, $R(z) = (z - z_0) + c(z - z_0)^k + \cdots$, with $c \neq 0$, we have Abel's equation, $F[R(z)] = F(z) + c$.

Julia proved the following two results.

- When two permutable rational functions R_1 and R_2 share an attracting fixed point z_0, then the Schröder equations relating to R_1 and R_2 and to the point z_0, have the same solution; the Böttcher equations, related to R_1 and R_2 and to the point z_0, also have the same solution.
- When two permutable rational functions R_1 and R_2 share a common indifferent fixed point, then the corresponding Abel equations have the same solution.

This theorem has the following interesting corollary.

COROLLARY J.1. *If f is a rational function with degree at least two, then there exist at most countably many rational functions which are permutable with f.*

J.2. Permutability and iteration (II): The Fatou version

In [**1923-24**] Fatou considered three functions f, g, and h. If

$$f(g) = h(f), \tag{J.1}$$

then we say that f is permutable with (g, h). We also say that g is conjugate to h. In particular, if $g = h$, then we say that f is permutable with g. We have the following graph:

$$\begin{array}{ccc} z & \xrightarrow{g} & g(z) \\ \downarrow{f} & & \downarrow{f} \\ f(z) & \xrightarrow{h} & f(g(z)) = g(f(z)). \end{array}$$

THEOREM J.2. *Let P and Q be two polynomials. If the entire function f is permutable with (P, Q), that is,*

$$f(P) = Q(f), \tag{J.2}$$

then f cannot be transcendental.

Fatou investigated permutable rational functions in neighborhoods of repelling, indifferent, and attracting fixed points.

Repelling fixed points. Consider a complex function f. The first derivative of f at a fixed point $z^* = f(z^*)$,

$$\lambda = f'(z^*),$$

is a complex number called a multiplier (or eigenvalue) of the point. The fixed point z^* is either attracting or repelling or indifferent (neutral) according to whether its multiplier satisfies

$$|\lambda| < 1, \qquad \text{or} \qquad |\lambda| > 1, \qquad \text{or} \qquad |\lambda| = 1.$$

In particular, z^* is a superattracting point (critical point) if $f'(z^*) = 0$.

Let $z = a$ be a fixed point of f and $\lambda = f'(a) \neq 0$. We consider the Schröder functional equation,

$$\psi(f(z)) = \lambda\psi(z) \tag{J.3}$$

and assume that $u = \psi(z)$ has the inverse $z = \phi(u)$. Then

$$f_t(z) = \phi[\lambda^t \psi(z)]$$

is permutable with f:

$$f_t(f) = f(f_t),$$

where t is a parameter. Therefore the study of the permutable function f_t is related to the solution of (J.3).

For convenience, we define the Fatou set of f, $\mathcal{F} = \mathcal{F}(f)$, to be the set of points for which there exists a neighborhood on which the family $\{f_n = f \circ \cdots \circ f : n \in \mathbb{N}\}$ is normal in the sense of Montel [**1927**]. The complement set of $\mathcal{F}(f)$, $\mathcal{J} = \mathcal{J}(f) = \mathbb{C} - \mathcal{F}(f)$, is called the Julia set.

For a rational function R we consider the Poincaré equation,

$$\phi(su) = R[\phi(u)] \quad (|s| > 1).$$

Then $\phi(u)$ is a single-valued map, in general meromorphic or, if R is a polynomial, entire. The exceptional values of $\phi(u)$, in the sense of Picard, fall into three different cases:

- If $\phi(u)$ has two exceptional values, then R is conjugate to $z^{\pm m}$, and $\phi(u)$ is conjugate to e^u, by the correspondence $u = 0$ to the repelling fixed point $z = 1$.
- If $\phi(u)$ has only one exceptional value, then R could be taken as a polynomial and $\phi(u)$ as a transcendental entire function, because the only exceptional value is ∞.
- If $\phi(u)$ has no exceptional values, then, by a linear transformation, we may assume that ∞ is not a critical point of R_n. Thus $\phi(u)$ is meromorphic with no multiple poles. The algebraic critical points of $\phi^{-1}(z)$ are the images of the roots of $R'(z)$. The transcendental critical points (if they exist) may be some of the points of the derived set of such images.

THEOREM J.3. *Given a rational function R, if ϕ is a periodic entire map satisfying*

$$\phi(su) = R[\phi(u)], \quad |s| > 1,$$

then $\phi(u)$ is conjugate to $\cos(u + h)$ *or e^u, where h is a constant.*

Indifferent fixed points. We consider the case where a rational function has indifferent fixed points. Assume that $z = 1$ is an indifferent fixed point of R. Then R can be expanded as

$$R(z) = a + \frac{b}{z} + h(z), \tag{J.4}$$

where h is analytic near $z = 1$ and

$$|h(z)| < C\,|z|^{-k}, \quad |h'(z)| < Ck\,|z|^{-k-1},$$

where $k > 1$ and C is a positive constant.

THEOREM J.4. *Assume that* $a \neq 0$. *Then any function which is permutable with* R *will be the Abel function*

$$A(z) = z - \frac{b}{a}\log\frac{z}{a} + O\left(\frac{1}{z^{\beta-1}}\right), \tag{J.5}$$

where $\beta = \min\{k, 2\}$. *It satisfies the Abel equation,*

$$A(R(z)) = A(z) + a. \tag{J.6}$$

REMARK J.5. If $a = 0$, then

$$R(z) = z + \frac{b}{z^{m-1}} + \cdots, \quad b \neq 0, \quad m > 1.$$

This case may be turned into the case $a \neq 0$ by transforming the given map by two auxiliary conformal maps, where the first map is analytic in the basin of ∞,

$$w = z + c_0 + \frac{c_1}{z} + \cdots + \frac{c_{m-2}}{z^{m-2}},$$

and the second map is of the form

$$u = w^m.$$

Attracting fixed points. If z^* is an attracting fixed point of a function R, then there is a branch $D_{z^*} \subset \mathcal{F}(R)$ which contains z^*. This branch is called the immediate attracting basin of z^*.

THEOREM J.6. *Let* R *and* S *be two permutable rational functions. Assume* $z = 0$ *is an attracting (nonsuperattracting) fixed point of them. Then there exist two nonzero integers* m *and* n *such that* $R_m = S_n$.

REMARK J.7. The origin $z = 0$ can be changed to any other point.

In the special case where z^* is a superattracting fixed point, that is, $\lambda = 0$, we may assume that $z^* = 0$. Then

$$R(z) = z^m + az^{m+1} + \cdots \quad (m > 1),$$

and there exists an analytic function $\Phi(z)$ in D_0 satisfying the functional equation

$$\Phi[R(z)] = [\Phi(z)]^m.$$

The function $\Phi(z)$ has infinitely many algebraic critical points, all converging to the boundary points of D_0.

Case in which the Julia set is not $\overline{\mathbb{C}}$. We consider two nonlinear rational maps, R and S, which are permutable and

$$R_n \not\equiv S_m$$

for any integer values of n and m.[3]

[3] Fatou wrote about the opposite case, when $J \equiv \mathbb{C}$ [**1923-24**, p. 380]: "One does not know, in this case, the general form of the maps R and S, but only examples coming from the multiplication formulas for elliptic functions".

THEOREM J.8. *If R has a fixed point or an attracting cycle, or a multiplier in the form $e^{i2\pi p/q}$ (p and q are two integers), then $\mathcal{J}(R) \neq \mathbb{C}$. Furthermore, R and S may be transformed, by the same linear map, into the form*

$$R(z) \sim z^m, \quad S(z) \sim z^n$$

or

$$R(z) \sim u\cos(m \arccos z), \quad S(z) \sim v\cos(n \arccos z),$$

where $u^2 = v^2 = 1$ and m and n are two integers.

In conclusion, Fatou stated that

THEOREM J.9. *Suppose that R and S are permutable rational functions of degree at least two. Then at least one of the following four cases holds:*

(1) *There exist positive integers m and n which satisfy*

$$R_m = S_n;$$

(2) $\mathcal{J}(R) = \mathbb{C}$;
(3) *R is conjugate to the function z^n for some integer n;*
(4) *R is conjugate to the Chebyshev polynomial T_n for some positive integer n, where T_n is the polynomial satisfying $\cos(nz) = T_n(\cos z)$.*

This theorem has the following interesting corollary.

COROLLARY J.10. *If R is a rational function of degree at least two, then there exist at most countably many rational functions which permute with R.*

THEOREM J.11. *Let R and S be two permutable rational functions. If S is linear, then S is an elliptic linear map.*

Like Julia, Fatou discovered all *entire* maps with two multiplication theorems for rational maps. The next open question was to find permutable *meromorphic* maps:

> *It is clear that the Poincaré entire maps with two distinct theorems of multiplication are essentially the functions e^u and $\cos u$, so there is no room to insist further on the meaning of the word "essentially".*
>
> *The search for meromorphic maps enjoying the same property which are, like $\tan u$, linear transformations of an entire map, appears difficult, and I do not know of any others besides the elliptic functions.*[4] [**1923-24**, p. 374]

[4]Here Fatou inserted this footnote: *From Ritt's researches, published after the redaction of this memoir, it seems that only the elliptic functions answer this question. Ritt's proofs, of an algebraic nature, have no link with the methods applied here.* With his naturally gentle temper, Fatou stated the theoretical independence of Ritt's works. In the second part of this memoir, published in 1924, Fatou returned to Ritt again, observing they effectively showed that only "independent and permutable rational maps" can be obtained via the multiplication formulas of Weierstrass $\wp$ elliptic functions. It is interesting that here he was aware of results from America, unlike the case with the center problem. See Section 9.6.

J.3. Permutability and iteration (III): Ritt's approach

In [**1815**, **1820**] Babbage studied the real functional equation,

$$f^n(x) = x, \tag{J.7}$$

where $f^n(x) = f[f^{n-1}(x)]$. Babbage proved that, if $f(x)$ is a solution of (J.7), then

$$\phi^{-1}[f(\phi(x))]$$

will also be a solution for any function ϕ, hence f and ϕ permute.

In [**1915**] Ritt first proved that, when $n > 2$,

$$\phi^{-1}[f^p(\phi(x))]$$

will also be a solution of equation (J.7), where p is an integer relatively prime to n. He also studied the case $n = 2$.

In [**1920**] Ritt studied the iteration of rational functions. By using a theorem[5] of Łucyan Böttcher in [**1904b**], he obtained some results which are not covered by Fatou, Julia, or other researchers. Ritt used the concept of *associate points of rational functions*.[6]

DEFINITION J.12. Let f be a function. If $f^p(z_1) = f^p(z_2)$ for some positive integer p, then we say that z_1 and z_2 are associates of each other. If $z_1 = f^p(z_2)$, then z_1 is called a *forward image* of z_2, and z_2 a *preimage* of z_1.

THEOREM J.13. *Let f be a rational function of degree greater than* 1. *Assume that f cannot be transformed into a polynomial or a power of z. Then every point has an infinite number of associates.*

THEOREM J.14. *For a rational function f, let $A(c)$ be the set of associates of c. Then:*

- *The set of limit points of $A(c)$ contains a perfect set.*
- *$A(c)$ cannot accumulate about a point of attraction.*
- *The forward images of a limit point of $A(c)$ cannot approach a point of attraction, z, for which $f'(z) \neq 0$, even if the forward images of c approach z.*
- *If the forward images of c tend toward a point of attraction at which $f'(z) \neq 0$, then $A(c)$ forms an isolated set.*
- *If c lies within a sufficiently small neighborhood of a point of attraction p and $f'(p) = 0$, then there exists a simple closed analytic curve, passing through c, and around p, on which $A(c)$ is dense everywhere. If f has a pole of order greater than* 1, *the associates of a point q in the neighborhood of ∞ are everywhere dense on a simple closed analytic curve which passes through q.*

Using the concept of associate points, Ritt proved the following important existence and uniqueness theorem for two polynomials.

[5]This theorem is as follows: *If $z = 0$ is an attracting point of f of order $m > 1$, then there exists, in a neighborhood of $z = 0$, a sufficiently small circle D centered at the origin and an analytic function ϕ such that $\phi(f(z)) = [\phi(z)]^m$, where $z \in D$.*

[6]Fatou also used the concept of an associate point, [**1919-1920**, p. 189, Chap II].

THEOREM J.15. *Let f and g be two polynomials of degree m and n, respectively. If*

$$f^p(z) = g^q(z),$$

then there exists a polynomial, h, in which the exponents form an arithmetic progression with ratio d, such that

$$f(z) = ah^s(z), \quad g(z) = bh^t(z),$$

for two constants a and b satisfying

$$a^d = 1, \quad b^d = 1, \quad a^{(m^p-1)/(m-1)} = b^{(n^q-1)/(n-1)}.$$

Now let f be a simply or doubly periodic and uniform analytic function with no essential singularity in the finite plane.

For a rational function R with repelling fixed point $z = 0$ (i.e., $R(0) = 0$, $|R'(0)| > 1$), Poincaré proved in [**1890a**] that there exists a meromorphic function f such that

$$f(cz) = R[f(z)], \quad c = R'(0). \tag{J.8}$$

In [**1922**], Ritt tried to determine the function f when it is periodic.

THEOREM J.16. *Let R and f be the functions appearing in* (J.8).

- *If f is simply periodic, then f is a linear function of* $\cos(\alpha z + \beta)$ *or* $e^{\alpha z}$, *where α and β are constants.*
- *If f is doubly periodic and not of the lemniscatic or equianharmonic types, then $f(z) = \wp(z + \alpha)$ for a certain constant α.*
- *For the lemniscatic case, there are functions of the form $\wp^2(z + \alpha)$ in addition to those of the form $\wp(z+\alpha)$, while for the equianharmonic case the forms $\wp(z + \alpha)$ and $\wp^2(z + \alpha)$ are possible.*

Ritt also considered the case where $|c| = 1$. Two rational functions, R_1 and R_2, are said to be *permutable* if

$$R_1(R_2(z)) = R_2(R_1(z)).$$

As we saw in the first two sections of this appendix, Fatou and Julia studied permutable polynomials, as well. We remark here that they both proved that if P_1 and P_2 are permutable polynomials such that no iterate of one is identical with any iterate of the other, then they can be obtained from the formulas for the multiplication of the argument in the functions e^z and $\cos z$. This prompted the following remark from Ritt about Julia's work:

> *His other results are mainly of a qualitative nature, and deal with the manner in which $\Phi(z)$ and $\Psi(z)$ behave*[7] *when iterated.* [**1923a**, p. 399]

In [**1923a**, pp. 399–400] Ritt considered this question again and distinguished his own approach:

> *The method used in the present paper differs radically from that of Julia and Fatou, and leads to results of much greater precision.* [**1923a**, p. 399]

[7] Φ and Ψ are R_1 and R_2 in our notation, respectively.

In Ritt's mind, this article was the obvious follow-up of the path started with [**1922**], when the function f was restricted to being a polynomial. His goal was in fact to solve the same problem in the more general case when f is fractional. The main result is as follows.

THEOREM J.17. *If two nonlinear permutable rational functions, R_1 and R_2, are such that no iterate of one is identical with any iterate of the other, then there exist a periodic function f and four constants, a, b, c and d, such that*

$$f(az+b) = R_1[f(z)], \quad f(cz+d) = R_2[f(z)]. \tag{J.9}$$

Ritt pointed out that the only possibilities for f are:

(1) any linear function of e^z, $\cos z$, and $\wp(z)$;
(2) in the lemniscatic case, $\wp^2(z)$;
(3) in the equianharmonic case, $\wp'(z)$ or $\wp^3(z)$.

As Ritt mentioned, his method to prove the existence of f in Theorem J.17 is based upon an algebraic approach. He also wondered (p. 400) whether such a question can be solved by means of the Poincaré function which Julia applied to the problems in his memoir [**1921a**]. In particular, for two nonlinear permutable polynomials, P_1 and P_2, Ritt proved the following result.

THEOREM J.18. *If two permutable polynomials do not come from the multiplication theorems of e^z and $\cos z$, then there exist a linear integral function $\lambda(z)$ and a polynomial $P(z) = zQ(z^r)$, where Q is a polynomial, such that*

$$P_1(z) = \lambda^{-1}\{uP^n[\lambda(z)]\}, \quad P_2(z) = \lambda^{-1}\{vP^m[\lambda(z)]\},$$

where u and v are rth roots of unity, and n and m are positive integers.

Ritt gave the following example, which satisfies the hypotheses of Theorem J.18 [**1923a**, p. 447]

$$P_1(z) = \frac{\varepsilon z^2+2}{\varepsilon z+1}, \quad P_2(z) = \frac{z^2+2}{\varepsilon z+1}, \quad P(z) = \frac{z^2-4}{z-1},$$

where ε is a primitive third root of unity.

Since Ritt did not initially use Montel's theory of normal families, he did not address the basic question in complex dynamics: How is the plane partitioned into regions of attraction, and what is the behavior of the boundaries of the regions?

APPENDIX K

The Last 60 Years in Permutable Maps

Xinhou Hua and Rémi Vaillancourt

Department of Mathematics and Statistics
University of Ottawa, Ontario, Canada

K.1. Introduction

Between 1918 and 1926, Pierre Fatou and Gaston Julia independently developed the theory of complex dynamics. They discovered the dichotomy of the Riemann sphere into the sets which now bear their names (see, e.g., Fatou [**1919-1920, 1926**] and Julia [**1918a**]).

Let f be a nonlinear meromorphic function. The sequence of iterates of f is denoted by

$$f^0 = \mathrm{id}, \quad f^1 = f, \quad \ldots, \quad f^{n+1} = f^n(f), \quad \ldots.$$

There should be no confusion with ordinary powers, which will be explicitly written as $(f(z))^n$ if necessary. We define

$$F = F(f) = \{z \in \overline{\mathbb{C}} : \text{the sequence } \{f^n\} \text{ is well defined and normal at } z\}$$

and

$$J = J(f) = \overline{\mathbb{C}} - F(f),$$

where $\overline{\mathbb{C}} = \mathbb{C} \cup \{\infty\}$, and the concept *normal* is in the sense of Montel. The sets $F(f)$ and $J(f)$ are called the Fatou and Julia sets of f, respectively. When there is no confusion, we briefly write F and J instead of $F(f)$ and $J(f)$. Each component of F will be called a Fatou component. Let D be a component of F. If the components $f^n(D)$ $(n = 0, 1, 2, \ldots)$ are disjoint, then D is called a *wandering domain.*

According to the definition, it can be easily verified that F is open (possibly empty) and J is closed.

Let f and g be two nonconstant meromorphic functions. If

$$f(g) = g(f),$$

then we say that f and g are *permutable.*

In general, for three nonconstant meromorphic functions, f, g, and h, it is more interesting to consider whether we have the relation

$$f(g) = h(f).$$

This goes back to Schröder's consideration.

K.2. Schröder's functional equation

A function g is said to be *analytically conjugate* to a function h on a domain $D \subset \mathbb{C}$ if there exists an analytic function f such that the following diagram commutes.

$$\begin{array}{ccc} D & \xrightarrow{g} & g(D) \\ \downarrow f & & \downarrow f \\ f(D) & \xrightarrow{h} & f(g(D)) \end{array}$$

Many important iterative properties are preserved by conjugation. To study the dynamics of a given function g, we select an analytically conjugate function h whose dynamics is easier to study. Thus, we transfer the dynamics of g to the dynamics of h.

K.3. Permutable rational functions

For two permutable rational functions, the relation between their dynamics is clear:

- Fatou [**1923-24**] and Julia [**1922d**] proved that if two rational functions f and g permute, then $F(f) = F(g)$.

An interesting case is the inverse: If two rational functions, R_1 and R_2, have the same Julia set, do they have special relations?

- Baker and Eremenko [**1987**]. For two polynomials p_1 and p_2, when the group of rotational symmetries of $J(p_1)$ is trivial, then

$$J(p_1) = J(p_2) \Longrightarrow p_1 = p_2.$$

- Fernández [**1989**]. If p_1 and p_2 have the same degree and same leading coefficient, then

$$J(p_1) = J(p_2) \Longrightarrow p_1 = p_2.$$

- Beardon [**1992**]. If p_1 and p_2 have the same degree, then

$$J(p_1) = J(p_2) \Longrightarrow p_1 = \sigma(p_2),$$

  where σ is a rotational symmetry of $J(p_1)$.
- Atela and Hu [**1996**]. For two polynomials p_1 and p_2 with degrees n and m, respectively, such that both of them have no second highest terms, if they have the same Julia set J, then:
  (1) If J is a circle, then there exist a linear map L and a rotational symmetry σ of J such that

$$p_1 = L \circ z^n \circ L^{-1}, \quad p_2 = L \circ \sigma \circ z^m \circ L^{-1}.$$

  (2) If J is an interval, then there exist a linear map L and a rotational symmetry $\sigma \in \{-1, 1\}$ such that

$$p_1 = L \circ T_n \circ L^{-1}, \quad p_2 = L \circ \sigma \circ T_m \circ L^{-1},$$

  where T_n and T_m are Chebyshev polynomials.

(3) In all other cases, there is a polynomial p and positive integers s and t such that

$$p_1 = \sigma_1 \circ p^s, \quad p_2 = \sigma_2 \circ p^t,$$

where σ_1 and σ_2 are two rotational symmetries of J.

- Schmidt and Steinmetz [**1995**]. To any Julia set (of a polynomial), J, which is not a circle or a straight line segment, there exists a polynomial p such that any polynomial with the same Julia set J can be written in the form σp^n, where σ is a rotation which maps J onto itself, and n is a positive integer.

K.4. Permutable transcendental entire functions

The study of the dynamics of permutable transcendental entire functions is more difficult than the study of the dynamics of rational functions. It is not easy to construct two permutable transcendental functions except for the function and its iterates.

EXAMPLE 1 (Hua and Yang [**2003**]). Let

$$\begin{aligned} f(z) &= ic\left[\exp\left(\frac{(4k+3)\pi}{8c^2}iz^2\right) + \exp\left(-\frac{(4k+3)\pi}{8c^2}iz^2\right)\right], \\ g(z) &= c\left[\exp\left(\frac{(4k+3)\pi}{8c^2}iz^2\right) - \exp\left(-\frac{(4k+3)\pi}{8c^2}iz^2\right)\right]. \end{aligned}$$

Then $f(g) = g(f)$.

Based on the results for rational functions, the following question is natural (see Baker [**1984**]):

> Given two permutable transcendental entire functions f and g, does it follow that $F(f) = F(g)$?

This question is still open. However, for some particular cases, we have an affirmative answer.

- Baker [**1984**]. Suppose f and g are permutable transcendental entire functions. If $g(z) = af(z) + b$, where a and b are constants and $a \neq 0$, then $F(f) = F(g)$.
- Poon and Yang [**1988**]. Suppose that f and g are two permutable transcendental entire functions. If both $\operatorname{sing}(f^{-1})$ and $\operatorname{sing}(g^{-1})$ are isolated in the finite complex plane, then $F(f) = F(g)$.
- Wang and Hua [**1997**]. If f and g are permutable and $\operatorname{sing}(f^{-1})$ is bounded, then $F(f) = F(g)$.
- Langley [**1999**]. Let f and g be permutable transcendental entire functions of finite order without wandering domains. Then $J(f) = J(g)$.

REMARK 1. If $g = af + b$ permute with f, where a and b are constants, then by the Hadamard three circle theorem, it is easy to conclude that $f(az+b) = af(z)+b$ with $|a| = 1$.

K.5. Permutable transcendental meromorphic functions

Let f and g be two permutable transcendental meromorphic functions.

- Hua and Vaillancourt [**2005**]. Let p be a nonconstant polynomial and q be a nonconstant entire function. Suppose that
$$p(f) = q(g).$$
Then $J(f) = J(g)$.

As an application, we obtain the following interesting conclusions.

- Let g be a transcendental meromorphic function. Then $J(g) = J(g^n)$.
- Let f and g be two permutable transcendental meromorphic functions, and a and b be two constants. If $f = ag + b$, then $J(f) = J(g)$.

K.6. Other dynamical properties of permutable functions

K.6.1. The classification theorem of rational functions. According to Sullivan [**1985**], rational functions have no wandering domains. Hence all the Fatou components are (pre)periodic. Possible types of such components are:

- One containing an attracting or superattracting fixed point.
- One whose boundary contains a fixed point where the derivative of the function equals 1 (called a parabolic fixed point—Leau domain).
- The component is simply connected (called a Siegel disk) or doubly connected (called a Herman ring).

K.6.2. The classification theorem of transcendental functions. The classification theorem for transcendental meromorphic functions was first stated by Baker, Kotus, and Lu [**1991**]. The proof is essentially due to Cremer [**1932a**] and Fatou [**1919-1920**].

Let f be a transcendental meromorphic function, and let U be a periodic component of period m. Then we have the following possibilities:

(1) U contains an attracting periodic point z_0 of period m. Then $f^{nm}(z) \to z_0$ for $z \in U$ as $n \to \infty$, and U is called the immediate attracting basin of z_0. Furthermore, U is called a Böttcher domain or a Schröder domain provided z_0 is superattracting or not, respectively.

(2) ∂U contains a periodic point z_0 of period m and $f^{nm}(z) \to z_0$ for $z \in U$ as $n \to \infty$. Then $(f^m)'(z_0) = 1$. In this case, U is called a Leau domain (or parabolic domain).

(3) There exists an analytic homeomorphism $\phi : U \to \Delta$ such that $\phi \circ f^m \circ \phi^{-1}(z) = e^{2\pi\alpha i}z$ for some $\alpha \in \mathbb{R}\backslash\mathbb{Q}$. In this case, U is called a Siegel disk. We have the following commutative diagram.

$$\begin{array}{ccc} U & \xrightarrow{f^m} & U \\ \downarrow \phi & & \downarrow \phi \\ \Delta & \xrightarrow{e^{2\pi i\alpha}} & \Delta \end{array}$$

(4) U is doubly connected and f^m is conjugate to either a rotation on an annulus or a rotation followed by an inversion. This U is called a Herman

ring (or an Arnold–Herman ring). Siegel disks, and Herman rings are known as *rotation domains*.

(5) $f^{nm}(z) \to z_0 \in \partial U$ for $z \in U$ as $n \to \infty$ but f^m is not holomorphic at z_0, and U is called a *Baker domain* (or infinite Fatou component, or essentially parabolic domain, or domains at ∞). In particular, if $m = 1$, then the only possible case is $z_0 = \infty$.

K.6.3. Other dynamical properties of permutable rational functions. For two permutable rational functions,

- Schmidt [**1992**] proved that they have the same attracting domains, Leau domains, Siegel disks, and Herman rings.

K.6.4. Other dynamical properties of permutable transcendental entire functions. For transcendental entire functions, the following two questions are natural.

Question 6.4.1. *Suppose that f and g are permutable transcendental entire functions and U is a common component of $F(f)$ and $F(g)$. Does U have the same/similar dynamical property for f and g?*

Question 6.4.2. *Suppose that f and g are permutable transcendental entire functions. If f has no wandering domain, does g also have no wandering domain?*

Next, we construct a counterexample to Questions 6.4.1 and 6.4.2.

Example 2. Let $f(z) = -\sin z + z$ and $g(z) = f(z) + 2\pi$. It is easy to check that $g(f) = f(g)$. By Baker's result, $F(f) = F(g)$. Obviously, for any $k \in \mathbb{N}$, $2k\pi$ is an attracting fixed point of f. Let U_{2k} be the attracting stable domain containing the attracting fixed point $2k\pi$. Then it is a subset of $F(g)$. Note that $U_{2k} \cap U_{2k'} = \emptyset$ for $k \neq k'$ and

$$g(2k\pi) = 2k\pi + 2\pi = 2(k+1)\pi.$$

Thus

$$g(U_{2k}) = U_{2(k+1)}.$$

Hence $(U_{2k})_{k\in\mathbb{N}}$ are wandering domains of g.

Note that $f(z) = -\sin z + z$ has finitely many multiple zeros and the super-attracting fixed points of f are zeros of f' with finitely many exceptions. By Bergweiler [**1993**], f has no wandering domain.

- Poon and Yang [**1988**]. For two nonlinear entire functions f and g, $f(g)$ has no wandering domain if and only if $g(f)$ has no wandering domain.

Remark 2. Baker and Singh [**1995**] proved the same result under the additional condition that f and g are of finite order.

Example 3. We know that $e^{P(z)}$ has no wandering domain for any nonconstant polynomial $P(z)$. The above result implies that $P(e^z)$ also has no wandering domain. Obviously, $P(e^z)$ is not of finite type.

- Wang and Yang [**2003**]. Let f and g be two distinct permutable transcendental entire functions, and let q be a nonconstant polynomial. Suppose that $q(g) = aq(f) + b$, where $a(\neq 0)$ and b are constants. Then $J(f) = J(g)$.

K.6.5. Other dynamical properties of permutable transcendental meromorphic functions. For permutable transcendental meromorphic functions,

- Hua and Vaillancourt [**2005**]. There exist two permutable meromorphic functions f and g such that $J(f) = J(g)$, and some wandering domains of f need not be wandering domains of g.

Example 4. Take two constants a and b such that

$$\left|1 + 2\pi i + \frac{4b\pi^2}{a}\right| < 1. \tag{K.1}$$

Let

$$f(z) = z - \frac{a}{e^z - b} \tag{K.2}$$

and

$$g(z) = f(z) + 2\pi i. \tag{K.3}$$

It is easy to check that $g(f) = f(g)$. It follows that

$$F(f) = F(g). \tag{K.4}$$

Let z_0 be a solution of the equation

$$e^z = b + \frac{a}{2\pi i}, \tag{K.5}$$

and let

$$z_k = z_0 - 2k\pi i, \qquad k = 1, 2, \ldots .$$

We deduce that

$$g(z_k) = z_k$$

and

$$g'(z_k) = 1 + 2\pi i + \frac{4b\pi^2}{a}.$$

Combining these two equations and (K.1), we see that z_k, for $k = 1, 2, \ldots$, are attracting fixed points of g. Let U_k be the Fatou components of $F(g)$ such that $z_k \in U_k$ $(k = 1, 2, \ldots)$. Obviously

$$g(U_k) \subset U_k$$

and

$$g^n(U_k) \to z_k$$

as $n \to \infty$. Note that

$$f(z_k) = g(z_k) - 2\pi i = z_k - 2\pi i = z_{k+1}.$$

This yields

$$f(U_k) \subset U_{k+1}.$$

By (K.4), we see that U_k $(k = 1, 2, \ldots)$ are also Fatou components of $F(f)$. Therefore, U_k $(k = 1, 2, \ldots)$ are wandering domains of f.

APPENDIX L

Understanding Julia Sets of Entire Maps

Dierk Schleicher
School of Engineering and Science,
Jacobs University Bremen, Bremen, Germany

ABSTRACT. We discuss similarities and differences in the understanding of the dynamics of specific examples of polynomial and transcendental entire functions. In particular, we show some of the surprising properties arising in the iteration theory of transcendental entire maps for which there are no (known) counterparts in polynomial dynamics.[1]

L.1. Introduction

Pierre Fatou, in the article [**1926**], *Sur l'itération des fonctions transcendantes entières*, extended his theory of holomorphic dynamics of rational maps to the case of transcendental functions.

The goal of his theory was of course to understand the dynamics of iterated holomorphic functions. In this text, we explore possible meanings of "understanding" dynamics, and then we compare the situation of iterated polynomial maps to that of transcendental entire maps from the point of view of what we know today. We now have quite a substantial theory of iterated polynomial maps (with a number of important questions still open), while much less is known about iterated transcendental entire maps, particularly concerning the detailed study of specific maps. It turns out that transcendental iteration theory features a number of surprising properties from the points of view of topology, geometry, or dimension theory; some of these will be discussed here.

What does it mean to "understand" a dynamical system, such as the iteration of a map $f\colon X \to X$ on a space X? Every point $x \in X$ has its orbit $x, f(x), f(f(x)), f(f(f(x))), \ldots, f^{\circ n}(x), \ldots$, and one may ask questions such as the following:

- What are the possibilities for the long-term behavior of the orbit of $x \in X$?
- Which of these possibilities are typical?
- Which possibilities are interesting?
- What can be said about the set of points which realizes a certain kind of dynamics?

[1]This essay was written in 2003 during the planning stages of this book, *Early Days in Complex Dynamics*. Since then, the field of transcendental dynamics has witnessed rapid progress. Doing justice to these developments would require a new essay. Still, the essay has a considerable historical component, and thus is not entirely inappropriate for inclusion in this book. A few references and an occasional remark were updated when *Early Days in Complex Dynamics* was completed in 2010.

Opinions may of course vary on what kind of dynamics is most interesting (thus contributing to the richness of the subject), or what should be asked about the set of points realizing certain dynamics. For example:

- **Topology:** Is a dynamically interesting set dense in X? Is it open, compact, connected, locally connected, pathwise connected?
- **Measure theory:** Are there measures which are preserved by the dynamics? What is the behavior of typical orbits with respect to these measures? What are the ergodic components of these measures?
- **Dimension theory:** What is the Hausdorff dimension (and the Hausdorff measure) of interesting dynamical subsets?

Even the precise meaning of "typical orbits" often allows for contradictory interpretations: "typical orbits in the sense of Lebesgue" describes a set of positive or full Lebesgue measure, while "typical orbits in the sense of Baire" describes a topologically generic (or residual) set, i.e., a set containing the nested intersection of dense open sets. Sometimes in mathematics, subsets $X_L, X_B \subset X$ are artfully constructed such that X_L is Lebesgue-typical for X, while X_B is Baire-typical and $X_L \cap X_B = \emptyset$; however, a frequent observation in dynamics is that such "paradoxical" situations happen quite naturally in many situations (see for example Milnor [**2006a**, §11]). We will describe a particularly striking apparently "paradoxical" example (of a different kind) at the end of this appendix.

An important step in the understanding of a dynamical system such as $f\colon X \to X$ is achieved when a dynamically meaningful decomposition $X = X_1 \dot\cup X_2 \dot\cup \cdots$ is found such that $f(X_i) \subset X_i$ for all i. In such a case, the points in X_1 are dynamically different from those in X_2 etc.

From now on, we will discuss the case of a holomorphic function $f\colon \mathbb{C} \to \mathbb{C}$ (or analogously, $f\colon \overline{\mathbb{C}} \to \overline{\mathbb{C}}$). For general background, we recommend Milnor [**2006a**] from the mathematical point of view, and Peitgen and Richter [**1986**] for a richly illustrated overview. Fatou and Julia started from the topological point of view, constructing the sets

$$\begin{aligned} F(f) &:= \left\{ \begin{matrix} z \in \mathbb{C}\colon z \text{ has a neighborhood on which the iterates of } f \text{ form a} \\ \text{normal family} \end{matrix} \right\}, \\ J(f) &:= \mathbb{C} \setminus F(f), \end{aligned}$$

which today are called the *Fatou set* and the *Julia set* of f. By definition, the Fatou set is open, the Julia set is closed, and both are forward and backward invariant under f. Excluding the trivial cases of rational maps of degree 0 and 1 from now on, the Julia set is never empty, while the Fatou set might be empty; an important special case when the Fatou set is never empty is the case of polynomials.

A connected component U of the Fatou set is called a *Fatou component*; it is called *periodic* of period n if $f^{\circ n}(U) \subset U$ (for rational maps, this implies $f^{\circ n}(U) = U$). A periodic Fatou component of a (holomorphic or meromorphic) map $f\colon \mathbb{C} \to \mathbb{C}$ is one of the following:

- **Attracting:** U contains an attracting periodic point a, and $f^{\circ kn}(z) \to a$ as $k \to \infty$ for all $z \in U$ uniformly on compact sets;
- **Parabolic:** there is a parabolic periodic point $a \in \partial U$, and $f^{\circ kn}(z) \to a$ as $k \to \infty$ for all $z \in U$ uniformly on compact sets;
- **Siegel disk:** there is an indifferent periodic point $a \in U$ and a conformal isomorphism $\varphi\colon U \to \mathbb{D}$ with $\varphi(a) = 0$ which conjugates the first return

dynamics on U to an irrational rotation: for all $w \in \mathbb{D}$, $(\varphi \circ f^{\circ n} \circ \varphi^{-1})(w) = e^{2\pi i \vartheta} w$ with $\vartheta \in \mathbb{R} \setminus \mathbb{Q}$ (in particular, U is simply connected);

- **Herman ring:** there is a conformal isomorphism $\varphi\colon U \to A_r = \{z \in \mathbb{C}\colon 1 < |z| < r\}$ (for a well-defined $r > 1$) which conjugates the first return dynamics on U to an irrational rotation: for all $w \in A_r$, $(\varphi \circ f^{\circ n} \circ \varphi^{-1})(w) = e^{2\pi i \vartheta} w$ with $\vartheta \in \mathbb{R} \setminus \mathbb{Q}$ (in particular, U is doubly connected and contains no periodic points);
- **Baker domain:** there is an $a \in \partial U \subset \overline{\mathbb{C}}$ at which f has an essential singularity, and $f^{\circ kn}(z) \to a$ as $k \to \infty$ for all $z \in U$ uniformly on compact sets.

Between Fatou and Julia, these possibilities were considered, although neither explicitly referred to what are now known as Herman rings. Moreover, in [**1919a**], Julia asserted that centers do not exist for rational functions (see Section 9.3 of this book). By the maximum modulus principle, entire functions (polynomial or transcendental) cannot have Herman rings, and clearly rational functions cannot have Baker domains. If a Fatou component is not periodic, then it may be preperiodic (so it is covered by the list above); if it is not, then the last case is:

- **Wandering domain:** all iterates $f^{\circ n}(U)$ are disjoint.

A question which naturally arises from the work of Fatou and Julia is the possibility of a wandering domain for rational functions. It was finally proved by Sullivan around 1980 that rational functions never have wandering domains. This proof initiated the use of quasi-conformal maps in complex dynamics and was an essential motor for the rebirth of the theory in the early 1980s, allowing for progress beyond what Fatou, Julia, and their contemporaries had achieved. One of the characteristic differences between the theories of iterated rational and transcendental functions is that the latter can have wandering domains and Baker domains. Eremenko and Lyubich showed that transcendental functions of "finite type" (see below) never have wandering domains or Baker domains, so dynamically they are the closest relatives to rational maps.

The Julia set $J(f)$ clearly contains all repelling periodic points; Fatou and Julia both showed that for rational maps f, the Julia set is the closure of the repelling periodic points, and Baker [**1968**] proved the same for transcendental maps. In general, the dynamics on the Fatou set is usually quite well understood, while the dynamics on the Julia set is "wild and chaotic". Thus, one of the guiding principles in complex dynamics has been to understand the Julia set using easier properties of the Fatou set.

A most important observation is that many dynamical properties of an iterated holomorphic map $f\colon \overline{\mathbb{C}} \to \overline{\mathbb{C}}$ are determined by the singular orbits: a *critical point* z of f is a point which has no neighborhood on which f is injective (if $z \in \mathbb{C}$, then this is equivalent to $f'(z) = 0$), and $f(z)$ is called a *critical value*; an *asymptotic value* is a value v for which there exists an essential singularity a and a curve $\gamma\colon [0,1] \to \overline{\mathbb{C}}$ with $\gamma(1) = a$ and $\gamma(t) \neq a$ for $t < 1$ such that $\lim_{t\to 1} f(\gamma(t)) = v$. Critical and asymptotic values are jointly known as *singular values*. An example of the decisive importance of singular orbits is the fact that every attracting or parabolic periodic orbit has at least one singular orbit which converges to the attracting or parabolic orbit; and the boundary of every Siegel disk or Herman ring is in the closure of the orbits of the singular values (see Milnor [**2006a**]). Similarly, wandering domains or Baker domains can exist only if there are infinitely many singular orbits, which implies that they cannot exist for rational maps (see Eremenko and Lyubich [**1984, 1992**]).

Not surprisingly, the best understood holomorphic dynamical systems are those with only few singular orbits, or where the singular orbits have special properties. Since rational maps of degree 1 are dynamically uninteresting, the story starts with degree 2: such maps have two critical points. The simplest case is when one of them is a fixed point, so its dynamics is particularly easy: such maps are conjugate to quadratic polynomials, which can always be written in the form $z \mapsto z^2 + c$ with a complex parameter c. A rational map of degree $d \geq 2$ has $2d - 2$ critical points, up to $d - 1$ of which may coincide. A particularly simple case is when the maximal number of $d - 1$ critical points coincides, and if in addition this multiple critical point is a fixed point, necessarily superattracting. Choosing coordinates of $\overline{\mathbb{C}}$ in which this superattracting fixed point is at ∞, we obtain a polynomial of degree d. The transcendental analogue of polynomials are entire functions. Here, the best-understood prototypical examples are exponential maps $z \mapsto E_\lambda(z) = \lambda \exp(z)$ (with a parameter $\lambda \in \mathbb{C}^*$) and cosine maps $z \mapsto C_{a,b}(z) = a \exp(z) + b \exp(-z)$ (with two complex parameters $a, b \in \mathbb{C}^*$). Exponential maps have a single asymptotic value 0 and no critical values, while cosine maps have no asymptotic values and two critical values $\pm 2\sqrt{ab}$. In the remainder of this text, we discuss what it may mean to understand certain polynomial and transcendental entire Julia sets and how radically different these two cases can be.

L.2. Dynamics of Polynomials

For an entire holomorphic function $f\colon \mathbb{C} \to \mathbb{C}$, we set

$$I := \{z \in \mathbb{C}\colon f^{\circ n}(z) \to \infty \text{ as } n \to \infty\} \cup \{\infty\}$$

(the set of points converging to *I*nfinity). This is motivated by the fact that for a polynomial p of degree $d \geq 2$, the point ∞ is always a superattracting fixed point, so I is always a nonempty connected open neighborhood of ∞. Restricting to the case of polynomials in this section, the complement

$$K := \mathbb{C} \setminus I$$

is known as the *filled-in Julia set* of p; all orbits in K are uniformly bounded. Since K contains all periodic points of p in $\mathbb{C}$, it is nonempty. The common boundary $J := \partial K = \partial I$ is the *Julia set* of p, while the *Fatou set* of p consists of the open set I, together with any components of the interior of K. (Note that we are not using Fatou's and Julia's definition of the Julia set as the locus of failing local normality of the iterates; our definition is equivalent for polynomials.)

The simplest case is that of polynomials $p_0\colon z \mapsto z^d$ (all critical points in $\mathbb{C}$ coincide, and they are fixed by p_0). Here $K = \overline{\mathbb{D}}$ and $J = \partial\mathbb{D}$ are the closed unit disk and the unit circle. All orbits in $\mathbb{D}$ converge to 0, all orbits in $\mathbb{C} \setminus \overline{\mathbb{D}}$ converge to ∞, while the restriction of p to J is a topological cover of degree d. Topology and measure of J are trivial; an easy but not obvious property of J is that there are dense orbits.

Small perturbations of p_0, like $p_\varepsilon\colon z \mapsto z^d + \varepsilon$ with $|\varepsilon|$ small, yield Julia sets with identical topological properties, but the Hausdorff dimension of J increases from 1 to values strictly greater than 1.

More generally, suppose only that all the $d - 1$ finite critical points of p have bounded orbits. In this case, both J and K are connected (see Douady and Hubbard

[**1984-85**] and Milnor [**2006a**]). The Fatou set is particularly easy to understand: without loss of generality, we may assume that we have coordinates in which p is monic and centered, i.e., $p(z) = z^d + a_{d-2}z^{d-2} + \cdots + a_1 z + a_0$. Then there is a unique conformal isomorphism $\varphi\colon I \to \overline{\mathbb{C}} \setminus \overline{\mathbb{D}}$ with $\varphi(\infty) = \infty$ and $\lim_{z\to\infty} \varphi(z)/z = 1$. The dynamics of p on I is particularly easy after a coordinate change by φ: the map $(\varphi \circ p \circ \varphi^{-1})\colon (\overline{\mathbb{C}} \setminus \overline{\mathbb{D}}) \to (\overline{\mathbb{C}} \setminus \overline{\mathbb{D}})$ takes the simple form $w \mapsto w^d$. Using the guiding principle mentioned above, we may now try to understand the dynamics on J in terms of the limiting behavior of the dynamics on I. Whether or not this is possible depends on the boundary behavior of φ or φ^{-1}.

For $\vartheta \in \mathbb{R}/\mathbb{Z}$, the *dynamic ray of p at angle* ϑ is the set

$$R(\vartheta) := \varphi^{-1}\left(\{re^{2\pi i\vartheta}\colon r > 1\}\right).$$

The *limit set* of the dynamic ray $R(\vartheta)$ is the set of possible limits

$$L(\vartheta) := \left\{a \in \mathbb{C}\colon a = \lim_{r_n \searrow 1} \varphi^{-1}(r_n e^{2\pi i\vartheta}) \text{ for some sequence } r_n \searrow 1\right\} ;$$

the ray $R(\vartheta)$ *lands* at some $z \in J$ if $L(\vartheta)$ consists of the point z alone, i.e., if $\lim_{r\searrow 1} \varphi^{-1}(re^{2\pi i\vartheta}) = z$. In general, not every dynamic ray lands, but for fixed p, the set of $\vartheta \in \mathbb{R}/\mathbb{Z}$ for which $R(\vartheta)$ does not land has measure 0.

In order to say more, we need more control on the set of critical orbits. Suppose all critical points in $\mathbb{C}$ are strictly preperiodic: all have finite orbits, but none is periodic. Examples of such are $p_{-2}\colon z \mapsto z^2 - 2$ and $p_i\colon z \mapsto z^2 + i$. In this case, the Fatou set is connected: $F = I$ (by the classification above, any bounded component of F would require a critical orbit of a kind which is not available). Therefore, $J = K$: the filled-in Julia set equals the Julia set, and both are dendrites, that is, compact, uniquely arcwise connected, and locally arcwise connected. The Hausdorff dimension of the Julia set is less than 2 and the Hausdorff measure in the appropriate dimension is finite and positive. In this case, the conformal isomorphism $\varphi^{-1}\colon \overline{\mathbb{C}} \setminus \overline{\mathbb{D}} \to I$ extends continuously to the boundaries and yields a continuous surjection $\mathbb{S}^1 \to J$ (sometimes known as the Carathéodory loop). This surjection associates to every angle ϑ the unique landing point of $R(\vartheta)$: every dynamic ray lands, and the landing points depend continuously on the angle (by Carathéodory's theorem, this property holds whenever the Julia set is locally connected). Since $J = \mathbb{C} \setminus I$ in our case, every point in $\mathbb{C}$ is either on a unique dynamic ray or on the landing point of one or several dynamic rays.

As a result, the dynamics of all points in $\mathbb{C}$ can be described quite simply: if $z \in I$, then $p^{\circ n}(z) \to \infty$ as $n \to \infty$; more precisely, $z = \varphi^{-1}(re^{2\pi i\vartheta})$ for a unique pair (r, ϑ) of coordinates with $r > 1$, and $p^{\circ n}(z) = r^{d^n} e^{2\pi i d^n \vartheta} + o(1)$ as $n \to \infty$. If $z \in J$ is the landing point of $R(\vartheta)$, then z is periodic under p if and only if ϑ is periodic under multiplication by d on $\mathbb{R}/\mathbb{Z}$; the orbit of z is dense in J iff the orbit of ϑ is dense in $\mathbb{R}/\mathbb{Z}$ (under multiplication of angles by d), etc. We also obtain a topological model of the Julia set (see Douday [**1992**]): $J \approx (\mathbb{S}^1/\sim)$, where $\mathbb{S}^1 = \mathbb{R}/\mathbb{Z}$ and $\vartheta \sim \vartheta'$ iff the dynamic rays $R(\vartheta)$ and $R(\vartheta')$ land at a common point. In other words, the Julia set J is homeomorphic to the quotient of the circle $\mathbb{S}^1$ by an equivalence relation $\sim$. This equivalence relation is quite easy to understand: for $\vartheta, \vartheta' \in \mathbb{S}^1$, there is a combinatorial algorithm to tell whether or not $\vartheta \sim \vartheta'$. It needs the finitely many dynamic rays landing at the finitely many critical points: clearly, for $R(\vartheta)$ and $R(\vartheta)$ to land at the same point, no pair of angles $(d^n\vartheta, d^n\vartheta')$ must be separated on $\mathbb{S}^1$ by the angles η and η' of two dynamic rays $R(\eta)$ and

$R(\eta')$ landing at the same critical point; i.e., for every $n \geq 0$, the angles $d^n\vartheta$ and $d^n\vartheta'$ must be in the same connected component of $\mathbb{S}^1 \setminus \{\eta, \eta'\}$. As it turns out, this necessary condition is also sufficient, yielding the promised easy algorithm. For details, see Douday and Hubbard [**1992**] or Thurston [**2009**].

As a rather simple example, the polynomial $p_{-2}\colon z \mapsto z^2 - 2$ has a single critical value at -2 which is the landing point of the dynamic ray at angle $1/2$, and the only critical point 0 is the landing point of the two dynamic rays at angles $1/4$ and $3/4$. It is an easy exercise to show that for given $\vartheta \neq \vartheta' \in \mathbb{S}^1$, there is an n such that $2^n\vartheta$ and $2^n\vartheta'$ are in different components of $\mathbb{S}^1 \setminus \{1/4, 3/4\}$ unless $\vartheta' = 1 - \vartheta$. Therefore, J is homeomorphic to the quotient of $\mathbb{S}^1 = \mathbb{R}/\mathbb{Z}$ with ϑ and $-\vartheta = 1 - \vartheta$ identified, i.e., the upper and the lower half circles are glued, yielding an interval. Indeed, the Julia set of p_{-2} is the interval $[-2, 2] \subset \mathbb{R}$.

A more interesting example is given by the polynomial $p_i\colon z \mapsto z^2 + i$. The critical value is the landing point of the dynamic ray $R(1/6)$, so the critical point is the landing point of $R(1/12)$ and $R(7/12)$. The three angles $1/7$, $2/7$, $4/7$ are mapped to each other, and they are all in the same component $(1/12, 7/12)$ of $\mathbb{S}^1 \setminus \{1/12, 7/12\}$, so the three rays $R(1/7)$, $R(2/7)$, and $R(4/7)$ land together at the same point α, say (which is a fixed point of p_i): this is the landing point of three rays, and accordingly it is a branch point of the Julia set which disconnects J into three parts. Similarly, the three angles $1/14$, $9/14$, and $11/14$ land together at the point $-\alpha$, which is the unique preimage of α other than α itself. Note that $1/7$ is periodic under angle doubling in $\mathbb{S}^1$, and $R(1/7)$ lands at a periodic point $\alpha \in J$, while $R(1/14)$ is strictly preperiodic, and its landing point $-\alpha$ is strictly preperiodic too.

This example $p_i\colon z \mapsto z^2 + i$ is already fairly typical in many ways for polynomials with all critical points preperiodic: the Julia set is a dendrite with infinitely many branch points; it turns out that all branch points are periodic or preperiodic.

Of course, the assumption that all critical points be preperiodic is rather restrictive; we now have a good understanding of the dynamics in many further cases, but this would go beyond the scope of this text. We recommend Milnor's book [**2006a**] for further reading.

L.3. Dynamics of Transcendental Entire Functions

For the understanding of polynomial dynamics, the understanding of the set I of escaping points was extremely helpful. Is there an analogous approach for transcendental entire functions?

To begin with, note that every transcendental entire function $f\colon \mathbb{C} \to \mathbb{C}$ has an essential singularity at ∞, so there can never be a neighborhood of ∞ which converges to ∞. If there was an interior point of I, then it would have to be in a Baker domain or in a wandering domain, and such do not exist in many cases: Eremenko and Lyubich [**1984, 1992**] have ruled out this possibility for entire maps with only finitely many singular values (called entire maps of *finite type*). However, a nontrivial theorem of Eremenko [**1989**] asserts that for every transcendental entire function, the set $I \cap J$ is nonempty: there are points which converge to ∞, but these are in the Julia set. In fact, for many transcendental entire maps, the Fatou set is empty, so the topological decomposition $\mathbb{C} = F \dot{\cup} J$ of Fatou and Julia is no longer useful in all cases.

In this section, we will discuss examples of dynamical systems given by iterated entire transcendental maps on $\mathbb{C}$, chosen in analogy to the easiest polynomial examples: exponential maps with an attracting fixed point, exponential maps with a strictly preperiodic asymptotic value, and cosine maps with strictly preperiodic critical values. It will turn out that all three examples are vastly more complicated than the corresponding polynomial examples, and each example features a different complication: explosion sets, indecomposable continua, and surprises with the Hausdorff dimension. A good survey on known results in transcendental dynamics is Bergweiler [**1993**], and an introduction specifically to the dynamics of transcendental entire functions can be found in Schleicher [**1998**].

For the first example, suppose that the exponential map $E_\lambda\colon z \mapsto \lambda\exp(z)$ has an attracting fixed point $a \in \mathbb{C}$. This is the case if and only if $\lambda = \mu\exp(-\mu)$ for some $\mu \in \mathbb{D}$. In this case, the asymptotic value 0 converges to a, and so does a neighborhood of 0 by continuity. Since the basin of a is invariant under E_λ, the E_λ-preimage of this neighborhood is contained in the basin too: this contains a left half-plane. The preimage of a half-plane consists of countably many horizontal strips spaced at integer multiples of $2\pi i$, and the preimages of each of those consist of countably many strips, etc. The Fatou set consists of a single connected component which is the basin of a. The Julia set has Hausdorff dimension 2 but measure 0. It has a rich topological structure (see Aarts and Oversteegen [**1993**], Devaney [**2001**], and Schleicher [**2007b**]): it consists of uncountably many connected components, each of which is a curve $\gamma\colon [0,\infty) \to \mathbb{C}$ with $\lim_{t\to\infty}\gamma(t) = \infty$. This curve joins the "endpoint" $\gamma(0)$ to ∞. We can write $J = E\dot{\cup}R$, where

$$E := \bigcup_{\gamma\subset J}\gamma(0) \qquad \text{and} \qquad R := \bigcup_{\gamma\subset J}\gamma((0,\infty)).$$

Here, E is the set of endpoints for all the Julia components, and R is the union of the remaining parts of the curves.

These points have a number of surprising properties. The set E is totally disconnected (i.e., every connected component of E is a single point), while $E\cup\{\infty\}$ (as a subset of $\overline{\mathbb{C}}$) is connected! In other words, starting from the uncountable connected set $E\cup\{\infty\}$, removing the single point ∞ yields a totally disconnected set: the set E "explodes" when ∞ is removed (see Mayer [**1990**]). For this reason, the point ∞ is called an *explosion point*. Moreover, it is known that the set E has Hausdorff dimension 2, while R has Hausdorff dimension 1 (see Karpińska [**1999**]): every point from the 2-dimensional totally disconnected set E is joined to ∞ by its own curve such that the union of all these curves has dimension 1! We will see an even more extreme situation below.

The previous example had nonempty Fatou and Julia sets, so the understanding of the dynamics could begin with the partition $\mathbb{C} = F\dot{\cup}J$. However, in many cases of transcendental entire functions, the Fatou set is empty. Saying that $J = \mathbb{C}$ allows for an easy description of the topology of J ("everything"), but it is dynamically meaningless. Guided by the experience from polynomials, one may wonder whether the set I of points "escaping" to ∞ yields interesting information even though it is no longer distinguished as a connected component of the Fatou set, as it is for polynomials. After having shown that I is always nonempty, Eremenko has asked whether $I\cup\{\infty\}$ is always connected or even path connected. This is in the spirit of the initial discussion: the set I is dynamically interesting, so what are its properties from the point of view of topology, geometry, dimension theory, etc.? It was shown very recently by Rippon and Stallard [**2005**] that I contains unbounded connected components for every transcendental entire function, and $I \cup \{\infty\}$ is connected if there is a wandering domain of a certain special type. For the case of exponential functions $z \mapsto \lambda \exp(z)$ and $z \mapsto a\exp(z) + b\exp(-z)$, it is known that $I \cup \{\infty\}$ is path connected, and there is a complete description of I (see Schleicher and Zimmer [**2003**] and Rottenfußer and Schleicher [**2008**]): every path connected component of I is a curve $\gamma\colon (0,\infty) \to \mathbb{C}$ or $\gamma\colon [0,\infty) \to \mathbb{C}$ with $\lim_{t\to\infty} \gamma(t) = \infty$. These curves are called *dynamic rays* in analogy to polynomials (sometimes, the name "hair" is used too). Note the second possibility: if $\gamma\colon [0,\infty) \to \mathbb{C}$ parametrizes a connected component of I, then this curve is a ray which lands at $\gamma(0)$, but the landing point of the curve is in I itself! (For polynomials, the landing points of rays always have bounded orbits in $\mathbb{C}$.)

In the years since the first version of this text was written, there has been some progress in this direction, in particular on Eremenko's question whether $I \cup \{\infty\}$ was connected. Barański [**2007**] proved this for certain entire functions. Based on Rottenfußer's thesis, it was discovered that this conjecture is false for certain entire functions, but it holds for a large class of entire functions, including those for which the singular values are contained in a bounded set, and the function does not grow faster than $e^{|z|^d}$ for some $d \in \mathbb{N}$ (see Rottenfußer, Rückert, Rempe, and Schleicher [**2011**]).

In analogy with the polynomial situation, one may wonder whether there are particularly nice transcendental entire maps for which every point in $\mathbb{C}$ is either on a dynamic ray or the landing point of a finite positive number of dynamic rays. The most promising example might seem to be the case of exponential maps $z \mapsto E_\lambda(z) = \lambda\exp(z)$ in which the singular orbit is strictly preperiodic (such as $\lambda = i\pi$). However, after trying hard to prove this for a long time, we realized that this is false: there are dynamic rays which "land" at an indecomposable continuum X, that is, a compact set X is a continuum if it is connected, and it is indecomposable if it is not the union of two compact connected proper subsets (i.e., of two proper subcontinua), not necessarily disjoint. There can even be uncountably many indecomposable continua, Devaney, Jarque, and Rocha [**2005**]. As with explosion sets, it might be hard to imagine that such sets exist, but they do, and they come up quite naturally in dynamical systems. For every postsingularly preperiodic exponential map, there are uncountably many rays for which the limit set is an indecomposable continuum. In fact, Lasse Rempe [**2003, 2007**] has generalized this observation to many further exponential maps with $J = \mathbb{C}$, and one may wonder whether there is *any* exponential map with empty Fatou set for which no dynamic ray has an indecomposable continuum as its limit set. It has been known for a long time that maps of the form $z \mapsto \lambda\exp(z)$ with $\lambda > 1/e$ have dynamically distinguished indecomposable continua (see Devaney [**2001**]).

As a final example, we discuss maps of the type $z \mapsto C_{a,b}(z) = a\exp(z) + b\exp(-z)$ with $a, b \in \mathbb{C}^*$. For all such maps, it is known that the set I is contained in J and has positive measure (McMullen [**1987**]) and there is a similar classification of I as for exponential maps (Rottenfußer and Schleicher [**2008**]): every path connected component of I is a curve (a dynamic ray) $\gamma\colon (0,\infty) \to \mathbb{C}$ or $\gamma\colon [0,\infty) \to \mathbb{C}$ with $\lim_{t\to\infty} \gamma(t) = \infty$. Again, it may happen that dynamic rays land at escaping points. Similarly, as in the first example, we set

$$E_I := \bigcup_{\gamma \subset I} \gamma(0) \cap I \qquad \text{and} \qquad R := \bigcup_{\gamma \subset I} \gamma((0, \infty)).$$

The set E_I consists of the endpoints of only those dynamic rays which land and for which the landing point is in I. For every path connected component of I (which is a curve), this is at most one point. All the remaining points of I are in R. Then every point in E_I is connected to ∞ by its own curve in R, and there are many further curves in R unrelated to points in E_I (and all these curves are disjoint). As before, the Hausdorff dimension of R equals 1, while the Hausdorff dimension of E_I equals 2, and E_I has positive (even infinite) Lebesgue measure in $\mathbb{C}$.

Let us now restrict to the special case when the two critical values $\pm 2\sqrt{ab}$ have strictly preperiodic orbits, such as when $a = -b = 2k\pi i$ for $k \in \mathbb{Z} \setminus \{0\}$ (see Schleicher [**2007a, 2007b**]). In this case, the set $E := \mathbb{C} \setminus R$ has the property that every point in E is connected to ∞ by its own ray (or sometimes by several rays), different points in E have different rays, and all the rays are disjoint! This is essentially the same situation as before, where E had Hausdorff dimension 2 or even infinite Lebesgue measure, only that in this case E contains every point in $\mathbb{C}$ except those in the one-dimensional set R! As far as the dynamical system is concerned, this is what we want. The dynamics for points in R is quite easily understood (they all escape to ∞ at an incredible speed, the asymptotics of which can be written down rather easily in terms of combinatorial details beyond the scope of this article), and the dynamics of the remaining points in $E = \mathbb{C} \setminus R$ can be described in terms of R: whether or not a point $z \in E$ has a periodic, preperiodic, dense, recurrent orbit etc. is described combinatorially in terms of the dynamic rays landing at z. All points in E can be distinguished dynamically.

In the beginning of this text, we mentioned that the meaning of "typical points" could be interpreted differently. In this situation, the situation is similar if not worse. If every point in E is connected to ∞ by one or several curves in R, then it seems fair to say that "most" points are in R (the topological point of view). However, since R has Hausdorff dimension 1, while $E = \mathbb{C} \setminus R$ is everything else in $\mathbb{C}$, the geometric point of view is that the points in R are extremely rare, and "most" points are in E.

How easy might it be, using bare hands, to construct a decomposition $\mathbb{C} = E \dot\cup R$ such that R is a disjoint union of curves (even infinitely differentiable) so that R has Hausdorff dimension 1, and every point in the "big" set E is connected to ∞ by one or several curves in R? We know of no example where someone had constructed such sets, while suddenly this arises quite naturally in the context of an easy dynamical system $z \mapsto 2\pi i \sin(z)$.

What is the message from examples like this? On the one hand, this is the first time that a transcendental dynamical system has been described at this level of detail, while on the other hand, even the simplest transcendental dynamical systems seem to have very complicated and surprising properties in terms of topology or geometry. We believe that these complications are no exceptions for the particular maps at hand, but rather illustrate the kind of difficulty that one has to expect in the study of transcendental dynamics—and who knows what further surprises might be waiting for us out there?

L.4. A Transcendental Function Which is Not Understood

Having described a number of holomorphic functions for which many important aspects of the dynamics are understood, we conclude this text by an interesting transcendental entire function of which the dynamics is *not* understood: an example whose origins lie in an entirely different field altogether (we should stress of course that, despite substantial recent progress, there are many transcendental maps whose dynamics is rather poorly understood).

A famous problem of elementary mathematics is known under many names: the $3n + 1$ problem, the Collatz problem, the Syracuse problem, etc. It is easy to describe: Start with any positive integer n. If it is even, then divide n by two; if it is odd, replace it by $3n + 1$. Do the same with the number thus obtained, and continue. Example: $7 \mapsto 22 \mapsto 11 \mapsto 34 \mapsto 17 \mapsto 52 \mapsto 26 \mapsto 13 \mapsto 40 \mapsto 20 \mapsto 10 \mapsto 5 \mapsto 16 \mapsto 8 \mapsto 4 \mapsto 2 \mapsto 1 \mapsto 4 \mapsto \cdots$. This is indeed a dynamical system $f\colon \mathbb{N} \to \mathbb{N}$, given by

$$f\colon n \mapsto \begin{cases} n/2 & \text{if } n \text{ is even,} \\ (3n+1)/2 & \text{if } n \text{ is odd}. \end{cases}$$

(Here, we have simplified things slightly: if n is odd, then its image $3n + 1$ is even and divided by 2 in the next step.) The question is, What happens in the long run to any $n \in \mathbb{N}$ under iteration of f? Will it eventually map onto 1 and then become periodic? Will it fall into any other periodic cycle? Or will it never fall onto any periodic cycle (which would necessarily mean that it tends to ∞)? Many billion positive integers have been tried out on a computer, and all have eventually reached the value 1. Nobody has been able to prove this, or even show that every integer is eventually periodic—and many people have tried hard (see Lagarias [**1985**]).

What does this have to do with holomorphic dynamics? Consider the map $g\colon \mathbb{C} \to \mathbb{C}$,

$$g(z) = \frac{z}{2} + \left(z + \frac{1}{2}\right)\frac{1 - \cos \pi z}{2} + \frac{1}{\pi}\left(\frac{1}{2} - \cos \pi z\right)\sin \pi z\ .$$

This is an entire holomorphic function, and it is quite easy to show that every integer $n \in \mathbb{Z}$ is in the Fatou set of g. Moreover, for integer values n, we have $g(n) = f(n)$; in other words, g interpolates the $3n + 1$ problem. Show that g has no wandering domains; it would then follow that every integer is in a (pre-) periodic Fatou component, so every integer would eventually become periodic (see Letherman, Schleicher, and Wood [**1999**]) this would be major progress on the $3n + 1$ problem. (The last term in the definition of g vanishes for every integer, so it does not contribute to the values on the integers. The reason for this term is that it makes the integers into critical points: recall that we want control on the critical points.) This is an interesting transcendental entire function, the dynamics of which "wants to be" understood, in particular the set of points escaping to ∞. These questions are beyond reach of the existing theory—there is always more to do!

APPENDIX M

Fatou: A Biographical Sketch

We thank Bertrand Fatou, who granted us access to his father Robert Fatou's recollections of his uncle Pierre Fatou [**F1**]. We also drew from two necrologies, one by his friend, the physicist (and philosopher) Leon Bloch [**1931**], and the other by Jean Chazy [**1932**], an eminent French astronomer who was a member of the Académie, as well as a eulogy given at Fatou's funeral by Henri Mineur [**1929**], his younger colleague at the Observatoire de Paris, where Fatou worked from from 1901 until his death at age 51. The eulogy was also printed in a local newspaper [**A11**]. We direct the reader wishing to know more about Fatou to the richly detailed treatment of Fatou's life in Audin's book, [**2009**, Chap V].

Shortly after Fatou's death, a portion of his personal papers was donated to the libraries at the Observatoire de Paris and the École Normale Supérieure (ENS). The Observatoire's collection seems to consist mostly of copies of Fatou's published work. The Fatou donation to the ENS was evidently not inventoried upon its arrival, and so the extent of its collection is unknown. In any event, it seems his personal effects—including correspondence and any photographs—are for the most part lost.[1]

M.1. Fatou the mathematician

FATOU, PIERRE JOSEPH LOUIS (1878–1929). Pierre Fatou was born on Thursday February 28, 1878, at 7:30 a.m., to Prosper Ernest Fatou and Louise Eulalie Courbet, in the seaside town of Lorient (department of Finistère), at rue des Fontaines 33. He was the last of four sons. His father died when he was a boy of 13; his mother died 20 years later on April 8, 1911.

Fatou's father was a captain in the French Navy and his brother Louis an admiral. Mineur's eulogy suggested that at one point his family had hoped Fatou might follow his father's path as well.[2] Had he wanted to join the Navy, health issues which plagued him his entire life would have most likely have precluded it.

Fatou attended a private Catholic school, Collège Stanislas, and the Lycée[3] Saint-Louis of Paris, where he placed first in the *concours général* (Mineur [**1929**]).[4] The Lycée Saint-Louis is a *grandes lycée*, the name given to a school which prepares students for a *grandes école*, which include the École Polytechnique (often referred to as "X" because, depending on who tells the story, of the crossed rifles on the

[1]See Audin [**2009**, pp. 172–174] for a more detailed account of these collections.

[2]Mineur observed that "... Fatou's health did not allow him to embrace the Navy and enjoy the quality and charm of the people of the sea" [**1929**].

[3]When referring to a child's education, "*collège*" refers to the first stage of secondary school and "*lycée*" to secondary school.

[4]This prize, started by the Université de Paris in 1747, was awarded to the top students of the *lycée* of Paris. See http://concoursgeneral.free.fr/.

school's insignia, or the strong influence of mathematics at the school), École Normale Supérieure (ENS), and the École des Mines, among others. Although there are private higher education institutions in France, students wishing to study at the university level typically go to school in the extensive public university system or to one of the elite *grandes écoles.*

Fatou was a superb student and ranked first in the entrance exams of 1898 for the École Normale Supérieure, attending from 1898 to 1901, when he placed second in the agrégation exams[5] which qualified him to teach as a temporary alternate, either in a *collège* or a *lycée.* He probably could have attained a full-time post at a secondary school, or perhaps sought a university level position in the provinces, but decided to work at the Observatoire de Paris as astronomer.[6]

The Observatoire had fallen a bit from its golden age in the second half of 19th century and lacked talented astronomers during the early 1900s. According to Léon Bloch, Jules Tannery believed that recruiting talented mathematicians to become theoretical astronomers would strengthen the Observatoire, and he actively recruited the best students for the Observatoire de Paris [**1931**, p. 55].

In November 1901 Fatou was hired as a "staging astronomer" at the Observatoire. In January 1904, he became an "astronomer's aide" and was promoted to "adjoint astronomer" in April. Later that same year he joined the Société Mathématique de France.[7] Under the direction of Maurice Lœwy,[8] and later, Édouard Benjamin Baillaud,[9] Fatou's work at the Observatoire initially involved the celestial meridian as well as determining the absolute position of the fundamental stars and the location of other celestial bodies [**F1**].

On February 14, 1907, Fatou officially earned his doctorate for his thesis entitled *Séries trigonométriques et séries de Taylor* [**1906a**], which represents an early application of measure theory to complex functions. It contained two famous propositions from complex analysis which bear his name, Fatou's Theorem,

> *Let F be an analytic function bounded inside the unit disc. Almost all radial limits*
>
> $$\lim_{r\to 1} F(re^{i\vartheta})$$
>
> *exist*

and Fatou's Lemma, a fundamental theorem in measure theory,

> *Let* $\{f_n\}$ *be Lebesgue measurable, nonnegative functions. Then*
>
> $$\int \liminf f_n \leq \liminf \int f_n.$$

---

[5]An *agrégation* is a competitive, qualifying exam taken by graduates of institutes of higher education interested in a teaching a career. They are given in various subjects throughout the humanities and sciences including mathematics. Until the end of World War I there were separate rankings for males and females. See Chervel [**1994**] for more about the agrégation exams.

[6]See Audin's book [**2009**, pp. 140–141] for an interesting discussion of Fatou's decision to work at the Observatoire. She also details frictions between Fatou and his superiors during his early years there.

[7]Fatou was presented on December 15th 1904 by Desiré André, who was its president at the time, and Émile Borel. Fatou served the SMF in many capacities, as under-secretary (1907–1917), archivist (1918–1921), vice-president (1922–1925), president in 1926 and a member of the council (1927–1928). See [**A4**], Chazy [**1932**], Fatou [**1929a**], and Audin [**2009**, pp. 168–169].

[8]Lœwy was the Director of the Observatoire de Paris during 1896–1907. He founded the department of astrophysics.

[9]Baillaud was director of the Observatoire de Paris during 1908–1926.

Fatou's Theorem states that radial limits exist almost everywhere on the circle except for a set of zero Lebesgue measure.

Fatou's health was often weak, and the astronomical observations tested his strength. In both 1906 and in 1912, he applied for extended sick leave, respectively, May 1–August 31, 1906, and March 1–April 30, 1912. He suffered from symptoms often associated with anxiety or depression, such as insomnia and heart palpitations, and he also had stomach problems. According to Bloch, Fatou had difficulty mustering enough energy to teach or speak at length in public (which may explain why he did not initially teach), but at the same time he enjoyed dancing and hiking when he was able [**1931**, pp. 52–53]. Most likely due to his poor health, he remained at the Observatoire during World War I and, unlike many in the French scientific community, did not join the in the military.

Audin asserts that he suffered from neurasthenia, which is a term—used much less frequently now, at least in the U.S., than it was during Fatou's lifetime—that covers a wide range of symptoms including those Fatou suffered from as well as chronic fatigue.[10]

During his first several years at the observatory, Fatou published only a few mathematical papers including his groundbreaking 1906 *Comptes Rendus* note on the iteration of the complex functions [**1906b**], and he evidently continued his investigation in the years following his notice.[11] This investigation came to fruition during the run-up to the 1918 *Grand Prix* when he published several *Comptes Rendus* notices, [**1917a, 1917b, 1918a, 1918b**]. Ultimately, he did not enter the competition, perhaps because the priority judgement that Julia instigated went against him. After the announcement that Julia was awarded the *Grand Prix*, Fatou published a long monograph on the iteration of rational functions in three parts [**1919-1920**], which we surmise he prepared in anticipation of the prize. These events are recounted in Chapter 6. The mathematics from [**1919-1920**] is discussed in Chapter 7.

Although we could find nothing regarding Fatou's reaction to the judgement against him, we have Mineur's comment about the priority of Fatou's work in iteration, delivered at his funeral:

> *To one of his most remarkable discoveries, i.e., to the iteration of rational functions, the name of Fatou is inalterably tied: he was the first to dare to attack the problem, the first one to solve it.* [**A11**]

Fatou's work in complex dynamics seems to have triggered a creative burst. Between 1921 and 1924 he followed his monograph on the iteration of rational functions with a two-part, 90-page paper on permutable functions, [**1923-24**] and

[10]The term seems to have originated in the U.S. in the 1870s in the work of the neurologist George Miller Beard. He thought it to be a disease brought upon by urbanization and modern life that tended to strike active professionals, and believed it to be a physical affliction of the nervous and emotional systems that could be treated though mental therapy. It was sometimes referred to "Americanitis", a term evidently popularized by the American philosopher and psychologist, William James, who was diagnosed with it. In the present-day U.S., someone who presented the symptoms associated with neurasthenia would probably be diagnosed with a more specific condition such as chronic fatigue syndrome, depression, or an anxiety disorder. Sigmund Freud had different ideas on the disease and evidently thought it had is roots in sexual neurosis. It is unclear whose interpretation of neurasthenia Audin presumed.

[11]See Section 2.2 for details about his notice; Audin [**2009**, p. 17] discusses a letter to Maurice Fréchet about Fatou's subsequent work on the subject.

a 75-page paper on the iteration functions of two complex variables [**1924**], as well as over ten shorter articles and notices concerning the iteration of complex functions, See the first four sections of Chapter 10 for a discussion of his research in the 1920s.

Around the same time, Fatou also developed a research program in celestial mechanics,[12] publishing several papers, including [**1921e, 1923c, 1928b**]. He continued with this line of research until his death; this includes two posthumous papers in which he examined the stability of a body in a fixed gravitational field, a short note in *Comptes Rendus* [**1929b**], and 64-page paper in *Acta Astronomica* [**1931**].[13]

Fatou extended his dynamical program in the mid-1920s to include the iteration of algebraic functions [**1922a**] and transcendental functions [**1926**] (see Sections 10.3 and 10.4). He was also elected as Président of Sociètè Mathèmatique de France in 1926.

After the death of Georges Marie Humbert in 1921, the chair in *Mathématiques* at the Collège de France became open.[14] Both Lebesgue and Fatou were candidates for the election. The Académie was charged with recommending candidates and by ballot vote awarded Lebesgue the first position on the ballot, with 35 votes, and Fatou the second, with 29. Lebesgue was elected later that year.[15]

Fatou's work at the Observatoire in the 1920s included observations of the celestial equator, as well the investigation of comets, planets, stars hidden by the moon, and the measurement of binary stars. In 1927 the position of titular astronomer became open. There were three candidates, and the first vote was won by Pierre Salet[16] with 36 votes; Lambert had 13, and Fatou 2. In the second vote the tally was, respectively, 36, 19, and 30.[17] In 1928, Fatou and Lambert competed again for the same position, and this time Fatou won.

Just one year later, Fatou died prematurely on the evening of Friday, August 9, 1929, at 8:00 p.m. in his room at the hotel Villa Brise Lames on Avenue Léon Dubas, in Pornichet (Loire-Inférieure), where he was vacationing not too far from his birthplace, Lorient. The exact circumstances of his death are not known.[18] No

---

[12]For a more detailed analysis of Fatou's work in celestial mechanics see Chazy's necrology of Fatou [**1932**].

[13]Aurel Witner, a Hungarian mathematician who emigrated to the U.S. to teach at John's Hopkins University in 1930 described an error he found in these papers in his article [**1946**].

[14]The Collège de France is a unique institution. It was founded in 1530, although its roots as a place of learning go back further. Its faculty give courses on their current research which are free, open to the public, and have no entrance requirements. Many of these lectures are available on podcast. Its faculty elect new members, and past chairs in *Mathématiques* include Gilles de Roberval (from 1634–1675), Joseph Liouville (1851–1882), and Edmund Laguerre (1883–1886). Currently, Jean-Christophe Yoccoz holds the chair in *equations différentielles et systèmes dynamiques.*

[15]Audin [**2009**, pp. 162–163] discusses Fatou's unconfirmed candidacies for chairs at the Sorbonne in 1923 as well as in 1927.

[16]Salet's career at the Observatoire de Paris started as a staging astronomer in 1899; later, in 1906, he became an astronomer's aide and, in 1912, adjoint astronomer. In 1927 he became a titular astronomer (see Le Gars [**2007-W**]). His main interests focused on polarization and the application of meridian instruments. He also published a book on the work of Omar Khayyam in [**1927**].

[17]See *Comptes Rendus*, 184 (1927), pp. 1153, 1394, and 187 (1928), pp. 1503, 1792.

[18]The certificate, signed by Florian Claude Marie La Porte, chief hydrographic engineer for the Navy, commander of the Legion of Honor, residing at rue Pont Carré in Lorient (close to Pornichet), and countersigned by a local officer, Auguste Jarniou, was filed the next day at 3:00

cause for the death was given on his death certificate, but Audin makes a persuasive case that he died from an unattended stomach ulcer [**2009**, pp. 170–172]. His nephew Robert noted

> *Having never thought it useful during his life to consult a doctor, my dear uncle died suddenly in a hotel room at Pornichet.* [**F1**]

Fatou's career was on an upward arc at the time of his death, and according to the Lorient daily newspaper, the *Nouvelliste du Morbihan*, he was soon going to enter the Académie des Sciences [**A11**]. This newspaper was probably mistaken: Audin quotes Fatou's brother Louis as saying that Picard assured him that Pierre would have been elected to the Académie at the end of the year. As there was an open seat in the Astronomy Section, this is most likely what Picard was referring to. But Audin's analysis makes it seem very unlikely that Fatou would have been elected [**2009**, pp. 165–166]. Perhaps Picard was being kind—or perhaps he was simply telling Admiral Louis Fatou what he thought the admiral wanted to hear.

Two death announcements were published in the Lorient daily newspaper, the *Nouvelliste* on August 11 and 13, as was a memorial article on Thursday, August 15, 1929 [**A11**]. After a religious service in the church of Saint-Louis, the funeral was held at 8:45 a.m. on August 14th.

Besides three family members—his brother Louis, his nephew Paul, and a brother in law—a report of the event lists the attendance of several dignitaries, including Léon Guillet, director of the École Centrale.[19] The eulogy was given by his colleague Mineur and included memorials from his colleagues at the Observatoire de Paris [**A11**]. At the time of his death, Fatou lived at 172 Boulevard Montparnasse, Paris [**A3**]. Pierre Fatou was buried at the lot 38/16 in the "Carnel" cemetery of Lorient.

M.2. The Private Fatou

As we mentioned, there are only a few sources for information about Fatou's life outside of mathematics. The necrology by Fatou's good friend Bloch [**1931**] offers a few glimpses into Fatou's private life, but the richest details are to be found in the fond recollections of Fatou's nephew Robert, who seemed to possess a deep affection for his uncle, even though he only knew him during two periods in his life. The first was in 1911, when he attended the Lycée Saint-Louis (which his uncle had also attended), and in 1920, when, while waiting for his naval assignment, he would sometimes take a few days leave in Paris.[20]

---

p.m. and contains a mistake: It lists the deceased as "*René Joseph Louis Fatou, born on February 28th 1878, son of Prosper Ernest Fatou and of Louise Eulalie Courbet*". Relatives verified that it was Pierre Joseph Louis Fatou [**F1**, p. 7].

[19] An old French institution which had become an engineering school.

[20] Robert Fatou had a long career in the French Navy, and became *Capitaine de vaisseau* (Commander) of the French training ship *Jeanne d'Arc*, on which he served during World War II. As France fell to the Germans, the *Jeanne d'Arc* helped transport France's gold reserves to Halifax, Nova Scotia, Canada. The ship then went to the island of Guadeloupe, where it was under the service of the Vichy government until 1943. Fatou was evidently an officer on the ship during this time, but not commander, and according to one source, chafed under the ship's service to Vichy (see `http://www.lescrutateur.over-blog.com/article-13030762.html`). After 1943 the ship sailed with the Free French, later participating in the Allied invasion of France. The website `http://www.netmarine.net/bat/croiseur/jeannedarc/histoire.htm` contains a photograph of Fatou as commander, dated 1946. Fatou also wrote a translation of a book about an incident

Certainly Bloch and Robert Fatou's accounts display a survivor's gilded recollections of someone they held dear, but they are remarkably consistent, and the same man emerges from both their portraits, something that does not always happen when one reads two accounts of a person's life. They leave the impression that Fatou was cordial and modest, but distrustful of what he perceived as vulgarity and especially of those who thought overly much of their accomplishments. Indeed, his colleague Mineur described him as rather the opposite of self-important, and recounts hearing Fatou remark shortly before his death that,

> *When I wonder about what Science would be had I never existed, I feel myself become quite modest.* [**A12**]

He evidently tended to avoid situations that put him at the center of attention. According to Bloch,

> *From the first glance, Fatou is a gentleman, a man of quality, whose distinction is absolutely evident; he is the epitome of a rare and precious man.* [**1931**, p. 53]

His nephew wrote,

> *...I started to see, under the extreme originality of my uncle, a heart of gold with an imperturbable frankness and freedom of judgment, hindered by neither convention nor the opinions of others. Nevertheless, his natural gentleness forbade him to offend his neighbor, and he preferred to remain quiet rather than to display contrary feelings to others.* [**F1**]

Music was a passion of his, perhaps developed while listening to his sister and brother play the piano as a child [**F1**]. Both [**1931**] and [**F1**] reveal that Fatou thoroughly enjoyed concerts and would sometimes travel to Bayreuth[21] for the annual Wagner festival. He also liked Russian and Spanish music.

Despite his love of music, he was not an accomplished musician, although according to his nephew he sometimes played piano for his own enjoyment. He also liked to read the scores of his favorite music, and at concerts, he often focused on the techniques of the musicians. Fatou also took delight in the singing of his Brazilian parrot Lorito, who he inherited from his father. Robert describes how they would sing a song with the words

> *When I drank claret,*
> *All turns, all turns.*
> *When I drank claret,*
> *All turns at the cabaret.*[22]

Often the bird would begin the song unbidden, which Robert found quite comical.

Both Bloch and Robert Fatou report that in addition to his great passion for music—a passion evidently only eclipsed by his love for mathematics—Fatou enjoyed photography. His goal seemed to be capturing moments of everyday life rather than the making of art. Bloch compared it to photojournalism and noted, "To him his subjects took on the most spontaneous and charming shapes." Fatou's

in the Chinese Civil War involving the *H.M.S. Amethyst* [**1972**], and an article on French Naval history [**1949**].

[21] A town in central Germany that was home to Richard Wagner for the last years of his life.

[22] The original French: *Quand j'ai bu du vin clairet, Tout tourne, tout tourne. Quand j'ai bu du vin clairet, tout tourne au cabaret.*

best work, Bloch said, was "artful" and possessed a "surprising delicacy". Robert Fatou did not share this high opinion and judged them to be rather ordinary, although he understood their value to his uncle both as a source of humor (for example, his photographs of Parisians in snobbish displays of outlandish dress) and as a way to remember things he held dear, such as his childhood in Bretagne.

Fatou also enjoyed dancing. On his second visit with his uncle, in 1920, his nephew was astonished to discover Fatou's fashionable side:

> *Indeed, I could not have been more surprised than to be invited by my uncle, on a beautiful Sunday, to come with him to Luna Park where he was going to amuse himself with the delights of the Tango with his dance teacher, who was a very beautiful girl, charming, but not conceited at all and escorted by her father. Thus—I thought—this scientist hovering onto the highest spheres of mathematics, this man indifferent to the vulgar joys of the world and disdaining the worldly meetings, used a professor's expense to start learning the secrets of the fashionable dance.*
>
> *I have seen it all!*
>
> *And so I stood together with the dancer's father for several hours, drinking beer and watching while my uncle oscillated into the dancer's arms that conscientiously guided her pupil's steps.*
>
> *I never doubted the perfect honesty of the relationships between my uncle and his dance teacher, always chaperoned by her father.* [**F1**]

In his astonishment, Robert Fatou mused upon whether the joy Fatou was experiencing was from the music alone, or whether the movements of the dance itself hypnotized him, as might a *fakir* or whirling dervish:

> *Whatever it was, I could not bring myself to ask my uncle about it.*

Fatou was interested in politics, although it is not clear how active he was. He also liked to spend evenings out with friends and to travel. Bloch recalls excursions along the beaches of Bretagne, the mountains of Tyrol, and the lake regions of northern Italy.

In Paris, he sometimes frequented an establishment near the Observatoire de Paris called *Le Bal Bullier*, where one could eat, dance the tango, listen to music, watch, or—in Fatou's case at least—do mathematics: According to Bloch, part of Fatou's thesis was written there [**1931**].[23] Bloch offers a view of Fatou shortly before his death as a man who finally obtained the position, notice, and, presumably, salary he had earned:

> *In the last months of his life, when he gained independence and authority, he took care of applying these two forces to the renewal of astronomy. The long evenings during mid-July 1929 were the last I spent with him, and had been devoted to the discussion of future plans, to the improvement of research programs, to the exchange of*

[23] *Le Bal Bullier*, whose roots go back to the French Revolution, has a colorful history, replete with several closings and reopenings. *Le Bal Bullier* was also a haunt of Parisian artists and writers, such as Alfred Jarry, Guillaume Apollinaire, and Andre Gidè and Sonia Delaunay, whose 1913 painting, *Le Bal Bullier* captures its dance hall with an almost fractal intensity. *Le Bal Bullier* was a rival of sorts to Luna Park, where Fatou danced the tango.

viewpoints about the education of a new generation of astronomers. [**1931**, p. 56]

One of the more touching passages in the nephew's remembrance is a spirited defense of his uncle against remarks made by the philosopher Alain several years after his death.[24] At one point Fatou notes that an unnamed but "very eminent" philosopher quipped that Fatou's thesis "was probably understood by only two people in all the world" [**F1**]. Near the end of his piece Robert quoted—and this time used his name—Alain's waggish remark that his uncle died from "*l'ennui mathématique*".[25] Robert's response: "I would hate to think so. Nothing seems less certain to me than that."

Robert found out about his uncle's death when he was in Australia, quite a while after it happened and experienced what he referred to as a deep sorrow. The news came to Bloch while he was vacationing in Switzerland, near Zermatt, with his brother Eugène, also a physicist. Coming back to their hotel, they unexpectedly ran into Paul Montel who said [**1931**],

> *"Do you know the news from Paris?"*
> *"What news?"*
> *"Fatou is dead."*

M.3. A Fatou genealogy[25]

The Fatou branch: Fatou's grandfather, Jean Baptiste Ambroise, was born on November 7, 1786, at Château de Versailles in what is now called the department of Yvelines, formerly known as Seine et Oise, to Jean Louis Fatou, a pharmacist, and Marie Anne Simonnot. On September 16, 1817, Fatou became a military pharmacist in Brest, department of Finistère. After moving to Quimper, he met his future wife, Jeanne Marie Marguerite Mougeat, whose parents were Pierre Louis Mougeat, also a pharmacist, and Anne Marie La Franc. They married on June 22, 1816, and died at Quimper; their five sons were

• Prosper Ernest Fatou, Pierre's father, who married Louise Eulalie Courbet on July 25, 1866, at 10:15 a.m. At Pierre's birth, his current rank was *Lieutenant de Vaisseau de première classe* (roughly equivalent to *Lieutenant Junior grade* in the U.S. Navy, *Sub-lieutenant* in the Royal Navy, and *Tenente di Vascello* in the Italian Navy). Since his parents had already died, the authorization to marry was given on July 5, 1866, by the current prefect Rear Admiral ad interim. In 1878 Prosper Ernest retired with the final rank of *Capitaine de Frégate.* He was a Cavalier of the Legion of Honor. He died in Lorient at rue de la Corderie $n^\circ 1$ at 3:00 a.m. on March 2, 1891, age 58. His mother died in 1911.

---

[24] Alain was the pseudonym of the philosopher Émile Auguste Chartier, who taught philosophy in Lorient between 1893–1900 where he had Pierre Fatou as a student. Alain generally wrote in a short, constrained form he called *propos* that he published in French daily newspapers. Although they often concerned political topics, he wrote sometimes about mathematics, and displayed a suspicion towards non-Euclidean geometry and a fourth-dimension, see *propos* XX, XXI, XXXIV in [**1942**]. His work is evidently not well known in the U.S., and a English speaking reader wishing to know more about him might consult Howell [**1965**].

[25] Alain's remarks are from a passage in his book *Histoire de mes pensées*, quoted in full in Audin [**2009**, p. 128], that also includes the quip about Fatou's thesis.

[25] Much of the information below is from the Archives in Quimper, the capital of the Finistère department in Bretagne, [**A14**].

• Louis Ambroise was appointed first class pharmacist in July 1849 in Paris after becoming a doctor of medicine on January 25, 1849. He married Elisabeth Marie Joubert (born in Paimboeuf on September 27, 1827, in what is now the Loire-Atlantique department) on April 12, 1856. They had four children: Louise Elisabeth Marie (born on January 1, 1857), Louis Émile Eugène (August 28, 1859), Amélie Anne Aimée (January 23, 1863), and George Ernest Joseph (September 23, 1867).

• Pierre Eugène, a pharmacist, who married Adelaïde Jan de Lagillardaie on August 5, 1872. They had a daughter, Anny Marie (1858–1874) and a son Paul Julien (1861–1942).

• Émile Auguste, also a *Capitaine de Frégate*, married Joséphine Félicité Philomèle Septans on August 5, 1872. They had one son, Émile Henri (1873–1957).

• Louis Eugène Alfred, a pharmacist, married Marie Apolline Rivet. Their only known son was Guillaume Fatou (lawyer, 1892–?).

The Courbet branch: Fatou's maternal grandfather was Pierre Joseph Courbet. Upon retirement his rank was *chef d'escadron de gendarmerie maritime* (a commandant in the national maritime police). He died at Brest. His wife Félicie Farvet, was the daughter of Claude François Farvet, a landlord, and Claude Françoise Pierrecy in Lorient.

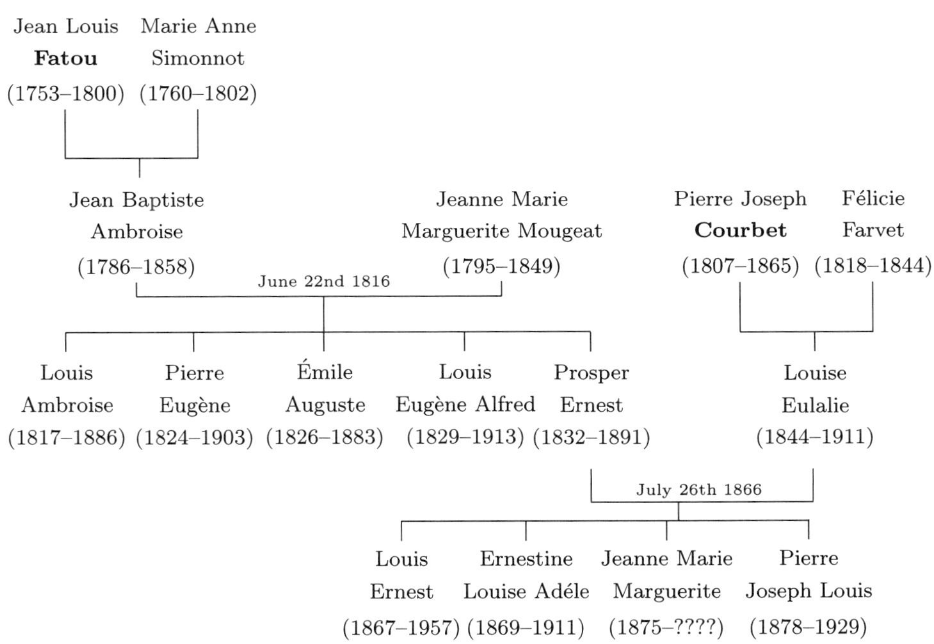

FIGURE M.1. **Fatou's family tree.**

APPENDIX N

Gaston Julia: A Biographical Sketch

A detailed narrative of Julia's life has yet to be written.[1] While we obviously cannot remedy that in such a short space, we hope to sketch his life in human terms, especially as it was lived prior to the horrific events that resulted in his becoming an iconic war hero. We rely on four kinds of sources: published accounts of his life and mathematics by others that are readily available in the literature and online,[2] archival documents, such as those found in his dossier at the French *Archives Nationales* [**A6, A7**] and Julia's own *Souvenirs sur Georges Humbert et Camille Jordan* [**A8**]; notices published in *Comptes Rendus* and *L'Enseignement Mathématique*; and finally, several *Allocutions* by Julia and others found in his Œuvres [**1968-1970**]. We do not analyze his mathematics here since we discuss it in Chapters 7, 9, and 10. Those wishing an overview of his entire mathematical career might want to look at Hervé [**1963**] or Garnier [**1978**].

N.1. Julia's early years

JULIA, GASTON MAURICE (1893–1978). Julia was born to Joseph Julia and Dolorès Benavent Julia on February 3, 1978, at 3:00 a.m. in Sidi Bel Abbès, in northern Algeria, not far from the Mediterranean Sea. According to Garnier's necrology in *Comptes Rendus* [**1978**, p. 126], Julia was raised in a deeply religious Catholic environment,[3] by a family of very modest means. Julia often worked to help his family, and his father, who was an agricultural mechanic, worked very hard to provide for his son's education.

As a young child, Julia attended a small Christian Brothers' school in Sidi Bel Abbès, and his unusual intelligence came to light early. By the time Julia was 11, the school's director, Fr. Attale, evidently decided that Julia's talents would soon eclipse the school's ability to nurture them. In his *Œuvres*, Julia recounted Fr. Attale saying to him,

> *My little boy, we have nothing more to teach you here.* [**1968-1970**, p. 13]

[1] Audin argued in [**2009**] that due to Julia's iconic stature one perhaps cannot write such a work. We are not so sure, although we make no pretense that we are getting to the essence of the private Julia. But we think that an examination of Julia's public life in the context of his involvement in two wars, as a hero in one and briefly investigated (and exonerated) for collaboration in the other, as well as his stature as a prince of mathematics for over 50 years is not only possible but it would make very interesting reading—or viewing.

[2] For example, Coulomb [**1978**], Garnier [**1978**], Hervé [**1963, 1981, 1992**], and Mendès-France [**1988**].

[3] His grandmother venerated Saint Theresa of Avila and of Saint John of the Cross, according to Garnier [**1978** p. 126].

Since Sidi Bel Abbès did not have a school appropriate for him, Julia and his family moved to Oran on the Mediterranean Sea in 1904, around 35 miles away. A door to a new world opened to him. Again, from his *Œuvres*:

> *...from only one tiny classroom with a single teacher to a place with many classes and many teachers.* [**1968-1970**, p. 13]

However, new obstacles arose. For one, he was not offered a full scholarship. The school also had a language requirement, but he had no knowledge of Latin or German. Julia, however convinced the German teacher to set Christmas as the deadline for demonstrating his proficiency in German. He beat this deadline handily and passed the exam within a month or so.[4]

He successfully completed both *collège* and the *lycée* and passed his *baccalauréat* exam in 1910. Around this time, he also published his first work, a solution to a problem posed on an *agrégation* exam, a qualifying exam for prospective teachers of mathematics (see footnote 5 in Appendix M), in the *Journal de Mathématiques Élémentaires.*

Julia was clearly ready to leave Oran. He was a very distinguished student with extraordinary talents in the sciences, especially mathematics and physics, and his goal was to eventually attend one of the *grandes écoles* (see page 341). However gaining entrance was quite competitive and required additional preparation.

As is the case in the present day, to prepare for the entrance exams to the *grande écoles*, students during Julia's time attended preparatory classes known in French as *Classes Préparatoires aux Grandes Écoles* (CPGE). Admission to preparatory classes required the student to have earned a *bachelier* and was subject to the judgement of a commission. One could choose preparatory classes in either literature or science, and Julia, not surprisingly chose the latter, the *Classe Préparatoire Scientifique.*[5] Obviously, the best CPGE were in Paris, at the *grandes lycées*, but Julia's family lacked the means to send him France, so he begin his preparation under the tutelage of his teachers of mathematics and physics at Oran—M. Caron and M. Robert [**A1-1911**, p. 3].[6]

The financial barriers to attending school in Paris were resolved when, after making a very strong impression on an inspector general who was visiting Oran, he received a fellowship to attend the *Janson-de-Sailly* in Paris (Garnier [**1978**, p. 127]).

The opportunity to attend the Lycée Janson was especially fortuitous since, along with the Lycée Saint-Louis (which Fatou had attended over 10 years earlier) and Louis-le-Grand, its CPGE had an excellent reputation. However, before he could leave Oran, another impediment arose. Julia fell seriously ill with typhoid fever in late summer 1910, soon after his *baccalauréat* exam, and had to delay his departure for France. Upon his recovery that fall, he finally left Oran on the *Duc de Bragance*, bound for Marseille (Julia [**1968-1970**, p. 13]).

---

[4]During a speech he gave at the 50th anniversary of the Lycée d'Oran, Julia recalled how he was worried that he might be refused admittance because of this, and expressed regret for not having spent the summer before coming to Oran more productively (Julia [**1968-1970**, p. 13]).

[5]This typically lasts two-years, the first year consisting of *mathématiques supérieures* and the second, *mathématiques spéciales*. Classes were strictly propaedeutic (which still holds today) and confer no degree. "None is admitted to the école but through the exams ..." [**A1-1911**, p. 3].

[6]In his *Œuvres* Julia did not use the first names of his teachers, always using the honorific "M." for *Monsieur*.

Despite the interruption occasioned by his illness, Julia's talent and preparation enabled him to skip the first year of the CPGE and move directly to *Mathèmatiques Spéciales*, and he completed his CPGE studies a year early. He took his entrance exams for the *grandes écoles* in 1911, and in July, his teacher of mathematics at the Lycée Janson, to whom Julia referred as "M. Cor", sent a letter to the director of the École Normale Supérieure, warmly recommending Julia together with two other students, both for his academic virtues and and his temperament,

> *Julia: A first year student who did not attend the special preparatory classes. An especially distinguished spirit. He longs to enter the* école. [**A7**]

Shortly thereafter, Julia took the entrance exams[7] (*concours d'entrée*) at both the École Polytechnique (X) and the École Normale Supérieure (ENS),[8] arguably the two most celebrated *grandes écoles*. He placed first on both,[9] a remarkable achievement, last accomplished by Gaston Darboux. Moreover, he dominated the tests. The gap between Julia and the second place at X was 131 points, while the gap between the second place finisher and the last was 111. He distinguished himself as well on the oral portion of the test at X, where Ernest Vessiot and Émile Borel were his examiners (Garnier [**1978**, p. 127]). Because of his rankings he received the title of *Cacique* (chief) at the École Normale Supérieure and *Major* at the École Polytechnique.

Garnier speculates that due to the urgings of Caron, his former teacher of mathematics at Oran, and his desire to teach [**1978**, p. 127], Julia decided to enter the École Normale Supérieure to study mathematics, but also to take advantage of the access to the culture it provided.

During his early years as a *Normaliens*, he began to develop ties to a number of eminent French mathematicians, including Georges Humbert, Émile Borel, Émile Picard, Camille Jordan, and Élie Cartan [**A8**].

Julia published his first research paper in 1913 at age 20, *Sur les lignes singulières de certaines fonctions analytiques* [**1913**], devoted to monogenic functions which could not be continued through a set of singularities (see Appendix C). The study of continuation of analytic functions was a popular topic at the time and, interestingly enough, Fatou also published a paper on the same topic in the same year [**1913a**]. In the fall of 1913 Julia helped care for his old teacher Caron during a period of convalescence following liver surgery.

His education continued until the outbreak of World War I. During the summer of 1914 the imminence of war forced a total mobilization of the French military. On

---

[7]The entrance exams at the École Polytechnique consist of three tests: 1st, compositions (numerical calculus, graphical drawing, imitation drawing); 2nd, first level test (Algebra and Trigonometry, Analytic, Mechanic and Descriptive Geometry, Physics and Chemistry, Calculus); 3rd, second level test (listed in order of importance, Mathematics, Physics, Chemistry, German and Physical education (fencing, gymnastics and horsemanship). A poor performance on the first level test excludes one from the second level, which is used to determine the ranking, [**A1-1911**, p. 5]. Over 220 students took the test.

[8]An examination of the list of candidates for the 1911 exams reveals that many students took both.

[9]The scores ranged from 0 to 20 points per question, [**A1-1911**, p. 8]. At X, Julia scored 565 points at the compositions/first level tests and 1513.50 at the second level. He also scored 1829 for the ENS exam.

August 2, 1914, after a brief meeting at the École Normale Supérieure, the *Normaliens* decided they should go to the front (Plantefol [**1961**, p. 312]). Julia, now 21, jointed the 57th infantry regiment and was soon promoted to sublieutenant[10] of the 144th Infantry Regiment and finally, on December 1914, assigned to the 5th company of the 34th regiment, which was headed for the northeastern front where Julia was severely wounded. His was shot in the face and disfigured, and for the rest of his life wore a patch to cover his nose. He was awarded as a *Chevalier de la Légion d'honneur* on February 2, 1915.[11]

Julia's recovery was long and arduous, and required multiple operations. He convalesced first at the hospital *Val-de-Grâce*, a military hospital in Paris, and later at a temporary military hospital set up at the École Normale Supérieure to treat the large influx of wounded soldiers (Plantefol [**1961**, p. 312]) where, according to [**A8**], he received visits from several mathematicians, including Paul Dupuy, who was an official at the ENS and authored a book on Évariste Galois, and Julia's teachers Picard and Humbert, the latter of whom who encouraged Julia's return mathematics. Julia also met a nurse, Marianne Chausson, daughter of the composer Ernest Chausson, and they fell in love, marrying in 1918.

Julia's wounds pained him his entire life, and as he aged, he suffered from a variety of ailments. His dossier [**A6**] lists several extended health-related absences. When his wife Marianne died in 1971, he moved to the *Hotel des Invalides* in Paris, a home for injured and disabled soldiers founded by Louis XIV. Julia died there on March 19, 1978, and his funeral was held on March 23 at the church of Saint-Louis des Invalides.

N.2. Julia's academic career

As his wounds healed and his pain lessened, he returned to mathematics. Julia referred to this time as his time of "relearning" analysis (Planetfol [**1961**, p. 314]). He attended Humbert's course in early 1916 and became intrigued by the work of Charles Hermite. It was during this time that he found the topic for his thesis.[12]

In 1917 he finished his thesis and submitted it for the *Prix Bordin*, which he won, along with 3000 francs. The theme for this contest was the improvement of some points in the arithmetic theory of nonquadratic forms ("*Perfectionner en quelque point important la théorie arithmétique des formes non quadratiques*" L'Enseignement Mathématique [**1916**, p. 134]). That same year he wrote his long monograph on the iteration of rational complex functions [**1918a**], which earned him the 1918 *Grand Prix des Sciences Mathématiques* (see Sections 6.4, 6.7, and 7.2).

Like Fatou, Julia extended his research on the iteration of holomorphic functions well into the 1920s, investigating, among other things, permutable rational functions (see Appendix J), the iteration of rational functions of two variables, and functional equations, which he did not discuss in great detail in his 1918 monograph. The last subject related to iteration that Julia researched was series of iterates around 1930–31, that is, series of the form $\sum_{k=0}^{\infty} a_k f^k(z)$, where f is a

[10] The promotion of ENS students to the rank of officer was quite common.

[11] See Sections 6.2 and 6.3 for a longer account of his military service and his recuperation from his wounds.

[12] See [**A8**, p. 88] and Garnier [**1978**, p. 127]. We discuss this period—as well as the gestation of his thesis—in Section 6.3.

rational function with an attractive or rationally indifferent fixed point. He also published a three-part memoir [**1919-21**] that included a generalization of Picard's theorems involving the existence of so-called "curves of Julia" (see Appendix E). Table 3 in Section 10.1 and Section 10.5 summarize his research in the 1920s related to the iteration complex functions.

In the late 1930s his interest shifted to Hilbert spaces, and of the final 59 papers listed under the heading "*Travaux de recherches*" (1938–1954) in his *Œuvres*, over 30 explicitly referenced Hilbert spaces in the title.

With his second major mathematical prize in as many years—and being a war hero to boot—Julia's career at the end of World War I was in ascendance. He found a teaching position in Paris shortly after he obtained his thesis, becoming *maître de conférences*[13] at the ENS in 1919 and at the Sorbonne in 1922.

In 1925 he was appointed *professeur titulaire* at the Sorbonne, and then to its chair of *Mathématiques générales* replacing Montel who was awarded the chair of *Mécanique rationnelle* where he served until 1934; this was followed with a succession of appointments within the Faculté des Sciences at the Sorbonne: *Calcul différentiel et calcul intégral* (1931–33), *Application de l'analyse à la géométrie* (1933–35), *Mécanique analytique et mécanique céleste* (1935–1941), and finally *Analyse supérieure* from 1941 until his retirement on February 3, 1964 [**A6**].

He had a parallel teaching career at the l'École Polytechnique. In 1919 he became a *Répétiteur d'analyse*,[14] initially under Hadamard (Garnier [**1978**, p. 131]), and remained in this role until 1937, when he was named to the *Géométrie* chair, succeeding Maurice d'Ocagne, a position Julia held until his retirement.

Julia was awarded numerous honors in the 1920s and 1930s. In 1926 and 1928, he won the *Françoeur* and *Poncelet* prizes, respectively, for his body of his mathematical works and for his work analysis and dynamics. In 1931 he was awarded the prize *Petit d'Ormoy* and was appointed Research Director at the *Caisse Nationale des Sciences*, which funds scientific research. In 1932, he was elected Président of the Société Mathématique du France. The next year he founded the *Cahiers Scientifiques*. These accomplishments culminated with his election to the Académie in 1934, filling the seat vacated with Paul Painlevé's death. The brief announcement of his election in *L'Enseignement Mathématique* [**1933**, p. 261] does not mention his work in iteration by name, noting instead the "points of Julia" stemming from his research on the Picard theorems, as well as his "remarkable work in analysis and geometry".

Julia remained an active researcher through the late 1940s. From the 1920s through the late 40s he published over 120 research papers and notes, and (counted with multiplicity) over 25 books including collaborations with Picard, Humbert, Jules Drach, and Henri Béghin. Julia's academic service included a stint on the editorial committee of *Annales Scientifiques de l'École Normale Supérieure* and organizing a subscription campaign to raise funds for the completion of Poincaré's collected works. In 1950 Julia was elected president of the Académie.

---

[13]In the French system there are generally two ranks of full-time professorship, *maître de conférences* and *professeur des universités* (the equivalent of full professor), rather than the three common to the U.S. In the sense of being the first step in the academic ladder, *maître de conférences* bears some similarity to the position of assistant professor, but it is probably closer to an associate professor in terms of prestige and job security.

[14]*Répétiteurs* assisted professors, and many famous mathematicians first served as a *répétiteur* at the École Polytechnique, for example André-Marie Ampere, Joseph Fourier, and Hermite.

Befitting a mathematician with his academic resume, Julia travelled quite a bit. He spoke at several international conferences, and he also visited Germany. In 1932 he gave seminars at the universities in Berlin and Göttingen on iteration and quasi-analytic functions (Jahresbericht [**1933**]).[15] Julia also travelled to Germany in late June 1937 to speak at the second centenary of the University of Göttingen (see Julia [**1937**]). He served on the international editorial board of *Zentralblatt für Mathematik* before World War II. These ties to Germany bred controversy after the end of the German occupation of France.

N.3. Julia under suspicion[15]

Julia travelled to Germany in 1942 to attend conferences in Berlin and Göttingen that had evidently been scheduled before the war. To call them conferences may be a misnomer since most academics were involved in the war effort. He also continued to serve on the editorial board of *Zentralblatt für Mathematik*. For the French provisional government formed at the end of the German occupation in 1944, this raised the possibility of collaboration, and an investigation ensued.

During the investigation, Julia was temporarily suspended, and his pay was cut in half. On November 10, 1944, the Superior Council of Inquiry of the Ministry of National Education met to hear Julia's case and recommended that the sanctions be lifted. While citing Julia's travels to Germany and his association with *Zentralblatt*, it came to the conclusion that Julia had travelled to Germany to use his influence with the Germans to seek the release of French mathematicians who were prisoners of war. Even "if M. Julia had agreed to collaborate with the Germans", the commission report said, it was with good, if "imprudent", intent, and he did not seek to support the enemy. It also said that his heroism during World War I, as well as his "grave mutilation", should be noted. René Capitant, the Minister of National Education, decreed on November 23, 1944, that the suspension be lifted and the charges dropped, and with that, the official investigation against Julia closed.

N.4. A Julia Genealogy

By the 1600s Julia's forebears had settled in Saint-Julien-de-Briola and Villasavary, two very small villages in the Department of Aude in the south of France whose economy revolved around the production of grains and grapes. Julia's grandfather, Antoine, was born in 1823 in Villasavary and worked as a wheelwright.

Years of both economic and military tension, exacerbated by European imperialism, finally boiled over in the late 1820s, and France began its conquest of what is now known as Algeria with a blockade of the city of Algiers in 1827. It invaded Algeria in 1830 and colonization began soon afterwards.

The French government encouraged migration of skilled workers to Algeria, which perhaps explains why Julia's grandfather moved to Sidi Bel Abbès, Algeria. The French colonists were joined by other Europeans, and on August 27, 1853, Antoine Julia married Josephine Sepulcre, whose family came from the Spanish

[15]There was a long-standing German interest in the iteration of complex functions, dating back to Ernst Schröder's groundbreaking papers on the iteration of complex functions [**1870, 1871**] (see Chapter 1). Humbert Cremer wrote a short monograph summarizing the works of Fatou and Julia [**1925**], that grew out of a presentation he gave at the University of Berlin in the summer of 1924, which introduced Germany to their research. Beginning in 1927, Cremer made significant progress on the center problem in series of papers written over a ten-year period (see Chapter 11).

[15]This section is based on documents found in Julia's dossier at the *Archives Nationales* [**A6**].

town of Novelda in the province of Alicante. Their oldest son Joseph, Gaston Julia's father, married Dolorès Benavent of Valencia, Spain, in Sidi Bel Abbès on May 17, 1890.

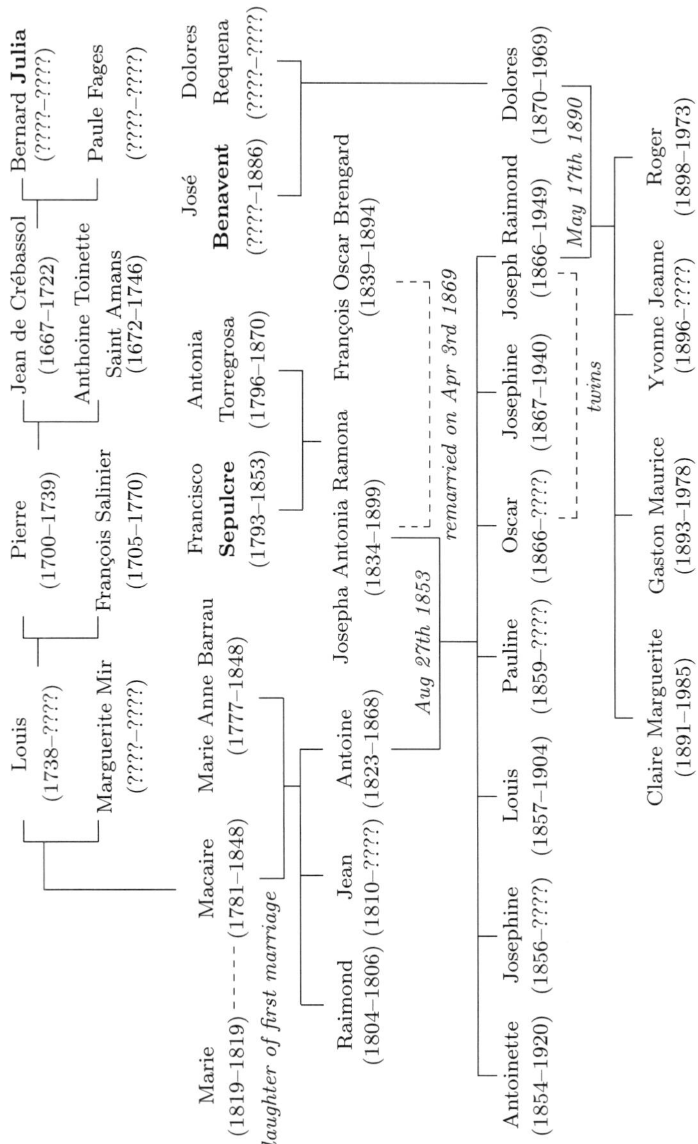

FIGURE N.1. Julia's family tree.

APPENDIX O

Selected Biographies

BĂDESCU, RADU LAZAR (1904–1988). After graduating from the Lyceum Matei Basarab in 1922, Bădescu entered the Technische Hochschule in Charlottenburg in Berlin to study engineering. Shortly thereafter he gave up engineering for mathematics, and graduated from the University of Cluj in Cluj-Napoca, Romania, in 1926. That same year he started his academic career as a teaching assistant at the University of Cluj. After passing his exams, he became a provisional full professor of mathematics at the Normal School of Piteşti in May 1927. Soon thereafter he was awarded a fellowship and moved to Geneva, Switzerland, where he obtained his Ph.D. in mathematics in 1930 and became an instructor before returning to the University of Cluj as an associate professor in the mid-1930s. He was appointed to full professor 1941, and remained at the University of Cluj until his retirement in the early 1970s.

His dissertation concerned linear integral equations, and several notices related to it appeared in *Comptes Rendus*, some of which are discussed in Section 13.3. He also visited Paris in the 1930s where he met Picard and Goursat. Bădescu's research mainly concerns functional analysis, in particular, work related to the Fredholm and Volterra equations. See Andonie [**1967**, pp. 53–60] for more information about his life and work. Our sources included Gabriel T. Pripoæ and Doru Stefanescu from the University of Bucharest, Romania.

BÖTTCHER, ŁUCYAN (1872–1937). Böttcher was a Polish mathematician born in Warsaw, which was then under Russian control. In 1893 he attended the University of Warsaw, at the time a Russian-language university. Shortly thereafter, Böttcher became involved in protests against Russia and was evidently forced to relocate (Domoradzki [**2010**, p. 62]). He continued his university studies at the Polytechnical School of Lwów[1] but soon resettled in Germany where he attended the University of Leipzig, earning his Ph.D. in 1898 with his dissertation [**1898**]. While at Leipzig, he attended courses taught by Sophus Lie and Felix Hausdorff. According to Domoradzki [**2010**, p. 64], Lie found Böttcher to be intelligent, hard working, and independent. Upon his graduation, Böttcher returned to Lwów and found a teaching position at his former university, the Polytechnical School of Lwów, where he taught until 1935. Böttcher published papers in several languages, and his chief mathematical interests were the iteration of functions as well as the foundations of mathematics. However, in 1912 he became interested in spiritualism and the occult, and his mathematical output subsequently diminished. Domoradzki [**2010**, p. 62] speculates that perhaps these interests minimized his interactions with the

[1]The city of Lwów (now referred to as Lviv) was a part of the Austrian-Hungarian Empire in the 1800s. Currently, it is part of Ukraine.

generation of Lwów mathematicians, including Stefan Banach, that emerged after World War I.

CREMER, HUBERT (1897–1983). Hubert Cremer was born to an academic family in Munich. His father Max was a physicist who is considered by many to be the inventor of the glass electrode. His brother Lothar was a professor at the Technical University of Berlin who studied acoustics, and his sister Erika was a chemist who made significant contributions to the study of gas chromatography.[2]

Cremer was drafted into the German army in 1916 and discharged after World War I. Like Julia, he suffered a serious facial wound: His jaw was evidently shattered by a bullet and replaced by a shin-bone. Cremer initially studied medicine at the University of Berlin, but soon turned to mathematics and astronomy. His teachers at Berlin included such eminent scientists and mathematicians as Richard von Mises, Max Planck, Albert Einstein, Erhard Schmidt, and Ludwig Bieberbach, who was the advisor of his dissertation, *Zum Zentrumproblem* (The Center Problem), which he obtained on July 29, 1927. His thesis was the basis for his groundbreaking work of the same name, discussed in Chapter 11.

Cremer taught at the universities of Münster and Leipzig. At Leipzig he served as Paul Koebe's assistant and become close to him. In the early 1930s he moved to the University of Köln, where he earned his *Habilitation*[3] in 1931. In 1940 he became a full professor at Breslau University of Technology.[4]

Cremer remained in Germany during World War II and was characterized by Sanford Segal as a "tacit anti-Nazi". Segal includes an excerpt from a post-war letter from Cremer to Friedrich Levi, who had resettled in Calcutta after he was forced to leave Germany before the war, where Cremer railed against the butchery of the Nazis, noting that while he "knew nothing of these horrors ... we dimly suspected them" [**2003**, p. 491].

In 1947 he was appointed as visiting professor at Aachen University of Technology, and from 1949 until his retirement in 1966 he served as a full professor. He became the director of the institute of mathematics as well as the computer science center. Under his direction Aachen developed a strong computer science center. It had a Zuse Z22 in 1957, a Siemens 2002 in 1959, and a CDC 6400 in 1965; the latter was evidently the largest computer in Germany at the time.

Cremer's research included function theory and the center problem before the war, and theoretical acoustics, stability problems in control theory, and computer science, after it. He and his brother Lothar, a professor at the Technical University of Berlin, published a joint work entitled "Theoretical derivations of the reverberation laws" [**1937**]. Cremer was the author of *Carmina Mathematica*, a very popular collection of humorous poems about mathematics, one of which is called "*Die Ballade von armen Epsilon*" ("The ballad of the poor epsilon").

Cremer died on February 26, 1983 (see [**A2-W**]).

---

[2]See Bobleter [**1990**] for more information on Erika Cremer.

[3]There is no equivalent to the *Habilitation* in the U.S. It is akin to a second doctorate, except that the research one does to qualify for it is to be independent (as opposed to being under the direction of a dissertation supervisor) and of a higher quality than what is usually required for a Ph.D. In some respects, the process one undergoes for the *Habilitation* is similar to a tenure review.

[4]Breslau was the name of Wrocław prior to World War II when it was part of Germany.

KASNER, EDWARD (1878–1955). Kasner was born in New York City and educated at the City College of New York, Columbia University, and at the University of Göttingen, Germany. He received his B.S. from the City College and an A.M. from Columbia in 1897, where he also received a Ph.D. in 1899. Felix Klein and David Hilbert are listed at the Mathematics Genealogy Project as his first and second advisors.

In 1900 Kasner started his career as a mathematics tutor at Barnard College, then affiliated with Columbia, and, in 1906, became adjunct professor at Barnard. In 1910, he was appointed full professor at Columbia University, and thereby became one of its first Jewish professors. Kasner was awarded the Adrain Professorship at Columbia in 1942 and retired in 1949.

His principal research interest was differential geometry. He passionately believed in making mathematics accessible to nonmathematicians, and wrote several books directed towards a general audience, such as as *Einstein's Theory of Gravitation* in 1920, *New Names in Mathematics* in 1937, and *Mathematics and the Imagination* (together with James Roy Newman) in 1940.

Kasner is also remembered for coining the terms "googol" and "googol-plex" (Newman [**1956**]). Actually the honor may belong to his nephews Milton and Edwin Sirotta: during a walk with them when they were quite young, Kasner was ruminating about names for a very large number, and the story is that Milton suggested "googol".

KŒNIGS, GABRIEL (1858–1931). Kœnigs was born in Toulouse, France. He studied at the École Normale Supérieure, earning his Ph.D. in 1882. He was *chargé de cours* for mechanics at the University of Besaçon from 1883 to 1885, held a chair in analysis at the University of Toulouse the next year, and then returned to Paris as *maitre de conferences* at the École Normale Supérieure in 1886. He became a *professeur* at the Sorbonne in 1896.

There are three principal strands to his research. The first is his interest in the iteration of complex functions, where he initiated the French study of the field and developed a robust local theory in the 1880s, which was extended by his students Leau and Grévy through the 1890s (see references to the "Kœnigs school" in the index). The second is geometry, where his work is often associated with Darboux. The third is applied mechanics, which coincided with the creation of the Laboratoire de Mécanique de la Faculté des Sciences de Paris in 1897. He directed the lab for many years, and according to Buhl's necrology, his name "is indelibly attached" to it [**1931**, p. 286]. During World War I, the laboratory made an important contribution to the war effort under his supervision, and he was officially commended by the French government.

Taton was of the opinion that Kœnigs' interest in mechanics was a loss for geometry:

> *Despite his success in these areas it is perhaps regrettable that this disciple of Darboux thus abandoned his initial approach, the originality of which appeared potentially more fruitful.* [**1980**, p. 446]

He won award several honors from the Académie, including its *Prix Bordin* in 1891. He was listed among the second line of candidates for election to the Académie in 1892 to fill the chair in mathematics vacated with the death of Ossian

Bonnet but he lost out to Paul Appell. However, he was elected to the mechanics section of the Académie in 1918.

Following World War I, Kœnigs became embroiled in the issues stemming from the International Research Council's (IRC) refusal to allow scientists from Germany or the other Central Powers[5] to participate in the IRC or any of its constituent scientific organizations, such as International Mathematical Union (IMU).

This embargo against the Central Powers had been controversial from its inception at the end of World War I. Godfrey Harold Hardy was vocal in his opposition to it, as were many U.S. mathematicians. Lehto also suggests that the French mathematicians Appell and Paul Painlevé were sympathetic towards the notion of reconciliation [**1998**, p. 32]. In 1926 the embargo was lifted. However, due to lingering hard feelings over their exclusion, many German scientists initiated their own boycott and decided not to attend the 1928 International Congress of Mathematicians in Bologna. Meanwhile, not everyone in the Allied side was happy that the embargo was officially, if not actually, over, and many French mathematicians, including Picard and Kœnigs, who in 1920 had been elected Secretary General of the IMU (part of whose responsibility was to organize the Congresses) urged French mathematicians not to participate—and many did not.

It may be that the controversy took a lot out of Kœnigs. In 1928 his health began to decline, and according to Lehto he became distant, and often did not respond to official IMU correspondence. In 1931 he died.

KOLMOGOROV, ANDREI (1903–1987). Kolmogorov was born in western Russia. His father abandoned his mother prior to his birth, and she died very soon after he was born. He was raised by his aunts, who were relatively young, on the estate of his nobleman grandfather in the Yaroslal' province. His grandfater was evidently was in political opposition to Czarist Russia and was paid regular visits by the police. When Kolmorogov was very young, as one story about him goes, a forbidden book was safely hidden beneath him during a police search of his grandfather's estate (see Nikol'skii [**2005**, p. 290]). Kolmogorov moved to Moscow with his aunts, and in the 1920s he attended Moscow University, earning his Ph.D. in 1929.

Early interests included probability theory, where he did both foundational and fundamental work, and logic. Around 1940 his interests shifted to dynamical problems, and he studied turbulence and eventually celestial mechanics. His address to the 1954 International Congress of Mathematicians helped give rise to the body of work called KAM theory (discussed in Sections 4.4 and 12.5). With the advent of the computer, he became interested in some of the mathematical issues surrounding computation, including the theory of algorithms and complexity theory.

He is generally credited, along with Vladimir Arnold, with solving, at least in part, Hilbert's 13th problem, which is to show the impossibility of solving the general 7th degree equation via functions of two variables. He and Arnold showed that every continuous function in k variables can be viewed as a composite of functions of two variables. In the eyes of some observers, this constitutes a complete solution; in the eyes of others (see, for example, Vitushkin [**2004**]), the problem

[5]The Central Powers were The German Empire, the Austro-Hungarian Empire, the Kingdom of Bulgaria, and the Ottoman Empire (Turkey). Among the Allies, or Triple Entente, were the British Empire, France, and the Russian Empire (the original Triple Entente), as well as Greece, Italy, the Empire of Japan, Romania, Serbia, and (after 1917) the U.S.

remains open. In 1980 he won the Wolf prize for his life's work. That Vladimir Arnold and Jürgen Moser also won the Wolf prize gives the "K", "A", and "M" of KAM theory an interesting, and perhaps unique, trifecta.

Towards the end of his life, he suffered from Parkinson's disease and poor eyesight. Yet, according to O'Connor and Robertson [**2010W-a**] he remained mathematically active until his death in 1987.

LATTÈS, SAMUEL (1873–1918). Lattès was born in Nice, France, on February 21, 1873, to Israël Vita Lattès, and he was educated at *lycées* in Nice. According to Montel's necrology [**1918a**], Lattès was precocious child. In 1888 his father died, which evidently brought hardships to his family. In 1892 after one year of *mathématiques spéciales* at the *lycée* of Marseille, Lattès entered the École Normale Supérieure. In 1895 he placed ninth in the mathematics *agrégation* exam (see footnote 5 on page 342 for more on the *agrégation* exams).

He became a professor of special mathematics at *lycées* in Nice, Alger, and Dijon. In 1906 Lattès published his dissertation thesis. Shortly thereafter, he was appointed professor of special mathematics at the Lycée Montpellier and, in November 1908, became a lecturer in the Faculty of Sciences at the University of Montpellier. On August 17, 1910, he married Jeanne Ferrier in Montpellier, who had been a student of his and who became a physicist of some renown.[6] Their only child, Florence Marguerite, was born on May 10, 1914.

In 1911 Lattès was teaching at the Faculty of Sciences in Besaçon when the chair of mechanics in Toulouse became vacant. Lattès was appointed to it with the unanimous consent of the Faculty Council.

Lattès' research interests focused on analysis, functional equations, iteration of functions, and related geometric transformations. In 1917, along with Pincherle and Julia, he submitted an entry in the competition for the 1918 *Grand Prix des Sciences Mathématiques* that was subsequently lost.[7] Lattès became ill and died of typhoid fever[8] on July 5, 1918, at 11 am, at rue de la Trinité 10, in Toulouse. When the results of the *Grand Prix* were announced in December 1918, he was awarded a posthumous honorable mention and 2000 francs.

MOSER, JÜRGEN (1928–1999). Moser was born in Königsburg, Germany. He was five when the Nazis were elected to power, and his father, an eminent neuroscientist, and his mother, who helped foster a life-long interest in music, were evidently not sympathetic to the Nazi regime. According to Lax [**2002**], Moser's parents were encouraged to send him to a school designed to produce the elite of the next generation of Nazis, but they refused, sending him instead to the nearby public school (one that David Hilbert graduated from). At age 15 he was pressed into service, ostensibly to compute the trajectories of enemy aircraft but ended up on the front lines in the war against the Soviets, where his unit was decimated.

Amidst the chaos of post-war Germany, he located his parents in Königsburg, now in the Russian zone of Germany, which would become East Germany. When

---

[6]See Audin [**2009**, p. 76] for a brief account of her life.

[7]See Sections 6.6 and 7.5 for more about Lattès' involvement with the competition.

[8]During World War II hygienic conditions in France deteriorated (Weber [**1986**, p. 58]), and this may have contributed to his becoming ill. One can contract typhoid fever from infected foods such as meat, poultry, eggs, seafood, vegetables, or any food contaminated by waste water.

he attempted to cross the border to join them, he was imprisoned, but soon escaped and made his way to Göttingen in 1947 where he enrolled in the university. His parents remained in East Germany, where they lived for quite some time.

His mathematical talent was recognized by Franz Rellich, and Moser began to study the spectral theory of differential equations. By 1951, Carl Ludwig Siegel had returned to Germany from the U.S., where he had spent the war at the Institute for Advanced Study at Princeton University, and he became close to Moser, influencing the direction of his study toward number theory and celestial mechanics (see Section 12.5). He received his Ph.D. from the University of Göttingen in 1952.

Moser was awarded a Fulbright fellowship, and he left Göttingen to study at at the Courant Institute at New York University (NYU) in 1953–54 where he met Richard Courant's daughter, Gertrude,[9] a biologist whom he later married. He briefly returned to Göttingen in 1954 where he attended Siegel's lecture on celestial mechanics. His notes for this course formed the basis for a text jointly authored with Siegel, which was published in 1956 and later translated to English as [**1971**].

He emigrated to the U.S. in 1955, and taught first at New York University, then, from 1960 through 1980, at the Massachusetts Institute of Technology. He moved back to Europe in 1980 where lived until his death. He gave several prestigious lectures, including three invited addresses at the International Congresses of Mathematicians, and he won a number of major mathematical prizes during the course of lifetime, including the Wolf Prize in 1995.

OKA, KIYOSHI (1901–1978). Oka's doctorate is from Kyoto Imperial University in 1925. Evidently, he was initially interested in physics, but decided upon mathematics in 1923. After receiving his degree, he stayed on for a while as a lecturer at the university. Between 1929 and 1932 he spent time studying mathematics, and it is thought he met with Julia.

He did fundamental work in the complex functions of several variables. Henri Cartan described his work thusly:[10]

> *The publication in 1934 of a monograph by Behnke-Thullen marked a crucial stage in the development of the theory of analytic functions of several complex variables. By giving a list of the open problems in the area, this work played an important role in deciding the direction of Oka's research. He set himself the almost super-human task of solving these difficult problems. One could say that he was successful, overcoming one after the other the obstacles he encountered on the way.*
>
> *However, the technical aspects of his proofs and the way he presents his results make his work difficult to read. It is only with considerable effort that one can appreciate the considerable strength of his results. This is why, even today, it is worth collecting his work as a tribute to its creator, Kiyoshi Oka.* (Oka, Cartan, Remmert and Narasimhan [**1984**, p. xiii])

---

[9] She was a granddaughter of Carl Runge and a grandniece of Emil Du Bois-Reymond.

[10] This quotation is a translation of a passage from a collection of his works (Oka, Cartan, Remmert and Narasimhan [**1984**]) found at the website of O'Connor and Robertson [**2010W-c**].

Oka addressed these open problems in a series of papers collectively referred to as *Sur les fonctions analytiques de plusieurs variables* which appear, translated to English, in the volume by Oka, Cartan, Remmert and Narasimhan [**1984**].

Upon his return from France, Oka taught at Hiroshima University. Evidently, during much of World War II, he lived in the village of Kimi (now Hashimoto) in Wakayama Prefecture, where he grew up. He became a professor at Nara Women's University in 1949, and retired from there in 1964. A few years later in 1969, he was appointed professor at Kyoto Sangyo University.

PFEIFFER, GEORGE ADAM (1889–1943). Born in New York City, he graduated from the High School of Commerce. In 1910, he received a masters degree in engineering from Stevens Institute of Technology, and in 1911 joined the faculty in mathematics. He eventually enrolled in the graduate program in mathematics at Columbia University and received his Ph.D. in 1914 with a dissertation entitled *Contribution to the conformal geometry of analytic arcs.*

During World War I, Pfeiffer served in the U.S. Army and was stationed at Princeton University as an instructor in meteorology for aviation students. After the war he taught at Columbia and was appointed assistant professor in 1924. In 1931 he was promoted to associate professor. He also served on the editorial board of the *Annals of Mathematics.* On December 28, 1943, Pfeiffer died following surgery.

In an interview with Melvin Hendriksen, the mathematician Leonard Gillman recounted an anecdote from a class he took from Pfeiffer as a graduate student:

> *The topology course with Pfeiffer was based on Max Newman's* Topology of plane sets. *Once during a discussion of continuity, one of the students (a very good one) became momentarily confused and asked a question that Pfeiffer did not understand. After they had talked at cross purposes for a while, the student finally blurted out, "But suppose I choose epsilon large?" "Ah," Pfeiffer replied with a twinkle, "but you don't choose epsilon. I choose epsilon."* [**1967-W**]

PINCHERLE, SALVATORE (1853–1936). Pincherle was born in Trieste (then part of Austria) into a Jewish family who soon thereafter moved to Marseille, France, where he completed his schooling. He attended the University of Pisa, where he studied under Enrico Betti and Ulisse Dini, and received his Ph.D. in 1874.

He received a scholarship to travel to Berlin, where he attended lectures in complex analysis by Weierstrass in 1877–78, which he used as the basis of an expository article he published the *Giornale di Matematiche* that was one of the early Italian language introductions to Weierstrass' work [**1880**]. That same year he obtained a chair in mathematics at the University of Palermo, and the following year, was appointed to a position at the University of Bologna, where he remained throughout his career.

Pincherle is considered to be one of the founders of the Italian school of functional analysis. Reinhard Siegmund-Schultze in [**2003**] divined three major directions in the developmental phase of functional analysis: abstract operator theory, abstract theory of spaces and functionals, and the theory of concrete functionals. He placed Pincherle, along with Carlo Bourlet, at the forefront of the first. Reflective of his status was the assignment to write the article on functional operations and equations in the *Enzyklopädie der Mathematischen Wissenschaften* [**1906**] and the *Encyclopédie des Sciences Mathématiques Pures et Appliquées* [**1912**].

As recounted in Section 8.6, Pincherle submitted an entry for the 1918 *Grand Prix des Sciences Mathématiques* that was returned to him without having been considered for the prize.

After World War I, Pincherle became involved with the International Research Council (IRC) and was elected president of the International Union of Mathematicians (IMU) in 1924 where he soon became involved in the controversy surrounding the embargo on German membership in the IRC and IMU. These events are discussed in our biographical sketch of Kœnigs above. At roughly the same time, Pincherle publicly supported the goals of the fascist movement in Italy, and in 1925 became one of the few mathematicians to sign the Manifesto Gentile, which called on Italian intellectuals to support the aims of Italian fascism.

REMOUNDOS, GEORGE J. (1878–1928). A Greek mathematician with strong ties to the French mathematical community, Remondous published over 100 articles. He studied at the University of Athens and received a scholarship that allowed him to attend the University of Paris in the early 1900s, where he encountered Appell, Painlevé, Picard, and Poincaré. In 1903, Remoundos joined the Société de Mathématiques de France, presented by Painlevé and Louis Raffy. While in France he published his first works in Picard theory, for example, [**1903, 1905, 1906**], which he extended to algebroid maps. His work was cited by Montel, for example, in his book [**1927**], and Zoretti in [**1911**]. One of Remondous' last publications is the book [**1927**], which summarized much his research into Picard theory.

He returned to Greece in 1905 and obtained a position at the University of Athens, becoming full professor of analysis in 1916. He obtained a chair in analysis at the National Technical University of Athens in 1918, evidently with the assistance of Carathéodory. In 1926 the Academy of Athens was founded, and he was appointed First Secretary.

His principal mathematical interest was analysis, where he investigated the zeros of transcendental maps, entire maps of finite order, numbers theory, growth theory, as well as algebroid maps. Along with with Carathédory, N. I. Hadjidakis (1872–1942), Stephanos (1857–1917), and Zervos, he brought modern mathematics to Greece. See Phili [**2003**] for more about Remoundos' mathematics.

RITT, JOSEPH FELS (1893–1951). Ritt was born six months after Julia. In 1908, after attending the City College of New York, he entered George Washington University, and he received his A.B. in 1913. He attended graduate school at Columbia University in New York City and wrote his doctoral thesis on differential operators under the supervision of Kasner. He was awarded his Ph.D. in 1917, and he served in the U.S. Army in the Ordnance Department as a master computer during the U.S. involvement in the World War I.

Following the war, he joined the Columbia faculty in 1919 as an instructor in mathematics. He was promoted to assistant professor in 1921, associate professor in 1927, and full professor in 1931. He taught at Columbia until his death. Ritt was also a member of the National Research Council and from 1938–41, a member of the National Academy of Sciences. He was vice president of the American Mathematical Society from 1938 to 1940, and served on the editorial board of the *Annals of*

Mathematics. In addition to his work involving the iteration of complex functions, described in Section 13.2 and Appendix J.3, Ritt developed a novel approach to differential equations, sometimes called differential algebra.

SCHRÖDER, ERNST (1841–1902). Schröder was born in Mannheim, Germany. His father was a teacher and also an amateur scientist of minor renown, according to Dipert's biographical essay [**1991**], the source for much of our own biographical sketch. Schröder was home-schooled by his grandfather.

He studied mathematics at the University of Heidelberg with Ludwig Otto Hesse, who is listed as his dissertation advisor, and he received his doctorate in 1862. He also enjoyed science, and took chemistry from Robert Bunsen of the eponymous burner. He continued with his studies at the University of Königsberg and taught in various *Gymnasien* (secondary schools) in Germany and Switzerland, where he earned his *Habilitation* in 1868, until he obtained a university position at the Technical Institute of Darmstadt as an *Ordinariat* professor in mathematics. In 1876 he was appointed to what came to be called the Technical Institute of Karlsruhe, where he remained until his death.

Schröder is perhaps best known outside of the study of complex analysis for his work in mathematical logic, in particular, its algebra. Evidently, his interest in the subject was piqued when he read Robert Graßmann's *Formenlehre* in 1872, which may account for the fact that he did not return to study the iteration of functions. He published widely and is the Schröder of the Cantor–Schröder–Bernstein theorem, which states that if there is a one-to-one mapping from set A into B and one from B into A, that the cardinality of the two sets is the same. One of his most famous works is the *Vorlesungen über die Algebra der Logik*, only small portions of which have been translated into English and which is regarded as a classic. He was an early adopter of Georg Cantor's ideas and came to correspond with Charles Sanders Pierce, with whom he had a somewhat complicated relationship. He evidently venerated Pierce (according to Dipert), but Pierce, while generally praising Schröder, could also be very critical of him.

Very of few of Schröder's private and professional unpublished writings remain. His professional writings were stored for possible posthumous publication, and his private correspondence was given to his nephew. In the 1920s the Deutschen Mathematiker-Vereinigung appointed a committee headed by Friedrich Schur to produce a volume from Schröder's unpublished papers and correspondence (which included exchanges with Pierce, Felix Klein, Giuseppe Peano, and Bertrand Russell, among others). Ludwig Bieberbach, who had taken over the project after Schur's death, tried to interest an editor, but to no avail. During the war, his papers were deposited at the University of Münster along with Gottlob Frege's papers. The building in which these papers were stored was destroyed during an Allied bombing raid on Palm Sunday in 1945, less than two months before the end of the European war.

Dipert observed that all of Schröder's biographers describe him as "even tempered and gentle", and that these qualities are evident in his correspondence [**1991**, p. 124]. Schröder was evidently very active physically, and enjoyed being out of doors. In his youth he went on unguided climbs in the Alps and was an avid cyclist up until the time of his death. When he turned 60, he took up skiing. He died at age 62, not long after coming down with a cold following an extended bicycle trip in February 1902.

Siegel, Carl Ludwig (1896–1981). Siegel is one of the mathematical luminaries of the 20th century. He made fundamental contributions in analytic number theory, celestial mechanics, and complex dynamics, and he won the Wolf Prize in 1978. Siegel's wide-ranging interests were perhaps a reflection of advice from Felix Klein not to specialize too early [**1985**, p. 77].[11]

Siegel was born in Berlin and entered the University of Berlin in 1916. He was evidently interested in mathematics, astronomy, and physics, but, perhaps due to the influence of Ferdinand Georg Frobenius, he chose mathematics, eventually earning his Ph.D. in Göttingen in 1920 under the direction of Edmund Landau.

Siegel's politics went against the German mainstream. Although drafted in 1917, he was discharged as being unsuited for military service. His health, evidently not robust in the best of times, suffered markedly. Landau's father, a physician, cared for Siegel during his service and helped obtain his release. Siegel encouraged Emmy Noether while she was at Göttingen, and, according to Davenport, his misgivings about the future of Germany [**1985**, p. 77] made him decide not to join the Deutschen Mathematiker-Vereinigun in 1921.

Siegel became a professor at the University of Frankfurt in 1922 and took part in the famous Frankfurt history of mathematics seminar (see Siegel [**1977**]). Siegel was the opposite of a Nazi sympathizer. In 1935 he left Frankfurt to spend a year at the Institute for Advanced Study at Princeton University. According to Sanford Segal [**2003**, pp. 67–68], he returned to Frankfurt the next year despite the increasing inevitability of Nazi brutality, in part to try and protect Jewish colleagues (he was not Jewish) but also to secure his pension. However, in 1940 Siegel could no longer bear to live in Nazi Germany and left for Norway, then at war with Germany, where, after receiving another appointment to the Institute for Advanced Study he caught one of the last boats to the U.S. to leave Norway before the German take-over.

He remained at Princeton for 11 years, returned to Germany in 1951, and worked at the University of Göttingen for the rest of his career.

Shimizu, Tatsujiro (1897–1992). Shimizu received his Ph.D. from Tokyo Imperial University in 1924 and served on the staff there until 1932, when he became a professor of mathematics at Osaka Imperial University. After World War II, he taught at Kobe University, the University of Osaka, and the Science University of Tokyo.

His list of publications [**A15-W**] suggests that as his career unfolded, his interests broadened from function theory (he is the Shimizu of the Shimizu–Ahlfors Theorem) to diverse areas of applied mathematics, including artificial intelligence, computational machines and devices, and numerical analysis. He evidently played a significant role in founding the Japanese Association of Mathematical Sciences after the end of the war and helped inaugurate its journal, *Mathematica Japonica*, in 1948.

[11] Davenport's article, written in 1985, recounts conversations he had with Siegel in 1966.

TAMBS-LYCHE, RALPH (1890–1991). Born in the U.S. to a Norwegian father and American mother, his family moved to Norway when he was two years old. In 1916 Tambs-Lyche graduated from the *Royal Frederick University* with a degree in mathematics and married Else Rasmussen. He began his academic career at the Norwegian Institute of Technology, where he taught for most of his professional career. He obtained his Ph.D. at the University of Strasbourg in 1927.

Tambs-Lyche was a politically active leftist, which contributed to his arrest by the Germans in the spring 1942. He was interned at the Falstad concentration camp for the duration of the German Occupation with other political prisoners, many of whom were subsequently executed.

He contributed to analysis, differential equations, and algebra and number theory. In 1957, Tambs-Lyche gave an example of a real function $f(x)$ that is differentiable for all $x \in \mathbb{R}\backslash\mathbb{Q}$ but not for $x \in \mathbb{Q}$. Moreover, $f(x)$ is of bounded variation, so its graph $(x, f(x))$ has an arc length center problem.

His other interests included botany, nature, and hiking in forests, and he evidently remained fit up until his death. See Selmer [**1991**] for more about his life.

VALIRON, GEORGES (1884–1954). Born in Lyon, France, Valiron entered the École Normale Supérieure, earning the top scores at the exams for the mathematics *agrégation* in 1908. During 1908–1912, Valiron was a professor at a *lycée* in Besançon in Eastern France, where André Bloch was among his students. Valiron received a scholarship to study at Commercy in eastern France from 1912–1914, and, in June 1914, he earned his doctorate from the University of Paris. His thesis advisor of record was Émile Borel.

Drafted into World War I, he was dismissed in 1919 with the rank of sublieutenant of Artillery. That same year, Valiron took a position at the University of Strasbourg, where he taught until was appointed to the Faculty of Sciences at the University of Paris in the mid-1930s. Valiron was awarded several prizes from the Académie des Sciences, among them the prestigious *Grand Prix des Sciences Mathématiques*. In 1954 he was conferred the honor of Commander of the Legion of Honor.

WOLFF, JULIUS (1882–1944). Wolff was born in Nijmegen, Netherlands. He took his Ph.D. under Diederik Korteweg in 1908. His principal research interests were analysis and function theory, and he became one of the leading mathematicians in The Netherlands in the period between the two world wars. Wolff's research in function theory is linked to that of Arnaud Denjoy, whom he knew from the time of Denjoy's professorship in Utrecht. It is no exaggeration to say that he represented the "Paris School of Analysis" in The Netherlands.

He held a chair at Groningen in 1917–1922 and in Utrecht from 1922 onwards. He was dismissed from his position for being Jewish in 1940, but evidently he and Otto Blumenthal established a two-person mathematical colloquium that lasted until 1942. Soon thereafter he was deported to Germany, where he was killed at the Bergen-Belsen concentration camp.

APPENDIX P

Remarks on Computer Graphics

(The main computer of the 'Enterprise' starship
is under the control of a cold, evil creature).

Mister Spock: Computer! This is a class-A compulsory directive.
Compute, to the last digit, the value of π.
[... *to captain Kirk*]
As we know, the value of π is
a transcendental figure without resolution.
The computer banks will work on this program
to the exclusion of all else until we order it to stop.

Captain Kirk: Yes, ... that should keep that thing busy for a while!

From the STAR TREK episode "Wolf in the fold."

P.1. Introduction

Most of the digital images in this book, included as Color Plates 1–11, have been computed using one of the following methods:

(1) the escape time method, in *indexed colors mode*;
(2) domain coloring, in RGB mode;[1] or
(3) grid-maps.

We describe each method below; see also Rosa [**2004, 2005b**] as well as Appendix H in Milnor [**2006a**]

P.2. Escape time method and its variants

P.2.1. Computation. Escape time is perhaps the easiest and most commonly used approach to display a graphical model to represent the dynamics generated by the iteration of a complex function. It is especially useful in distinguishing bounded from unbounded orbits, so it is often used when the point at infinity is an attracting or superattracting fixed point, the latter of which occurs when

$$f(z) = \frac{p(z)}{q(z)}, \qquad \text{with } d \equiv \deg(p) - \deg(q) \geq 2, \tag{P.1}$$

where p and q are polynomials. The escape time method is often used to depict the Mandelbrot set, M, as well as the associated Julia sets for $z^2 + c$ where $c \in \mathbb{C}$.

The escape time method[2] and its variants discussed below rely on the property that for a rational function f of degree at least two, any complete domain of

[1] In the *indexed colors mode* colors are applied from a finite palette; in RGB mode the color range is in theory infinite.

[2] Also called *boundary scanning method.*

attraction D of an attracting point or periodic orbit admits the Julia set J of f as its boundary.

In particular, this allows us to view the Julia set of a function satisfying (P.1) as the boundary of the complete basin of attraction for infinity, which we will denote Ω. The complement of Ω is known as the filled-in Julia set, and it consists of the set of points whose orbits remain bounded. Since the connected component of Ω containing infinity contains a neighborhood of infinity (a set of the form $\{z : |z| > r\}$ for some constant r), the Julia set for functions with an attracting fixed point at infinity, such as those of the form (P.1), is bounded. This circumstance suggests an obvious algorithm to approximate J which we will discuss below.

Escape time refers to the fact that we choose a trigger, or *stopping criteria* to end computation. First of all, determine a closed disc $\overline{D}$ of radius r centered at a point w that contains the Julia set. By definition, the complement of $\overline{D}$ is contained in Ω.

Let P be a portion of the domain of f, completely or partially enclosed in D, whose dynamics we wish to portray. We introduce a grid Γ of points covering P and fix a sufficiently large *stopping criteria* $k \in \mathbb{N}$. Using the r from the paragraph above, we compute the forward orbit $O^+(z) = \{z_n = f^n(z),\ n = 1, 2, \dots\}$ individually for each point z from the grid Γ, stopping when either of the following occurs,

$$\text{(P.2)} \qquad |z_{\nu(z)} - w| > r \quad \text{or} \quad \nu(z) = k,$$

and storing the couple $(\nu(z), z)$ for later use. We try to a choose a k so that if the second condition is reached, we can be reasonably sure that $O^+(z)$ will remain bounded for all n.

For the purposes of our explanation, let us assume that $P \subseteq D$, and we are initially trying to determine which orbits remain bounded (i.e., in the complement of Ω) and which escape to infinity. If the first condition in (P.2) holds, the orbit of z leaves D after $\nu(z) < k$ iterations and enters Ω, escaping to infinity. The integer $\nu(z)$ is referred to as the *iteration number* and measures the escape rate of the orbit originating at z.

Alternatively, if the orbit survives for at least k iterations, we presume that $f^n(z)$ will remain in $\overline{\mathbb{C}} \setminus \Omega$ for all n. In this case we set $\nu(z) = k$, to indicate that the orbit remains bounded since the maximum number of iterations for escape has been reached.

When each orbit computation concludes, the algorithm uses the stored information $(\nu(z), z)$ to associate a specific color to each point $z \in \Gamma$, depending on the escape rate $\nu(z)$.[3] The simplest configuration is obtained by using two colors, such as black if $\nu(z) = k$ (i.e., the point z does not escape), and white if $\nu(z) < k$ (z does escape). The boundary of the black region will approximate the Julia set of f with an accuracy that can be improved by either refining the grid Γ or increasing the stopping criteria k, or both.

[3]When z escapes to the basin of infinity, as it does when the first condition in (P.2) holds, we can obtain a continuous coloring by extending the iteration number to a real value, the so-called *real iteration number*, $\hat{\nu}(z)$. This is accomplished by subtracting a suitable number in the interval $[0, 1)$ from $\nu(z)$ and setting $\hat{\nu}(z) = \nu(z) - \log\left(\frac{\log z_{\nu(z)}}{\log r}\right) / \log d$, where the latter quantity is obtained by observing that, for ν large, the assumption (P.1) implies the fixed point at infinity is superattracting and $z_\nu \simeq z_0^{d^\nu}$. Assuming also that $|z_\nu| \geq r$ and $|z_{\nu-1}| < r$, our computation yields the exponent $\hat{\nu} \in (\nu - 1, \nu]$ satisfying $z_0^{d^{\hat{\nu}}} = r$.

P.2.2. Color indexing. A robust application of colors uses a palette to illustrate the dynamics of the orbits according to their escape rates $\nu(z)$. This is accomplished by mapping the discrete interval $[\min \nu(z), \max \nu(z)]$ to a given finite color map or to a portion of it, which is indexed according to the numerical order.[4] For example, in Color Plates 3 and 6, we choose a blue palette for the basin of infinity, where lighter colors indicate that a larger number of iterations is needed to escape the disc D. There are many choices of palettes, as our examples illustrate, but we generally require colors to change through a small spectrum in order to depict varying escape rates.

P.2.3. Generalizations. The region $\overline{\mathbb{C}} \setminus \overline{D}$ may be interpreted as a neighborhood $\hat{N}$ of infinity, in which case the escape time method measures whether a point enters $\hat{N}$ or not. We can trivially generalize this notion to locate orbits originating at points z in a given region P that converge to a finite periodic orbit. To do this, we need to detect the portion of P which belongs to a particular component of the Fatou set of f. Suppose, for example, that we know that λ is an attracting fixed point of f. We consider a small disc N centered at λ with radius r and, for each orbit $O^+(z)$, use the following stopping criterion in place of (P.2); that is, instead of checking when (if at all) a point enters a neighborhood $\hat{N}$ of infinity, we are checking to see when a point enters a neighborhood N of an attracting fixed point:

$$|z_{\nu(z)} - \lambda| < r \quad \text{or} \quad \nu(z) = k. \tag{P.3}$$

We can then color z using methods analogous to those we described above. Notice that in this setting, the scanned region P and the neighborhood N may be exterior to each other, or $N \subset P$.

We can also use this procedure to locate the various basins of attraction associated to different attracting fixed points, $\lambda_1, \lambda_2, \ldots, \lambda_s$. To this end, we store the couples

$$\{(\nu(z), \lambda_i) \ : \ |z_{\nu(z)} - \lambda_i| < r_i, \ z \in \Gamma\}$$

and associate a different color C_i, or perhaps a range of shades, to all z whose orbits converge to λ_i.

Color Plates 5 and 8 show examples of this method. In the first, the three fundamental colors, red, green and blue, are used to highlight the three basins of attraction of the Newton method applied to the cubic $z^3 - 1$. This is referred to as a *fixed color technique.* We could have instead also shaded the three colors to indicate the rates of convergence within these domains, as we did for the basin of infinity in Color Plate 8.

P.2.4. Additional considerations. Issues of scale can arise if we use our methods on generic higher-degree rational functions. For example, if we want to find the fixed points of a function of degree higher than four, we must solve $f(z) = z$, which can result in an equation that cannot be solved by exact methods. Instead we would need to use some root finding method to numerically approximate the roots of $f(z) = z$, and then compute derivatives to see which fixed points are attracting, before applying the escape time method, all of which will increase computation time.

[4] When using the real iteration number, one should instead consider a theoretically infinite range of colors. See Section P.3 for a discussion of this point of view.

We can also take another route and, slightly modifying (P.3), focus on the difference between successive iterates z_ν and $z_{\nu+1}$ by testing for

$$|z_{\nu+1} - z_\nu| < \varepsilon, \tag{P.4}$$

where ε is the accuracy required. The stopping criterion (P.4) avoids the preliminary computation of the attracting fixed points λ_i, and it is based upon the fact that one can use condition (P.4) to turn the function f itself into a root finding method by constructing various orbits which simultaneously approximate the attracting fixed points.

In other words, (P.4) is tantamount to (P.3) for a certain attracting fixed point. In fact, assuming that the sequence $\{z_n\}$ converges to a fixed point λ, we have

$$\begin{aligned} z_{n+1} - z_n &= z_{n+1} - \lambda + \lambda - z_n = f(z_n) - f(\lambda) - (z_n - \lambda) \\ &= f'(\lambda)(z_n - \lambda) - (z_n - \lambda) + O(|z_n - \lambda|^2) \simeq (f'(\lambda) - 1)(z_n - \lambda), \end{aligned}$$

where, in the last step, we neglect the higher order terms in $|z_n - \lambda|$. It follows then that

$$|z_\nu - \lambda| \le \frac{1}{|1 - f'(\lambda)|} |z_{\nu+1} - z_\nu|,$$

which means that (P.4) implies (P.3) with $r = \frac{1}{|1-f'(\lambda)|}\varepsilon$.

Each of the above methods can be modified for functions with attracting periodic orbits: if we determine that there is an attracting orbit of order p, we can, for example, use (P.3) or (P.4) on f^p instead of f.

P.3. Domain coloring

P.3.1. Introduction. Domain coloring takes an interdisciplinary approach, mixing complex analysis and color theory, and it connects each to Euclidean geometry.

The notion of domain coloring arose in complex dynamics from the need to develop sharper numerical estimates than the ones used in the escape time method, which relies on a limited palatte set. Domain coloring relies on the fact that color can be represented by mixing proportions of red, green, and blue. If we denote the relative proportion of each by a number between 0 and 1, then in there we have an infinite number of colors to choose from, and colors can be continuously blended into each other. The question then becomes how to match this panoply of color to points.

Motivation to represent complex functions with an infinite array of colors came from several sources. One was the visualization of elementary complex functions in William Kahan [**1987**]. John Richardson developed a method for visual complex wave mappings that arise in Schrödinger's equation and its scattering process by introducing the application of a color model associated to points in $\mathbb{C}$; see [**1991**].

Frank Farris is often given credit for coining the term "domain coloring" in [**1998**], but the technique was used earlier, and perhaps originates with Lawrence Crone. A few years after Crone, Douglas N. Arnold, lamenting the lack of available methods to produce detailed visualizations of complex mappings, developed his own approach. He assigned colors to predetermined grids of points based on the values of f on these grids and created rich images which demonstrated the power of a chromatic approach [**2008-W**]. Cleve Moler in [**1998-W**] and George Abdo and Paul Godfrey in [**1999-W**] assigned a continuous coloring scheme to the mapping of complex functions and obtained highly detailed pictures. Abdo and Godfrey's methods were an advance over those of Arnold, resulting in richer images.

P.3.2. Puzzling out an initial strategy. Plotting complex points is challenging, and we will see that this question needs to be attacked from two directions. We shall consider this fundamental fact: just as the graph (x, y) of a mapping $f : \mathbb{R} \to \mathbb{R}$ is plotted with two real coordinates, the graph of a complex map $f : \mathbb{C} \to \mathbb{C}$ requires two complex points, $(z, f(z))$ since we are associating one complex surface, P, to another, W, via $f : P \to W$.

Clearly, we want a means to distinguish points in P depending on how they are mapped to W, and colors are ideal. Rendering a mapping of complex surfaces in graphical terms cannot be achieved in the same manner used with real functions of a single variable because the graph of a complex function is four dimensional. One solution is to drop down one dimension, that is, move from a four-dimensional space to a three-dimensional model (akin to the graph of a two-variable real valued function, $(x, y, f(x, y))$, through some sort of projection.

The concomitant loss of data is a sacrifice we must make if we are to reach a solution to the problem of representing a complex mapping with a single image. We can condense our problem into a single sentence: "We need to represent the graph of a complex function in three-dimensional space in a way that uses color meaningfully."

P.3.3. 3-D models and projections. The kind of three-dimensional model we are discussing is generally referred to as a *color model*, and color theory allows us to use geometric parameters to locate colors. The color of a point on the screen is coded by a three-dimensional vector in the RGB (red, green, blue) additive model, where each vector component is a primary color. All other colors are coded by shading each component using an interval, usually from zero to one, ranging from darker to lighter tones. In other words, the domains of R, G, and B are each $[0, 1]$.

This gives rise to a conceptualization of the RGB space as a unit cube composed of the triples (R,G,B) that satisfy the following properties:

- R is the unit vector $(1, 0, 0)$, G $= (0, 1, 0)$, and B $= (0, 0, 1)$;
- white $=$ R+G+B $= (1, 1, 1)$ and black $= (0, 0, 0)$;
- the four corners of the base of the cube are $(0, 0, 0)$, $(1, 0, 0)$, $(0, 1, 0)$, and $(1, 1, 0)$.

Colors are thus vectors of the form (a, b, c) which, as suggested above, follow the usual additive rules. For example, cyan equals $(0, 1, 1)$ since it is composed by mixing fully saturated G and B; likewise, magenta $=$ R+B $= (1, 0, 1)$, and, finally, yellow $= (1, 1, 0)$.

The RGB color cube is then projected to a planar hexagon whose vertexes are the three primary colors R, G, B plus cyan, magenta and yellow. The center of the hexagon could be a black or white point, depending on the orientation of RGB projection—exactly which point is mapped to the center determines the overall color shading scheme for our final image, a point we will come back to. Finally, the hexagon is warped to a disc (much as a circle can be approximated by a hexagon). This is a fundamental expedient since the the disc becomes a representation of the complex plane (or the Riemann sphere).

P.3.4. Coding colors and complex points. When we project the three-dimensional model down to two dimensions to represent it, we compensate for the loss of information by shading from dark (i.e., black) to light (white). We can visualize this new two-dimensional space as either a cone or a cylinder whose base is the disc itself: The top of the object could correspond to either black or to white, depending on the modeler's inclinations. This gives us the means to represent colors: we have the two dimensions of the original projection of the cube, while the third dimension is shaded, running from white to black.

These models are thus based upon hue, saturation, and lightness, which we can think of brightness or intensity. For convention, we will refer to this model as HSL.

Caption for Color Plate 11: (See p. 66.) In the left we see the square $[-1, 1]$ by $[-1, 1]$, and in the middle, its complement, including the point at infinity. The projection used takes black $(0, 0, 0)$ to the center of the hexagon, thus, black represents zero, and the lighter the color, the larger the modulus is. One can visualize the distribution over the sphere by joining the two images in the obvious way. These two images together serve as a guide to link colors to points. For example, as the positive real axis stretches towards infinity, it remains red but lightens, and the point at infinity is white.

In the right, we see the image of the domain plane under $f(z) = z^2 - 1$ in the viewing window $[-1.25, 1.25]$ by $[-1.25, 1.25]$. The zeros of the mapping at $z = \pm 1$ are evident, and one can also see that f wraps the regions around the origin in the domain in the picture in the left around $z = \pm 1$.

P.3.5. Complex analysis enters in. The Uniformization Theorem assures us that $\mathbb{C}$ (which can be viewed as a punctured sphere) is conformally equivalent to the unit disc $\mathbb{D}$. To assign colors to points in our model, we use the map given by the conformal equivalence to associate a point $z \in \mathbb{C}$ to one in unit disc $\mathbb{D}$, and then to the HSL model. We thereby obtain a unique triplet of colors for each complex point on the punctured sphere.

The last stage, mapping from $\mathbb{D}$ to the HSL model, results in three-component vectors with two components corresponding to a complex point on the sphere and a third corresponding to a modulus, which we interpret depending on the shading. We can add a point at infinity if we wish.

Recall that there are two ways to project from the RGB cube to the hexagon depending on whether $(0, 0, 0)$ in the RGB cube or $(1, 1, 1)$ is mapped to the center of the hexagon. The choice we make determines how we interpret the shading. If $(0, 0, 0)$ maps to the center, then the modulus is a real parameter where darker shades have lower modulus and lighter shades, higher. The color plates in this book use domain coloring that follows this convention. Thus, the origin of the sphere is black and the point at infinity white. This means that if we are representing a mapping of $f(z)$, its roots would be black and its poles white (see Color Plate 11). The convention of mapping black to the center of the hexagon is not universal; for example, in the 1990s it was common to map white to the center, hence poles were often black.

P.4. Grid-maps

Grid maps yield a useful representation of the dynamics near a rationally neutral point because they are well suited to plotting local vector fields. A grid is essentially a localized coordinate system and can be orthogonal, radial, or some other configuration. Different grid systems highlight different topological properties when the orbit of a point in the domain grid is mapped onto an image of the source grid. In Figure P.1 we show an example of the same image plotted with orthogonal and radial grids.

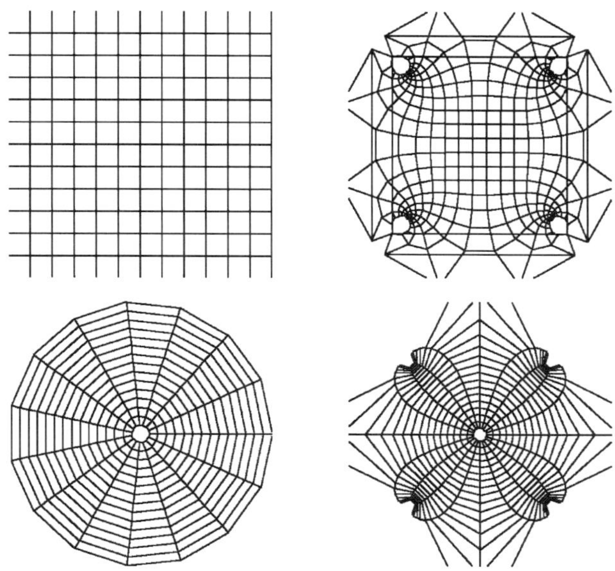

FIGURE P.1. Grid Maps. Here we iterate points from orthogonal and radial grids under the mapping $f(z) = z + z^5$ which produces a flower. The properties that one sees is affected by the choice of the grid. A radial grid highlights the structure of the flower. In the orthogonal grid, note that the image is quite close to the domain near the origin while, near its edges, it starts to resemble a flower. That the portion of the grid near the origin is almost preserved under iteration suggests that dynamics near the fixed point changes slowly. In other words, near the origin, it may take a relatively high index of iteration to see the nature of the orbits.

Glossary

Note: Italicized terms in the glossary indicate another glossary entry.

$\overline{\mathbb{C}}$: The extended complex plane or the Riemann sphere.
$\mathbb{D}$: The open unit disc: $\{z : |z| < 1\}$.
$\mathbb{H}$: The upper half-plane: $\{z : \Re(z) < 1\}$.
$\mathcal{I}$: The iterative family of f, that is, the set $\mathcal{I} = \{f^n : n = 0, 1, \dots\}$.

Abel functional equation: The functional equation $A \circ f = A + c$, where f is known, and c is a complex constant. In the canonical Abel functional equation, $c = 1$.
antecedent: An image of a point z under the iteration of f, that is, $f^n(z)$, where $n \in \mathbb{N}$.
attracting cycle (or orbit): An *attracting fixed point* or *attracting periodic orbit.* We will refer generically to an *attracting periodic orbit* or an *attracting fixed point* as an *attracting orbit*, *attracting cycle* or attractor.
attracting fixed point: Given a function f, a *fixed point* λ satisfying $|f'(\lambda)| < 1$. If $f'(\lambda) = 0$ then the fixed point λ is a *superattracting fixed point.*
attracting periodic orbit of order p**:** A *periodic orbit of order* p such that

$$\left|\frac{d}{dz} f^p(\lambda_k)\right| = \left|\prod_{k=0}^{p-1} f'(\lambda_k)\right| < 1.$$

Also called an *attracting cycle.*

backward orbit: The set $O^-(z) = \{f^{-n}(z) : n = 0, 1, 2, \dots\}$, where f^{-1} is the *total inverse* of f. One can also consider the orbit of a point under a branch of the inverse.
Baker domain: A region D in $\overline{\mathbb{C}}$, *forward invariant* under f, with an essential singularity $a \in \partial D$, with the property that while no orbits $O^+(z)$ originating in D have an accumulation point in D, there is a $p \in \mathbb{N}$ such that for all $z \in D$, $f^{pk}(z) \to a$ as $k \to \infty$.
basin of infinity: If the point at infinity is an *attracting fixed point*, then its *domain of convergence* is referred to as the basin of infinity.
Böttcher functional equation: The functional equation $B \circ f = (B)^k$ where f is known, and k is the order of the first nonzero term in the Taylor series of f expanded about a point c.

Cantor set: A totally disconnected, perfect set.
center: Given a function f, a point λ satisfying $f(\lambda) = \lambda$ where iteration is conformally equivalent to a rotation of $\mathbb{D}$. Alternatively, λ is a center if it is satisfies the canonical *Schröder functional equation* $S \circ f = f'(\lambda) \cdot S$, on a neighborhood surrounding λ.
completely invariant domain: A connected component D of the *Fatou set* such that $f(D) \subseteq D$ and $f^{-1}(D) \subseteq D$ for all branches of the inverse of f.
consequent: A *preimage of a point* z under the iteration of f, that is, a point w such that $z = f^n(w)$, where $n \in \mathbb{N}$.
critical point: A point z such that $f'(z) = 0$.
critical value: The point $w = f(z)$, where z is a *critical point* of f.

degree of a rational function: If $f = p/q$ is a *rational function*, the maximum of $\{\deg f, \deg q\}$, written $\deg f$.
derived set: The set S' of limit points of a given set S.

domain: An open, connected set in $\overline{\mathbb{C}}$. Sometimes this will be part of the name of a particular set and is not intended to imply that the set is open or connected; for example, the domain of nonnormality is closed, and the domain of normality can be an infinite union of open, connected sets.

domain of convergence: The set of all z such that $f^k(z)$ converges to a fixed point λ, often denoted $A(\lambda)$. It is possible that this set has infinitely many connected components in which case the component containing λ is called the *immediate domain of convergence*.

domain of nonnormality: The set of points z from $\overline{\mathbb{C}}$ on which the *iterative family* $\mathcal{I}$ is not a *normal family* on neighborhoods of z. Also called the *Julia set* of f.

domain of normality: The set of points z from $\overline{\mathbb{C}}$ on which the *iterative family* $\mathcal{I}$ is a *normal family* on some neighborhood of z. Also called the *Fatou set* of f.

entire function: An analytic function which is single valued at all finite points of the plane.

exceptional value: Let $\mathcal{F}$ be a family of analytic (or *meromorphic*) functions on a *domain* D. If $w \notin \bigcup_{f\in\mathcal{F}} f(D)$, then then w is an exceptional value for $\mathcal{F}$ on D.

Fatou set: For a given function f the *domain of normality* for the *iterative family* $\mathcal{I}$. Since this term did not come into common usage until the 1980s, our use of it is somewhat anachronistic.

filled-in Julia set: The union of the *Julia set* J and the set of all $z \in \mathbb{C}$ for which the *forward orbit* of z, under f, is bounded. This is usually applied when J is a closed curve in which case the filled-in Julia set equals the union of J and its interior.

fixed point: A point λ satisfying $f(\lambda) = \lambda$ for a given function f.

forward invariant: A connected component D of the *Fatou set* such that $f(D) \subseteq D$. It is also referred to as an *invariant domain*. If a forward invariant domain is also invariant for the *total inverse* of f, that is, if $f^{-1}(D) \subseteq D$ for all branches of the inverse, it is said to be *completely invariant*.

forward orbit: The set $O^+(z) = \{f^n(z) : n = 0, 1, 2, \ldots\}$.

fundamental circle: A circle C in $\overline{\mathbb{C}}$ such that $f(C) = C$ and $f(\mathrm{Int}(C)) = \mathrm{Int}(C)$.

Herman ring: A doubly connected component D of the *Fatou set* of f which is conformally isomorphic to an annulus A on which iteration by f (or some iterate of f) is conformally equivalent to a rotation of A. It is an example of a *rotation domain*.

immediate domain of convergence: The component D of the *domain of convergence* of an *attracting fixed point* λ that contains λ. If the domain of convergence only consists of one component, then the domain of convergence coincides with the immediate domain of convergence.

invariant domain: A connected component D of the *Fatou set* such that $f(D) \subseteq D$. An invariant domain is sometimes called *forward invariant*. If an invariant domain is also invariant under all branches of the *total inverse* of f, that is, if $f^{-1}(D) \subseteq D$ for all branches of the inverse, it is *completely invariant*.

irrationally neutral or indifferent fixed point: Given a function f, a *fixed point* λ where $|f'(\lambda)| = 1$ but is not a root of unity.

iterative family: The set $\mathcal{I} = \{f^n : n = 0, 1, \ldots\}$.

Julia set: For a given function f, the *domain of nonnormality* for the iterative family $\mathcal{I}$. For *rational functions* of degree two or more, this is is equivalent to the closure of the set of *repelling periodic points*. Since this term was not used until at least the 1920s and did not come into common usage until the 1980s, our use of it is somewhat anachronistic.

Kœnigs' linearization theorem: Let f be analytic on a neighborhood of λ. If λ is a *fixed point* of f satisfying $0 < |\lambda| < 1$, then there is a conformal mapping S from a neighborhood of λ onto a neighborhood of the origin satisfying the canonical *Schröder functional equation*, $S \circ f = f'(\lambda)S$.

Lattès function: A *rational function* l whose *Julia set* equals the entire Riemann sphere.

limit function: A function L on a component D of the *Fatou set* of f, such that a subsequence of the *iterative family* $\mathcal{I}$ converges uniformly to L on compact subsets of D.

meromorphic function: A function which has a pole at one or more points z in a *domain* D is said to be meromorphic on D.

Montel's normality criterion: Let $\mathcal{F}$ be a family of *meromorphic* functions on a *domain* D. If $\mathcal{F}$ has at least three *exceptional values* on D, then it is a *normal family* on D. If the family is analytic, only two exceptional values are sufficient.

multiplicity of a fixed point λ**:** The multiplicity of a *fixed point* λ viewed as a root of $f(z) = z$. Since $f(z) = z$ reduces to

$$0 = (\lambda - z) + a_1(z - \lambda) + a_{n+1}(z - \lambda)^{n+1} + \cdots ,$$

near λ, its multiplicity is greater than one only when the *multiplier* $f'(\lambda) = a_1$ is one, in which case the multiplicity equals $n + 1$.

multiplier of a fixed point: The quantity $f'(\lambda)$ where λ is a fixed point of f.

neighborhood of infinity: The set $\{z : |z| > R\}$, usually for large R.

neutral (or indifferent) fixed point: Given a function f, a *fixed point* λ with $|f'(\lambda)| = 1$. A neutral fixed point is *rationally neutral or indifferent* if $f'(\lambda)$ is a root of unity; if it is not, λ is an *irrationally neutral or indifferent* fixed point.

normal family: A family of functions $\mathcal{F}$ is normal on a *domain* D if every sequence of functions from $\mathcal{F}$ contains a subsequence f_n which converges uniformly on compact subsets of D, or converges uniformly to ∞ on D.

periodic orbit of order p**:** The set $O^+(z_0) = \{z_0, z_1 = f(z), \ldots, z_{p-1} = f^{p-1}(z)\}$, where p is a *periodic point of order* p.

periodic point of order p**:** A point z satisfying $f^p(z_0) = z_0$, for $p \in \mathbb{N}$, with $f^k(z_0) \neq z_0$ for $0 < k < p - 1$.

Picard's Big Theorem: In any neighborhood of an isolated essential singularity, an analytic function takes on every value of $\mathbb{C}$ except at most one. Each nonomitted value is taken on an infinite number of times in this neighborhood. The omitted value is called an *exceptional value*. For the case of an essential singularity of a *meromorphic function*, two values are excluded.

Picard's Little Theorem: An *entire function* from $\mathbb{C}$ to $\mathbb{C}$ that omits two or more values is constant.

Poincaré functional equation: The functional equation $f \circ \theta = \theta \circ (s \cdot I)$, where f is known, and s is a complex constant. Usually, $s = f'(\lambda)$ where λ is a *fixed point* of f with $|s| \geq 1$.

post-critical set: The union of the sets $O^+(w)$, as w ranges over the set of *critical values* of f.

preimage of a point z**:** A point w such that $z = f^n(w)$, where $n \in \mathbb{N}$.

properly discontinuous group: Our usage is intended to suggest how the term was used when those mathematicians we wrote about used it. The action of a group G is discontinuous on a set D if G contains no infinitesimal transformations; that is, there is no element $g \neq I$ of G such that $g(z)$ is infinitesimally close to z for all $z \in D$. If there are points in D where an infinitesimal transformation exists, a discontinuous group is improperly discontinuous at that point, a term that is generally out of use nowadays; otherwise, its action is properly discontinuous. See Forsyth [**1918**, pp. 717–719].

rational function: A complex function $f = p/q$, where p and q are polynomials.

rationally neutral or indifferent fixed point: Given a function f, a *fixed point* λ where $|f'(\lambda)|$ is a root of unity.

repelling cycle (or orbit): A *repelling fixed point* or *repelling periodic orbit.*

repelling fixed point: A *fixed point* λ of a function f with $|f'(\lambda)| > 1$.

repelling periodic orbit of order p**:** A *periodic orbit of order* p such that

$$\left|\frac{d}{dz} f^p(\lambda_k)\right| = \left|\prod_{k=0}^{p-1} f'(\lambda_k)\right| > 1.$$

Also called a *repelling cycle.*

rotation domain: The collective name for a *Siegel disc* or a *Herman ring.*

Schröder functional equation: The functional equation $S \circ f = c \cdot S$, where f is known and c is a complex constant. In the canonical Schröder functional equation, $c = f'(\lambda)$ where λ is a *fixed point* of f.

***d'un seul tenant*:** Used by Fatou and Julia to describe path connectedness. Julia gave the following definition: a set is *d'un seul tenant* if, given $\epsilon > 0$, "one can find in the set a succession of points beginning and ending with the two points, such that the distance of each to the one following it is less than ϵ" [**1918a**, p. 58].

Siegel disc: A singly connected region D of the *Fatou set* of f which is conformally isomorphic to $\mathbb{D}$ on which f (or some iterate of f) is conformally equivalent to a rotation of $\mathbb{D}$. It is an example of a *rotation domain.* The *fixed point* is called a *center.*

singular domain: A component of the *Fatou set* of f such that the *iterative family* $\mathcal{I}$ has nonconstant *limit functions.*

superattracting fixed point: Given a function f, a point λ satisfying $f(\lambda) = \lambda$ and $|f'(\lambda)| = 0$.

total inverse: The multifunction defined by $f^{-1}(z) = \{w : f(z) = w\}$.

total orbit: The union of the *forward orbit* $O^+(z)$ and the *backward orbit* $O^-(z)$.

wandering domain: A component of the *Fatou set* D such that the infinite sequence $\{f^n(D)\}$ is pairwise disjoint.

Bibliography

We list archival material first, followed by a list publications by author, publisher, or journal. In the case of joint authorhship we have ordered the names alphabetically. Our citation key is the year of publication, followed by a lower-case letter (e.g., Fatou [**1906b**]) when there are multiple publications for the author in a given year. An upper-case "W" following a year indicates the work is available only on the World Wide Web. A paper that is accepted for publication but has yet to appear is denoted by "appear" and a paper in preparation but not submitted is denoted by "prep".

Archivial material

[A1-1911] *Arrête pour l'admission à l'École Polytechnique*, Ministère de la Guerre, Direction du Génie, Bureau du Personnel, 5, 9 Mars 1911, pp. 1–15. Also in Bull. Officiel, 32, Édition Méthodique, p. 73.

[A2-W] *Zur Erinnerung an Professor Dr. Hubert Cremer, Professor für Mathematik an der Rheinisch-Westfälischen-Technischen Hochschule Aachen.* Initiator und Erster Leiter des Rechenzentrums der RWTH Aachen; Zusammengestellt durch das Rechenzentrum der RWTH Aachen anl. eines Kolloquiums zur Ehrung von Prof. Dr. Hubert Cremer u. zur Feier des 25jährigen Bestehens desvRechenzentrums. Aachen, den 16 Mai 1984 [25 pp.]. `http://www.archiv.rwth-aachen.de/Zuse%20Z22.pdf`

[A3] *Death certificate, Pierre Fatou*, Archives de la Mairie de Pornichet, August 9th 1929.

[A4] F/17/25674, *Dossier de carrière de Pierre Joseph Louis Fatou*, Centre Historique des Archives Nationales, Paris.

[A5] l'Illutration, *Le plus jeune membre de l'institut Grand Mathématicien Français*, n° 4751, 24 Mars 1934.

[A6] F/17/28319, *Dossier, Gaston Maurice Julia,* Centre Historique des Archives Nationales, Paris.

[A7] 61/AJ/170, *Concours d'entrée à l'Éc. Norm. Sup.*, Centre Historique des Archives Nationales, Paris.

[A8] *Souvenirs sur Georges Humbert et Camille Jordan*, Archives du Collège de France, CDF 16/24, pp. 87–89.

[A9] F/17/25827, *Dossier, Samuel Isaac Lattès*, Centre Historique des Archives Nationales, Paris.

[A10] *Dossier, Samuel Isaac Lattès*, University Paul Sabatier, Toulouse.

[A11] *(Le) Nouvelliste de Morbihan*, daily journal: n° 187, August 11th 1929, p. 4; n° 188, August 13th 1929, p. 4; n° 190, August 15th 1929, p. 3.

[A12] *(Les) Nouvelles de Lorient et de Morbihan*, weekly journal: n° 469, August 17th 1929, p. 3.

[A13] *Plis Cachetés*, Annuaire de l'Académie des Sciences, 1918, pp. 335–336.

[A14] Quimper, Town archives, *État des médecins, pharmaciens et sages femmes excerçant dans la ville de Quimper*, September 7th 1836, filed 46—1.

[A15-W] List of Tatsujiro Shimuzi's publications, `http://www.jams.or.jp/shimizu/lshimz-e.html`

Books and Journal Articles

Aarts, J. and Oversteegen, L.

[1993] *The geometry of Julia sets*, Trans. AMS, 338, 1993, pp. 897–918.

Abate, M.

[1988] *Horospheres and iterates of holomorphic maps*, Math. Zeitschrift, 198, 1988, pp. 225–238.

[1989] *Iteration theory of holomorphic maps on Taut domains*, Mediterranean Press, 1989.

[1990] *The Lindelöf principle and the angular derivative in strongly convex domains*, J. Analyse Math., 54, 1990, pp. 189–228.

[1991] *Angular derivatives in strong pseudoconvex manifolds*, Several Variables and Complex Geometry, Part 2 (Santa Cruz, CA, 1989), pp. 23–40; Proc. Sympos. Pure Math., 52, Part 2, AMS, Providence, RI, 1991.

[1998] *The Julia-Wolff-Carathéodory theorem in polydisks*, J. Analyse Math., 74, 1998, pp. 275–306.

Abate, M. and Tauraso, R.

[1999] *The Julia-Wolff-Carathéodory theorem(s)*, Complex Geom. Analysis in Pohang, 1997, pp. 161–172; Contemp. Math., 222, AMS, Providence, 1999.

Abdo, G. and Godfrey, P.

[1999-W] *Table of conformal mappings using continuous coloring*, 1999, `http://my.fit.edu/~gabdo/`.

Abel, N.H.

[1881] *Œuvres complètes d'Abel*, 2 vols., Johnson Reprint Corporation, I, 1964, pp. 61–65.

Ageron, P.

[2003] *Histoire d'un mathématicien de Caen: Ludovic Zoretti*, `http://www.math.unicaen.fr/~ageron/histoire-philo.html`, 2003.

Ahlfors, L.V.

[1932] *Quelques propriétés des surfaces de Riemann correspondant aux fonctions méromorphes*, Bull. SMF, 60, 1932, pp. 197–207.

[1935] *Zur Theorie der Überlagerungsflächen*, Acta Math., 65, 1935, pp. 157–194.

[1966] *Complex Analysis*, 2nd edition, McGraw-Hill, New York, 1966.

Alain (Émile August Chartier)

[1936] *Histoire de mes pensées*, NRF, Gallimard, 1936.

[1942] *Vigiles de l'esprit*, NRF, Gallimard, 1942.

Alexander, D.S.

[1994a] *A history of complex dynamics: From Schröder to Fatou and Julia*, Vieweg, 1994.

[1994b] *Civilized Mathematics*, MAA Focus, June 1994.

[1995] *Gaston Darboux and the history of complex dynamics*, Historia Mathematica, 22, 1995, pp. 179–185.

[1996a] *Newton's Method — Or is it?*, MAA Focus, October 1996.

[1996b] *An Episodic History of Complex Dynamics from Schröder to Fatou and Julia*, Suppl. Rend. Circ. Mat. Palermo, Ser. II, 44, 1996, pp. 57–83.

Amaldi, U.

[1954] *Della vita e delle opere di Salvatore Pincherle*, in *Opere Scelte di Salvatore Pincherle*, UMI, Edizioni Cremonese, 1954.

Ando, T. and Fan, K.

[1979] *Pick-Julia theorems for operators*, Math. Zeitschrift, 168, 1979, pp. 23–34.

Andonie, G.S.

[1967] *Istoria Matematicii în România*, Editura Ştiinţifică, Bucureşti, Volume III, 1967.

Andreoli, G.

[1916] *Sull' iterazione delle funzioni di variabili reali*, Rend. Accad. Lincei Roma, (5), 25, 2, 1916, pp. 129–133.

[1919] *Sull' iterazione di una speciale funzione*, Rend. Accad. Lincei Roma, (5) 28, 1, 1919, pp. 420–424.

[1937] *Sull' iterazione delle funzioni monotone*, Rend. Accad. Sci. Fis. Mat., Napoli, (4), 7, 1937, pp. 10–18.

Appell, P.

[1891] *Sur des équations différentielles linéaires transformables en elles-mêmes par un changement de fonction et de variable*, Acta Math., 15, 1891, pp. 281–315.

[1897] *Observations sur la Communications Précédente*, C.R. Acad. Sci. Paris, 124, 1897, pp. 1433–1434.

Appell, P., Goursat, É. and Fatou, P.

[1930] *Théorie des fonctions algébriques d'une variable et des transcendantes qui s'y rattachent, tome II: Fonctions automorphes*, Gauthier-Villars, 1930, Deuxième édition, revue et augementée.

Arago, F.

[1957] *Joseph Fourier*, Biographies of Distinguished Scientific Men, London, 1957, pp. 242–286.

Arnold, D.N.

[2008-W] *Graphics for complex analysis*, `http://www.ima.umn.edu/~arnold/complex.html`

Arnold, V.I.

[1963] *Proof of a theorem by A.N. Kolmogorov on the invariance of quasi-periodic motions under small perturbations of the Hamiltonian*, Russian Math. Survey 18, 1963, pp. 13–40.

[1965] *Small Denominators I: On the mappings of the circumference onto itself*, AMS Translations, 46, 1965, pp. 213–284.

[1978] *Mathematical methods of classical mechanics*, 1st edition, Graduate Texts in Mathematics, 60, Springer-Verlag, New York, 1978.

Arnold, V.I., Kozlov, V.V. and Neishtadt, A.I.

[2006] *Mathematical aspects of classical and celestial mechanics*, Dynamical Systems III, Series: Encyclopaedia of Mathematical Sciences, Volume 3. Springer-Verlag 3rd ed., Berlin, 2006

Atela, P. and Hu, J.

[1996] *Commuting polynomials and polynomials with the same Julia set*, Internat. J. Bif. Chaos Appl. Engrg., Vol. 6, No. 12A, 1996, pp. 2427–2432.

Athreya, K.B. and Ney, P.E.

[1972] *Branching processes*, Springer-Verlag, Berlin, 1972.

Audin, M.

[2009] *Fatou, Julia, Montel, le grand prix des sciences mathématiques de 1918, et après, ...*, Springer-Verlag, Berlin, 2009.

Babbage, C.

[1815] *Phil. Trans. Royal Soc. London*, 105, 1815, pp. 389–423.

[1820] *Examples of the solutions of functional equations*, Cambridge, 1820.

Bădescu, R.

[1929a] *Sur l'équation intégrale d'Abel généralisée*, C.R. Acad. Sci. Paris, 188 1929, pp. 217–219.

[1929b] *Sur l'équation intégrale d'Abel généralisée*, C.R. Acad. Sci. Paris, 188 1929, pp. 851–853.

[1929c] *Distribution des singularités. De la solution d'une équation intégrale linéaire*, C.R. Acad. Sci. Paris, 189, 1929, pp. 831–834.

[1929c] *Sur une équation fonctionnelle*, Bull. Acad. Bruxelles, (5), 15, 1929, pp. 1062–1072.

[1930a] *Sur une équation fonctionnelle*, C.R. Acad. Sci. Paris, 191, 1930, pp. 480–482.

[1930b] *Solutions logarithmiques d'une équation intégrale*, C.R. Acad. Sci. Paris, 191, 1930, pp. 428–431.

[1930c] *Sur l'équation intégrale de Volterra*, Mathesis, 44, 1930, pp. 71–74.

[1931] *Résolution d'une équation fonctionnelle et fonctions itératives généralisées*, C.R. Acad. Sci. Paris, 192, 1931, pp. 599–602.

[1932-36] *Sur l'équation intégrale de Fredholm dans le domaine complexe*, Verhandlungen Kongress Zürich, 2, 1932, p. 116; Bull. Acad. Sci. Roumaine, 16, 1933, pp. 108–110; Acad. Roumaine, Bull. Sect. Sci., 17, 1936, pp. 3–6.

Bagnera, G. and De Franchis, M.

[1906] *Sur les surfaces hyperelliptiques*, C.R. Acad. Sci. Paris, 145, 1906, pp. 747–749.

Baire, R.

[1990] *Lettres de René Baire à Emile Borel.* Cahiers du Séminaire d'Histoire des Mathématiques, 11, 1990, pp. 33–120.

Baker, I.N.

[1955] *The iteration of entire transcendental functions and the solution of the functional equation $f(f(z)) = F(z)$*, Math. Ann., 129, 1955, pp 174–180.

[1968] *Repulsive fixed points of entire functions*, Math. Zeitschrift, 104, 1968 pp. 252–256.

[1970] *Completely invariant domains of entire functions*, in Shankar [**1970**], 1970, pp. 33-35.

[1975] *The domains of normality of an entire function,* Ann. Acad. Sci. Fenn. Ser. A Math., 1, 1975, pp. 277–283.

[1976] *An entire function which has wandering domains*, J. Austral. Mat. Soc., 22, 1976, pp. 173–176.

[1984] *Wandering domains in the iteration of entire functions*, Proc. LMS, (3), 49, 1984, pp. 563–576.

Baker, I.N. and Eremenko, E.

[1987] *A problem on Julia sets*, Ann. Acad. Sci. Fenn, 12, 1987, pp. 229–236.

Baker, I.N., Kotus, J. and Lu, Y.

[1991] *Iterates of meromorphic functions I*, Ergod. Th. Dynam. Sys., 11, 1991, pp. 241–248.

Baker, I.N. and Singh, A.P.

[1995] *Wandering domains in the iteration of compositions of entire functions*, Ann. Acad. Sci. Fenn. Ser. A Math., 20, 1995, pp. 149–153.

Barański, K.

[2007] *Trees and hairs for some hyperbolic entire maps of finite order.* Math. Zeitschrift, 257, 1, 2007, pp. 33–59.

Barba, G.

[1929] *Un problema d'iterazione razionale relativo ad un fascio sizigetico di forme biquadratiche*, Atti Catania, (5), 16, Nr. 18′, 1929.

[1936] *Sulla iterazione di una classe di funzioni*, Atti Accad. Naz. Lincei, Cl. Sci. Fis. Mat. Nat, (6), 23, 1936, pp. 473–477.

Bargmann, D., Bonk, M., Hinkkanen, A. and Martin, G.J.

[1999] *Families of meromorphic functions avoiding continuous functions*, J. Analyse Math., 79, 1999, pp. 379–387.

Barrow-Green, J.

[1997] *Poincaré and the Three Body Problem*, History of Mathematics, 11, AMS-LMS, 1997.

[appear] *An American goes to Europe. Three letters from Oswald Veblen to George Birkhoff in 1913/1914*, Math. Intelligencer, to appear.

Bartocci, C.

[2000] *Henri Poincaré: geometria e caso, scritti di matematica e fisica*, Bollati Boringhieri, 2000.

Beardon, A.F.

[1990] *Iteration of contractions and analytic maps*, J. LMS, 41, 1990, pp. 141–150.

[1991] *Iteration of rational functions*, Springer-Verlag, New York, 1991.

[1992] *Polynomials with identical sets*, Complex Variables, 17, 1992, pp. 195–200.

Bedding, S. and Briggs, K.

[1996] *Iterations of quaternion functions*, Amer. Math. Monthly, 103, 1996, pp. 654–664.

Bell, E.T.

[1986] *Men of Mathematics*, Simon & Schuster, New York, 1986.

Belna, C.K., Colwell, P. and Piranian, G.

[1985] *The radial behavior of Blaschke products*, Proc. AMS, Vol. 93, No. 2 (February 1985), pp. 267–271.

Beltrami, E.

[1978] *Memorie per la storia dell'Universitá di Pavia*, Ser. 1a, 1878.

Bennett, A.A.

[1915] *The iteration of functions of one variable*, Ann. Math., (2), 17, 1915, pp. 23–60.

[1916a] *A case of iteration in several variables*, Ann. Math., (2), 17, 1916, pp. 188–196.

[1916b] *A case of iteration in several variables*, Bull. AMS, 22, 1916, pp. 487–488.

Bergweiler, W.

[1993] *Iteration of meromorphic functions*, Bull. AMS, 29, 2, 1993, pp. 151–188.

[1998] *A new proof of the Ahlfors five islands theorem*, J. Analyse Math., 76, 1998, pp. 337–347.

[2000] *The role of the Ahlfors five islands theorem in complex dynamics*, Conf. Geo. and Dyn. 4, 2000, pp. 22–34.

Bergweiler, W. and Eremenko, A.

[1995] *On the singularity of the inverse to a meromorphic function of finite order*, Rev. Mat. Iberoamericana, 11, 1995, pp. 355–373.

Bergweiler, W., Haruta, M., Kriete, H., Meier, H.G. and Terglane, N.

[1993] *On the limit functions of iterates in wandering domains*, Ann. Acad. Sci. Fenn. Math., 18, 1993, pp. 369–375.

Berkson, E. and Porta, H.

[1978] *Semigroups of analytic functions and composition operators*, Michigan Math. J., 25, 1978, pp. 101–115.

Bernoulli, J.

[1710] *Problème inverse des forces centrales, extrait de la réponse de Monsieur Bernoulli à Monsieur Herman*, Mém. de l'Acad. R. des Sciences de Paris, 1710, pp. 521. *Opera Omnia I*, pp. 470-480.

Bernstein, V.

[1932a] *Sur les directions de Julia de certaines fonctions entières*, J. Éc. Polytechnique, (2), 30, 1932, pp. 191–219.

[1932b] *Sur les directions de Julia et de Borel des fonctions entières d'ordre fini*, C.R. Acad. Sci. Paris, 195, 1932, pp. 356–358.

[1932c] *Sur l'analogie entre la distribution des droites de Julia des fonctions holomorphes et celle des points singuliers des fonctions analytiques*, C.R. Acad. Sci. Paris, 194, 1932, pp. 1629–1631.

Berthon, P.

[1986] *Les plis cachetés de l'Académie des Sciences*, Revue d'Histoire des Sciences, 1986, XXXIX/1, pp. 71–78.

Bertrand, M.J.

[1877] *Sur la possibilité de déduire d'une seule des lois de Kepler le principe de l'attraction*, C. R. Acad. Sci. Paris Sér. I Math., 84, 1877, 671–674.

Bhattacharya, P.

[1969] *Iteration of analytic functions*, Ph.D. dissertation, University of London, 1969.

[1971] *On the domain of normality of an attractive fixpoint*, Trans. AMS, 153, 1971, pp. 89–98.

Bieberbach, L.

[1933] *Beispiel zweier ganzer Funktionen zweier komplexer Variablen, welche eine schlichte volumtreue Abbildung des $\mathbb{R}^4$ auf einen Teil seiner selbst vermitteln*, Pr. Akad. Wiss. Sitzungsber, 1933, pp. 476–479.

[1953] *Conformal Mapping*, Chelsea, 1953. Translation of 4th ed., *Einführung in die Konforme Abbildung*, Berlin, 1949.

Biernacki, M.

[1928a] *Sur les droites de Julia des fonctions entières*, C.R. Acad. Sci. Paris, 186, 1928, pp. 1410–1412.

[1928b] *Sur les droites de Julia des fonctions entières*, C.R. Acad. Sci. Paris, 186, 1928, pp. 1260–1262.

Blanchard, P.

[1984] *Complex analytic dynamics on the Riemann sphere*, Bull. AMS, 11, 1984, pp. 85-141.

Blaschke, W.

[1915] *Eine Erweiterung des Satzes von Vitali über Folgen analytischer Funktionen*, Leipziger Berichte, 67, 1915, pp. 194–200.

Bloch, L.

[1925] *Les théorèmes de M. Valiron sur les fonctions entières et la théorie de l'uniformisation*, Ann. Fac. Sci. Toulouse, Ser. 3, 17, 1925, pp. 1–22.

[1926] *Les fonctions holomorphes et méromorphes dans le cercle unité*, Mémorial des Sciences Mathématiques, fasc. 20, Gauthier-Villars Paris, 1926, p. 61.

[1931] *Fatou*, Notices de l'Ass. des Anciens Élèves de l'Éc. Norm. Sup., 1931, pp. 52–58.

Boas, H.

[2010] *Julius and Julia: Mastering the Art of the Schwarz Lemma*, MAA Monthly, 117, 2010, pp. 770–785.

Bobleter, O.

[1990] *Professor Erika Cremer — A pioneer in gas chromatography*, Chromatographia, 30, 1990, pp. 471–476.

Boehm, C.

[1937] *Mathematische Statistik in Wirtschaft und Technik*, Jahresbericht DMV, 47, 1937, pp. 239–242.

Bogush, A.A., Gazizov, A.Z., Kurochkin, Y.A. and Stosui, V.T.

[2001] *On symmetry properties of Quaternionic Analogs of Julia sets*, Proc. 9th Annual Seminar NPCS-2000, Minsk, Belarus (L. Babichev and V. Kuvshivov, eds.), 2001, pp. 304–309.

Bohr, H. and Landau, E.

[1910] *Über das Verhalten von $\zeta(s)$ und $\zeta_k(s)$ in der Nähe der Geraden $\sigma = 1$*, Gött. Nachr., 1910, pp. 303–330.

Boltzmann, L.

[1982] *On the methods of theoretical physics*, Proc. Phys. Soc. London, Vol. No. 12, Issue 1, 1892, pp. 336–345.

Borel, É.

[1894] *Sur quelques points de la théorie des fonctions*, C.R. Acad. Sci. Paris, 118, 1894, pp. 340–342.

[1895] *Sur quelques points de la théorie des fonctions*, Ann. Sci. Éc. Norm. Sup., Ser. 3, 12, 1895, pp. 9–55.

[1896a] *Fondements de la théorie des séries divergentes sommables*, J. Math. Pure Appl., (5), 2, 1896, pp. 103–122.

[1896b] *Sur la région de sommabilité d'un développement de Taylor*, C.R. Acad. Sci. Paris, 123, 1896, pp. 548–549.

[1896c] *Applications de la théorie des séries divergentes sommables*, C.R. Acad. Sci. Paris, 122, 1896, pp. 805–807.

[1896d] *Sur les séries de Taylor*, C.R. Acad. Sci. Paris, 123, 1896, pp. 1051–1052.

[1896e] *Démonstration élémentaire d'un théorème de M. Picard sur les fonctions entières*, C.R. Acad. Sci. Paris, 122, 1896, pp. 1045–1048.

[1897] *Sur les séries de Taylor*, Acta Math., 21, 1897, pp. 243–248.

[1898] *Sur la recherche des singularités d'une fonction définie par un développement de Taylor*, C.R. Acad. Sci. Paris, 127, 1898, pp. 1001–1003.

[1899] *Sur le prolongement des fonctions analytiques*, C.R. Acad. Sci. Paris, 128, 1899, pp. 283–284.

[1900a] *Sur les séries de fractions rationnelles*, C.R. Acad. Sci. Paris, 130, 1900, pp. 1061–1064.

[1900b] *Sur la généralisation du prolongement analytique*, C.R. Acad. Sci. Paris, 130, 1900, pp. 1115–1118.

[1914] *Introduction géométrique à quelques théories physiques*, Gauthier-Villars, Paris, VII, 8°, 1914.

[1917] *Leçons sur les fonctions monogènes uniformes d'une variable complexe*, Gauthier-Villars, 1917. Gaston Julia, ed.

[1919] *L'intégration des fonctions non bornées*, Ann. Sci. Éc. Norm. Sup., Ser. 3, 36, 1919, pp. 71–92.

[1926] *Notice sur la vie et les travaux de Georges Humbert* (1859–1921). Mémoires Acad. Inst. Fr., (2), 58, No. I, 1926, pp. I–XIX.

Bortolotti, E.

[1898] *Sulla convergenza degli algoritmi periodici e sulla risoluzione approssimata delle equazioni algebriche*, Bologna: Civelli, 8°, 1898.

Bottazzini, U.

[1999] *Poincaré, il cervello delle scienze razionali*, 'I grandi della scienza', Le Scienze, 7, 1999.

Böttcher, Ł.

[1898] *Beiträge zu der Theorie der Iterationsrechnung*, doctoral dissertation, Leipzig University, 1898.

[1899] *Zasady Rachunki Iteracyjnego* (Principles of Iteration), Prace Matematyczno-Fizyczne, Warsaw, 10, 1899–1900, pp. 65–101.

[1904a] *Über die Iteration der linearen Funktionen*, Wiad. Mat., 8, 1904, pp. 291–307; 9, pp. 77–86.

[1904b] *The principal laws of convergence of iterates and their application to analysis*, Kasan Ges., (2), 13; Nr. 1, pp. 1–37; 14, Nr. 2, pp. 155–200; Nr. 3, 1904, pp. 201–234.

[1909] *Nouvelle méthode d'intégration d'un système de n équations fonctionnelles linéaires du premier ordre de la forme* $U_i(z) = \sum_{j=1}^{j=n} A_{ij}(z)U_jF(z)$, Ann. Sci. Éc. Norm. Sup., Ser. 3, 26, 1909, pp. 519–543.

Bourbaki, N.

[1960] *Éléments d'histoire des mathématiques*, Hermann, 1960.

Bourlet, C.

[1897a] *Sur les opérations en général et les équations différentielles linéaires d'ordre infini*, 3, 14, Ann. Sci. Éc. Norm. Sup., Ser. 3, 14, 1897, pp. 133–190.

[1897b] *Sur certaines équations analogues aux équations différentielles*, C.R. Acad. Sci. Paris, 124, 1897, pp. 1431–1433.

[1899] *Problème de l'itération*, Ann. Fac. Sci. Toulouse, Ser. 1, 12,1899, pp. C1–C12.

Boutroux, P.

[1907a] *Sur les fonctions inverses des fonctions entières*, C.R. Acad. Sci. Paris, 145, 1907, pp. 1406–1409.

[1907b] *Sur les points critiques transcendants et sur les fonctions inverses des fonctions entières*, C.R. Acad. Sci. Paris, 145, 1907, pp. 708–710.

[1909a] *Sur les singularités transcendantes des fonctions inverses de fonctions entières*, C.R. Acad. Sci. Paris, 149, 1909, pp. 255–258.

[1909b] *L'inversion des fonctions entières*, Rome 4th Math. Congr., 2, 1909, pp 31–35.

Branner, B.

[1989] *The Mandelbrot set*, in *Chaos and fractals; the mathematics behind the computer graphics*, Devaney, R. and Keen, L., eds., 1989, pp. 75–105. See Devaney and Keen [**1989**].

Brian, E. and Demeulenaere-Douyère, C.

[1996] *Histoire et mémoire de l'Académie des Sciences. Guide de recherches*, Paris : Technique & Documentation, 1996.

Bricard, R.

[1913] *Carlo Bourlet*, Nouv. Ann., (4), 13, 1913, pp. 433–438.

Briot, Ch. and Bouquet, J.-K.

[1856] *Recherches sur les propriétés des fonctions définies par des équations différentielles*, J. Éc. Polytechnique, Cachier, 36, 1856, pp. 133–198.

Brjuno, A.D.

[1965] *Convergence of transformations of differential equations to normal forms* Dokl. Akad. Nauk USSR, 165, 1965, pp. 987–989.

Broer, H.W., Huitema, G.B. and Sevryuk, M.B.

[1996] *Quasiperiodic tori in families of dynamical systems: order amidst chaos*, Lecture Notes in Math. 1645, Springer-Verlag 1996.

Brolin, H.

[1965] *Invariant sets under iteration of rational functions*, Arkiv för Matematik, 6, 1965, pp. 103–144.

Brooks, R. and Matelski, J.P.

[1978] *The dynamics of 2-generator subgroups of* PSL(2, $\mathbb{C}$), Proc. 1978 Stony Brook Conference of "Riemann surfaces and related topics", Kra and Maskit eds., pp. 65–71.

Buff, X.

[2000] *Fibonacci fixed point of renormalization*, Erg. Th. Dynam. Systems, 20, 2000, pp. 1287–1317.

Buff, X. and Chéritat, A.

[2005] *Ensembles de Julia quadratiques de mesure de Lebesgue strictement positive*, C.R. Acad. Sci. Paris, Ser. I, 340, 2005.

Buhl, A.

[1921] *Éloge de Samuel Lattès*, Extrait des Mémoires Acad. Sc. Inscriptions et Belles-Lettres de Toulouse, IX, 1921, pp. 1–13.

[1931] *Nécrologie: Gabriel Kœnigs*, L'Enseign. Math., 30, 19310, pp. 286–287.

Bulletin de Société de Mathématiques de France

[1921] Comptes Rendus des Séances, 49, 1921, pp. 1–59.

[1922] Comptes Rendus des Séances, 50, 1922, pp. 1–34.

[1926] Comptes Rendus des Séances, 54, 1926, pp. 1–45.

Burckel, R.B.

[1979] *An introduction to classical complex analysis*, v. 1, Birkhäuser Basel, 1979.

[1981] *Iterating analytic self-maps of discs*, Amer. Math. Monthly, 88, 1981, pp. 396–407.

Butzer, P.L., Jansen, M. and Zilles, H.

[1982] *Johann Peter Gustav Lejeune Dirichlet (1805–1859): Genealogie und Werdegang*, Dürerner Geschichtsblätter, 71, 1982, pp. 31–56.

Cahen, E.

[1894] *Sur la fonction $\zeta(s)$ et sur des fonctions analogues*, Ann. Sci. Éc. Norm. Sup., Ser. 3, 11, 1894, pp. 75–164.

Camacho, C.

[1978] *On the local structure of conformal mappings and holomorphic vector fields in* $\mathbb{C}^2$, Astérisque, 1978, pp. 59–60; pp. 83–94.

Cantor, G.

[1884] *On the power of perfect sets of points*, in *Classics on Fractals*, Gerald A. Edgar, ed., Addison-Wesley, Boston, 1993. Translation of *De la puissance des ensembles parfait de points*, Acta Math. 4, 1884, pp. 381–392.

Carathéodory, C.

[1912] *Untersuchungen über die konformen Abbildungen von festen veränderlichen Gebieten*, Math. Ann., 72, 1912, pp. 107–144.

[1913a] *Über die gegenseitige Beziehung der Ränder bei der konformen Abbildung des Inneren einer Jordan schen Kurve auf einen Kreis*, Math. Ann., 73, 1913, pp. 305–320.

[1913b] *Über die Begrenzung einfach zusammenhängender Gebiete*, Math. Ann., 73, 1913, pp. 323–370.

[1926] *Über das Schwarzsche Lemma bei analytischen Funktionen von zwei komplexen Veränderlichen*, Math. Ann., 97, 1926, pp. 76–98.

[1929a] *Über die Winkelderivierten von beschränkten analytischen Funktionen*, Sitzungsberichte Preuss Akad. Wiss. Berlin, Phys.-Math Kl., 1929, pp. 39–54.

[1929b] *Stetige Konvergenz und normale Familien von Funktionen*, Math. Ann., 101, 1929, pp. 515–533.

Carathéodory, C. and Landau, E.

[1911] *Beiträge zur Konvergenz von Funktionenfolgen*, Sitz. Kön. Preuss. Akad. Wiss. Berlin, 1911, pp. 587–613.

Carleson, L. and Gamelin, T.W.

[1993] *Complex Dynamics*, Springer-Verlag, New York, 1993.

Cauchy, A.-L.

[1842] *Mémoire sur un théorème fondamental, dans le calcul intégral*, C.R. Acad. Sci. Paris, 14, 1842, pp. 1020–1028.

Cayley, A.

[1879a] *Applications of the Newton–Fourier Method to an imaginary root of an equation*, Quarterly J. Pure and Appl. Math., 16, 1879, pp. 179–185 = *Collected Papers*, **XI**, pp. 114–121.

[1879b] *The Newton–Fourier imaginary problem*, Amer. J. of Math., 2, 1879, p 97 = *Collected Papers*, **X**, p. 405.

[1880] *On the Newton–Fourier imaginary problem*, Proc. Camb. Phil. Soc., 3, 1880, pp. 231–232 = *Collected Papers*, **XI**, p. 143.

[1889–1897] *The Collected Papers of Arthur Cayley*, Cambridge University Press 1889–1897.

Celletti, A.

[2009] *Perturbation Theory in Celestial Mechanics*, Encyclopedia of Complexity and System Science, R.A. Meyer eds., Springer, 2009.

Charles Scribner's Sons (publisher)

[D1] *Dictionary of Scientific Biography*, Charles Scribner's Sons, New York, 1970–1980.

Chazy, J.

[1932] *Pierre Fatou*, Bull. Astronomique, 8, 1932, pp. 379–384.

Chen, H.H. and Fang, M.L.

[1995] *On the value distribution of $f^n f'$*, Sci. Sinica, Ser. A, 38, 7, 1995, pp. 787–798.

Chen, H.H. and Gu, Y.

[1993] *An improvement on Marty's criterion and its applications*, Sci. China, Ser. A, 36, 6, 1993, pp. 674–681.

Chéritat, A.

[2001] *Recherche d'ensembles de Julia de mesure de Lebesgue positive*, Thesis, Orsay, 2001.

Cherry, T.M.

[1964] *A singular case of iteration of analytic functions: a contribution to the small divisors problem*, Non-linear problems of Engineering, Academic Press, New York, 1964, pp. 29–50.

Chervel, A.

[1994] *Brève histoire de l'agrégation de mathématiques*, Gazette des Mathématiciens, 59, Janvier 1994, pp. 3–9.

Chierchia, L. and Falcolini, C.

[1994] *A direct proof of a theorem by Kolmogorov in Hamiltonian systems*, Ann. Scuola Norm. Sup. Pisa Cl. Sci. (4) 21, 1994, no. 4, pp. 541–593.

Chi-Tai, C.

[1993] *Normal families of meromorphic functions*, World Scientific, 1993.

Chu, C.H. and Mellon, P.

[1997] *Iteration of compact holomorphic maps on a Hilbert ball*, Proc. Amer. Math. Soc., 125, 1997, 1771–77.

Chuang, C.T.

[1937] *Quelques théorèmes sur les directions de Julia et de Borel des fonctions méromorphes*, C.R. Acad. Sci. Paris, 204, 1937, pp. 404–405.

Cigala, A.R.

[1904] *Sopra un criterio di instabilità*, Annali di Mat., (3), 11, 1904, pp. 67–78.

Cinquini, S.

[1933] *Sulle successioni di funzioni convergenti verso una funzione olomorfa*, Rend. Accad. Lincei Roma, (6) 17, 1933, pp. 505–506.

Cohen, A.

[1983] *A cache of 18th-century string*, The Galpin Society Jour., Vol. 36, 1983, pp. 37–48.

Comptes Rendus Hebdomadaires des Séances de l'Académie des sciences. Vie Académique (journal)

[CR-1904] C.R. Acad. Sci. Paris, 139, 1904.
[CR-1906] C.R. Acad. Sci. Paris, 143, 1906.
[CR-1913] C.R. Acad. Sci. Paris, 157, 1913, p. 1350.
[CR-1914] C.R. Acad. Sci. Paris, 159, 1914, pp. 156 & 948.
[CR-1915] C.R. Acad. Sci. Paris, 161, 1915.
[CR-1918] C.R. Acad. Sci. Paris, 167, 1918.
[CR-1918-Rapport] *Rapport de Grand Prix des sciences mathématiques*, C.R. Acad. Sci. Paris, 167, 1918, pp. 811–814.
[CR-173] C.R. Acad. Sci. Paris, 173, 192.

Cooke, R.

[2001-W] *The Cauchy-Kovalevskaya Theorem*, unpublished manuscript. See `http://www.cem.uvm.edu/~cooke/ckthm.pdf`

[2010-W] *Andrey Nikolayevich Kolmogorov*, *Encyclopedia Britannica* Online, `http://www.britannica.com/EBchecked/topic/321441/Andrey-Nikolayevich-Kolmogorov`, 2010.

Costabel, P.

[1984] *Les plis cachetés et la mémoire confiée à l'Académie*, La Vie des Sciences, Comptes Rendus, série générale, tome 1, n° 2, 1984, pp. 143–144.

Cotter, J.F.

[2000] *The Divine Comedy*, Revised Edition, Forum Italicum Publications, 2000.

Coulomb, J.

[1978] *Necrology of Gaston Julia*, C.R. Acad. Sci. Paris, 287, 1978, pp. 1–2.

Couturat, L.

[1903] *Opuscules et fragments inédits de Leibniz*, 1903, Felix Alcan, ed.

Cowen, C.C.

[1981] *Iteration and the solution of functional equations for functions analytic on the unit disk*, Trans. AMS, 265, 1981, pp. 69–95.

Cowen, C.C. and MacCluer, B.D.

[1995] *Composition operators on spaces of analytic functions*, CRC Press, Boca Raton, FL, 1995.

Cremer, H.

[1925] *Über die Iteration rationaler Funktionen*, Jahresbericht DMV, 33, 1925, pp. 185–210.

[1927] *Zum Zentrumproblem*, Math. Ann., 98, 1927, pp. 151–163.

[1930a] *Über das Zentrumproblem (mit besonderer Berücksichtigung der Lückenreihen)*, Ber. Math.-Phys. Akad. Leipzig, 32, 1930, pp. 243–250.

[1930b] *Ein Existenzbeweis der Kreisringabbildung zweifach zusammenhängender schlichter Bereiche*, Berichte der Mathematisch.-Phys. Klasse der Sächsischen Akademie der Wissenschaften zu Leipzig, 82, 1930, pp. 190–192.

[1932a] *Über die Schrödersche Funktionalgleichung und das Schwarzsche Eckenabbildungsproblem*, Leipziger Berichte, 84, 1932, pp. 291–324.

[1932b] *Zur Frage der Existenz nichtkonstanter Grenzfunktionen der Folgen der Iterierten rationaler Funktionen*, Jahresbericht DMV, 42, 1932, 2, p. 131.

[1932c] *Über die Schrödersche Funktionalgleichung*, Jahresbericht DMV, 41, 2, 1932, p. 78.

[1934] *Uber eine Eigenschaft der rationalen Funktionen*, Jahresbericht DMV, 44, 1934, pp. 276–278.

[1935] *Überkonvergenz und Zentrumproblem*, Math. Ann., 110, 1935, pp. 739–744.

[1938] *Über die Häufigkeit der Nichtzentren*, Math. Ann., 115, 1938, pp. 573–580.

[1948] *Dreikreisesatz und Zentrumproblem*, Commentarii Mathematici Helvetici, 21, 1948, pp. 185–188.

[1951] *Über das Iterationsproblem der analytischen Funktionen einer komplexen Veränderlichen*, Jahresbericht DMV, 54, 2, 1951, p. 31.

[1968] *Erinnerungen an Paul Koebe*, Jahresbericht DMV, Bd. 70, 1968, pp. 158–161.

Cremer, H. and Cremer, L.

[1937] *Über die theoretischen Ableitungen der Nachhallgesetze*, Akustische Zeitschrift, 1937, pp. 224ff.

Dalloz (publisher)

[1908] *Dictionnaire de Droit. Complet en deux volumes*, Dalloz, 1908.

Dang, Y. and Kauffman, L.H.

[1997] *Hypercomplex Fractal Distance Estimation*, in *Fractal Frontiers*, Novak M.M. and Dewey T.G., eds., World Scientific, 1997, pp. 117–130.

Dang, Y., Kauffman, L.H and Sandin, D.J.

[2002] *Hypercomplex Iterations Distance Estimation and Higher Dimensional Fractals*, Series on Knots and Everything, 17, World Scientific, 2002.

Davenport, H.

[1985] *Reminiscences of conversations with Carl Ludwig Siegel*, Math. Intelligencer, **7**, 2, 1985, pp. 76–79

de la Llave, R.

[2001] *A tutorial on KAM theory*, Smooth Ergodic Theory and Its Applications (Seattle, WA, 1999), pp. 175–292, Proc. Sympos. Pure Math., 69, Amer. Math. Soc., Providence, RI, 2001.

Delaunay, C.E.

[1846] *Nouvbelle théorie analytique du mouvement de la lune*, Comptes Rendus 23, 968–970.

[1860] *Théorie du mouvement de la lune I*, Mémoire de l'Academie des Sciences 28, 1–883.

[1867] *Théorie du mouvement de la lune II*, Mémoire de l'Academie des Sciences 29, 1–931.

Dell'Agnola, C.A.

[1908a] *Sulla funzione limite di una successione di funzioni continue*, Lomb. Ist. Rend., (2) 41, 1908, pp. 676–684.

[1908b] *Le successioni di funzioni continue e il teorema di Arzelà*, Lomb. Ist. Rend., (2) 41, 1908, pp. 287–307.

[1909] *Sulla convergenza uniforme di una successione di funzioni continue*, Ven. Ist. Atti, 69 [(8) 12], 1909, pp. 151–159.

de Longchamps, G.G.

[1902] *Sui radicali sovrapposti*, Suppl. al Period., 5, 1902, pp. 81–83.

Denjoy, A.

[1907] *Sur les fonctions entières de genre fini*, C.R. Acad. Sci. Paris, 145, 1907, pp. 106–108.

[1926] *Sur l'itération des fonctions analytiques*, C.R. Acad. Sci. Paris, 166, 1926, pp. 255–257.

[1980] *Lettres à Paul Lévy*, Cahiers du Séminaire d'Historie des Mathématiques, tome 1, 1980, pp. 51–67.

Devaney, R.

[1989] *An introduction to chaotic dynamical systems,*, 2nd edition, Addison-Wesley, 1989.

[2001] *e^z: Dynamics, Topology, and Bifurcations of Complex Exponentials*, Topology and Its Applications, 110, 2001, pp. 133–161.

Devaney, R., Jarque, X., and Rocha, M.M.

[2005] *Indecomposable continua and Misiurewicz points in exponential dynamics*, Internat. J. Bifur. Chaos Appl. Engrg., 15, 2005, pp. 3281–3293.

Devaney, R. and Keen, L., eds.

[1989] *Chaos and fractals; the mathematics behind the computer graphics,* Proc. Sympos. Pure and Appl. Math., AMS, Providence, RI, 1989.

[1996] **Diacu, F. and Holmes, P.**

Celestial encounters. The origins of chaos and stability, Princeton University Press, Princeton, NJ, 1996.

Dickson, L.E.

[1919] *Mathematics in war perspective*, Bull. AMS, 25, 1919, pp. 289–311.

Dinghas, A.

[1938] *Über das Phragmén-Lindelöfsche Prinzip und den Julia-Carathéodoryschen Satz*, S. B. Preuss. Akad. Wiss. Phys.-math. Kl., 1938, pp. 32–48.

Dipert, R.

[1991] *The life and work of Ernst Schröder*, Modern Logic, **1**, 2-3, 1990-1991, pp. 117–139.

Dirichlet, J.

[1837] *Sur les séries dont le terme général dépend de deux angles, et qui servent à exprimer des fonctions arbitraires entre des limites données*, J. Reine Angew. Math., 17, 1837, pp. 35–56.

Domoradzki, S.

[2010] *Mathematics in Lwów before the Lwów Mathematical School*, in *Mathematics in the Austrian-Hungarian Empire*, M. Bečvăová and C. Binder, eds., Matfyzpress, Praha, 2010, pp. 55–73.

Douady, A. and Hubbard, J.H.

[1984-85] *Étude dynamique des polynômes complexes.* Prépublications Mathématiques d'Orsay, 2, 1984 and 4, 1985.

[1992] *Compacts in* $\mathbb{C}$, Topological methods in modern mathematics, Publish or Perish, Houston, 1992, pp. 429–465.

Drasin, D.

[1969] *Normal families and the Nevanlinna theory*, Acta Math., 122, 1969, pp. 231–263.

du Bois-Reymond, P.

[1883] *Über den Gültigkeitsbereich der Taylor'schen Reihenentwicklung*, Münch. Ber., 1876, pp. 225–237; and, with the same title, Klein Ann. XXI, 1883, pp. 253–254 and Klein Ann, XXI, 1883, pp. 109–117.

[1888] *Sur les caractères de convergence et de divergence des séries à termes positifs*, C.R. Acad. Sci. Paris, CVI, 1888, pp. 941–944.

Dugué, D.

[1952] *Vers un théorème de Picard global*, Ann. Sci. Éc. Norm. Sup., Ser., Ser. 3, 69, 1952, pp. 65–81.

Dulac, H.

[1903-1904] *Recherches sur les points singuliers des équations différentielles*. Thesis, Gauthier-Villars, Paris, 4°, 1903 and J. de l'Éc. Pol., (2), 9, 1904, pp. 5–125.

Dupuy, P.

[1896] *La vie d'Evariste Galois*, Ann. Sci. Éc. Norm. Sup., Ser. 3, 13, 1896, pp. 197–266.

Écalle, J.

[1975] *Théorie itérative: Introduction a la théorie des invariants holomorphes*, J. Math. Pure Appl., 54, 1975, pp 183–258.

[1981-1985] *Le fonctions résurgentes et leurs applications*, Publ. Math. Orsay, y. I, no 81-05; t. II, no 81-06; t. III, no 85-05., 1981–1985

Eggers, H.

[1876a] *Calculation of radicals*, The Analyst, 3, 1876, pp. 100-102.

[1876b] *A new method of solving equations*, The Analyst, 3, 1876, pp. 149-150.

Eliasson, L.H.

[1988-96] *Absolutely convergent series expansion for quasi-periodic motions*, Math. Phys. Electron. J. 2, 1996, paper 4, 33 pp. (electronic); see also: Report 2-88, Dept. of Math., University of Stockholm, 1988.

Elin, M. and Shoikhet, D.

[2001] *Dynamic extension of the Julia-Wolff-Carathéodory theorem*, Dynam. Systems Appl., 10, 2001, pp. 421–438.

Elin, M., Reich, S. and Shoikhet, D.

[2002] *Asymptotic behavior of semigroups of ρ-nonexpansive and holomorphic mappings on the Hilbert ball,* Ann. Mat. Pura Appl., 181, 2002, pp. 501–526

[2008] *A Julia-Carathéodory theorem for hyperbolically monotone mappings in the Hilbert ball* Israel J. Math., 164, 2008, pp.397-411.

Elin, M., Harris, L.A., Reich, S. and Shoikhet, D.

[2002] *Evolution equations and geometric function theory in J^*-algebras*, J. Nonlinear Convex Analysis 3, 2002, pp. 81–121.

Eljoseph, N.

[1968] *On the iteration of linear fractional transformations*, Amer. Math. Monthly, 75, 1968, 362-366

Encyclopediae

[E1] *Encyclopedia of Mathematics*, Kluwer Academic Publishers, London, 1987–1994.

[E2] *Encyclopedic Dictionary of Mathematics*, 2nd edition, Mathematical Society of Japan, 1977

Enriques, F. and Severi, F.

[1909] *Mémoire sur les surfaces hyperelliptiques*, Acta Math., 32 (1909), pp. 283–392.

L'Enseignement Mathématique (journal)

[1908] L'Enseign. Math., 10, 1908.
[1914] L'Enseign. Math., 16, 1914.
[1916] L'Enseign. Math., 18, 1916.
[1918] L'Enseign. Math., 20, 1918.
[1923] L'Enseign. Math., 23, 1923.
[1924] L'Enseign. Math., 24, 1924–1925.
[1928] L'Enseign. Math., 27, 1928.
[1929] L'Enseign. Math., 28, 1929.
[1933] L'Enseign. Math., 32, 1933.

Eremenko, A.

[1989] *On the iteration of entire functions*, Dynamical Systems and Ergodic Theory, Banach Center Publications, Polish Scientific Publications, 23, 1989, pp. 339–345.

Eremenko, A. and Lyubich, M.

[1984] *Iterates of entire functions*, USSR Math. Dokl., 30, 1984, pp. 592–594.
[1992] *Dynamical properties of some classes of entire functions.* Ann. Inst. Fourier, Grenoble, 42, 4, 1992, pp. 989–1020.

Fabry, E.

[1896] *Sur les points singuliers d'une fonction donnée par son développement en série et l'impossibilité du prolongement analytique dans des cas très généraux*, Ann. Sci. Éc. Norm. Sup., Ser. 3 13, 1896, pp. 367–399.
[1897] *Sur les séries de Taylor*, C.R. Acad. Sci. Paris, 124, 1897, pp. 142–143.
[1898a] *Sur les séries de Taylor qui ont une infinité de points singuliers*, Acta Math., 22, 1898, pp. 35–88.
[1898b] *Sur les points singuliers d'une série de Taylor*, J. Math., (5), 4, 1898, pp. 317–358.
[1899] *Généralisation du prolongement analytique d'une fonction*, C.R. Acad. Sci. Paris, 128, 1899, pp. 78–80.

Fan, K.

[1979] *Julia's lemma for operators*, Math. Ann., 239, 1979, pp. 241–245.
[1982] *Iterations of analytic functions of operators*, Math. Zeitschrift, 179, 1982, pp. 293–298.
[1983] *Iterations of analytic functions of operators II*, Linear and Multilinear Algebra, 12, 1983, pp. 295–304.
[1986] *The angular derivative of an operator-valued analytic function*, Pacific J. Math. 121, 1986, pp. 67–72.

Farkas, J.

[1884] *Sur les fonctions itératives*, J. Math. Pure Appl., (3), 10, 1884, pp. 101–108.

Farris, F.

[1998] *Review of* Visual complex analysis, *by Tristan Needham,* Amer. Math. Monthly, 105, 6, 1998, pp: 570–576. See `http://www.maa.org/pubs/amm\_complements/complex.html` for supplementary material.

Fatou, P.

[1904] *Sur les séries entières à coefficients entiers*, C.R. Acad. Sci. Paris, 138, 1904, pp. 342–344.
[1906a] *Séries trigonométriques et séries de Taylor*, Acta Math., 30, 1906, pp. 335–400.
[1906b] *Sur les solutions uniformes de certaines équations fonctionnelles*, C.R. Acad. Sci. Paris, 143, 1906, pp. 546–548.

[1910] *Sur une classe remarquable de séries de Taylor*, Ann. Sci. Éc. Norm. Sup., (3), 27, 1910, pp. 43–53.

[1913a] *Sur les lignes singulières des fonctions analytiques*, Bull. SMF, 41, 1913, pp. 113–119.

[1913b] *Sur la convergence absolue des séries trigonométriques*, Bull. SMF, 41, 1913, pp. 47–53.

[1917a] *Sur les substitutions rationnelles*, C.R. Acad. Sci. Paris, 164, 1917, pp. 806–808.

[1917b] *Sur les substitutions rationnelles*, C.R. Acad. Sci. Paris, 165, 1917, pp. 992–995.

[1918a] *Sur les équations fonctionnelles et les propriétés de certaines frontières*, C.R. Acad. Sci. Paris, 166, 1918, pp. 204–206.

[1918b] *Sur les suites de fonctions analytiques*, C.R. Acad. Sci. Paris, 167, 1918, pp. 1024–1026.

[1919-1920] *Sur les équations fonctionnelles*, Bull. SMF: 47, 1919, pp. 161–271 (Chap. I to III) and 48, 1920, pp. 33–94 (Chap. IV and V) and 208–314 (Chap. VI and VII). [Because of duplication of pages numbers, chapter numbers are included in citations.]

[1921a] *Sur les domaines d'existence de certaines fonctions uniformes*, C.R. Acad. Sci. Paris, 1921, pp. 344–346.

[1921b] *Sur les fonctions qui admettent plusieurs théorèmes de multiplication*, C.R. Acad. Sci. Paris, 173, 1921, pp. 571–573.

[1921c] *Sur un groupe discontinu de substitutions algébriques*, C.R. Acad. Sci. Paris, 173, 1921, pp. 694–696.

[1921d] *Sur l'evanouissement d'une branche de fonction uniforme aux points d'une singulière*, Darboux Bull., (2), 45, 1921, pp. 65–81.

[1921e] *Sur le mouvement d'une planète dans un milieu résistant*, Bull. des Pub. Astr., Tome 1, VI, 1921.

[1922a] *Sur l'itération de certaines fonctions algébriques*, Darboux Bull., (2), 46, 1922, pp. 188–198.

[1922b] *Sur les fonctions invariantes par une substitution rationnèlle*, Bull. SMF, 50, 1922, pp. 37–41.

[1922c] *Sur certaines fonctions uniformes de deux variables*, C.R. Acad. Sci. Paris, 175, 1922, pp. 862–865.

[1922d] *Sur les fonctions méromorphes de deux variables*, C.R. Acad. Sci. Paris, 175, 1922 pp. 1030–1033.

[1923a] *Sur les fonctions holomorphes et bornées à l'intérieur d'un cercle*, Bull. SMF, 51, 1923, pp. 191–202.

[1923b] *Sur les frontières de certains domaines*, Bull. SMF, 51, 1923, pp. 16-22.

[1923c] *Sur le mouvement d'une planète dans un milieu résistant*, Darboux Bull. (2), 47, 1923, p. 19-40.

[1923-24] *Sur l'itération analytique et les substitutions permutables*, J. Math. Pures Appl., 2, 1923, pp. 343–384; 3, 1924, pp. 1–49.

[1924] *Substitutions analytiques et équations fonctionnelles à deux variables* Ann. Sci. Éc. Norm. Sup., Ser. 3, 41, 1924, pp. 67–142.

[1926] *Sur l'itération des fonctions transcendantes entières*, Acta Math., 47, 1926, pp. 337–370.

[1927] *Observations d'étoiles doubles*, J. des Observateurs, 10, 1927. p.96–101.

[1928a] *Observations d'étoiles doubles*, J. des Observateurs, 11, 1928, p.76–84.

[1928b] *Sur le mouvement d'un système soumis à des forces à courte période*, Bull. SMF, 56, 1928, pp. 98-139.

[1929a] *Notice sur les travaux scientifiques*, Imp. Tequi, Paris, 1929.

[1929b] *Sue un critère de stabilité*, C.R. Acad. Sci. Paris, 189, 1929, pp. 967–969.

[1931] *Sur le mouvemelnt d'un point matériel dans un champ de gravitation fixe*, Acta Astron., Ser. a, 2, 1931, 101-164

Fatou, R.

[F1] *Souvenir relatifs à Pierre Fatou: Astronome à l'observatoire de Paris.* Private recollections.

[1949] *La marine de guerre en temps de paix*, Les Cahiers de l'Est, 4, 1[a] ser, 1946, pp 117–124.

[1972] *L'évasion de Chine de la frégate Amethyst*, Editions France Empire, 1972. Translation of *The Escape of the Amethyst*, by C. E. Phillips, Heinmann, London, 1957.

Fehr, H.

[1913] *Carlo Bourlet*, L'Enseign. Math., 15, 1913, pp. 417–418.

Fejer, L. and Riesz, F.

[1921] *Über einige funktionentheoretische Ungleichungen*, Math. Zeitschrift, 11, 1921, pp. 305–314.

Fernández, J.

[1989] *A note on the Julia sets of polynomials*, Complex Variables, 12, 1989, pp. 83–85.

Flanigan, F.J.

[1983] *Complex variables: harmonic and analytic functions*, Dover, New York, 1983.

Forman, P.

[1973] *Scientific internationalism and the Weimar Physicists: The ideology and its manipulation in Germany after World War I.*, Isis, 64, 2, 1973, pp. 151–180.

Formenti, C.

[1875] *Su alcuni problemi di Abel*, Rendiconti Ist. Lombardo di Scienze e Lettere, Milano, 8, 1875, pp. 276–282.

Fornaess, J.E.

[2002] *Real methods in complex dynamics*, in *Real methods in complex and CR geometry*, CIME, Martina Franca (Taranto), 2002 CIME Lecture Notes.

Forsyth, A.R.

[1918] *Theory of Functions*, Vol 1 and 2, 3rd edition, Cambridge University Press, 1918 (rprt. by Dover, 1965).

Fourier, J.B.J.

[1822] *Théorie de la chaleur*, Paris, Firmin Didot 1822. Grand in-4°.

[1831] *Analyse des équations déterminées*, 1831.

de Franqueville, A. Ch.

[1896] *Le premier siècle de l'Institut de France: 1795–1895*, Ed. Rothschild: Tome I, 1895; Tome II, 1896.

Fréchet, M.

[1965] *La vie et l'oeuvre d'Émile Borel*, L'Enseign. Math., 11, 1965, pp. 1–95.

Frederick, C. and Schwartz, E.L.

[1990] *Conformal Image Warping* IEEE Comp. Graph. Appl., March, 1990, pp; 54–61. March 1990.

Fricke, R. and Klein, F.

[1911] *Vorlesungen über die Theorie der automorphen Funktionen*, B.G. Teubner, Leipzig und Berlin, 1911.

Friedland, S. and Milnor, J.

[1989] *Dynamical properties of plane polynomial automorphisms*, Ergod. Th. & Dynam., 9, 1989, pp. 67–99.

Fulbrook, M.

[1991] *A Coincise History of Germany*, Cambridge University Press, 1991.

Fuller, F. and Ward, M.

[1934] *The continuous iteration of real functions*, Bull. AMS, 40, 1934, pp. 393–396.

Gallois, J.

[1994] *Ernest Chausson*, Fayard, 1994.

Galois, É.

[1830] *Note sur la résolution des équations numériques*, Bull. Sci. Math., Physiques et Chimiques, XIII, §216, Juin 1830, pp. 413–414.

Galton, F.

[1869] *Hereditary Genius: An inquiry into its laws and consequences*, MacMillan, London, 1869.

Galton, F. and Watson, H.W.

[1874] *On the probability of extinction of families,* J. Anthropol. Inst. Great Britain and Ireland, 4, 1874, pp. 138–144.

Galuzzi, M.

[2001] *Galois' note on the approximative solution of numerical equations (1830)*, Arch. Hist. Exact Sci., 2001, Vol. 56, No. 1, pp. 29–37.

Gamelin, T.W. and Greene, R.E.

[1999] *Introduction to Topology*, 2nd edition, Dover, 1999.

Garibian, V.S.

[1952] *Pensions and Rising Prices*, P. J. D. Wiles Oxford Economic Papers, New Series, Vol. 4, No. 2 (Jul., 1952), pp. 131–148.

Garnier, R.

[1978] *Notice nécrologique sur Gaston Julia*, C.R. Acad. Sci. Paris, 286, 1978, pp. 126–133.

Gauja, P.

[1934] *L'Académie des sciences de l'Institut de France*, Gauthier-Villars, Paris, 1934.

Gersevanov, N.M.

[1941] *Über einige Anwendungen der Operationen der gebrochenen und der komplexen Iteration*, Doklady Akad. Nauk USSR, (2), 31, 1941, pp. 835–836.

Gispert, H.

[1991] *La France Mathématique: La Société Mathématique de France* (1872–1914). Société Française d'Histoire des Sciences et des Techniques, Paris, 1991.

Gispert, H. and Hulin, N.

[2000] *L'enseignement des mathématiques dans ses liens à d'autres disciplines. Une perspective historique.* Bull. de l'Union des Professeurs de Spéciales, 192, October 2000.

Gleick, J.

[1987] *Chaos: The making of a new science*, Viking Penguin, New York, 1987.

Glicksberg, I.

[1976] *Julia's lemma for function algebras*, Duke Math. J., 43, 1976, pp. 277–284.

Goebel, K.

[1982] *Fixed points and invariant domains of holomorphic mappings of the Hilbert ball*, Nonlinear Analysis, 6, 1982, pp. 1327–1334.

Goebel, K. and Reich, S.

[1982] *Iterating holomorphic self-mappings of the Hilbert ball*, Proc. Japan Acad., 58, 1982, pp. 349-352.

[1984] *Uniform convexity, hyperbolic geometry and non-expansive mappings*, Marcel Dekker, 1984.

Golusin, G.M.

[1939] *Iterationsprozesse für konforme Abbildungen mehrfach zusammenhängender Bereiche*, Rec Math., Moscou, (2), 6, 1939, pp. 377–382.

Gontcharoff, W.

[1927a] *Über ganze Funktionen und die Geraden von G. Julia*, Charikov, Comm. Soc. Math., (4), 1, 1927, pp. 94–107.

[1927b] *Sur les fonctions entières et les droites de Julia*, C.R. Acad. Sci. Paris, 185, 1927, pp. 1100–1102.

Götz, O.

[1967] *Über das funktionentheoretische Zentrumproblem*, Math. Ann., 172, 1967, pp. 211–216.

Goursat, É.

[1887] *Sur les fonctions à espaces lacunaires*, Bull. S.M.F., 2, 11, 1887, pp. 109–114.

[1893] *Sur une fonction à espace lacunaire*, Bull. S.M.F., 2, 17, 1893, pp. 247–248.

Granata, E. and Lanzani, A.

[2006] *Quando la periferia é sinonimo di città pubblica*, Animazione Sociale, 8/9, 2006.

Gräser, E.

[1930] *Über konforme Abbildung des allgemeinen zweifach zusammenhängenden schlichten Bereiches auf die Fläche eines Kreisrings*, Leipzig, Philos. Diss., 1930.

Greene, R.E. and Krantz, S.G.

[1997] *Function theory of one complex variable*, Wiley, 1997.

Grevy, A.

[1892] *Sur les équations fonctionnelles*, Bull. Sci. Math. Astr., Series 2, 16, Part 1, 1892, pp. 311–313.

[1894] *Étude sur les équations fonctionnelles*, Ann. Sci. Éc. Norm. Sup., Ser 3, 11, 1894, pp. 249–323.

[1896] *Étude sur les équations fonctionnelles*, Ann. Sci. Éc. Norm. Sup., Ser. 3, 13, 1896, pp. 295–338.

[1897] *Équations fonctionnelles avec second membre*, Bull. SMF, 25, 1897, pp. 57–63.

Griffin, C.J. and Joshi, G.

[1991] *Octonionic Julia sets*, UM-P-91-79, July, 1991, 61 pp.

[1992] *Octonionic Julia sets*, Chaos, Solitons & Fractals 2, 1992, pp. 11–24.

[1993] *Transition Points in Octonionic Julia Sets*, Chaos, Solitons & Fractals 3, 1993, pp. 67–88.

Gu, Y.X.

[1979] *A criterion for normality of families of meromorphic functions* (Chinese), Science Sinica Special Issue on Math., 1, 1979, pp. 267–274.

Guerraggio, A. and Nastasi, P.

[1993] *Gentile e i matematici italiani — lettere* 1907–1943, Bollati Boringhieri, 1993.

Guichard, C.

[1883] *Théorie des points singuliers essentiels*, Ann. Sci. Éc. Norm. Sup., (2), XII, 1883, pp. 301–395.

Guiraldenq, P.

[1999] *Émile Borel* (1871–1956). *L'espace et le temps d'une vie sur deux siècles*, 1999.

Hadamard, J.

[1888] *Sur le rayon de convergence des séries ordonnées suivant les puissances d'une variable*, C.R. Acad. Sci. Paris, 106, 1888, pp. 259–262.

[1892] *Essai sur l'étude des fonctions données par leur développement de Taylor*, J. Math., 4, VIII, 1892, pp. 101–186.

[1893] *Sur les caractères de convergence des séries*, C.R. Acad. Sci. Paris, 117, 1893, pp. 844–845.

[1896] *Sur la distribution des zéros de la fonction $\zeta(s)$ et ses conséquences arithmétiques*, Bull. SMF, 24, 1896, pp. 199–220.

[1901] *Sur l'itération et les substitutions asymptotiques des équations différentielles*, Bull. SMF, 29, 1901, pp. 224–228.

[1944] *Two works on iteration and related questions*, Bull. AMS, (2), 50, 1944, pp. 67–75.

Hahn, R.

[1993] *L'anatomie d'une institution scientifique*, l'Académie des sciences de Paris 1666–1803, Ed. Archives Contemporaines, 1993.

Hamilton, E.J.

[1977] *The role of war in modern inflation*, J. Economic History, Vol. 37, No. 1, The Tasks of Economic History (Mar., 1977), pp. 13–19.

Handt, T. and Kneser, H.

[1930] *Beispiele zur Iteration analytischer Funktionen*, Mitt. der Naturwiss. Ver. für Neuvorpommern und Rügen, 57, Greifswald, 1930.

Hardy, G.H. and Littlewood, J.E.

[1917] *Sur la convergence des séries de Fourier et des séries de Taylor*, C.R. Acad. Sci. Paris, 165, 1917, pp. 1047–1049.

Harnack, A.

[1878] *Über eine Eigenschaft der Koeffizienten der Taylor'schen Reihe*, Clebsch Ann., 1878, XIII, pp. 555–559.

Harris, L.A.

[1979] *Schwarz-Pick systems of pseudometrics for domains in normed linear spaces*, in *Advances in Holomorphy*, J.A. Borroso, ed., North Holland, Amsterdam, 1979, pp. 345–406.

Harris, T.E.

[1963] *The theory of branching processes*, Springer-Verlag, Berlin Göttingen-Heidelberg, 1963.

Hart, J.C., Francis, G.K. and Kauffman, L.H.

[1994] *Visualizing quaternion rotation*, ACM Trans. on Graphics, 13(3), July 1994, pp. 256–276.

Hart, J.C. Kauffman, L.H. and Sandin, D.J.

[1989] *Ray tracing deterministic 3-D fractals*, Computer Graphics 23 (3), (Proc. SIGGRAPH 89,) July 1989, pp. 289–296.

[1990] *Interactive visualization of quaternion Julia sets*, Proc. of Visualization, IEEE Computer Society Press, Oct. 1990, pp. 209–218.

Hartogs, F.

[1927] *Über die Grenzfunktionen beschränkter Folgen von analytischen Funktionen*, Math. Ann., 98, 1927, pp. 164–178.

Hartogs, F. and Rosenthal, A.

[1928] *Über Folgen analytischer Funktionen*, Math. Ann., 100, 1928, pp. 212–263.

van Haselen, H.

[1929] *Van asymptotische ontwikkeling van holomorfe functies in een halfvlak*, Proefschrift Utrecht. VIII + 36 + II blz. Groningen, P. Noordhoff, 1929.

[1932] *Sur l'itération de* $\log(1+z)$, L'Enseign. Math., 30, 1932, pp. 269–271.

Hasselblatt, B. and Katok, A.

[2002] *The development of dynamics in the 20th century and the contribution of Jürgen Moser*, Ergod. Theory Dynam. Systems, 22, 2002, pp. 1343–1364.

Hayman, W.K.

[1959] *Picard values of meromorphic functions and their derivatives*, Ann. Math., 70, 1959, pp. 9–42.

[1967] *Research problems in function theory*, The Athlone Press, 1967.

Heins, M.H.

[1941] *On the iteration of functions which are analytic and single-valued in a given multiply-connected region*, Amer. J. Math., 63, 1941, pp. 461–480.

Hendricksen, M.

[1967-W] *Leonard Gillman; an interview*, Topology Atlas, 37, `http://at.yorku.ca/t/o/p/c/37.htm`, 1967

Herman, M.

[1979] *Sur la conjugaison différentiable des difféomorphismes du cercle à des rotations*, Publ. Math. IHES, 49, 1979, pp. 5–234. MR 81h:58039.

[1984] *Exemples de fractions rationnelles ayant une orbite dense sur la sphère de Riemann*, Bull. SMF, 112, 1984, pp. 93–142.

Hermann, J.

[1710a] *Extrait d'une Lettre de M. Herman à M. Bernoulli*, datée de Padoüe le 12. Juillet 1710, Histoire de l'Academie Royale des Sciences (Paris), pp. 519–521.

[1710b] *Metodo d'investigare l'Orbite de' Pianeti, nell'ipotesi che le forze centrali o pure le gravità delgi stessi Pianeti sono in ragione reciproca de' quadrati delle distanze, che i medesimi tengono dal Centro, a cui si dirigono le forze stesse*, Giornale de Letterati D'Italia 2, pp. 447—467.

Hermite, C.

[1985] *Lettres de Charles Hermite à Gösta Mittag-Leffler*, Cahiers du Séminaire d'Histoire des Mathématiques, 6, 1985, pp. 79–217.

Hervé, M.

[1963] *Quelques propriétés des applications analytiques d'une boulé à m dimensions dans elle-même*, J. Math. Pures Appl., 42, 1963, pp. 117–147.

[1981] *L'œuvre de Gaston Julia*, Cahiers du Séminaire d'Histoire des Mathématiques, 2, 1981, pp. 1–8.

[1992] From *Supplement to The Dictionary of Scientific Biography*, Charles Scribner's Sons, Preprint.

Hoidn, H.

[1982] *Osculation methods for the conformal mapping of doubly connected regions*, J. Appl. Math. and Phy. (ZAMP), 33, 1982, pp. 640–652.

Holbrook, J.A.R.

[1983] *Quaternionic astroids and starfields*, Applied Mathematical Notes, 8 (2), 1983, pp. 1–34.

[1987] *Quaternionic Fatou-Julia sets*, Ann. Sci. Math. Quebec, (1), 1987, pp. 79–94.

Howell, R.F.

[1965] "The philosopher Alain and French classical radicalism", *The Western Political Quarterly*, 18, 3, September 1965, pp. 594–614.

Hua, X.H. and Vaillancourt, R.

[2005] *Dynamics of permutable meromorphic functions*, Scientific Proc. Riga Technical University, 26, Boundary Field Problems and Computer Simulation, 47th issue, 2005, pp. 25–31; CRM–3190, *Centre de Recherches Mathématique de l'Université de Montréal*, 08/2005.

Hua, X.H., Wang, X.L. and Yang, C.C.

[2000] *Dynamics of transcendental functions*, Proc. of the Second ISAAC Congress, Vol. 1, Kluwer, 2000, pp. 311–316.

Hua, X.H. and Yang, C.C.

[1998] *Dynamics of transcendental functions*, Gordon and Breach Science Publishers, 1998.

[2003] *Dynamics of permutable transcendental entire functions*, J. Math. Anal. Appl., 278, 2003, pp. 512–526.

Humbert, G.

[1917a] *Sur une Communication de M. Gaston Julia, intitulée 'Sur les substitutions rationnelles'*, C.R. Acad. Sci. Paris, 165, 1917, pp. 1096–97.

[1917b] *Rapport sur le mémoire, présenté comme thèse, par M. Julia*, C.R. Acad. Sci. Paris, 165, 1917, pp. 819–823.

[1918] *Rapport Humbert*, Internal report on entries to the 1918 *Grand Prix des Sciences mathématiques*, Archives, Académie des Sciences, Institute de France, Paris.

Hurwitz, A.

[1906] *Sur les points critiques des fonctions inverses*, C.R. Acad. Sci. Paris, 143, 1906, pp. 877–879.

[1907] *Sur les points critiques des fonctions inverses*, C.R. Acad. Sci. Paris, 144, 1907, pp. 63–65.

[1914] *Sur les points critiques des fonctions inverses des fonctions entières*, C.R. Acad. Sci. Paris, 158, 1914, pp. 1007–1008.

Imaeda, K. and Imaeda, M.

[2000] *Sedenion: Algebra and analysis*, Applied Mathematics and Computation, Elsevier, Vol. 115, No. 2-3, pp. 77–88.

Inou, H. and Shishikura, M.

[2006] *Renormalization for near-parabolic fixed points of holomorphic maps*, RIMS Kokyuroku, Kyoto University, 1482, 2006, pp. 47–48.

Isenkrahe, C.

[1897] *Das Verfahren der Funktionswiederholung, Seine geometrische Veranschaulichung und algebraische Anwendung*, B. G. Teubner, Leipzig, 1897.

Israel, G. and Nurzia, L.

[1984] *The Poincaré-Volterra theorem: a significant event in the history of the theory of analytic functions*, Hist. Math., 11, 1984, pp. 161–192.

Iversen, F.

[1914] *Recherches sur les fonctions inverses des fonctions méromorphes*, Ph.D dissertation, l'Université de Helsingfors, 1914.

[1918] *Sur les valeurs asymptotiques des fonctions méromorphes et les singularités transcendantes de leurs inverses*, C.R. Acad. Sci. Paris, 166, 1918, pp. 156–159.

Jafari, F.

[1993] *Angular derivatives in polydisks*, Indiana J. Math., 35, 1993, pp. 197–212.

Jahresbericht DMV **(German Mathematical Society, publisher)**

[1925] Jahresbericht DMV, 33, Vol. 2, 1925, p. 119.
[1933] Jahresbericht DMV, 42, 1933, p. 87.
[1934] Jahresbericht DMV, 43, 1934, p. 116.

Jaisson, M.

[2003] *Prix et Subventions de l'Académie des Sciences* 1916–1996, Editions Brepols, 2003, 2 vols.

Jakobson, M.V.

[1968] *Structure of polynomial mappings on a singular set*, Math. Sbornik 77(119), No. 1, 1968, pp. 105–124, in Russian. English translation: Math. USSR Sb., Vol. 6, No. 1, 1968, pp. 97–114.
[1969] *On the problem of the classification of polynomial endomorphisms of the plane*, Vol. 80(122), No. 3(11), 1969, pp. 365–387, in Russian. English translation: Math. USSR Sb., Vol. 9, No. 3, 1969, pp. 345–364.

Jentzsch, R.

[1917] *Untersuchungen zur Theorie der Folgen analytischer Funktionen*, Acta Math., 41, 1917, pp. 219–251.

Johansen, P.

[1939] *Iteration von Funktionen zweier reellen Variablen und einer komplexen Variablen*, Skand. Aktuarietidskr., 22, 1939, pp. 101–113.

Jordan, C.

[1883] *Cours d'analyse de l'École Polytechnique*, 3 vols., Gauthier-Villars, Paris, 1883.

Joshi, G. and Kricker, A.

[1995] *Bifurcation phenomena of the non-associative octonionic quadratic*, Chaos, Solitons & Fractals, 5, 1995, pp. 761–782.

Julia, G.

[1913] *Sur les lignes singulières de certaines fonctions analytiques*, Bull. SMF, 41, 1913, pp. 351–366 = *Œuvres*, **IV**, article #1.
[1916a] *Sur les formes quadratiques binaires positives*, C.R. Acad. Sci. Paris, 162, 1916, pp. 151–154 = *Œuvres*, **V**, article #2.
[1916b] *Sur la réduction des formes quadratiques positives*, C.R. Acad. Sci. Paris, 162, 1916, pp. 320–322 = *Œuvres*, **V**, article #3.
[1916c] *Sur la réduction des formes quadratiques quaternaires positives*, C.R. Acad. Sci. Paris, 162, 1916, pp. 498–501 = *Œuvres*, **V**, article #4.
[1916d] *Sur quelques propriétés du groupe fuschsien formé des substitutions modulaires qui n'altèrent pas une forme d'Hermite indéfinie*, C.R. Acad. Sci. Paris, 163, 1916, pp. 599–600 = *Œuvres*, **V**, article #5.
[1916e] *Sur les formes de Dirichlet et sur les substitutions loxodromiques du groupe de Picard*, C.R. Acad. Sci. Paris, 163, 1916, pp. 691–694 = *Œuvres*, **V**, article #6.
[1917a] *Étude sur les formes binaires non quadratiques a indéterminées réelles ou complexes, ou a indéterminées conjuguées*, (*Prix Bordin* 1917), Mémoires de l'Académie des Sciences, 55, 1917, pp. 1–296 = *Œuvres*, **V**, article #14.
[1917b] *Sur la réduction des formes binaires à coefficients réels de degré quelconque*, C.R. Acad. Sci. Paris, 164, 1917, pp. 32–35 = *Œuvres*, **V**, article #7.
[1917c] *Sur les formes binaires à coefficients et indéterminées complexes, de degré quelconque*, C.R. Acad. Sci. Paris, 164, 1917, pp. 352–355 = *Œuvres*, **V**, article #8.

[1917d] *Sur la réduction des formes binaires de degré quelconque à coefficients et indéterminées réels ou complexes*, C.R. Acad. Sci. Paris, 164, 1917, p. 484 = *Œuvres*, **V**, article #9.

[1917e] *Sur la réduction des formes à indéterminées conjuguées non quadratiques*, C.R. Acad. Sci. Paris, 164, 1917, pp. 571–574, pp. 619–622 = *Œuvres*, **V**, article #10.

[1917f] *Sur les formes biquadratiques à indéterminées conjuguées et à coefficients entiers*, C.R. Acad. Sci. Paris, 164, 1917, pp. 910–913 = *Œuvres*, **V**, article #12.

[1917g] *Sur les formes binaires à indéterminées conjuguées qui restent invariantes par un groupe de substitutions linéaires*, C.R. Acad. Sci. Paris, 164, 1917, pp. 991–993 = *Œuvres*, **V**, article #13.

[1917h] *Sur les substitutions rationnelles*, C.R. Acad. Sci. Paris, 165, 1917, pp. 1098–1100 = *Œuvres*, **I**, article #15.

[1918a] *Mémoire sur l'itération des fonctions rationnelles*, *Grand Prix des Sciences Mathématiques*, J. Math. Pur. Appl., (8), 1, 1918, pp. 47–245 = *Œuvres*, **I**, article #22.

[1918b] *Sur l'itération des fractions rationnelles*, C.R. Acad. Sci. Paris, 166, 1918, pp. 61–64 = *Œuvres*, **I**, article #16.

[1918c] *Sur des problèmes concernant l'itération des fractions rationnelles*, C.R. Acad. Sci. Paris, 166, 1918, pp. 153–156 = *Œuvres*, **I**, article #17.

[1918d] *Sur les substitutions rationnelles*, C.R. Acad. Sci. Paris, 166, 1918, pp. 599–601 = *Œuvres*, **I**, article #18.

[1919a] *Sur quelques problèmes relatifs à l'itération des fractions rationnelles*, C.R. Acad. Sci. Paris, 168, 1919, pp. 147–149 = *Œuvres*, **I**, article #23.

[1919b] *Sur les fonctions entières ou méromorphes*, C.R. Acad. Sci. Paris, 168 1919, pp. 990–992 = *Œuvres*, **II**, article #29.

[1919-21] *Sur quelques propriétés nouvelles des fonctions entières ou méromorphes*, Ann. Sci. Éc. Norm. Sup., Ser. 3, 36, 1919, pp. 93–125; 37, 1920, pp. 165–218; 38, 1921, pp. 165–181 = *Œuvres*, **II**, article #31,39,40.

[1920] *Extension nouvelle d'un lemme de Schwarz*, Acta Math., 42, 1920, pp. 349–355 = *Œuvres*, **IV**, article #38.

[1921a] *Sur la permutabilité des substitutions rationnelles*, C.R. Acad. Sci. Paris, 173, 1921, pp. 690–693 = *Œuvres*, **I**, article #46.

[1921b] *Sur un classe d'équations fonctionnelles*, C.R. Acad. Sci. Paris, 173, 1921, pp. 813–816 = *Œuvres*, **I**, article #47.

[1921c] *Sur les fonctions entières ou méromorphes*, C.R. Acad. Sci. Paris, 174, 1921, pp. 964–967 = *Œuvres*, **II**, article #48

[1921d] *Sur les solutions méromorphes de certaines équations fonctionnelles*, C.R. Acad. Sci. Paris, 173, 1921, pp. 1149–1152 = *Œuvres*, **I**, article #49.

[1922a] *Les équations fonctionnelles et la représentation conforme*, C.R. Acad. Sci. Paris, 174, 1922, pp. 517–519 = *Œuvres*, **III**, article #51.

[1922b] *Nouvelles applications de la représentation conforme aux équations fonctionnelles*, C.R. Acad. Sci. Paris, 174, 1922, pp. 653–655 = *Œuvres*, **III**, article #52.

[1922c] *Sur la transformation des substitutions rationnelles en substitutions linéaires*, C.R. Acad. Sci. Paris, 174, 1922, pp. 800–802 = *Œuvres*, **III**, article #53.

[1922d] *Mémoire sur la permutabilité des fractions rationnelles*, Ann. Sci. Éc. Norm. Sup., Ser. 3, 39, 1922, pp. 131–215 = *Œuvres*, **I**, article #55.

[1922e] *Sur les substitutions rationnelles à deux variables*, C.R. Acad. Sci. Paris, 175, 1922, pp. 1182–85 = *Œuvres*, **I**, article #58.

[1923a] *Sur les substitutions rationnelles à deux variables*, C.R. Acad. Sci. Paris, 175, 1923, pp. 58–60 = *Œuvres*, **I**, article #59.

[1923b] *Sur une classe d'èquations fonctionnelles*, Ann. Sci. Éc. Norm. Sup., Ser. 3, 40, 1923, pp. 97-150 = *Œuvres*, **I**, article #60.

[1924] *Sur quelques applications de la représentation conforme à la résolution d'équations fonctionnelles*, Bull. SMF, 52, 1924, pp. 279–315 = *Œuvres*, **III**, article #61.

[1925a] *Séries de fractions rationnelles d'itération*, C.R. Acad. Sci. Paris, 180, 1925, pp. 563–566 = *Œuvres*, **II**, article #62.

[1925b] *Les séries d'itération et les fonctions quasi-analytiques*, C.R. Acad. Sci. Paris, 180, 1925, pp. 720–723 = *Œuvres*, **IV**, article #63.

[1925c] *Sur un type de fractions quasi-analytiques,*, C.R. Acad. Sci. Paris, 180, 1925, pp. 1150–1153 = *Œuvres*, **IV**, article #64.

[1925d] *Fonctions quasi-analytiques et fonctions entières d'ordre nul,* C.R. Acad. Sci. Paris, 180, 1925, pp. 1240–1242 = *Œuvres*, **IV**, article #65.

[1925e] *Sur les séries de fractions rationnelles itérées,* C.R. Acad. Sci. Paris, 181, 1925, pp. 1119–1122 = *Œuvres*, **II**, article #66.

[1926a] *Sur la représentation conforme des aires simplement connexes*, C.R. Acad. Sci. Paris, 182, 1926, pp. 1314–1316 = *Œuvres*, **III**, article #70.

[1926b] *Sur une série de polynômes liée à la représentation conforme des aires simplement connexes*, C.R. Acad. Sci. Paris, 183, 1926, pp. 10–12 = *Œuvres*, **III**, article #71.

[1927] *Sur une classe de polynômes*, C.R. Acad. Sci. Paris, 184, 1927, pp. 1227–1228 = *Œuvres*, **III**, article #75.

[1930a] *Sur la convergence des séries de fractions rationnelles itérées,* C.R. Acad. Sci. Paris, 191, 1930, pp. 987–989 = *Œuvres*, **II**, article #87.

[1930b] *Les séries de fractions rationnelles itérées et les fonctions quasi-analytiques*, Arkiv Mat. Ast. Fys., 22A, 1930, pp. 1–8 = *Œuvres*, **IV**, article #89.

[1931a] *Sur l'allure des séries d'itérées au voisinage des frontières de convergence*, C.R. Acad. Sci. Paris, 193, 1931, pp. 690–692 = *Œuvres*, **II**, article #91.

[1931b] *Mémoire sur la convergence des séries formées avec les itérées successives d'une fraction rationnelle*, Acta Math., 56, 1931, pp. 149–195 = *Œuvres*, **II**, article #92.

[1931c] *Fonctions continues sans dériveées formées avec les itérées d'une fraction rationalle*, Ann. Sci. Éc. Norm. Sup., Ser. 3, 48, 1931, pp. 1–14 = *Œuvres*, **II**, article #93.

[1931d] *Mémoire sur l'extension du théorème d'Abel aux séries d'itérées* $\sum a_n R_n(z)$, Ann. Sci. Éc. Norm. Sup., Ser. 3, 48, 1931, pp. 439–495 = *Œuvres*, **II**, article #95.

[1932] *Addition au Mémoire sur la convergence des séries formées avec les itérées successives d'une fraction rationnelle*, Acta Math., 58, 1932, pp. 407–12 = *Œuvres*, **III**, article #102.

[1934] *Notice de Candidature Académique*, *Œuvres*, **I**, pp.1–88.

[1937] *Allocution prononcée au* 2e *centenaire de l'Université de Göttingen (27 juin 1937)*, 1937 = *Œuvres*, **VI**, article #195.

[1965] *Discours prononcès à l'Ecole Polytechnique pour la cérémonie d'adieux au Professeur Gaston Julia*, discours du Général Cazelles, Commandant de l'École Polytechnique, 1965 = *Œuvres*, **VI**, article #231.

[1968-1970] *Œuvres*, 6 vols., Gauthier-Villars, Paris, 1968–1970.

Kahan, W.

[1987] *Branch cuts for complex elementary functions and much ado about nothing's sign it*, in *The state of the art in numerical analysis*, eds., Iserles and Powell, Clarendon Press, Oxford, 1987, pp. 165–211.

Kahane, J.P.

[1991] *Séries de Fourier, séries de Taylor, séries de Dirichlet; un aperçu de l'importance des travaux des mathématiciens français dans le période 1880–1910*, Cahiers d'Histoire & de Philosophie des Sciences, 34, 1991, p. 284.

Kakutani, S.

[1939] *Iteration of linear operations in complex Banach spaces*, Proc. Acad., Tokyo, 14, pp. 295–300. Collect. Papers Fac. Sci. Osaka Univ., A 6, 1939, Nr. 16, 6 pp.

Kapeluszny, J., Kuczumow, T. and Reich, S.

[1999] *The Denjoy-Wolff theorem in the open unit ball of a strictly convex Banach space*, Adv. Math. 143, 1999, pp. 111-123.

Karpinska, B.

[1999] *Hausdorff dimension of the hairs without endpoints for* $\lambda\exp(z)$, C.R. Acad. Sci. Paris, 328, Ser. I, 1999, pp. 1039–1044.

Kasner, E.

[1911a] *The subdivisions of curvilinear angles*, Bull. AMS, (2), 17, 1911, p. 527.

[1911b] *Conformal geometry and equilong invariants of horn angles*, Bull. AMS, (2) 17, 393, 1911.

[1913] *Conformal geometry*, Proc. 5th Int. Math. Cong., Cambridge, 2, 1913, pp. 81–87.

[1915] *Conformal classification of analytic arcs or elements: Poincaré's local problem of conformal geometry*, Trans. AMS, 16, 1915, pp. 333–349.

Kasner, E. and Harrison, I.

[1950] *Voltaire on mathematics and Horn angels*, Scripta Math., 16, 1950, pp. 13–21.

Kasner, E. and Newman, J.R.

[1956] *New Names for Old*, The World of Mathematics, Volume III, Simon and Schuster, New York, 1956, pp. 1996–2010.

Keegan, J.

[2003] *La première guerre mondiale*, Perrin Editions, 2003.

Kent, D.

[preprint] *Ambitious editors and amateur audiences: nineteenth-century mathematical periodicals in the United States*, preprint, 2011.

Khatskevich, V. and Senderov, V.

[2001] *The Abel-Schröder equations for linear fractional maps of operator balls*, Dokl. Ross. Akad. Nauk, 379, 2001.

Khatskevich V., Reich S. and Shoikhet D.

[2001] *Schröder's functional equation and the Koenigs embedding property*, J Nonlinear Analysis, 47, 2001, pp. 3977–3988.

Klingen, B.

[1962] *Wachstum und Periodeneigenschaften inverser Schrd̈erfunktionen*, Ph.D. dissertation, Universität zu Köln, 1962.

[1968] *Wachstumsvergleich bei ganzen analytischen Funktionen*, Math. Annalen, 175, 1968, pp. 50–80

von Koch, H.

[1906] *Une méthode géométrique élémentaire pour l'étude de certaines questions de la théorie des courbes planes*, Acta Math., XXX, 1906, pp. 145–176.

Koebe, P.

[1907a] *Über die Uniformisierung beliebiger analytischer Kurven I*, Math. Ann., 67, 1907, pp. 145–224.

[1907b] *Über die Uniformisierung beliebiger analytischer Kurven*, Gött. Nachr., 1907, pp. 191–210.

[1910] *Über die Uniformisierung der algebraischen Kurven II*, Math. Ann., 69, 1910, pp. 1–81.

[1915] *Abhandlungen zur Theorie der konformen Abbildung*, J. für Math., 145, 1915, pp. 177–225.

[1927] *Allgemeine Theorie der Riemannschen Mannifaltigkeiten*, Acta Math. 50, 1927, pp. 27–157.

[1941] *Zur allgemeinen Iterationstheorie der Uniformisierung algebraischer Funktionen*, Ber. Sächs. Akad. Wiss. Leipzig Math. Physische Kl., 93, 1941, pp. 43–66.

Kœnigs, G.

[1883] *Recherches sur les substitutions uniformes*, Bull. Soc. Math. Astr., Ser. 2, 7, Part I, 1883, pp. 340–357.

[1884a] *Sur une généralisation du théorème de Fermat, et ses rapports avec la théorie des substitutions uniformes*, Bull. Soc. Math. Astr., Ser. 2, 8, 1884, pp. 236–288.

[1884b] *Recherches sur les intégrales de certaines équations fonctionnelles*, Ann. Sci. Éc. Norm. Sup., Ser. 3, 1, 1884, pp. 1–41.

[1884c] *Sur les intégrales de certaines équations fonctionnelles*, C.R. Acad. Sci. Paris, 99, 1884, pp. 1016–1017.

[1885a] *Sur les conditions d'holomorphisme des intégrales de l'équation itérative et de quelques autres équations fonctionnelles*, C.R. Acad. Sci. Paris, 101, 1885, pp. 1137–1139.

[1885b] *Nouvelles recherches sur les équations fonctionnelles*, Ann. Sci. Éc. Norm. Sup., Ser. 3, 2, 1885, pp. 385–404.

Kojima, C.

[2004] *La Physique française avant Louis de Broglie*, Annales de la Fondation Louis de Broglie, 29, Hors serie 1, 2004, pp. 767–783.

Kollerstrom, N.

[1992] *Thomas Simpson and "Newton's method of approximation:" an enduring myth*, Brit. Jour. Hist. Sci., 3, 1992, pp. 347–354.

Kolmogorov, A.N.

[1953] *On dynamical systems with an integral invariant on the torus* (Russian), Doklady Akad. Nauk SSSR (N.S.) 93, 1953, 763–766.

[1954] *On conservation of conditionally periodic motions under small perturbations of the Hamiltonian*, Dokl. Akad. Nauk SSSR 98, 1954, pp. 527–530.

[1957] *Théorie générale des systèmes dynamiques et mécanique classique* (French), Proceedings of the International Congress of Mathematicians, Amsterdam, 1954, Vol. 1 pp. 315–333, Erven P. Noordhoff N.V., Groningen; North-Holland Publishing Co., Amsterdam, 1957. Engl. Transl., Abraham R., *Foundations of mechanics*, Benjamin, 1967, Appendix D.

Koppe, M.

[1909] *Die Iteration des Sinus und anderer Funktionen*, Progr. (Nr. 115) Luisenstädt. Realgymn. Berlin. 28 S., 4°. 3 Fig.-Taf, 1909.

Korkine, A.

[1882] *Sur un problème d'interpolation*, Bull. Soc. Math. Astr., (2), 6, 1882, pp. 228–242.

Kovalishina, I.V.

[1990] *The Carathéodory-Julia theorem for matrix functions*, Teor. Funktsii Funktsional. Analysis iPrilozhen. 43, 1985, pp. 70–82, in Russian. English translation J. Soviet Math. 48, 1990, pp. 176–186.

Kubota, Y.

[1983] *Iteration of holomorphic maps of the unit ball into itslef,* Proc. AMS, 88, 1983, pp. 476-480

Kuczumow, T., Reich, S. and Shoikhet, D.

[2001a] *The existence and non-existence of common fixed points for commuting families of holomorphic mappings,* Nonlinear Anal., 43, 2001, pp. 45-59.

[2001b] *Fixed points of holomorphic mappings: a metric approach*, Handbook of Metric Fixed Point Theory, W. A. Kirk and B. Sims, eds. Kluwer, Dordrecht, 2001, pp. 437–515.

Kuhn, T.S.

[1976] *The Copernican Revolution*, Harvard University Press, 1976.

Lagarias, J.

[1985] *The $3x + 1$ problem and its generalizations*, Amer. Math. Monthly, 92, 1, 1985, pp. 3–23.

Lagrange, J.L.

[1798] *De la résolution des équations numériques de tous les degrés*, Paris, Duprat, an VI, 1798. Grand in-4.

[1879] *Sur l'intégration d'une équation différentielle à differences finies, qui contient la théorie des suites récurrentes*, Œuvres, t. I; *Recherches sur les suites récurrentes*, Œuvres, t. IV; *Mémoire sur l'expression du terme général des séries récurrentes*, Œuvres, t. V, Gauthier-Villars, Paris, 1879.

Laisant, C.A. and Bricard, R.

[1913] *Décès de M. Carlo Bourlet*, Nouv. Ann. Math., (4) 13, 1913, p. 337.

Lakner, M. and Petek, P.

[1997b] *The one-equator property*, Experimental Math., 6, 2, 1997, pp. 109–115.

Lakshmikantham, V. and Trigiante, D.

[2002] *Theory of Difference Equations. Numerical Methods and Applications*, Second Edition, Marcel Dekker, New York, 2002.

Lakshmikantham, V., Leela, S. and Martynyuk, A.A.

[1989] *Stability Analysis Of Nonlinear Systems*, Marcel Dekker, Inc., New York, 1989.

Landau, E.

[1904] *Über eine Verallgemeinerung des Picardschen Satzes*, Berl. Berichte, 1904, pp. 1118–1133.

Landau, E. and Valiron, G.

[1929] *A deduction from Schwarz's lemma*, J. LMS, 4, 1929, pp. 162–163.

Langley, J.K.

[1984] *On normal families and a result of Drasin*, Proc. Roy. Soc. Edin., 98A, 1984, pp. 385–393.

[1999] *Permutable entire functions and Baker domains*, Math. Proc. Camb. Phil. Soc., 125, 1999, pp. 199–202.

Laplace, P.S.

[1840] *A philosophical essay on probabilities*, J. Wiley, New York, 1902. Translation of *Essai philosophique sur les probabilitiés*, 65h ed., Bachelier, Paris, 1840, by F.W. Truscott and F.L. Emory.

Lattès, S.

[1903] *Sur une classe d'équations fonctionnelles*, C.R. Acad. Sci. Paris, 136 1903, pp. 905–908.

[1906] *Sur les équations fonctionnelles qui définissent une courbe ou une surface invariante par une transformation*, Annali di Mat., 3, 13, 1906, pp. 1–137.

[1908] *Nouvelles recherches sur les courbes invariantes par une transformation* $(X, Y; x, y, y')$, Ann. Sci. Éc. Norm. Sup., Ser. 3, 25, 1908, pp. 221–254.

[1910a] *Sur les séries de Taylor à coefficients récurrents*, C.R. Acad. Sci. Paris, 150, 1910, pp. 1413–1415.

[1910b] *Sur la convergence des relations de récurrence*, C.R. Acad. Sci. Paris, 150, 1910, pp. 1106–1109.

[1911a] *Sur les suites récurrentes non linéaires et sur les fonctions génératrices de ces suites*, Ann. Fac. Sci. Toulouse, Ser. 3, 3, 1911, pp. 73–124.

[1911b] *Sur les formes réduites des transformations ponctuelles à deux variables. Application à une classe remarquable de séries de Taylor*, C.R. Acad. Sci. Paris 152, 1911, pp. 1566

[1911c] *Sur les formes réduites des transformations ponctuelles dans le domaine d'un point double*, Bull. SMF, 39, 1911, pp 309–345.

[1914] *Sur le prolongement analytique de certaines séries de Taylor*, Bull. SMF, 42, 1914, pp. 95–112.

[1918a] *Sur l'itération des substitutions rationnelles et les fonctions de Poincaré*, C.R. Acad. Sci. Paris, 166, 1918, pp. 26–28.

[1918b] *Sur l'itération des substitutions rationnelles à deux variables*, C.R. Acad. Sci. Paris, 166, 1918, pp. 151–153.

[1918c] *Sur l'itération des fractions rationnelles*, C.R. Acad. Sci. Paris, 166, 1918, pp. 486–489. This original title appeared as *Sur l'itération des fractions irrationnelles* but was latter corrected, C.R. Acad. Sci. Paris, 166, 1918, p. 580.

Lax, P.D.

[2002] *Jürgen Moser, 1928–1999*, Ergodic Theory Dynam. Systems, 22, 2002, pp. 1337–1342.

Leau, L.

[1897] *Étude sur les équations fonctionelles à une ou plusieurs variables*, Ann. Fac. Sci. Toulouse, Ser. 1, 11, 1897, pp. 1–110.

[1899] *Sur les fonctions définies par un développement de Taylor*, C.R. Acad. Sci. Paris, 128, 1899, pp. 804–805.

[1932] *Les suites de fonctions en general (domaine complexe)*, Mémorial des Sciences Mathématiques, LIX, Gauthier-Villars, Paris, 1932.

Lebesgue, H.

[1926] *Notice sur la vie et les travaux de Camille Jordan* (1838–1922), Mémoires Acad. Institut de France, (2), 58, No. I, 1926, pp. XXXIX–LXVI.

[2004] *Les lendemains de l'intégrale; lettres à Émile Borel*, Paris, Vuibert, 2004.

Lecornu, L.

[1887] *Sur les séries entières*, C.R. Acad. Sci. Paris, 104, 1887, pp. 349–352.

Lee, S.

[1974] *Origins of Marvel Comics*, Simon and Schuster/Fireside Books, New York, 1974.

Le Gars, S.

[2007-W] *L'émergence de l'astronomie physique en France (1860-1914): Aceurs et pratiques, Tome 1*, Doctoral Thesis, Universite de Nantes, 2007. `http://en.scientificcommons.org/47959478`

Legendre, A.M.

[1816] *Méthodes nouvelles pour la résolution approchée des équations numériques*, §III of *Supplément à l'Essai sur la théorie des nombres, seconde édition.* Bound with *Essai sur la théorie des nombers*, 2nd edition, Courcier, Paris, 1808.

Lehto, O.

[1998] *Mathematicians without Borders*, Springer-Verlag, New York, 1998.

Lémeray, E.

[1895] *Un théorème sur les fonctions itératives*, Bull. SMF, 23, 1895, pp. 255–263.

[1896] *Sur la convergence des substitutions uniformes*, C.R. Acad. Sci. Paris, 123, 1896, pp. 793–94.

[1897a] *Sur la convergence des substitutions uniformes*, Nouv. Ann. Math., 16, 1897, pp. 306–319.

[1897b] *Dérivée des fonctions itératives par rapport à l'indice d'itération*, Bull. SMF, 25, 1897 pp. 92–95.

[1898a] *Sur quelques algorithmes généraux et sur l'itération*, Bull. SMF, 26, 1898, pp. 10–15.

[1898b] *Sur le calcul des racines des équations par approximations successives*, Nouv. Ann. Math., (3) 17, 1898, pp. 534–539.

[1898c] *Sur la convergence des substitutions uniformes*, Nouv. Ann. Math., 17, 1898, pp. 75–80.

Le Roy, Éd.

[1898a] *Sur les points singuliers d'une fonction définie par un développement de Taylor*, C.R. Acad. Sci. Paris, 127, 1898, pp. 348–350.

[1898b] *Sur les séries divergentes et les fonctions définies par un développement de Taylor*, C.R. Acad. Sci. Paris, 127, 1898, pp. 654–657.

[1899a] *Sur les séries divergentes et les fonctions définies par un développement de Taylor*, C.R. Acad. Sci. Paris, 128, 1899, pp. 492–495.

[1899b] *Sur les séries divergentes et les fonctions définies par un développement de Taylor*, Ann. Fac. Sci. Toulouse, Ser. 2, 2, 1900, pp. 317–384 and pp. 385–430.

Letherman, S., Schleicher, D. and Wood, R.

[1999] *On the $3n+1$-Problem and Holomorphic dynamics*, Experimental Mathematics, 8, 3, 1999, pp. 241–251.

Levi, B. and Viola, T.

[1933] *Intorno ad un ragionamento fondamentale nella teoria delle famiglie normali di funzioni*, Bollettino U.M.I., 12, 1933, pp. 197–203.

Levi-Civita, T.

[1901] *Sopra alcuni criteri di instabilitá*, Annali di Matematica Pura ed Applicata, Serie III – Tomo V, 1901, pp. 221–307.

Lévy, P.

[1912a] *Sur une généralisation des théorèmes de MM. Picard, Landau et Schottky*, C.R. Acad. Sci. Paris, 153, 1912, pp. 658–660.

[1912b] *Remarques sur le théorème de M. Picard*, Bull. SMF, 40, 1912, pp. 25–39.

[1927a] *Sur l'itération de la fonction exponentielle*, C.R. Acad. Sci. Paris, 184, 1927, pp. 500–502.

[1927b] *Sur l'itération des fonctions et la notion de croissance régulière*, C.R. Acad. Sci. Paris, 184, 1927, pp. 663–665.

[1927c] *Sur l'itération de la fonction exponentielle*, C.R. Acad. Sci. Paris, 184, 1927, pp. 500–502.

[1928b] *Fonctions à croissance régulière et itération d'ordre fractionnaire*, Ann. di Mat. 4, 5, 1928, pp. 269–298.

Lewittes, J.

[1968] *An extension of hyperbolic geometry and Julia's theorem*, Duke Math. J., 35, 1968, pp. 777–782.

Li, S.

[1984] *The normality criterion of a class of the functions*, J. East China Normal Univ., 2, 1984, pp. 156–158.

Li, X.

[1985] *The proof of Hayman's conjecture on normal families*, Sci. Sinica, Ser. A , 28, 6, 1985, pp. 596–603.

Liebeck, H.

[1966] *The convergence of sequences with linear fractional recurrence relation*, Amer. Math. Monthly, 73, 1966, pp. 353–355.

Lindelöf, E.

[1898a] *Remarques sur un principe général de la théorie des fonctions analytiques*, Acta Soc. Sci. Fenn., 24, 4°, 1898.

[1898b] *Sur la transformation d'Euler et la détermination des points singuliers d'une fonction définie par son développement de Taylor*, C.R. Acad. Sci. Paris, 126, 1898, pp. 632–634.

[1915] *Sur un principe général de l'analyse et ses applications à la, théorie de la représentation conforme*, Acta Soc. Sci. Fenn., 46, pp. 1–35.

Lindstedt, A.

[1883] *Beitrag zur integration der differentialalgleichungen der störungstheorie*, Mémories de l'Académie Impériale des Sciences de St. Pétersbourg (7) **31** (no.4), 1883, pp. 1–19.

Liouville, J.

[1851] *Sur les classes très-étendues de quantités dont la valeur n'est ni algébrique, ni même réductible à des irrationnelles algébriques*, J. Math. Pures Appl., 16, 1851, pp. 131–142.

Litten, F.

[1996] *Ernst Mohr — Das Schicksal eines Mathematikers*, Jahresbericht DMV, 98, 4, 1996, pp. 192–212.

Liverpool, L.

[1971] *Analytic Functions and Iteration Theory*, Ph.D. dissertation, University of London, 1971

[1974] *On entire functions with infinite domains of normality*, Aequationes Math., Vol. 10, 1974, pp. 189–200.

Lombardo-Radice, L.

[1994] *Nikolaj Lobacevskij: Nuovi principi della geometria, con una teoria completa delle parallele*, Bollati Boringhieri, 1994.

Lorch, E.R.

[1951] *Joseph Fels Ritt*, Bull. AMS, 57, 1951, pp. 307–318.

[1980] *Joseph Fels Ritt* from *The Dictionary of Scientific Biography*, Volume 11, Charles Scribner's Sons, 1980, pp. 470–471.

Lorentzen, L.

[1999] *Convergence of composition of self-mappings*, XIIth Conference on Analytic Functions, Ann. Univ. Marie Curie-Sklodowka Sect., A, 53, 1999, pp. 121–145.

Lovett, E.O.

[1895] *The great inequality of Jupiter and Saturn*, Astronomical Journal, 15, 1895, pp. 113–127.

Luckert, H.-J.

[1937] *Der Mathematiker in Technik und Industrie*, Jahresbericht DMV, 47, 1937, pp. 242–250.

[1938] *Bemerkungen zur Ausbildung des Industriemathematikers*, Jahresbericht DMV, 48, 1938, pp. 258–260.

Lyapunov, A.M.

[1907] *Problème général de la stabilité du mouvement*, Ann. Fac. Sci. Toulouse, Ser. 2, 9, 1907 (tr. from Russian by E. Davaux), pp. 203–474.

MacCluer, B.D.

[1983] *Iterates of holomorphic self-maps of the open unit ball in C^n*, Michigan Math. J., 30, 1983, pp. 97–106.

Mackaay, E.

[1990] *Les droits intellectuels entre propriété et monopole*, J. Économ. Études Humaines, Vol. 1, no. 1, 1989/1990, pp. 61–100.

Mackey, M. and Mellon, P.

[2002] *On an angular derivative result of Fan*, Math. Proc. R. Ir. Acad., 102 A, 2002, pp. 115–126.

[2003] *Angular derivatives on bounded symmetric domains*, Israel J. Math., 138, 2003, pp. 291–315.

MacLane, G.R. and Ryan, F.B.

[1962] *On the radial limits of Blaschke products*, Pacific J. Math., 12, no. 3, 1962, pp. 993-998.

Maindron, E.

[1881] *Les fondations de prix à l'Académie des Sciences. Les Lauréats de l'Académie 1714–1880*, Gauthier-Villars, Paris, 1881.

Mandelbrojt, S. and Schwartz, L.

[1965] *Jacques Hadamard* (1865–1963), Bull. AMS, 71, 1965, pp. 107–129.

Mandebrot, B.

[2004] *Fractals and chaos: the Mandelbrot set and beyond: selecta*, Springer-Verlag, New York, 2004.

Marguet, F.

[1917] *L'histoire de la longitude à la mer*, Augustin Challamel, Paris, 1917.

Marie, M.

[1872] *Détermination du périmètre de la région de convergence de la série de Taylor et des portions des différentes conjuguées comprises dans cette région, ou construction du tableau général des valeurs d'une fonction que peut fournir le développement de cette fonction suivant la série de Taylor*, C.R. Acad. Sci. Paris, 75, 1872, pp. 469–472.

[1873] *Détermination du point critique où est limitée la convergence de la série de Taylor*, Liouville J., (2) XVIII, 1873, pp. 53–67.

Marmi, S.

[1999] *An Introduction to small divisors*, `arXiv:math/0009232v1`, 1999.

Marty, F.

[1931] *Recherches sur la répartition des valeurs d'une fonction méromorphe* Ann. Fac. Sci. Toulouse, Ser. 3, 23, 1931, pp. 183–261.

Mayer, J.

[1990] *An explosion point for the set of endpoints of the Julia set for* $\lambda\exp(z)$. Ergodic Theory Dynam. Systems, 10, 1990, pp. 177–184.

McMullen, C.

[1987] *Area and Hausdorff dimension of Julia sets of entire functions.*, Trans. AMS, 300, 1, 1987, pp. 329–342.

Mellon, P.

[1996] *Another look at results of Wolff and Julia type for J*-algebras*, J. Math. Anal. Appl., 198, 1996, pp. 444–457.

Mendès-France, M.

[1988] *Nevertheless*, Math. Intelligencer, **10**, 4, 1988, p. 35.

Mercer, P.R.

[1999] *On a strengthened Schwarz-Pick inequality*, J. Math. Anal. Appl., 234, 1999, pp. 735–739.

[2000] *Another look at Julia's lemma*, Complex Variables Theory Appl., 43, 2000, pp. 129–138.

Milloux, H.

[1926] *Sur le théorème de Picard*, Bull. SMF, 53, 1926, pp. 181–207.

Milnor, J.W.

[2004] "On Lattès' maps", in *Dynamics on the Riemann sphere. A Bodil Branner Festschrift*, P. Hjorth and C.L. Petersen, eds., European Math. Soc., 2006.

[2006a] *Dynamics in one complex variable*, 3rd edition, Princeton Univ. Press, 2006.

Minetti, S.

[1933] *Un teorema generale sulle successioni di funzioni convergenti verso una funzione olomorfa*, Rend. Accad. Lincei Roma, (6) 17, 1933, pp. 58–61.

Mineur, H.

[1929] *Les obsèques de Monsieur Pierre Fatou*, a Fatou family document which is a draft of the eulogy contained in the *Nouvelliste de Morbihan* (see entry [**A11**]).

Minialoff, A.

[1935] *Sur une propriété des transformations dans l'espace de deux variables complexes*, C.R. Acad. Sci. Paris, 200, 1935, pp. 711–717.

Miranda, C.

[1935] *Sur un nouveau critère de normalité pour les familles de fonctions holomorphes*, Bull. SMF, 63, 1935, pp. 185–196.

Misiurewicz, M.

[1981] *On iterates of* e^z, Ergodic Theory Dynam. Systems, 1, 1981, pp. 103–106.

Moler, C.

[1998-W] *Cleve's corner: trigonometry Is a complex subject*, MATLAB News & Notes, Summer, 1998, `http://www.mathworks.com/company/newsletters/news\_notes/clevescorner/sum98cleve.html`

Montel, P.

[1904] *Sur les suites des fonctions analytiques*, C.R. Acad. Sci. Paris, 138, 1904, pp. 469–471.

[1907] *Sur les suites infinies de fonctions*, Ann. Sci. Éc. Norm. Sup., 24, Ser. 3, 1907, pp. 233–334.

[1908] *Sur les points irréguliers des séries convergentes de fonctions analytiques*, C.R. Acad. Sci. Paris, 145, 1908, pp. 910–913.

[1911] *Sur les fonctions analytiques qui admettent deux values exceptionnelles dans un domain*, C.R. Acad. Sci. Paris, 153, 111, pp. 996–998.

[1912a] *Sur les familles de fonctions analytiques qui admettent des valeurs exceptionnelles dans un domaine*, Ann. Sci. Éc. Norm. Sup., Ser. 3, 29, 1912, pp. 487–535.

[1912b] *Sur quelques généralisations des théorèmes de M. Picard*, C.R. Acad. Sci. Paris, 155, 1912, pp. 1000–1003.

[1916] *Sur les familles normales de fonctions analytiques*, Ann. Sci. Éc. Norm. Sup., Ser. 3, 33, 1916, pp. 223–302.

[1917a] *Sur la représentation conforme*, J. Math. Pures Appl., Ser. 7, 3, 1917, pp. 1–54.

[1917b] *Sur la représentation conforme*, C.R. Acad. Sci. Paris, 164, 191, 1917, pp. 879–881.

[1918a] *Lattès*, Notices de l'Ass. des Anciens Élèves de l'Éc. Norm. Sup., 1918, pp. 79–81.

[1918b] *Sur les polynômes d'approximation*, Bull. SMF, 46, 1918, pp. 151–192.

[1922a] *Sur les familles quasi-normales*, C.R. Acad. Sci. Paris, 174, 1922, pp. 22–24.

[1922b] *Sur les familles quasi-normales de fonctions méromorphes*, C.R. Acad. Sci. Paris, 175, 1922, pp. 516–519.

[1924] *Sur les familles quasi-normales de fonctions analytiques*, Bull. SMF, 52, 1924, pp. 85–114.

[1926] *Sur le domaine correspondant aux valeurs d'une fonction analytique*, C.R. Acad. Sci. Paris, 183, 1926, pp. 940–942.

[1927] *Leçons sur les familles normales de fonctions analytiques et leurs applications. Recueillies et rédigées par J. Barbotte*, Gauthier-Villars, Paris, 1927. Reprinted in 1974 by Chelsea Publishing Company, New York.

[1929a] *Sur les domaines formés par les points représentant des valeurs d'une fonction analytique*, Ann. Sci. Éc. Norm. Sup., (3) 46, 1929, pp. 1–23.

[1929b] *Sur les familles de fonctions analytiques dont aucune fonction limite n'est constante*, Mathematica 1, pp. 1–4; Bull. Cluj., 4, pp. 125–128.

[1934] *Le rôle des familles normales*, L'Enseign. Math., 33, 1943, pp. 5–21.

Moser, J.

[1959] Review of Kolmorgorov's [**1957**], AMS Math. Reviews, MR0097598 (20 #4066), 1957.

[1962] *On invariant curves of area-preserving mappings of an annulus*, Nachr. Akad. Wiss. Göttingen Math.-Phys. Kl. II **1962**, 1962, 1–20

[1966] *A rapidly convergent iteration method and non-linear differential equation II*, Ann. Scuola Norm. Sup. Pisa, Vol. 20, No. 3, 1966, pp. 499–535.

Moser, J. and Siegel, C.L.

[1971] *Lectures on celestial mechanics*, Classics in Mathematics, Springer-Verlag, New York, 1971.

Mues, E.

[1979] *Über ein Problem von Hayman*, Math. Zeitschrift, 164, 1979, pp. 239–259.

Mumford, D., Series, C. and Wright, D.

[2002] *Indra's Pearls, The vision of Felix Klein*, Cambridge University Press, 2002.

Myller-Lébédeff, V.

[1929] *Sur les fonctions entières ayant un ensemble donné de droites de M. Julia*, Mathematica, 2, 1929, pp. 44–47.

Myrberg, P.

[1958] *Iteration von Quadratwurze1operationen,* Ann. Acad. Sci. Fenn. Ser. A, 259, 1958.

[1960] *Inversion der iteration für rationale Funktionen*, Ann. Acad. Sci. Fenn. Ser. A, 292, 1960.

[1964] *Iteration der polynome mit reellen Koeffizienten,* Ann. Acad. Sci. Fenn. Ser. A, 348, 1964.

[1965] *Iteration der polynome mit reellen Koeffizienten,* Ann. Acad. Sci. Fenn. Ser. A 374, 1965.

Nauenberg, M.

[2010] *The early application of the calculus to the inverse square force problem*, Arch. Hist. Exact Sci. 64, 2010, pp. 269–300.

Needham, T.

[2000] *Visual complex analysis*, Oxford University Press, 2000.

Nehari, Z.

[1952] *Conformal mapping*, Dover Publications, New York, 1952.

Netto, E.

[1887a] *Zur Theorie der iterirten Functionen*, Math. Ann., 29, 1887, pp. 148–153.

[1887b] *Ueber einen Algorithmus zur Auflösung numerischer algebraischer Gleichungen*, Math. Ann., 29, 1887, pp. 140–157.

[1887c] *Zur Theorie der iterirten Functionen*, Math. Ann., 29, 1887, pp. 148–153.

[1894] *Ueber Iterirung gebrochener Functionen*, Schlömilch Z., XXXIX, 1894, pp. 382–384.

[1896] *Vorlesungen über der Algebra,* Druck und Verlag von B. G. Teubner, Leipzig, 1869.

Nevanlinna, F. and Nevanlinna, R.

[1922] *Über die Eigenschaften einer analytischer Funktionen in der Umgebung einer singulären Stelle oder Linie*, Acta Soc. Sci. Fenn., 50, 5, 1922, pp. 1–46.

Nevanlinna, R.

[1929] *Über beschränkte analytische Funktionen*, Ann. Acad. Sci. Fenn., 32 (Lindelöf-Festschrift), 7, 1929.

[1970] *Analytic functions*, Die Grundlehren der Mathematischen Wissenschaften in Einzeldarstellungen (translated by Phillip Emig), Springer-Verlag, New York, 1970.

Nevo, S.

[2001a] *Applications of Zalcman's lemma to Q_m-normal families*, Analysis, 21, 2001, pp. 289–325.

[2001b] *Normality properties of the family $\{f(nz)\colon n \in \mathbb{N}\}$*, Comput. Methods Function Theory, 1, 2001, pp. 375–386.

[2003a] *Generating quasinormal families of arbitrary degree and order*, Analysis, 23, 2003, pp. 125–149.

[2003b] *Transfinite extension to Q_m-normality theory*, Results in Mathematics, 44, 2003, pp. 141–156.

Newton, I.

[1687] *Philosophiae Naturalis Principia Mathematica*, S. Pepys, Societatas Regalis, London, 1987.

[1713] *Philosophiae Naturalis Principia Mathematica*, 2nd edition, Cornelius Crawford, Cambridge, 1713.

Nicoletti, O.

[1906] *Sulla teoria dell' iterazione*, Mem. Soc. It. Sci., (3), 14, 1906, pp. 181–257.

[1907] *Sulla teoria della convergenza degli algoritmi di iterazione*, Annali di Mat., (3), 14, 1907, pp. 1–32.

[1917] *Su una classé di algoritmi di iterazione per l'approssimazione degli irrazionali quadratici*, Rend. Circ. Mat. Palermo, 42, 1917, pp. 73–79.

Nikol'skii, S.M.

[2005] "The Great Kolmogorov", transl. by Cooke, R., in *Mathematical events of the twentieth century*, by Bolibruch, A.A., Osipov Yu.S., and Sinai, Ya.G., Springer-Verlag, Berlin, 2005, pp. 283–296.

Norton, A.V.

[1982] *Generation and rendering of geometric fractals in 3-D*, Computer Graphics, 16 (3), 1982, pp. 61–67.

[1989] *Julia sets in the quaternions*, Computers and Graphics, 13 (2), 1989, pp. 267–278.

Noshiro, K.

[1938] *Contributions to the theory of meromorphic functions in the unit circle*, J. Fac. Sci. Hokkaido Univ., 7, 1938, pp. 149–159.

Nowicki, T. and van Strien, S.

[1994] *Polynomial maps with a Julia set of positive measure*, Stony Brook IMS preprint, #1994/3.

O'Connor, J.J. and Robertson, E.F.

[2010W-a] *Andrey Nikolaevich Kolmogorov*, The MacTutor History of Mathematics archive, `http://www-history.mcs.st-andrews.ac.uk/Biographies/Kolmogorov.html`

[2010W-b] *Jürgen Moser*, The MacTutor History of Mathematics archive, `http://www-history.mcs.st-and.ac.uk/Biographies/Moser\_Jurgen.html`

[2010W-c] *Kiyoshi Oka*, The MacTutor History of Mathematics archive, `http://www-history.mcs.st-andrews.ac.uk/Biographies/Oka.html`

Oka, K.

[1919-29-W] Unpublished writings, `http://www.lib.nara-wu.ac.jp/oka/moku/moku19.html`, 1919–1929.

[1930-34-W] Unpublished writings, http://www.lib.nara-wu.ac.jp/oka/moku/moku19.html, 1930–1934.

[1930-W] *Fonctions algébriques permutables avec une fonction rationnelle non-linéaire*, unpublished, http://www.lib.nara-wu.ac.jp/oka/ikou/s19/p000-1.html, 1930.

Oka, K., Cartan, H.P., Remmert, R., and Narasimhan, R.

[1984] *Kiyoshi Oka, Collected Papers*, Springer-Verlag, Berlin, 1984.

Osgood, W.F.

[1900] *On the existence of Green's function for the most general simply connected plane regions*, Trans. AMS, I, 1900, pp. 310–314.

Oshkin, I.B.

[1982] *On a test of the normality of families of holomorphic functions*, Uspekhi Mat. Nauk, 37, 2, 1982, pp. 221–222.

Ostrowski, A.

[1922] *Über vollständige Gebiete gleichmässiger Konvergenz von Folgen analytischer Funktionen*, Hamb. Abh., 1, 1922, pp. 327–350.

[1925] *Über Folgen analytischer Funktionen und einige Verschärfungen des Picardschen Satzes*, Math. Zeitschrift, 24, 1925, pp. 215–258.

[1931] *Studien über den Schottkyschen Satz*, Rektoratsprogramm der Universität Basel für das Jahr 1931. IV + 111 S. Basel, F. Reinhardt, 1931.

[1934] *Bemerkung zu meiner Abhandlung: Über Folgen analytischer Funktionen und einige Verschärfungen des Picardschen Satzes*, Math. Zeitschrift, 38, 1934, p. 642.

Outler, A.C.

[2006] *Augustine: Confessions and Enchiridion*, Westminster John Knox Press, Liousville, Kentucky, 2006.

Oya, Shin'ichi

[1983] *Japanese mathematics during the last 100 years*, Volume 1, Iwanami Shoten, Tokoyo, 1983 (in Japanese).

Painlevé, P.

[1887] *Sur les lignes singulières des fonctions analytiques*, Thesis, Paris, 4°, 1887.

[1897] *Leçons sur la théorie analytique des équations différentielles professées à Stockholm (Septembre, Octobre, Novembre 1895) sur l'invitation de S. M. le Roi de Suède et de Norwège*, Paris: A. Hermann. 19 + 6 + 589 S., 4°., lith., 1897.

[1898] *Sur la représentation des fonctions analytiques uniformes*, C.R. Acad. Sci. Paris, 126, 1898, pp. 200–202.

[1907] *Report of Fatou's thesis*, February 17th 1907, Ser. AJ^{16} 5539 des Centre Historique des Archives Nationales, Paris.

Pang, X.

[1989] *Bloch's principle and normal criterion*, Science in China, 32, 1989, pp. 782–791.

[1990] *On normal criterion of meromorphic functions*, Science in China, 33, 1990, pp. 521–527.

Pang, X., Nevo S., and Zalcman, L.

[2005] *Quasi-normal families of meromorphic functions*, Rev. Mat. Iberoamericana, 21, no. 1, 2005, pp. 249–262.

Parshall, K. and Rowe, D.

[1994] *The emergence of the American mathematical research community, 1876–1900: J. J. Sylvester, Felix Klein and E. H. Moore*, History of Mathematics, AMS and LMS, 1994.

Peitgen, H.-O. and Richter, P.

[1986] *The beauty of fractals*, Springer-Verlag, Berlin, 1986.

Peitgen, H.-O., Richter, P. and Saupe, D.

[2004] *Chaos and fractals*, Springer-Verlag, New York, 2004.

Peitgen, H.-O. and Saupe, D.

[1988] *The Science of Fractal Images*, Heinz-Otto Peitgen and Dietmar Saupe, eds., Springer-Verlag, New York, 1988.

Pelles, D.A.

[1974] *Iteration of endomorphisms of $P^n(C)$*, Univ. of North Carolina (mimeographed).

Pérès, J.

[1913] *Détermination de toutes les fonctions permutables de première espèce avec une fonction donnée*, C.R. Acad. Sci. Paris, 156, 1913, pp. 378–381.

Pérez, R.

[2011] *A brief but historic article of Siegel*, Notices AMS, 58, 2011, pp. 558–566.

Pérez-Marco, R.

[1992] *Solution complète au problème de Siegel de linéarisation d'une application holomorphe au voisinage d'un point fixe (d'apres J.Ch. Yoccoz)*, Sem. Bourbaki, 753: Astérisque: 206, 1992, pp. 273–310.

[1995a] *Sur une question de Dulac et Fatou*, C.R. Acad. Sci. Paris, 321, I, 1995, pp. 1045–1048.

[1995b] *Non linearizable holomorphic dynamics having an uncountable number of symmetries*, Invent. Math., 119, 1995, pp. 67–127.

[1997] *Fixed points and circle maps*, Acta Math., 179, 1997, pp. 243–294.

Petek, P.

[1997] *Circles and periodic points in the quaternic Julia sets*, Open Sys. Informat. Dyn., 4, 1997, pp. 487–492.

Petracovici, L.

[2006] *Non-Accessible critical points of certain rational functions with Cremer points*, Ann. Acad. Sci. Fenn. Math., 31, 2006, pp. 3–11.

Pettoello, R.

[1994] *Bernhard Riemann: Sulle ipotesi che stanno alla base della geometria e altri scritti scientifici e filosofici*, Universale Bollati Boringhieri, Torino, Italy, 1994.

Pfeiffer, G.A.

[1915a] *On the conformal geometry of analytic arcs*, Amer. J.. Math., 37, 1915, pp. 395–430.

[1915b] *On the conformal geometry of analytic arcs*, Bull. AMS, 21, 1915 p. 163.

[1916a] *On the conformal map of curvilinear angles*, Bull. AMS, 22, 1916, p. 491.

[1916b] *Existence of divergent solutions of the functional equations $\Phi[g(x)] = a\Phi(x), f[f(x)] = g(x)$, where $g(x)$ is a given analytic map, in the irrational case*, Bull. AMS, 22, 1916, p. 163.

[1917] *On the conformal mapping of curvilinear angles. The functional equation $\Phi[f(x)] = a_1\Phi(x)$*, Trans. AMS, 18, 1917, pp. 185–198.

[1918] *The functional equation $f[f(x)] = g(x)$*, Ann. Math., (2), 20, 1918, pp. 13–22.

Phili, Ch.

[2003] *Reflexions of Picard's theorem and Monge's problem on two Greek mathematicians: G. Remoundos and P. Zervos*, Acta Hist. Naturalium Necnon Technicarum, New Ser., 7, 2003, pp. 95–117.

Picard, É.

[1878] *Sur la forme des intégrales des équations différentielles du second ordre dans le voisinage de certains points critiques*, C.R. Acad. Sci. Paris, 87, 1878, pp. 430–432, pp. 743–745.

[1879] *Sur les fonctions entières*, C.R. Acad. Sci. Paris, 89, 1879, pp. 662–665.

[1881] *Sur la décomposition en facteurs primaires des fonctions uniformes ayant une ligne de points singuliers essentiels*, C.R. Acad. Sci. Paris, 92, 1881, pp. 690–692.

[1884] *Sur la forme des intégrales des équations différentielles du premier ordre dans le voisinage de certains points critiques*, Bull. SMF, XII, 1884, pp. 48–51.

[1887] *Report of Painlevé's thesis*, January 4th 1887, Ser. AJ16 5534 des Archives Nationales.

[1892] *Report of Hadamard's thesis*, May 18th 1892, Ser. AJ16 5535 des Archives Nationales.

[1896] *Traité d'analyse*, Tome III, Gauthier-Villars, Paris, 1896.

[1900] *Sur une classe de surfaces algébriques dont les coordonnées s'expriment par les fonctions uniformes des deux paramètres*, Ann. Sci. Éc. Norm. Sup., Ser. 3, 28, 1900, pp. 17–25.

[1904] *Sur certaines équations fonctionnelles et sur une classe de surfaces algébriques*, C.R. Acad. Sci. Paris, 139, 1904, pp. 5–9.

[1905] *Traité d'analyse*, Tome II, 2nd edition, Gauthier-Villars, 8°, 1905.

[1908] *Traité d'Analyse*, Tome III, Gauthier-Villars, Paris, 1908.

[1913] *Sur une classe de transcendantes*, Ann. Sci. Éc. Norm. Sup., Ser. 3, 30, 1913, pp. 247–253.

[1917] *Notice Historique sur Gaston Darboux*, Gauthier-Villars, Paris, 1917.

[1926] *La vie et l'œuvre de Jules Tannery*, Mémoires de l'Académie des Sciences, T. 58, 1926, pp. I–XXXII.

Pierpont, J.

[1900] *Mathematical instruction in France*, Bull. AMS, 6, 1900, pp. 225–249.

Pincherle, S.

[1880] *Saggio di una introduzione alla teoria delle funzioni analitiche secondo i principi del Prof. Weierstrass*, Giornale di Mat. 18, 1880, pp. 178–254; pp. 317–357.

[1894-95] *Sulle operazioni distributive commutabili con una operazione data*, Atti Reale Accad. Sci. Torino, 30 (1894–1895), pp. 820–844.

[1896a] *Della validità effettiva di alcuni sviluppi in serie di funzioni*, Atti Reale Accad. Lincei, Cl. Sc. Fis. Mat. Nat., Roma, (5), 5, 1896, pp. 27–33.

[1896b] *Le operazioni distributive e le omografie*, Reale Istit. Lombardo Scienze e Lettere, Classe Scienze Fisiche, Matematiche e Naturali, Milano, (2), 29, 1896, pp. 397–405.

[1896c] *Operazioni distributive: le equazioni differenziali non omogenee*, Atti Reale Accad. Lincei, Rend. Cl. Sc. Fis. Mat. Nat., Roma, (5), 5, 1896, pp. 301–306.

[1897a] *Cenno sulla geometria dello Spazio funzionale*, Rend. Reale Accad. Sci. Instit. Bologna (Nuova Serie), 1, 1896–1897, pp. 85–95.

[1897b] *Sulla generalizzazione del determinante wronskiano*, Atti Reale Accad. Lincei, Rend. Cl. Sc. Fis. Mat. Nat., Roma, (5), 6, 1897. pp. 301–307.

[1906] *Funktionaloperationen und Gleichungen* from *Encyklopädie der Mathematischen Wissenschaften*, Volume II, A, §11, Teubner, Leipzig, 1906, pp. 761–817.

[1912] *Equations et opérations functionnelles* from *Encyclopédie des sciences mathématiques pures et appliquées*, Volume V, Book II, §26, Gauthier-Villars, Paris, 1912, , pp. 1–81,

[1914] *Alcune osservazioni sulla iterata di una funzione data*, Rend. Reale Accad. Sci. Instit. Bologna (Nuova Serie), 18, (1913–1914), pp. 75–88.

[1916] *Sopra alcuni nuclei analitici*, Rend. Reale Accad. Sci. Instit. Bologna (Nuova Serie), 20, (1915–1916), pp. 85–100.

[1917a] *Appunti su alcuni problemi di iterazione*, Rend. Reale Accad. Sci. Instit. Bologna (Nuova Serie), 21, (1916–1917), pp. 86–97.

[1917b] *Sur l'itération de la substitution $x^2 - a$*, submission to French Academy of Sciences for its 1918 *Grand Prix des Sciences Mathématiques*, 1917.

[1918a] *Sulle catene di radicali quadratici*, Atti Reale Accad. Torino, 53, 1918, pp. 745-763.

[1918b] *Sulle catene di radicali quadratici*, Atti Reale Accad. Bologna, 53, 1918, pp. 437–455.

[1918c] *Sulle radici reali delle equazioni iterate di una equazione quadratica*, Rend. Accad. Lincei Roma, (5), 27, 1918m pp. 177-18.

[1918d] *Sull iterazione della funzione $x^2 - a$*, Rend. Accad. Lincei Roma, (5), 27, 2, 1918, pp. 337–343.

[1919] *Un teorema sull' iterazione della funzione quadratica*, Rend. Reale Accad. Sci. Instit. Bologna (Nuova Serie), 23, 1918–1919, pp. 65–70.

[1920a] *L'iterazione completa di $x^2 - 2$*, Rend. Accad. Lincei Roma, (5), 27, 1920, pp. 329–333.

[1920b] *Sulla funzione iterata di una razionale intera*, Rend. Accad. Lincei Roma, (5), 29, 1, 1920, pp. 403–407.

[1920c] *Sopra alcune equazioni funzionali*, Rend. Accad. Lincei Roma, (5), 29, 2, 1920, pp. 279–281.

[1925] *Notice sur les travaux*, Acta. Math., 46, 1925, pp. 341–362.

Plantefol, L.

[1961] *Allocution to Julia*, in Julia's Œuvres, VI, Gauthier-Villars, Paris, 1970, pp. 309–315.

Podetti, F.

[1897] *Sulle sostituzioni uniformi*, Giornale di Battaglini, 35, 1897, pp. 264–266.

Poincaré, H.

[1879] *Sur les propriétés des fonctions définies par les équations différentielles aux differences partielles*, Thesis presented to the Faculty of Sciences in Paris, August 1st 1879, Gauthier-Villars, Paris, = *Œuvres*, **I**, pp. XLIX–CXXIX.

[1881-82] *Mémoire sur les courbes définies par une équation différentielle*, Résal J., (3), VII, 1881, pp. 375–422; (3) VIII, 1882, pp. 251–296 = *Œuvres*, **I**, pp. 3-44, 44–84.

[1883a] *Sur un théorème général des fonctions*, Bull. SMF, 11, 1883, pp. 112–125 = *Œuvres*, **IV**, pp. 57–69.

[1883b] *Mémoire sur les groupes Kleiniéns*, Acta Math., III, 1883, pp. 49–92 = *Œuvres*, **II**, pp. 258–299.

[1884] *Sur les groupes des équations linéaires*, Acta Math., IV, 1884, pp. 201–312 = *Œuvres*, **II**, pp. 300–401.

[1885] *Sur les courbes définies par les équations différentielles*, Jordan J., (4), I, 1885, pp. 167–244 = *Œuvres*, **I**, pp. 90–161.

[1886] *Sur les courbes définies par une équation différentielle*, J. Math. Pures Appl., Quatrième Partie, 4, 2, 1886, pp. 151–217 = *Œuvres*, **I**, pp. 167–222.

[1890a] *Sur une classe nouvelle de transcendantes uniformes*, J. Math. Pures Appl., 6, 4, 1890, pp. 313–365 = *Œuvres*, **IV**, pp. 537–582.

[1890b] *Sur le problème des trois corps et les équations de la dynamique*, Acta Math., XIII, 1890, pp. 1–270 = *Œuvres*, **VII**, pp. 262–479.

[1892a] *Les méthodes nouvelles de la mécanique céleste. Tome I. Solutions périodiques. Non-existence des intégrales uniformes. Solutions asymptotiques.*, Gauthier-Villars et Fils., Paris, 8°, 1892, pp. 385.

[1892b] *Sur les fonctions à espaces lacunaires*, Amer. J.. Math., 14, 1892, pp. 201–221 = *Œuvres*, **IV**, pp. 36–55.

[1893] *Les méthodes nouvelles de la mécanique céleste.* Tome II. Gauthier-Villars et Fils, 1893.

[1899b] *Les méthodes nouvelles de la mécanique céleste, Tome III*, Gauthier-Villars et Fils., Paris, 8°, 1899.

[1908] *Sur l'uniformisation des fonctions analytiques*, Acta Math., 31, 1908, pp. 1–64 = *Œuvres*, **IV**, pp. 70–139.

[1921] *Analyse de ses travaux scientifiques*, Acta Math., 38, 1921, pp. 36–135.

[1915–1956] *Œuvres de Henri Poincaré*, 11 Volumes, Gauthier-Villars, Paris, 1915–1956.

Pólya, G.

[1927] *Sur les fonctions entières à série lacunaire*, C.R. Acad. Sci. Paris, 184, 1927, pp. 1526–1528.

Pólya, G. and Hurwitz, A.

[1916] *Zwei Beweise eines von Herrn Fatou vermuteten*, Acta. Math. 40, 1916, pp. 179–183.

Pommerenke, Ch.

[1981] *On asymptotic iteration of analytic functions in the disk*, Analysis, 1, 1981, pp. 45–61.

Pompeo Faracovi, O.

[2002] *Enriques e Severi: Matematici a confronto nella cultura del novecento*, Atti del Convegno di Livorno 24–25 Ottobre 2002, Agor Edizioni, 2004.

Poon, K.K. and Yang, C.C.

[1988] *Dynamical behavior of two permutable entire functions*, Ann. Polon. Math., 68, 1998, pp. 159–163.

Pringsheim, A.

[1890] *Allgemeine Theorie der Divergenz und Convergenz von Reihen mit positiven Gliedern*, Math. Ann., 35, 1890, pp. 297–394.

[1893] *Zur Theorie der Taylor'schen Reihe und der analytischen Funktionen mit beschränktem Existenzbereich*, Math. Ann., 42, 1893, pp. 153–184.

[1894a] *Ueber die notwendigen und hinreichenden Bedingungen des Taylor'schen Lehrssatzes für Funktionen einer reellen Variablen*, Math. Ann., 44, 1894, pp. 57–82.

[1894b] *Ueber Funktionen, welche in gewissen Punkten endliche Differentialquotienten jeder endlichen Ordnung, aber keine Taylorsche Reihenentwickelung besitzen*, Math. Ann., 49, 1894, pp. 41–56.

Rådström, H.

[1953] *On the iteration of analytic functions*,. Math. Scand. 1, 1953, 85-92.

Rassias, T.M.

[2004] *The Greek Mathematical Society*, EMS Newsletter, 2004, pp. 34–35.

Rausenberger, O.

[1881] *Theorie der allgemeinen Periodicität*, Klein Ann., XVIII, 1881, pp. 379–410.

Reich, S.

[1985] *Averaged mappings in the Hilbert ball.*, J. Math. Appl. Math., 109, 1985, pp. 199-206.

[1991] *The asymptotic behavior of a class of nonlinear semigroups in the Hilbert ball,* J. Math. Anal. Appl., 157, 1991, pp. 237-242.

Reich, S. and Shoikhet, D.

[1996] *Generation theory for semigroups of holomorphic mappings in Banach spaces*, Abstr. Appl. Analysis, 1, 1996, pp. 1–44.

[1997a] *Semigroups and generators on convex domain with the hyperbolic metric*, Atti Accad. Naz. Lincei, 8, 1997, pp. 231–250.

[1997b] *The Denjoy-Wolff theorem*, Proc. Workshop of Fixed Point Theory, Ann. Univ. Mariæ Curie-Sklodowska Sect., A, 51, 1997, pp. 219–240.

[2001] *The Denjoy-Wolff Theorem*, Encyclopedia of Mathematics, Suppl. III, Kluwer Academic Publishers, Dordrecht, 2001, pp. 121-123.

Reichenbächer, E.

[1908] *Über das Iterationsproblem*, Zs. f. Math. Naturw. Unterr., 39, 1908, pp. 233–246.

Reidemeister, M.

[1926a] *Elementare Begründung der Knotentheorie*, Abhandlungen Hamburg, 5, 1926, pp. 24–32.

[1926b] *Knoten und Gruppen*, Abhandlungen Hamburg, 5, 1926, pp. 7–23.

[1928] *Über Knotengruppen*, Abhandlungen Hamburg, 6, 1928, pp. 56–64.

Remmert, R.

[1990] *Theory of complex functions*, Springer-Verlag, 1990.
[1997] *Classical topics in complex function theory*, Springer-Verlag, New York, 1997.
[1998] *From Riemann surfaces to complex spaces*, Séminaires et Congrès SMF, 3, 1998

Remoundos, G.J.

[1903] *Une nouvelle généralisation du théorème de M. Picard sur les fonctions entières*, C.R. Acad. Sci. Paris, 136, 1903, pp. 953–955.
[1905] *Sur le cas d'exception de M. Picard et les fonctions multiformes*, Bull. SMF, 33, 1905, pp. 191–201.
[1906] *Sur les zéros d'une class de fonctions transcendantes*, Ann. Fac. Sci. Toulouse, Ser. 2, 9, 1906, pp. 1–72.
[1907] *Sur les points critiques d'une classe de fonctions*, C.R. Acad. Sci. Paris, 144, 1907, pp. 65–67.
[1913a] *Le théorème de M. Picard dans un cercle dont le centre est un point critique algébrique*, C.R. Acad. Sci. Paris, 167, 1913, pp. 694–697.
[1913b] *Sur les familles de fonctions multiformes admettant des valeurs exceptionnelles dans un domaine*, C.R. Acad. Sci. Paris, 167, 1913, pp. 542–545.
[1914a] *Sur les séries de fonctions multiformes dans un domaine*, C.R. Acad. Sci. Paris, 158, 1914, pp. 929–932.
[1914b] *Sur la convergence des séries de fonctions analytiques*, C.R. Acad. Sci. Paris, 158, 1914, pp. 248–250.
[1915] *Sur les familles et les séries de fonctions multiformes dans un domaine*, Annali di Mat., (3), 23, 1915, pp. 1–24.
[1917] *Sur la classification des points transcendants des inverses des fonctions entières ou méromorphes*, C.R. Acad. Sci. Paris, 165, 1917, pp. 331–334.
[1923] *Sur l'itération des fonctions multiformes*, C.R. Acad. Sci. Paris, 176, 1923, pp. 274–275.
[1927] *Extensions aux fonctions algébroïdes multiformes du théorème de M. Picard et des généralisations*, Gauthier-Villars, Paris, 1927.

Rempe, L.

[2003] *Dynamics of Exponential Maps*, doctoral dissertation, Christian-Albrechts-Universität Kiel, 2003.
[2007] *On nonlanding dynamic rays of exponential maps*, Ann. Acad. Sci. Fenn., 32, 2007, pp. 353–359.

von Renteln, M.

[1992] *Des Einflußder Lebesgueschen Integrationstheorie auf die komplexe Funktionentheorie zu Beginn dieses Jahrhunderts*, in Jahrbuch Äberblicke Mathematik, 1992 (Braunschweig, 1992), pp. 75–96.

Richardson, J.

[1991] *Visualizing quantum scattering on the CM-2 supercomputer*, Comp. Phys. Comm., **63**, 1-3, 1991, pp. 84–94.

Riemann, B.

[1859] *Über die Anzahl der Primzahlen unter einer gegebenen Grösse*, Monatsberichte Berl. Akad. 1859.
[1898] *Œuvres mathématiques. Traduites par L. Laugel. Avec une préface par Hermite et un discours par F. Klein*, Gauthier-Villars, Paris, 1898, XXXV + 953 p. gr. 8°.

Riesz, F. and Riesz, M.

[1916] *Ueber die Randwerte einer Analytischen Funktion*, 4° Congress of Scandinavian Mathematicians, Stockholm, 1916, pp. 27–44.

Rippon, P. and Stallard, G.

[2005] *On questions of Fatou and Eremenko*, manuscript, Proc. AMS, Vol. 133, 4, 2005, pp. 1119–1126.

Rippon, P. and Stallard, G., eds.

[2008] *Transcendental dynamics and complex analysis: a tribute to Noel Baker*, Cambridge University Press, 2008, pp. IX–XIX.

Ritt, J.F.

[1915] *On certain real solutions of Babbage's functional equation*, Annals of Maths, Ser. 2, 17, 1915, pp. 113–123.

[1918a] *Sur l'itération des fonctions rationnelles*, C.R. Acad. Sci. Paris, 166, 1918, pp. 380–381.

[1918b] *Polynomials with a common iterate*, Bull. AMS, 24, 1918, pp. 374.

[1918c] *On the iteration of polynomials*, Bull. AMS, 24, 1918, p. 467.

[1920] *On the iteration of rational functions*, Trans. AMS, 21, 1920, pp. 348–356.

[1921] *On the conformal mapping of a region into a part of itself*, Annals of Maths, Ser. 2, 22, 1921, pp. 157–160.

[1922] *Prime and composite polynomials*, Trans. AMS, 23, 1922, pp. 51–66.

[1923a] *Permutable rational maps*, Bull. AMS, 29, 1923, p. 147.

[1923b] *Sur les fonctions rationnelles permutables*, C.R. Acad. Sci. Paris, 176, 1923, pp. 60–61.

[1923c] *Permutable rational functions*, Trans. AMS, 25, 1923, pp. 399–448.

[1924] *Equivalent rational substitutions*, Trans. AMS, 25, 1924, pp. 221–229

Rosa, A.

[2004] *Inwards to chaos and complex mapper: two graphical interfaces for simulations in complex analysis and dynamics*, Electronic J. Diff. Eq. Control Proc., St. Petersburg, Russia, 3, 2004.

[2005a] *Hommage à Gaston Julia* (1893–1978), Quadrature, 58, 2005.

[2005b] *Advanced graphics in complex analysis and dynamics*, Electronic J. Diff. Eq. Control Proc., 1, 2005.

[2005c] *Methods and applications to display quaternion Julia sets*, Electronic J. Diff. Eq. Control Proc., St. Petersburg, 4, 2005.

[2006] *On the digital visualization of hedgehogs in holomorphic dynamics*, Electronic J. Diff. Eq. and Control Proc., St. Petersburg, Russia, 1, 2007, pp. 1–36.

Rosa, A. and Alexander, D.S.

[prep] *A history of graphical iteration*, in preparation.

Rosay, J.P. and Rudin, W.

[1988] *Holomorphic maps from $\mathbb{C}^n$ to $\mathbb{C}^n$*, Trans. AMS, 310, 1988, 1, pp. 47–86.

Rosenbloom, P.C.

[1948] *L'itération des fonctions entières*, C. R. Acad. Sci. Paris, 227, 1948, pp. 382-383.

[1952] *The fixed points of entire functions*, Medd. Lunds. Univ. Mat. Sem., Tome Suppl. 1952, pp. 186–192.

Rothman, T.

[1982] *Genius and biographers: The fictionalization of Evariste Galois*, Amer. Math. Monthly, 1982, pp. 89–84.

Rottenfußer, G. and Schleicher, D.

[2008] *Escaping points of cosine maps*, in *Transcendental dynamics and complex analysis: a tribute to Noel Baker*, P. Rippon and G. Stallard, eds., 2008, Cambridge University Press, pp. 396–424.

Rottenfußer, G., Rückert, J., Rempe, L., and Schleicher, D.

[2011] *Dynamic Rays of Entire Functions*, Annals of Math., to appear.

Rouche, N., Habets, P. and Laloy, M.

[1977] *Stability theory by Liapunov's direct method*, Applied Mathematical Series, 22, Springer-Verlag, New York, 1977.

Rubel, L.

[1986] *Four counter-examples to Bloch's principle*, Proc. AMS, 98, 1986, pp. 257–260.

Rudin, W.

[1980] *Function theory in the unit ball of* $\mathbb{C}^n$, Classics in Mathematics, Springer-Verlag, Berlin, 1980.

Runge, C.

[1885] *Zur Theorie der eindeutigen analytischen Functionen*, Acta Math., VI, 1885, pp. 229–244.

Rußman, H.

[1983] *Das Werk C. L. Siegels in der Funktionentheorie,* Jahresbericht DMV, 85, 1983, pp. 174–200.

Sandin, D.J. and Kauffman, L.H.

[2004] *A ray tracer to visualize higher dimensional Julia sets*, Proc. of the Fifth Interdisciplinary Conference of The International Society of The Arts, Mathematics, and Architecture, Chicago, 2004.

Salet, P.

[1927] *Omar Khayyam, savant et philosophe*, Maisonneuve Frères, Paris, 1927.

Salomon-Bayet, C.

[1978] *L'institution de la Science et l'expérience du vivant. Méthode et expérience á l'Académie Royale des Sciences 1666–1793*, Ed. Flammarion, Champs Sciences, Paris, 1978.

Sarason, D.

[1988] *Angular derivatives via Hilbert space*, Complex Variables Theory Appl., 10, 1988, pp. 1–10.

Saunders, E.S.

[1978] *The Archives of the Academy of Sciences*, French Historical Studies, 10, No. 4., 1978, pp. 696–702.

Saxer, W.

[1928a] *Über quasi-normale Funktionsscharen und eine Verschärfung des Picardschen Satzes*, Math. Ann., 99, 1928, pp. 707–737.

[1928b] *Sur la structure des familles normales de fonctions méromorphes*, C.R. Acad. Sci. Paris, 187, 1928, pp. 752–754.

[1930] *Über konvergente Folgen meromorpher Funktionen*, Comment. Math. Helvetici, 2, 1930, pp. 18–34.

Schapira, H.

[1887a] *Bemerkungen zu der Grenzfunction algebraischer Iterationen*, Schlömilch. Z., XXXII, 1887, pp. 310–314.

[1887b] *Ueber ein allgemeines Princip algebraischer Iterationen*, Vortrag. Heidelberg, 1887, 24 pp.

Schiff, J.L.

[1993] *Normal families*, Universitext, Springer-Verlag, New York, 1993.

Schleicher, D.

[1998] *Dynamics of entire functions*, in, *Holomorphic dynamical systems*, Graziano Gentili, Jacques Guenot, and Giorgio Patrizio, eds., Lecture Notes in Math., Springer-Verlag Berlin, 1988, pp. 295–339.

[2007a] *The dynamical fine structure of iterated cosine maps and a dimension paradox*, Duke Math. J., Vol. 136, No. 2, 2007, pp. 343–356.

[2007b] *Hausdorff dimension, its properties, and its surprises.* Amer. Math. Monthly, 114, 6, 2007, pp. 509–528.

Schleicher, D. and Zimmer, J.

[2003] *Escaping points of exponential maps*, J. LMS, 67, 2003, pp. 380–400.

Schmidt, W.

[1992] *On the periodic stable domains of permutable rational functions*, Complex Variables, 17, 1992, pp. 149–152.

Schmidt, W. and Steinmetz, N.

[1995] *The polynomials associated with a Julia set*, Bull. LMS, 27, 1995, pp. 239–241.

Schönflies, A.

[1908] *Die Entwicklung der Lehre von den Punktmannigfaltigkeiten, II*, Bericht erstattet der Deutschen Mathematiker-Vereinigung (Report to the German Math. Soc.), Teubner, Leipzig, 1908.

Schröder, E.

[1870] *Über unendlich viele Algorithmen zur Auflösung der Gleichungen*, Math Ann., 2, 1870, pp. 317–365.

[1871] *Über iterirte Funktionen*, Math. Ann., 3, 1871, pp. 296–322.

Schwarz, H.A.

[1890] *Gesammelte mathematische Abhandlungen*, II, Springer-Verlag, Berlin, 1890.

Schwick, W.

[1989] *Normality criteria for families of meromorphic functions*, J. Analyse Math., 52, 1989, pp. 241–289.

[1997] *Repelling periodic points in the Julia set*, Bull. LMS, 29, 1997, pp. 314–316.

Segal, S.L.

[1981] *Nine introductions in complex analysis*, Mathematical Studies, North Holland, Amsterdam, 1981.

[2003] *Mathematicians Under the Nazis*, Princeton University Press, 2003.

Seidel, W. and Littauer, S.B.

[1931] *On the lines of Julia of entire functions. I, II*, Bull. AMS, 37, 1931, 816

Selmer, E.S.

[1991] *Ralph Tambs-Lyche in memoriam*, Normat Nr., 39, (1), 1991, pp. 1–3.

Serrin, J.

[1956] *A note on harmonic function defined in a half-plane*, Duke Math J., 23, 1956, pp. 523–526.

Severini, C.

[1909] *Sulle successioni infinite di funzioni analitiche*, Roma 4th Mathematical Congress, 2, Accad. dei Lincei, Rome, 1909, pp. 183–193.

Shankar, H., ed.

[1970] *Essays dedicated to A. J. Macintyre*, Ohio Univ. Press, Athens, OH, 1970.

Shapiro, J.H.

[1993] *Composition operators and classical function theory*, Universitext: Tracts in Mathematics, Springer-Verlag, New York, 1993.

Shimizu, T.

[1931a] *On the iterations of algebraic functions I*, Proc. Phys.-Math. Soc. Japan, III, Ser. 13, 1931, pp. 255–266.

[1931b] *On the iterations of algebraic functions II*, Proc. Phys.-Math. Soc. Japan, III, Ser. 13, 1931, pp. 292–296.

Shishikura, M.

[1987] *On the quasiconformal surgery of rational functions*, Ann. Sci. Éc. Norm. Sup., 20, 1987, pp. 1–20.

Shoikhet, D.

[2001] *Semigroups in geometrical function theory*, Kluwer Academic Publishers, Dordrecht, 2001.

Siegel, C.L.

[1941] *On the integrals of canonical systems*, Annals of Math., 42, 1941, pp. 806–822.

[1942] *Iterations of analytic functions*, Annals of Math., 43, 1942, pp. 607–612.

[1977] *On the history of the Frankfurt Mathematics Seminar*, Math. Intelligencer, **1**, 4, 1977, pp. 223–230.

Siegmund-Schultze, R.

[2003] *The origins of functional analysis*, in *A History of Analysis*, H. N. Jahnke, History of Mathematics, 24, AMS-LMS, 2003, pp. 385–404.

[2004] *The ideology of applied mathematics within mathematics in Germany and the U.S. until the end of World War II*, Llull ISSN 0210–8615, 2004, Vol. 27, 60, pp. 791–811.

Sintzow (Sintsov), D.

[1904] Review of *The prinipal laws of convergence of iterates and their application to analysis* by L. Böttcher, JFM, 35.0398, 1904. See Böttcher [**1904b**].

Smith, H.J.S.

[1875] *On the integration of discontinuous funcitons*, Proc. LMS, 6, 1875, pp. 140–153.

Société de Mathmatiques de France **(publisher)**

[1905] *Vie de la Société*, Astérisque, Bull. SMF, 33, 1905.

[1906] *Vie de la Société*, Astérisque, Bull. SMF, 34, 1906.

[1907] *Vie de la Société*, Astérisque, Bull. SMF, 35, 1907.

[1908] *Vie de la Société*, Astérisque, Bull. SMF, 36, 1908.

[1913] *Vie de la Société*, Astérisque, Bull. SMF, 41, 1913.

[1915] *Vie de la Société*, Astérisque, Bull. SMF, 43, 1915, pp. 1–25.

[1916] *Vie de la Société*, Astérisque, Bull. SMF, 44, 1916, pp. 1–36.

[1917] *Vie de la Société*, Astérisque, Bull. SMF, 45, 1917, pp. 1–19.

[1918] *Vie de la Société*, Astérisque, Bull. SMF, 46, 1918, pp. 1–38.

[1919] *Vie de la Société*, Astérisque, Bull. SMF, 47, 1919, pp. 1–50.

[1920] *Vie de la Société*, Astérisque, Bull. SMF, 48, 1920, pp. 1–54.

[1929] *Vie de la Société*, Astérisque, Bull. SMF, 57, 1929, pp. 1–24, supplément spécial.

[1942] *Arnaud Denjoy – Évocation de l'homme et de l'œuvre*, Astérisque SMF, 1975, pp. 28–29.

Spiess, O.

[1902] *Die Grundbegriffe der Iterationsrechnung*, Bern. Mitt., pp. 106–137; sep. Basel. 34 S. 8°., 1902.

[1906] *Theorie der linearen Iteralgleichung mit konstanten Koeffizienten*, Math. Ann. 62, 1906, pp. 226–252.

Stachura, A.

[1985] *Iterates of holomorphic self-maps of the unit ball in Hilbert spaces,* Proc. AMS, 93, 1985, pp. 88-90.

Stavinschi M. and Mioc V.

[2005] *Astronomical researches in Poincaré's and Romanian works*, Journées 2004 - systèmes de référence spatio-temporels, Observatoire de Paris, Paris, September 2005, pp. 155–156.

Stefanescu, D.

[2005] *Bull. Mathématique—A brief history*, Bull. Soc. Sci. Math. Roumanie, 48(96), 1/2005, pp. 1–4.

Steinmetz, N.

[1993] *Rational Iteration: Complex Analytic Rational Systems*, de Gruyter Studies in Mathematics, de Gruyter, Berlin, 1993.

Sternberg, S.

[1969] *Celestial mechanics*, Part II, Mathematics Lecture Note Series, Benjamin, New York, 1969.

Stewart, G.

[1993] *Infinitely many algorithms for solving equations*, Computer Science Technical Report Series, Vol. CS-TR-2990, 1993. Translation of E. Schröder's *Über unendlich viele Algorithmen zur Auflösung der Gleichungen*. See Schröder [**1870**].

Sugiyama, H.

[1994] *A sketch of the life of Dr. Shimizu*, Mathematica Japonicæ, Vol. 40, No. 1, 66, 1994, p. 2.

Sullivan, D.

[1985] *Quasi-conformal homeomorphisms and dynamics I: Solution of the Fatou-Julia problem on wandering domains*, Ann. Maths, 122, 1985, pp. 402–418.

Tambs-Lyche, R.

[1927] *Une formule d'itération*, Bull. SMF, 55, 1927, pp. 102–113.

[1928a] *Sur la convergence de la série* $\sum \begin{bmatrix} n \\ r \end{bmatrix}$, C.R. Acad. Sci. Paris, 186, 1928, pp. 1810–1812.

[1928b] *Les fonctions limites d'une fonction à centre*, C.R. Acad. Sci. Paris, 187, 1928, pp. 102–104.

[1929a] *L'itération de la fonction* $z_1 = sz + z^2$, Forhandlinger Norske Selskab, 2, 18, 1929, pp. 62–65.

[1929b] *Remarques sur certains déterminants*, Forhandlinger Norske Selskab 2, 19, 1929, pp. 66–69.

[1930] *Sur le problème du centre*, VII-th Skand. Mathematikerkongress, 1930, pp. 158–169.

Tan Lei, ed.

[2000] *The Mandelbrot Set, Theme and Variations*, LMS Lecture Note Series, Cambridge University Press, 2000.

Taton, R.

[1980] *Gabriel Kœnigs* in *The Dictionary of Scientific Biography*, Charles Scribner's Sons, Vol. 7, 1980, p. 446.

Tazzioli, R.

[1999] *La matematica all'Universitá di Catania dall'Unità alla riforma Gentile*, Annali di Storia delle Universitá Italiane, Vol. 3, 1999.

Tao, Z.G.

[1988] *Julia's lemma for analytic operator functions*, Chinese Ann. Math., Ser. B, 9, 1988, pp. 156–160.

Thurston, W.

[2009] *On the geometry and dynamics of iterated rational maps*, D. Schleicher and N. Selinger, eds., with an appendix by Schleicher, in *Complex Dynamics*, D. Schleicher, ed., A. K. Peters, Wellesley, MA, 2009, pp. 3–137.

Tonelli, L.

[1910a] *Sull iterazione*, Giornale di Battaglini, 48, (3), 1, 1910, pp. 341–373.

[1910b] *Sull iterazione*, Rend. Accad. Lincei Roma, (6), 19, 1, 1910, pp. 676–681.

[1910c] *Su gli zeri del limite di una successione di funzioni analitiche*, Rend. Accad. Lincei Roma, 19, 1, 1910, pp. 5–10.

[1937] *Salvatore Pincherle*, Annali della Scuola Normale Superiore di Pisa, Classe di Scienze, 2, t. 6, 1, 1937, pp. 1–10.

Töpfer, H.

[1939] *Über die Iteration der ganzen transzendenten Funktionen insbesondere von* sin *z und* cos *z*, Math. Ann., 117, 1939.

[1949] *Komplexe Iterationsindizes ganzer und rationaler Funktionen*, Math. Ann., 121, 1949, pp. 191–222.

Tricomi, F.

[1916a] *Un teorema sulla convergenza delle successioni formate delle successive iterate di una funzione di una variabile reale*, Giornale di Battaglini, 54, (3) 7, 1916, pp. 1–9.

[1916b] *Sull iterazione delle funzioni di una variabile complessa*, Rend. Accad. Lincei Roma, (5) 25, 2, 1916, pp. 156–162

Tschebotaröw, N.

[1930] *Bemerkung über analytische Iterationen*, Bull. Soc. Phys.-Math. de Kazan, (3), 4, 1930, pp. 19–50.

Tucciarone, J.

[1973] *The development of the theory of summable divergent series from 1880 to 1925*, Arch. Hist. Exact Sci., Vol 10, number 12, 1973, pp. 1–40.

Valiron, G.

[1913] *Sur les fonctions entières d'ordre nul et d'ordre fini et en particulier les fonctions à correspondance régulière,* Ann. Sci. Fac. Sci. Toulouse, Ser. 3, 9, 1913, pp. 117–257

[1920] *Les théorèmes généraux de M. Borel dans la théorie des fonctions entières*, Ann. Sci. Éc. Norm. Sup., Ser. 3, 37, 1920, pp. 219–253.

[1925a] *Remarque sur un théorème de M. Julia*, Bull. Soc. Math., (2), 49, 1925, pp. 68–73.

[1925b] *Supplément à la note 'Remarque sur un théorème de M. Julia'*, Bull. Soc. Math., (2), 49, 1925, pp. 270–275.

[1925c] *Sur les fonctions méromorphes qui sont exceptionnelles relativement au théorème de M. Julia*, C.R. Acad. Sci. Paris, 180, 1925, pp. 1895–1896.

[1927] *Compléments au théorème de Picard-Julia*, Bull. Soc. Math., (2), 51, 1927, pp. 167–183.

[1928] *Le théorème de M. g et le complément de M. Julia*, J. Math., (9), 7, 1928, pp. 113–126.

[1930] *Sur les familles normales de fonctions analytiques*, Ann. Sci. Éc. Norm. Sup., Ser. 3, 47, 1930, pp. 79–92.

[1931] *Sur l'itération des fonctions holomorphes dans un demi-plan*, Bull. Soc. Math., (2), 55, 1931, pp. 105–128.

[1932] *Le théorème de Borel-Julia dans la théorie des fonctions méromorphes* Verhandlungen Kongress Zürich, 1, 1932, pp. 270–279.

Venkatachaliengar, K.

[1934] *A note on Julia's lemma*, Mathematics Student 2, 1934, pp. 149–150.

Viola, T.

[1932a] *Sui punti irregolari di una famiglia non normale di funzioni olomorfe*, Verhandlungen Kongress Zürich, 2, 1932, Orell Füssili, Zürich, pp. 36–37.

[1932b] *Una proprietà degli aggregati perfetti di punti, utile nello studio delle famiglie non normali di funzioni olomorfe*, Bollettino U.M.I., 11, 1932, pp. 287–290.

[1935] *Sulle famiglie di funzioni analitiche ammettenti valori eccezionali e sul comportamento di una funzione uniforme in prossimità di un punto singolare essenziale isolato*, Mem. R. Accad. Italiana, Cl. Sc. Fis. Mat. Nat., 6, 1935, pp. 1293–1307.

Vitali, G.

[1903] *Sopra le serie di funzioni analitiche*, Rendiconti del Real Istituto Lombardo, (2), 36, 1903 and Annali di Mat., (3), 10, 1904.

Vitushkin, A.G.

[2004] *On Hilbert's thirteenth problem and related questions*, Russ. Math. Surv., **59**, 1, pp. 11-25.

Volk, O.

[1976] *Miscellanea from the history of celestial mechanics*, Proceedings of the Fifth Conference on Mathematical Methods in Celestial Mechanics (Oberwolfach, 1975), Part II. Celestial Mech. 14 (3), 1976, pp. 365–382.

Volterra, V.

[1888] *Sulle funzioni analitiche polidrome*, Atti Reale Accad. dei Lincei, (IV), 4, 1888, pp. 355–361.

[1910] *Sopra le funzioni permutabili*, Rom. Acc. L. Rend., (5), 19, 1, 1910, pp. 425–437.

[1911a] *Contributo allo studio delle funzioni permutabili*, Rom. Acc. L. Rend., (5), 20, 1, 1911, pp. 296–304.

[1911b] *Sopra le funzioni permutabili di* 2^a *specie e le equazioni integrali*, Rom. Acc. L. Rend., (5), 20, 1, 1911, pp. 521–527.

Voronin, S.M.

[1981] *Analytic classification of germs of conformal mappings* $(\mathbb{C},0)\to(\mathbb{C},0)$ *with identity linear part*, Functional Analytic Appl., 15:1, 1981, pp. 1–17.

Wallis, J.

[1656] *De angulo contactus et semicirculi tractatus*, (1656) in vol. 2 of Wallis' *Opera Mathematica*, Georg Olms Verlag, New York, 1972.

Wang, X.L. and Hua, X.H.

[1997] *Fatou components of transcendental entire functions*, Nanjing Daxue Xuebao Shuxue Bannian Kan, 14, 1997, pp. 180–186.

Wang, X.L. and Yang, C.C.

[2003] *On the Fatou components of two permutable transcendental entire functions*, J. Math. Anal. Appl., 278, 2003, pp. 512–526.

Ward, M.

[1936] *Note on the iteration of one variable*, Bull. AMS, 42, 1936, pp. 688–690.

Weber, E.

[1986] *France, fin de Siècle*, Studies in Cultural History, Belknap Press, Cambridge, MA, 1986.

Wegner, U.

[1928] *Über die ganzzahligen Polynome, die für unendlich viele Primzahlmoduln Permutationen liefern*, Berlin, Philos. Diss, 1928.

Weyl, H.

[1913] *Die Idee der Riemannschen Fläche*, Teubner, 1913 (1955 edition).

Williams, D.R.

[1928] *Compléments au théorème de M. Julia*, Rend. Circ. Mat. Palermo, 52, 1928, pp. 373–415.

Wilson, C.

[1985] *The great inequality of Jupiter and Saturn: from Kepler to Laplace*, Arch. Hist. Exact Sci. 33 (no. 1–3), 1985, pp. 15–290.

Wintner, A.

[1946] *The adiabatic linear oscillator*, Amer. J. Math, 68, 1946, 383-397.

Włodarczyk, K.

[1986] *Pick-Julia theorems for holomorphic maps in J^*-algebras and Hilbert spaces*, J. Math. Anal. Appl., 120, 1986, pp. 567–571.

[1987] *Julia's Lemma and Wolff's Theorem for J^*-Algebras*, Proc. AMS, 99, 3, 1987.

Wolff, J.

[1920] *Über Folgen analytischer Funktionen*, Math. Ann., 81, 1920, pp. 48–51.

[1926a] *Sur l'itération des fonctions holomorphes dans une région et dont le valeurs appartiennent à cette région*, C.R. Acad. Sci. Paris, 182, 1926, pp. 42–43.

[1926b] *Sur l'itération des fonctions bornées*, C.R. Acad. Sci. Paris, 182, 1926, pp. 200–201.

[1926c] *Sur une généralisation d'un théorème de Schwarz*, C.R. Acad. Sci. Paris, 182, 1926, pp. 918–920.

[1926d] *Sur une généralisation d'un théorème de Schwarz*, C.R. Acad. Sci. Paris, 183, 1926, pp. 500–502.

Yang, L. and Chang, K.

[1965] *Recherches sur la normalité des familles de fonctions analytiques à des valeurs multiples. I. Un nouveau critère et quelques applications*, Sci. Sinica, 14, 1965, pp. 1258–1271; II. *Généralizations*, Sci. Sinica, 15, 1966, pp. 433–453.

Yang, P.

[1978] *Holomorphic curves and boundary regularity of biholomorphic maps of pseudoconvex domains*, unpublished, 1978.

Yoccoz, J. Ch.

[1988] *Linéarisation des germes de difféomorphismes holo- morphes de* $(\mathbb{C}, 0)$, C.R. Acad. Sci. Paris, 306, 1988, pp. 55–58.

Ypma, T.

[1995] *Historical development of the Newton-Raphson methods*, SIAM Rev., 37, 1995, pp. 531–551.

Yusuda, K.

[1998] *Contrasts and Analagues in Iteration Theory and Dynamical Systems*, Ph.D. dissertation, University of Jos, 1998.

Zalcman, L.

[1975] *A heuristic principle in complex function theory*, Amer. Math. Monthly, 82, 1975, pp. 813–817.

[1995] *On some questions of Hayman*, unpublished manuscript, Bar-Ilan University, 1995.

[1998] *Normal families: new perspectives*, Bull. AMS, 35, 1998, pp. 215–230.

Żorawski, K.

[1893] *Ozbieznośći iteracyi*, Krak. Ber, 24, 1993 pp. 271–288.

Zoretti, L.

[1911] *Leçons sur le prolongement analytique*, Collection de Monographs sur la Théorie des Fonctions, Gauthier-Villars, Paris, 1911.

Index